solar directory

edited by

CAROLYN PESKO

This book is dedicated to Dr. Jerry Plunkett and Richard Crowther A.I.A. who assisted in the initial conception of this work

Elizabeth Kingman, Albert Nuney Jr. and very specially to Karen George who through her long hours of support made this possible.

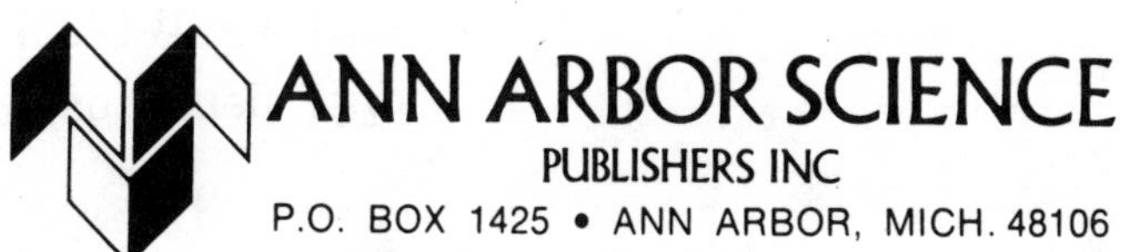

ANN ARBOR SCIENCE
PUBLISHERS INC
P.O. BOX 1425 • ANN ARBOR, MICH. 48106

First Published 1975

Second Printing 1975
Third Printing 1975
Fourth Printing 1976

Library of Congress Catalog Card Number 75-18574
ISBN 0-250-40109-6

Published by Ann Arbor Science Publishers, Inc.
P.O. Box 1425, Ann Arbor, Michigan 48106

through a special arrangement with
Carolyn Pesko

Formerly published by Environmental Action of Colorado

cg 4/6/77

Publisher's Preface

This work is designed to help the researcher and developer, the manufacturer and distributor communicate together on a mutually beneficial basis. Its content covers the wide scope of solar energy activity in the United States primarily but also in other countries, at the academic, governmental and industrial levels.

It remains substantially unchanged by the publisher, retaining all the personality and charm originally included by its editor, Carolyn Pesko.

this book was brought to you by...

EDITOR & PROJECT DIRECTOR
Carolyn Pesko

BIBLIOGRAPHIC RESEARCH
Karen George
Harry Burris
Pamala Engler
John Lefler
Carolyn Pesko
Bob Wolcott

COMPUTERIZED RESEARCH
Carolyn Pesko
Karen George
Albert Nunez

COMPUTER PROGRAM
Johnathan Castor

GRAPHICS
Carolyn Pesko
and friends

HELPING HANDS
John Claydon
Beverly Crispin
Ernie Perry
Cathy Regard
Larry Sainz
Carol Taylor
Bruce Williamson

KEY PUNCH
Colorado National Bank
Colorado State Dept of Highways
Sherry Enterline
Kathy Florez
Jo Shultz

LAY OUT
Carolyn Pesko
Mike Cooper
Albert Nunez
Al Quinn

SOLAR PROJECTS RESEARCH
Albert Nunez
Karen George
Elizabeth Kingman
Carolyn Pesko
Kathy Regard
Greg Spadafore

TYPING
Margy Eckles
Karen George
Debby Hoagie
Jiiva Kala (Janet James)
Carolyn Pesko

all the little people

all the people who shared information

and by everyone of you who
trusted
and believed in
what we were doing
and sent us your money
before it was published
you all made this possible

thank you

the preparation of this document was financed in part through a grant from National Science Foundation
Grant No. AER74-18503 A01

CAROLYN PESKO, Editor

The editor and principal compiler of this directory now resides in Pescadero, California, lecturing and consulting on solar energy, with emphasis on solar heating and passive systems.

The new concept of *low energy living*--solar energy is part of the integrated system--now draws her attention.

"Giving everyone aluminum flat plate collectors will not solve our energy crisis as there is not enough aluminum," says Pesko. "We have to start living a *low energy profile*. Dwellings must again be built of indigenous materials in harmony with their local climate."

"We will see a design revolution in the next few years--in multipurpose tools and appliances, such as utilization of 'waste heat' from refrigerators for domestic water heating. The possibilities are virtually limitless."

Those of you researching, building, manufacturing in the areas of energy conservation, low energy consumption, passive solar systems and/or total systems integration, please write to me:

Carolyn Pesko
Low Energy Living
P.O. Box 428
Pescadero, CA 94040

Describe your work, results and/or failures; send brochures, published works, photos and other pertinent information. Perhaps we will have another useful reference work in the future.

Photo by W. P. Davis,
Black Mountain

POEMS

QUOTATIONS

COVER DESIGN Carolyn Pesko

ILLUSTRATIONS*

*all illustrations without a name identification are the work of Carolyn Pesko

PREFACE

Sunbeams were gently filtering through the leaves as Kea stepped out of the sunborne transport unit. 'Such a nice period it's been for the annual family retreat. Sure feels good to be home again.'

Each retreat the family verbally recalled the history of the tribe. And somehow after refreshing memories, the day after always seemed more special. Kea felt a renewed sense of awe of what had come to pass in the elders life-time.

'To think, at one time the cities had to be washed to remove pollution from the outer walls. Pol·lu'tion. POL·LU'TION. A hard word-concept to grasp. The elders knew, they had seen it and now I must remember. Feel it, see it through the elders' eyes, so that my children yet to be may see it through mine. It must ramain a living history, a warning to our tribe so that this dream we live can remain a reality. And to think, people fought over parts of the tribe's spaceship. Sucking out its vital solids and liquids to burn and SQUAN·DER, blackening the skies and causing people to die painful, fearful deaths.' The film discs had it all on record and it certainly helped one who had never experienced such oppression.

'It always feels nice to visit the elders' house in the city, so clean and fresh. Trees, grass, and birds shoot up everywhere between the tall shiny buildings. The orange trees were in bloom this time of year and the city smelled so fine. AU·TO·MO·BILE. Why had they used such things? Machines of destruction Seth called them, and this year I think I have finally realized what that meant. To think they covered all land in cities with concrete just to maneuver the autos about.' Mya, an elder, had explained the difficulties she had had when sunborne transport units came to be. But she was already ancient then, the younger ones had had no trouble. But now Mya was glad for the change. She had said, "Now I know when I go for a journey I am not killing and robbing things for my pleasure."

Rob had recalled the horrors of the coal mines where his grandfather had labored. "How good it is to work now, for each other, in the sun works."

This year Kea had participated in the extra studies for those who carried the being yet to be. Conservation of all things was stressed as well as the impact of WASTE. It had struck an unconsious chord and she realized her mother had instructed her in the same ways she was now to teach her child. 'Our living heritage.' The thought brought a smile to Kea's lips as she glided through the airlock door into home.

'Home! So good to be back. Each year at this time you seem so much more special to me. To think, our forefathers used to dwell in thin walled boxes totally on the surface, burning enormous fires to warm their skin as well as the outside world. WĀSTE, yes that's what it had been, WĀSTE.'

Kea sunk slowly into the vibe chair as the sun streamed through the southern windows warming her face. 'So good to relax after the long journey. I've a few things to do before Yancy returns.' He stayed longer for the wind machine studies. It was his time now to help their neighbors with their wind machines. 'Brother Russell, how he's grown.' He had chosen the city to dwell. 'How exciting for him to work for the city's wind plant. Russell and Yancy had spent much time this year comparing jobs and remarking on the differences scale made.'

Kea activated the screen for todays reports.

DATE: 01:03:2029...TIME: 14:36 RMT...ALL SYSTEMS OPERATIONAL...HEAT STORAGE ESTIMATED AT 7 DAYS...ELECTRICAL STORAGE ESTIMATED AT 7.9 DAYS...METHANE PRODUCTION UP 6%...SLURRY SHOULD BE DRAWN OFF FOR GARDENS IN 3.6 DAYS (86.4 HOURS)... 60.75 GAL DOMESTIC WATER AT 88°C...SOYBEANS READY FOR HARVEST IN 2.5 DAYS (50 HOURS) GREENHOUSE CAVITY 2...WINDS FROM THE NORTH WEST AT........

Kea's mind drifted off, 'So good to be home.' The little one stirred within. 'Another generation yet to be, to carry the tribes living heritage into the future.'

It is our hope that this reference work will in some small way assist us all in taking one step closer to making this a reality.

in peace

carolyn

January 3, 1975
Black Mountain, Colorado

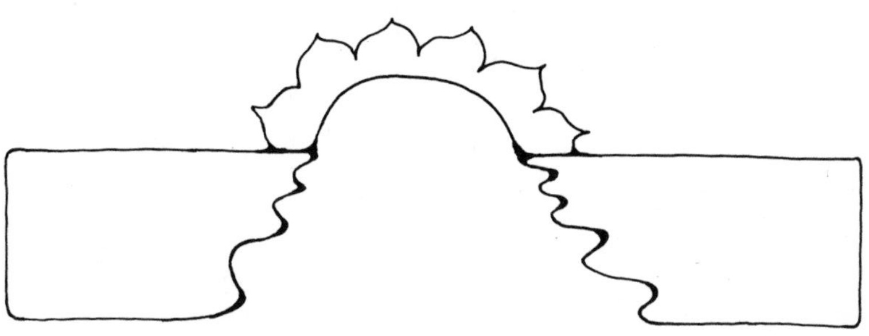

sun... you come
in the most beautiful packaging job
...you beat
CHEER and MR CLEAN and BRILLO boxes
all to hell
sun... you were
alizrin crimson today
.... tonight you will go out
in such style...
all the lights in LAS VEGAS
could never be you
ever changing ever right there

sun sun here comes the sun

and one day they'll realize...

Carolyn

INTRODUCTION

The Solar Directory was first conceived in early 1973, at which time it became apparent that a reference book of this kind was necessary to facilitate solar energy development. The project was accepted as a challenge and seen as a way to make a positive contribution to an effort that is vitally important to the survival of this spaceship. Research was initiated in August, 1973.

We began this project with the goal of providing an on-going information service to the solar community and the general public. We still maintain this goal, and urge your participation in creating the kind of publication that best serves your needs. We would appreciate feedback from you, the user, regarding this book's weak and strong points. Also, if you are engaged in solar activities and are not included, please send us full details, including brochures and other pertinent information so that we may update our files. Please do likewise if your data has changes.

HOW TO USE THIS BOOK

This survey consists of five major sections: Section 1, a compilation of our computerized data; Section 2, and inventory of solar projects; Section 3, an annotated bibliography; Section 4, appendicies; and Section 5 and 6, indices for Section 1. How to use each section is explained in detail in the following text.

GENERAL GUIDELINES:

1) Readers looking specifically for architects, engineers, and contractors involved in the actual design and construction of solar heated and/or cooled buildings should refer to Section 2, Index of Solar Projects.

2) Manufacturers of solar products are listed on pages 1.077-1.158 (Section 1). This list is divided into the following categories: Components Manufacturers, Solar Products Distributors, Solar Manufacturers, Solar Manufacturers-Educational Equipment, and Solar Potential Manufacturers-Established Business or New Business.

3) If you require sources of further information, please refer to: Section 1, Information Dissemination Groups, pages 1.017-1.061 and Section 3, Annotated Bibliography.

4) We request that you refrain from contacting researchers for information unless it states that they have published information available, or a particular individual who handles public inquiry (indicated by an asterisk next to an individual's name). In general, researchers are already overburdened. We have listed them here so others doing similar work would be aware of their activities and the public would be appraised of the types of projects being conducted.

5) Late additions, entries we were unable to include in Sections 1-3 prior to printing, are provided in Section 4, Appendix B, Late Additions.

6) The abbreviations used herein are listed in Section 4, Appendix A.

7) Universities are listed in the following fashion: California, University of.

8) Sections 5 and 6 are in alphabetical order, reading from left to right for each line of entries.

SECTION 1, By Classification Reports

This listing provides descriptions of the interests, activities, or products of particular individuals, companies, or organizations. Within each classification, entries are in alphabetical order with individuals (last name) without an affiliation listed first. Foreign entries follow U.S. listings and are in alphabetical order by country, and within country by company or affiliation.

SAMPLE LISTING

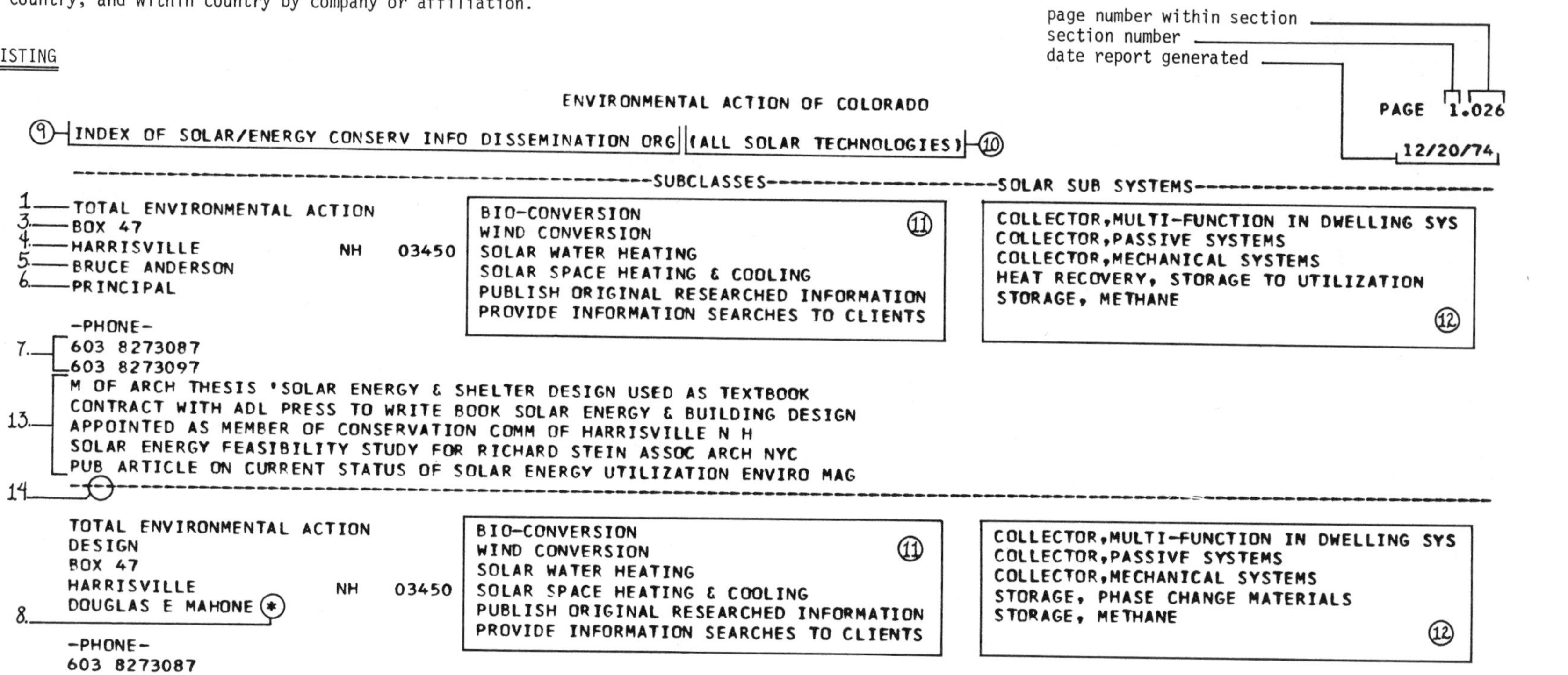

(1)...company or affiliation
(2)...department or division (not shown above)
(3)...street or mailing address
(4)...city, state, and zipcode-or country if other than U.S.
(5)...individual's name
(6)...individual's title
(7)...telephone number(s) note: if 2 are given the 1st is the primary number
(8)...indicates key individual who handles public inquiry (see explanation of use in Section 5 sample listing).
(9)...major classification note: these are included in the listings in the indexes (Sections 5 and 6) and can be used for locating an individual or a company/organization.
(10)..major subclassification see note for (9).
(11)..subclassifications indicate secondary areas of work note: listings have only been sorted by major area of work (9 and 10).
(12)..solar subsystems give supplemental detail on the areas of activity.
(13)..this is additional pertinent information such as published works, project description, product description, etc.
(14)..dotted line indicates end of a particular listing.

Citations in Section 1 are indexed in Sections 5 and 6. Section 5 is an index of the companies and organizations; Section 6 an index of individuals. Further details on these indexes are provided on the following page.

INTRODUCTION (con't)

SECTION 2, Index of Solar Projects

See introduction to Index of Solar Projects, page 2.001.

SECTION 3, Bibliography

See introduction to Bibliography, page 3.001.

SECTION 4, Appendices
These include:

- Appendix A...Abbreviations
- Appendix B...Late Additions
- Appendix C...NSF/RANN Solar Energy Research Grants & Contracts In FY 1974
- Appendix D...Solar Energy Legislation
- Appendix E...Solar Energy Courses
- Appendix F...Money
- Appendix G...Environmental Action of Colorado

SECTION 5, Index of Academic Institutions/Government Agencies/Industries

This is an index of all the companies and organizations listed in Section 1, and indicates the basic category by which they are listed. It is not to be used as an index of companies and organizations listed in the other sections.

SAMPLE LISTING

company or affiliation	BROOKHAVEN NATIONAL LAB
department or division	PUBLIC RELATIONS
street or mailing address	40 BROOKHAVEN AVE
city, state, zipcode-or country if other than U.S.	UPTON NY 11973
individual's name	CARL R THIEN (*)
individual's title	INFORMATION OFFICER
telephone number(s) note: if 2 given 1st is primary	-PHONE- 516 3452345
major classification*	SOLAR RESEARCH
major subclassification*	BIO-CONVERSION
key individual**	(*)

* These classifications will enable the reader to locate the complete detailed listing in Section 1. There are two methods which may be used to locate the listing:

1) Refer to Table of Contents. If using the example given above, you would locate Solar Research, (Bioconversion) and find the page numbers for this category (1.202-1.207). You could then find the entry by doing an alphabetical scan of those pages. or
2) Refer to the back cover for the easy access tab indicators for the particular section you desire. You can then scan the section until the proper category is found and do an alphabetical search.

** Represents an individual within an organization, if there is more than one listed, who handles public inquiry.

SECTION 6, Index of Individuals

This is a list of all the individuals listed in Section 1, and indicates the basic category in which you may find them in Section 1. It does not pertain to those listed in other sections.

SAMPLE LISTING

key individual**	(*)
individual's name (last name 1st)	HOGG (*) F G
individual's title	CHIEF
company or affiliation	C I R S O
department or division	DIV OF MECH ENGR
street or mailing address	P O BOX 26
city, state, zipcode-or country if other than U.S	HIGHETT, VICTORIA AUSTRALIA
telephone number(s) note: if 2 given 1st is primary	-PHONE- 03 950333
major classification*	SOLAR RESEARCH
major subclassification*	MEDIUM TEMPERATURE SOLAR COLLECTOR SYSTEMS

* For explanation of how to use these classifications to locate complete listing in Section 1, see how to use Section 5.
** For explanation of key individual, see how to use Section 5.

and now.........

.....................heeeeeeerrrrreeeeeeesssssssss the Directory

TABLE OF CONTENTS

TABLE OF CONTENTS (con't)

TABLE OF CONTENTS (con't)

----SUBCLASSES----SOLAR SUB SYSTEMS----

PRUDENTIAL INS CO OF AMERICA
REAL ESTATE INVESTNT
1180 RAYMOND BLVD
NEWARK NJ 07102
CHARLES F REYNOLDS JR
VICE PRES
-PHONE-
201 3364578
AS REAL ESTATE DEVELOP & INVEST PRIMARY INTEREST ENERGY-SAVINGS POTNL

	SUBCLASSES	SOLAR SUB SYSTEMS
DEPT OF HOUSING & URBAN DEVEL ENERGY AFF RM 10230 451 7TH STREET S W WASHINGTON DC 20410 GEORGE TAPPERT -PHONE- 202 7556480	PROVIDE INSURED LOANS GOVERNMENT AGENCIES INCENTIVES FOR SOLAR & ENERGY CONSERVATION	

```
------------------------------------------------SUBCLASSES-------------------SOLAR SUB SYSTEMS------------------------

COLORADO NATIONAL BANK
P O BOX 5168TA
DENVER                  CO   80217
ROGER WHITE *
ASST TO PRESIDENT

-PHONE-
303 8931862
-----------------------------------------------------------------------------------------------------------------------

FIRST NATIONAL CITY BANK            MEDIUM TEMPERATURE SOLAR COLLECTOR SYSTEMS
AREA 3 OPERATING GP                 NEW CONSTRUCTION, MECHANICAL SYSTEMS
PO BOX 939 CHURCH ST                OFFICE/BANK BUILDINGS
NEW YORK                NY   10008  1MILLION SQ FT + UP,UNDER CONSTRUCTION
HENRY DEFORD                        HIGH RISE COMPLEX,UNDER CONSTRUCTION
SENIOR VP
CONDUCTING STUDY WITH CONSOLIDATED EDISON ENERGY LAB MIT AND CUSHMAN &
WAKEFIELD INC ON WAYS TO ADAPT NEW BANK SKYSCRAPER TO USE SOLAR ENERGY
...SEE CONSOLIDATED EDISON
-----------------------------------------------------------------------------------------------------------------------

FULTON FEDERAL SAVINGS & LOAN       NEW CONSTRUCTION, RESEARCHING
P O BOX 1077                        PROVIDE INSURED LOANS
ATLANTA                 GA   30301  PROVIDE FINANCIAL CONSULTING SERVICES
ROBERT FREDERIC

-PHONE-
404 5222300
SPONSORING DEVELOPMENT OF ENERGY CONSERVING HOMES UTILIZING SUPPLEMENT
ARY SOLAR HEATING AND COOLING...SEE FRANK CLARKE GEORGIA INST OF TECH
-----------------------------------------------------------------------------------------------------------------------

UNITED BANK OF BOULDER              RESIDENTIAL OTHER
1300 WALNUT ST                      PROVIDE FINANCIAL CONSULTING SERVICES
BOULDER                 CO   80302  FINANCIAL OTHER
JACK MINNEMAN
V P

-PHONE-
303 4423734
PROVIDE FULL BANK SERVICES INCL HOME IMPROVEMENT
-----------------------------------------------------------------------------------------------------------------------
```

--------SUBCLASSES--------SOLAR SUB SYSTEMS--------

PERCY WILSON MORTGAGE&FINANCE
COMM & INDUST LOAN
221 N LASALLE ST
CHICAGO IL 60601
TODD J BROMS
ASST LOAN OFFICER
-PHONE-
312 8556726
PERCY WILSON-SUBSIDIARY OF US STEEL... T BROMS RESPONSIBLE FOR LOAN
SOLICITATION UNDERWRITING & APPRAISAL OF INCOME PROPERTY LOANS...P W
IS A MORTGAGE BANK

--------SUBCLASSES--------SOLAR SUB SYSTEMS--------

HEIZER CORPORATION
20 NORTH WACKER DR
CHICAGO IL 60606
PETER J GILLESPIE
VICE PRESIDENT

-PHONE-
312 6412200
INTERESTED IN IDENTIFYING & EVALUATING VENTURE CAPITAL INVESTMENTS IN ANY HIGH POTENTIAL NEW BUSINESS IN SOLAR ENERGY FIELD

FOR ADDITIONAL FINANCIAL INSTITUTIONS SEE:
INDEX OF SOLAR PROJECTS SECTION 2 PROJECT FUNDING SOURCES

----------SUBCLASSES----------SOLAR SUB SYSTEMS----------

OFFICE OF TECH ASSESSMENT 119 'D' STREET NE WASHINGTON DC 20510 RONAL W LARSON	ALL SOLAR TECHNOLOGIES SOLAR SPACE HEATING & COOLING

-PHONE-
202 2258996
IEEE CONGRESSIONAL FELLOW,1973-74, MAJOR RESPONSIBILITY WAS ON SOLAR HEATING AND COOLING DEMONSTRATION ACT (HR11864)

SEE APPENDICES FOR FURTHER FEDERAL LEGISLATION

INDEX OF SOLAR POLITICAL CONSIDERATIONS (ALL SOLAR TECHNOLOGIES)

------------------------------SUBCLASSES------------------SOLAR SUB SYSTEMS------------------

FLORIDA ENERGY COMMISSION GOVERNMENT AGENCIES
SENATE OFFICE BLDG LEGISLATIVE, STATE
TALLAHASSEE FL 32304
MARVIN YAROSH

-PHONE-
904 4881078
904 4881167

FRIENDS RESEARCH CENTER
ENVIRONMENTAL STUDYS
308 HILTON AVE
CATONSVILLE MD 21228
BRUCE L WELCH

-PHONE-
301 7472773

"The people, and the people alone, are the motive force in the making of world history."
Mao Tsetung.

		SUBCLASSES	SOLAR SUB SYSTEMS
BRADFORD,UNIV OF BRADFORD YORKSHIRE TOM T STONIER PROF OF PEACE STUD	ENGLAND	BIO-CONVERSION PUFLISH LAY PERIODICALS	

-PHONE-
33466

PROPOSAL FOR INTERNATIONAL SOLAR ENERGY DEVELOPMENT DECADE SUBMITTED TO GOVERNING COUNCIL U N ENVIRONMENT PROGRAM...ALSO PUB IN MAY 1972 ISSUE OF BULLETIN OF ATOMIC SCIENTIST

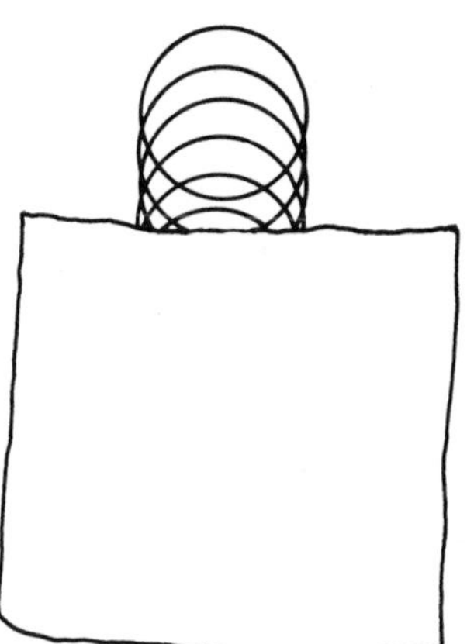

WASHINGTON UNIVERSITY
CTR BIOL OF NAT SYS
BOX 1126
ST LOUIS MO 63130
BARRY COMMONER
DIRECTOR
-PHONE-
314 8630100
314 8636768
PROMINENT ENVIRONMENTALIST ADVOCATING SOLAR ENERGY...SEE PLAYBOY INTER
VIEW JULY 1974

"...solar energy, unlike nuclear energy has no 'critical mass'; and indeed no hazards at all except the possibility of sunburn, which is seldom fatal. Nor are there any waste products to be disposed of. It does not take a Ph.D. degree in physics to make use of solar energy in one's backyard."

Halacy, The Coming Age of Solar Energy

------SUBCLASSES------SOLAR SUB SYSTEMS------

DESERT SUNSHINE EXPOSURE TESTS	SOLAR RADIATION MEASUREMENTS
CLIENT SERVICES	HIGH TEMPERATURE SOLAR COLLECTOR SYSTEMS

BOX185 BLACKCAN STGE
PHOENIX AZ 85020
JOHN L SCOTT *

-PHONE-
602 4657525
602 4657521

TEST HOW YOUR PRODUCT WEATHERS UNDER THE SUN...TEST FACILITIES INCLUDE STATIONARY 45DEGREE RACKS.FIXED ANGLE SPRAY RACK.EQUITORIAL MOUNT WITH & WITHOUT MIRRORS.OTHER TESTS ON REQUEST
VERY GOOD SUN TRACKING SYS

	SUBCLASSES	SOLAR SUB SYSTEMS
JEFFERSON CO PUBLIC SCHOOLS ENVIRONMENTAL EDUC 2192 COORS DRIVE GOLDEN CO 80401 BETTY JANE MEADOWS ENV EDUC SPECIALIST -PHONE- 303 2793346 303 2793347	ACADEMIC BUILDINGS GREENHOUSES LOBBY TO PROMOTE SOLAR LEGISLATION LEGISLATIVE, STATE LEGISLATIVE, LOCAL INCENTIVES FOR SOLAR & ENERGY CONSERVATION	MAGNETOHYDRONAMICS COLLECTOR,MULTI-FUNCTION IN DWELLING SYS HEAT RECOVERY, STORAGE TO UTILIZATION HEAT PUMPS STORAGE, MECHANICAL/PUMPED LIQUIDS STORAGE, MECHANICAL/COMPRESSED GAS STORAGE, FUEL CELLS

"If, over a 3-decade period, 2/3 of the residences in the U.S. were provided with solar equipment, more energy would be saved each year than is currently produced by all the electric power plants in the U.S."

J. Weingart

	SUBCLASSES	SOLAR SUB SYSTEMS
SOLAR ENVIRONMENTAL ENGR CO PO POX 1914 FORT COLLINS CO 80521 THOMAS E CORDER VICE PRESIDENT -PHONE- 303 4934480 303 4846386	SOLAR SPACE HEATING & COOLING NEW CONSTRUCTION, MECHANICAL SYSTEMS HOMES SINGLE FAMILY, PROPOSED INDUSTPIAL BUILDINGS PUBLISH TECHNICAL PERIODICALS	COLLECTOR,MECHANICAL SYSTEMS

COMPUTER AIDED DESIGN AND OPTIMIZATION OF SOLAR HEATING & COOLING SYS PUBLICATIONS INCLUDE SIMULATION MODEL OF SOLAR HEATED & COOLED BLDGS IN'JOURNAL OF SIMULATION & PRESENTATIONS AT INTERNATIONAL SOLAR ENERGY CONF, AMR SOCIETY OF MECH ENGINEERS WINTER ANNUAL MEETING AND 8TH ANNUAL SIMULATION SYMPOSIUM

	SUBCLASSES	SOLAR SUB SYSTEMS
TEXAS A&M UNIV CHEMICAL ENGRG DEPT COLLEGE STATION TX 77843 J H MARTIN	SOLAR PONDS HOMES	

-PHONE-
713 8453361

TECH & ECONOMIC STUDY OF SOLAR HEAT & COOL COMMUNITY USING FLOATING POND COLLECTORS...HEAT BY CIRCULATING WELL H2O THROUGH PONDS IN SUMMER & RETURN H2O TO UNDERGROUND AQUIFER & CIRCULATE HEATED H2O TO HOMES FOR WINTER HEAT...COOL BY CIRCULATING COOL H2O FROM AQUIFER TO HOMES

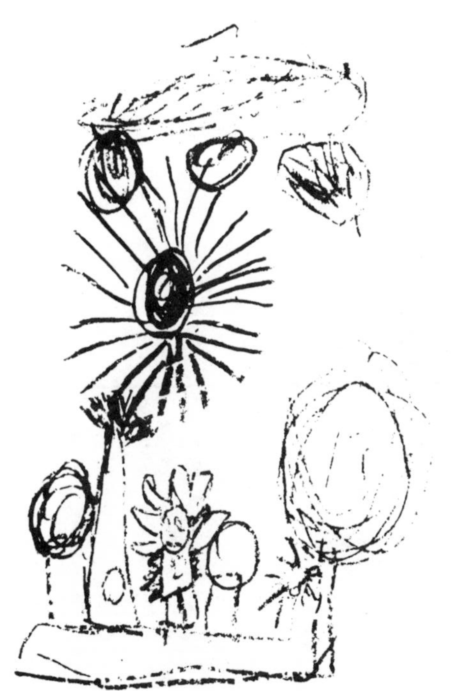

Yancy

----SUBCLASSES----SOLAR SUB SYSTEMS----

U S DEPT HOUSING & URBAN DEV   SOLAR SPACE HEATING & COOLING
OFF POLICY DEV&RESCH
451 7TH ST S W
WASHINGTON   DC   20410
JOSEPH SHERMAN
DIR BLDG TECH&SAFETY
-PHONE-
202 7555364
RESPONSIBLE FOR CRITERIA BY WHICH FUNDS ALLOCATED BY 74 SOLR HEAT&COOL
DEMO ACT WILL BE DISTRIBUTED...PROPOSAL SOLICITATION SCHEDULED FIRST
OF 1975...WRITE TO BE PLACED ON MAILING LIST

-------------------------------SUBCLASSES----------------SOLAR SUB SYSTEMS-------------------------------

UTAH STATE UNIVERSITY		SOLAR SPACE HEATING & COOLING
DEPT OF ECONOMICS		HOMES
LOGAN	UT 84322	INCENTIVES FOR SOLAR & ENERGY CONSERVATION
CRAIG H PETERSON		ZONING ORDINANCES FOR SOLAR UTILIZATION
ASST PROF ECONOMICS		

-PHONE-
801 7524100 X7307
ENVIRONMENTAL COST & INCOME DISTRIBUTION EFFECTS OF SOLAR ENERGY
GOVERNMENT POLICY REGARDING SOLAR ENERGY/CONSUMER&BUSINESS INCENTIVES

----------SUBCLASSES----------SOLAR SUB SYSTEMS----------

HELEN K VERMEERS
358 SW 27TH AVE #12
MIAMI FL 33135

-PHONE-
305 6423494
IMPACT OF SOLAR ON CULTURAL PATTERNS & ECONOMICS OF

CTR FOR RESEARCH-ACTS OF MAN
4025 CHESTNUT STREET
PHILADELPHIA PA 19104
MADELEINE S KLAUSNER
DIR OF ADMIN

-PHONE-
215 5946241
IMPACTS OF SOLAR ENERGY-RESEARCH ON ENERGY CONSV, RATIONING, STRUCTURE
OF ENGY INDUSTRIES & SOCIAL IMPACTS OF ENGY USE

U S WEATHER BUREAU
NATL WEATH RECOD CTR
FEDERAL BLDG
ASHEVILLE NC 28801

PUBLISH ORIGINAL RESEARCHED INFORMATION
PROVIDE INFORMATION SEARCHES TO CLIENTS

STATISTICAL WEATHER DATA FOR ALL OVER USA FOR PAST 50 YRS

"strangely enough, despite the many hours which the astronauts have spent on the moon, there have been no direct measurements of solar radiation intensity reported from that vantage point nor is there any longer the daily monitoring of the suns output as there was during Dr. Abbott's era at the Smithsonian Institution."

J. Yellott

----------SUBCLASSES----------SOLAR SUB SYSTEMS----------

JEROME A SILBERT MD SOLAR ELECTRIC POWER GENERATION
444 EAST 84TH ST
NEW YORK NY 10028

-PHONE-
212 2496487
PUBLICIZE CONCEPTS OF OCEAN THERMAL TO INCREASE GOVT FUNDING &INTEREST
BY PRIVATE INDUSTRY
CLARENCE ZENER,'SOLAR SEA POWER,' PHYSICS TODAY, VOL 26, NO 1, JAN 73,
PP 48-53.

SUBCLASSES — SOLAR SUB SYSTEMS

GOBAR GAS RESEARCH STATION
AJITMAL ETAWAH U P INDIA
RAM BUX SINGH
INCHARGE

STORAGE, METHANE

ANAEROBIC DECOMPOSITION OF ORGANIC WASTES FOR OBTAINING METHANE FUEL & ORGANIC MANURE WITH THE AID OF A BIO-GAS PLANT...HAVE INSTALLED OVER 200 PLANTS IN INDIA...2 RESEARCH MANUALS PUB IN ENGLISH(AVAIL 5DOLLARS EACH FROM ABOVE):'BIO-GAS PLANT GENERATING METHANE FROM ORGANIC WASTES &'BIO-GAS PLANT DESIGNS WITH SPECIFICATIONS'

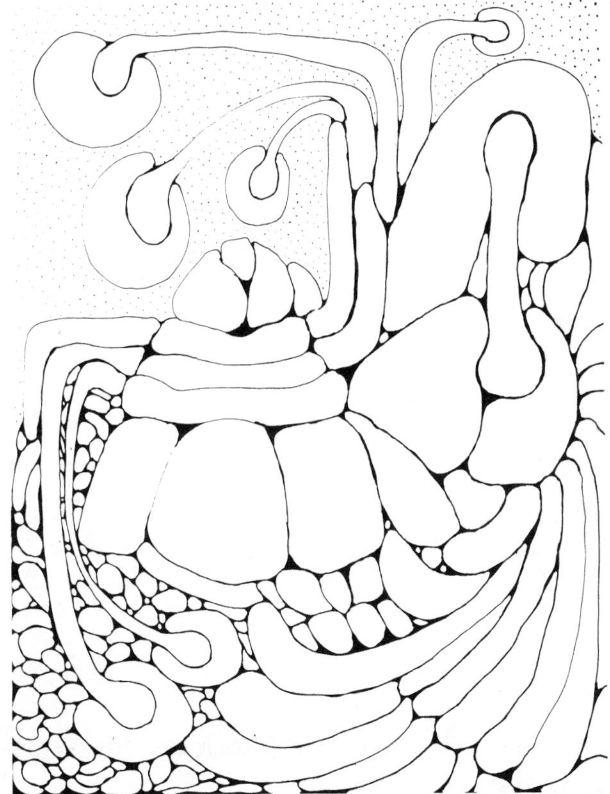

---SUBCLASSES---SOLAR SUB SYSTEMS---

BILL FLANNIGANS PLANS SOLAR ELECTRIC POWER GENERATION
20032 23RD ST
ASTORIA NY 11100
BILL FLANNIGAN

-PHONE-
212 7282083
SUPPLIES PLANS FOR HOMEMADE WINDMILL $10

EARTHMIND PUBLISH LAY PERIODICALS
JOSEL SAUGUS CA 91350
DAVID HOUSE

-PHONE-
805 2513053

WINDWORKS PUBLISH INFORMATIONAL REPRINTS
BOX 329 ROUTE 3
MUKWONAGO WI 53149

WIND ENERGY BIBLIOGRAPHY PREPARED & AVAILABLE...COVERS WIND WINDMILLS
AERODYNAMICS ELECTRICAL TOWERS STORAGE CONVERSION HYDROGEN PROD
CATALOGUES OF SUPPLIERS & GEN READING

	--SUBCLASSES--	--SOLAR SUB SYSTEMS--
AMERICAN INST AERONAUT&ASTRONT	PUBLISH LAY BOOKS	
MILWAUKEE STUDENT BR	PROVIDE AUDIO & VISUAL SERVICES	
3228 N 26TH STREET		
MILWAUKEE WI 53206		
DANIEL M LENTZ		

-PHONE-
414 8730133
414 9635168

ILLUSTRATE TO PUBLIC BENEFITS THAT HAVE COME THROUGH SPACE EXPLORATION INCLUD SOLAR & FUEL CELLS...AIAA STUDENT BRANCH HAS PREPARED 60 MIN SLIDE PRESENTATION & DISCUSSION...DELIVERED TO GRADE SCHOOL TO COLLEGE & GENERAL PUBLIC...GENERAL LITERATURE AVAIL SOON ON SOLAR CELLS & FUEL CELLS IN SPACE & ON EARTH

Kea

			SUBCLASSES	SOLAR SUB SYSTEMS
DUBIN-MINDELL-BLOOME ASSOC PE CONSULTING ENGINEERS 42 W39 ST NEW YORK FRED S DUBIN PRESIDENT -PHONE- 212 8689700	NY	10018	WIND CONVERSION SOLAR HEAT,COOL,&ELECTRIC POWER GENERATION RETROFITTING, TECHNOLOGIES NEW CONSTRUCTION, TECHNOLOGIES CLUSTER HOMES PROVIDE INFORMATION SEARCHES TO CLIENTS	COLLECTOR,MULTI-FUNCTION IN DWELLING SYS COLLECTOR,ALSO FUNCTION AS STRUCTURES SKIN COLLECTOR,MECHANICAL SYSTEMS HEAT RECOVERY, STORAGE TO UTILIZATION HEAT PUMPS STORAGE, LIQUIDS ELECTROLYSIS

SOLAR ENERGY-ENERGY CONSERVATION CONSULTANTS AND ENGR DESIGN FOR CARY ARBORTUM NY-BOTANICAL GARDENS-GSA DEMONSTRATION BLDG NH-GRASSY BROOK VILLAGE VT-NASA CAL RETROFIT FOR SOLAR COOLING-ARGONNE ENERGY CONSERVATION & SOLAR OFFICE BLDG-US HOME CORP ENGY CONSV PROTOTYPE RES-PINCHOT RES SOLAR HEATING CT-ANEGADA BWI TOTAL COMMUNITY USING SOLAR AND WIND ENERGY...AUTHOR GSA ENGY CONSV GUIDELINES FOR OFFICE BLDGS

MOTHER EARTH NEWS INC CORRESPONDENCE SECT P O BOX 70 HENDERSONVILLE * -PHONE- 704 6924256	NC	28739	MEDIUM TEMPERATURE SOLAR COLLECTOR SYSTEMS BIO-CONVERSION WIND CONVERSION PUBLISH LAY PERIODICALS PUBLISH INFORMATIONAL REPRINTS PUBLISH ORIGINAL RESEARCHED INFORMATION	

MOTHER EARTH NEWS INC RESEARCH DEPT P O BOX 70 HENDERSONVILLE JOHN A POLL DIRECTOR R&D -PHONE- 704 6924256	NC	28739	MEDIUM TEMPERATURE SOLAR COLLECTOR SYSTEMS BIO-CONVERSION WIND CONVERSION PUBLISH LAY PERIODICALS PUBLISH INFORMATIONAL REPRINTS PUBLISH ORIGINAL RESEARCHED INFORMATION	

RESEARCH IN GRASS ROOTS UTILIZATION OF SOLAR ENERGY INCLUDING BIO-CONVERSION AND METHANE PRODUCTION...PUBLISH THE MOTHER EARTH NEWS A BI-MONTHLY MAGAZINE WITH EMPHASIS ON ALTERNATIVE LIFESTYLES ECOLOGY WORKING WITH NATURE & DOING MORE WITH LESS SUBSCPIPTIONS 1YR 6 ISSUES 6 DOLL 2YRS 12 ISSUES 11 DOLL...WRITE FOR FULL LIST OF PUBLICATIONS

--------SUBCLASSES--------SOLAR SUB SYSTEMS--------

ARIZONA STATE UNIV — HAVE&MAINTAIN ENERGY DATA BANK
LIBRARY-SCI-ENGR SER
ROOM 257
TEMPE AZ 85281
MARY BEECHER
LIBRARIAN
-PHONE-
602 9657608
IN CHARGE LARGE COMPUTERIZED SOLAR COLLECTION...NUCLEUS GIFT OF ISES THEIR RESEARCH COLLECTION.....BOOKS JOURNALS GOVT DOC PATENTS PICTURES ETC...ONGOING UPDATED OCC...MATERIALS AVAIL INTERLIBRARY LOAN OR REPRODUCTION SER FOR FEE

BOULDER SOLAR ENERGY SOCIETY
P O BOX 3431
BOULDER CO 80303

GROUP OF RESEARCHERS AND OTHER INTERESTED PARTIES WHO MEET REGULARLY FOR DISCUSSION AND EXCHANGE OF IDEAS

COLORADO, UNIV OF-DENVER
ENVIRONMTL ACTION-CO
1100 14TH STREET
DENVER CO 80202
MOREY WOLFSON
DIRECTOR
-PHONE-
303 5341602 X10
303 8921117 X471

PUBLISH INFORMATIONAL REPRINTS
PUBLISH ORIGINAL RESEARCHED INFORMATION
PROVIDE VISUAL SERVICES
PROVIDE AUDIO SERVICES
PROVIDE AUDIO & VISUAL SERVICES
HAVE&MAINTAIN ENERGY DATA BANK

BRING ABOUT INCREASED RECOGITION & DEVELOP OF SOLAR ENGY:PROVIDE:SPEAKERS BUREAU TO DISCUSS ENERGY/ECOLOGY DILEMMA, EXTENSIVE GRAPHICS FILE ON ENERGY-ECOLOGY INCL SOLAR ENGY, MAINTAIN LIBRARY-CONTAIN WIDE RANGE ENERGY & ENVIRONMENT INFO,MAJOR EMPHASIS ON SOLAR & NUCLEAR INFO
SEE ALSO: E KINGMAN SOLAR EXHIBIT PROG & C PESKO SOLAR DIRECT PROJ DIR & BEN BILLINGS ENVIRO ACT REPRINT SERV & ALBERT NUNEZ SOLAR PROJ COORD

	SUBCLASSES	SOLAR SUB SYSTEMS
INTL SOLAR ENGY SOC/US SECT C/O SMITHSONIAN RBL 12441 PARKLAWN DR ROCKVILLE MD 20852 DR WILLIAM H KLEIN SECRETARY/TREASURER FOR GOALS & SERVICES OFFERED SEE INTL SOLAR ENERGY SOC AUSTRALIA MEMBERSHIP IN US SECT & INTL SECT ARE CONCURRENT WRITE THE ABOVE FOR DETAILS	PUBLISH TECHNICAL PERIODICALS	
METROPOLITAN ECOLOGY WORKSHOP 74 JOY STREET BOSTON MA 02114 ROBERT POLLUCK -PHONE- 617 7234699 DEVELOP & PROMOTE ALT ENGY SOURCES ESTAB LOCAL NETWORK & LIBRARY ALSO ASE PROJECTS GOAL TO CONVERT HOMES & OFFICES TO ASE	MEDIUM TEMPERATURE SOLAR COLLECTOR SYSTEMS HIGH TEMPERATURE SOLAR COLLECTOR SYSTEMS BIO-CONVERSION WIND CONVERSION	STORAGE, METHANE
MTN AREA DOMESTIC SOLAR ENGY TECH CONSORTIUM RR 2 BOX 274 EVERGREEN CO 80439 MALCOLM A LILLYWHITE COORDINATOR -PHONE- 303 6746633 PURPOSE OF MADSETC IS TO DEVELOP & DISSEMINATE DOMESTIC TECH EMPHASIS ON DOM SOLAR TEC TOWARDS ENERGY AUTONOMOUS FAMILY HOUSING&COMMUNITIES ADDITIONAL TECH APP: WATER HEATING REFRIGERATION HEATING&COOLING METHANE...OVER 800 MEMBERS OF MADSETC...FOR LIST OF PUBLICATIONS AVAIL ON DOMESTIC TECH SEND SELF ADDRESSED STAMPED ENVELOPE	TOTAL SYSTEMS INTEGRATION SOLAR AGRICULTURAL DRYING SOLAR WATER DISTILLATION SOLAR HEAT,COOL,&ELECTRIC POWER GENERATION PUBLISH INFORMATIONAL REPRINTS PUBLISH ORIGINAL RESEARCHED INFORMATION	STORAGE, METHANE COLLECTOR,MULTI-FUNCTION IN DWELLING SYS COLLECTOR,PASSIVE SYSTEMS COLLECTOR,MECHANICAL SYSTEMS STORAGE, SOLID MATERIALS STORAGE, PHASE CHANGE MATERIALS STORAGE, LIQUIDS

	SUBCLASSES	SOLAR SUB SYSTEMS
N Y SCI COMM FOR PUBLIC INFO 30 EAST 68TH ST NEW YORK NY 10021 CAROLYN S KONHEIM EXECUTIVE DIR	PUBLISH LAY PERIODICALS PROVIDE AUDIO & VISUAL SERVICES	

-PHONE-
212 7377302
SPONSORED 'ROUND TABLE ON SOLAR ENERGY' NY CITY JAN 21 1974...FOR 15 YRS SCPI HAS BEEN INTERPRETING FOR GENERAL PUBLIC TECHNICAL INFO WHICH HAS SOCIAL IMPLICATIONS...HAS VERY ACTIVE SPEAKERS BUREAU...PUBLISH BI-MONTHLY NEWSLETTER

	SUBCLASSES	SOLAR SUB SYSTEMS
NEW ENGLAND SIERRA CLUB 14 BEACON ST BOSTON MA 02108 RICHARD ELY	PROVIDE INFORMATION SEARCHES TO CLIENTS	

-PHONE-
617 2275339

NEW MEXICO SOLAR ENERGY ASSOC
602 1/2 CANYON RD
SANTA FE NM 87501
KEITH HAGGART
PRESIDENT

-PHONE-
505 9832861
SOCIETY FORMED TO ENCOURAGE EXCHANGE OF IDEAS AND KEEP THOSE IN STATE WELL-INFORMED

	SUBCLASSES	SOLAR SUB SYSTEMS
NORTHERN PLAINS RESOURCE CONSL 421 STAPLETON BLDG BILLINGS MT 59101 KYE COCHRAN	BIO-CONVERSION WIND CONVERSION	STORAGE, METHANE

-PHONE-
406 2596114
INFO ON ALL WAYS BY WHICH SOLAR ENERGY MAYBE UTILIZED...KEEPING UP TO DATE ON CURRENT ACTIVITIES IN SOLAR & RELATED ALTERNATIVE ENERGY FIELD ARTICLES PUBLISHED IN COUNCILS MONTHLY NEWSLETTER...COMPILING FACT BOOK ON SOLAR WIND GEOTHERMAL & METHANE GAS PRODUCTION

	SUBCLASSES	SOLAR SUB SYSTEMS
SOLAR ENERGY RESEARCH&INFO CTR	PUBLISH LAY PERIODICALS	
1001 CONN AVE NW	PROVIDE INFORMATION SEARCHES TO CLIENTS	
WASHINGTON DC 20036		
SOLAR ENERGY INDUSTRIES ASSOC		

ASSOC FORMED BY MEMBERS OF INDUSTRY TO ACCELERATE DEVELOP USE OF SOLAR ENERGY...TO ADVISE... GATHER & TABULATE GLOBAL STATISTICAL & MARKET DATA...DISSEMINATE REPORTS OF DEVELOPMENTS TO MEMBERS...RECOGNIZE ACH-IEVEMENT IN INDUSTRY

	SUBCLASSES	SOLAR SUB SYSTEMS
TOTAL ENVIRONMENTAL ACTION	BIO-CONVERSION	COLLECTOR,MULTI-FUNCTION IN DWELLING SYS
BOX 47	WIND CONVERSION	COLLECTOR,PASSIVE SYSTEMS
HARRISVILLE NH 03450	SOLAR WATER HEATING	COLLECTOR,MECHANICAL SYSTEMS
BRUCE ANDERSON	SOLAR SPACE HEATING & COOLING	HEAT RECOVERY, STORAGE TO UTILIZATION
PRINCIPAL	PUBLISH ORIGINAL RESEARCHED INFORMATION	STORAGE, METHANE
	PROVIDE INFORMATION SEARCHES TO CLIENTS	

-PHONE-
603 8273087
603 8273097
M OF ARCH THESIS 'SOLAR ENERGY & SHELTER DESIGN USED AS TEXTBOOK
CONTRACT WITH ADL PRESS TO WRITE BOOK SOLAR ENERGY & BUILDING DESIGN
APPOINTED AS MEMBER OF CONSERVATION COMM OF HARRISVILLE N H
SOLAR ENERGY FEASIBILITY STUDY FOR RICHARD STEIN ASSOC ARCH NYC
PUB ARTICLE ON CURRENT STATUS OF SOLAR ENERGY UTILIZATION ENVIRO MAG

	SUBCLASSES	SOLAR SUB SYSTEMS
TOTAL ENVIRONMENTAL ACTION	BIO-CONVERSION	COLLECTOR,MULTI-FUNCTION IN DWELLING SYS
DESIGN	WIND CONVERSION	COLLECTOR,PASSIVF SYSTEMS
BOX 47	SOLAR WATER HEATING	COLLECTOR,MECHANICAL SYSTEMS
HARRISVILLE NH 03450	SOLAR SPACE HEATING & COOLING	STORAGE, PHASE CHANGE MATERIALS
DOUGLAS E MAHONE *	PUBLISH ORIGINAL RESEARCHED INFORMATION	STORAGE, METHANE
	PROVIDE INFORMATION SEARCHES TO CLIENTS	

-PHONE-
603 8273087

INTL SOLAR ENERGY SOCIETY — PUBLISH TECHNICAL PERIODICALS
PO BOX 52
PARKVILLE VIC — AUSTRALIA
THE SECRETARY

INTERDISCIPLINARY ORG THAT PROVIDES A COMMON MEETING GROUND FOR THOSE CONCERNED WITH THE NATURE & USE OF THIS NATURAL RESOURCE...PUBLISH 'SOLAR ENERGY' JOURNAL OF SOLAR ENERGY SCI & TECH PUB QUARTERLY...PUB MONTHLY NEWSLETTER...HOLD ANNUAL MEETINGS IN US & ABROAD FOR MEMBERSHIP INFO PLEASE WRITE...US RESIDENTS SEE ISES US SECTION

MCGILL UNIV MACDONALD COLLEGE — WIND CONVERSION
BRACE RESEARCH INST — SOLAR COOKING
ST ANN DE BELEVUE800 — SOLAR WATER DISTILLATION
QUEBEC H9X 3MI — CANADA — SOLAR WATER HEATING
RON ALWARD — GREENHOUSES
PUBLISH ORIGINAL RESEARCHED INFORMATION

-PHONE-
514 4576580 X341

R&D OF ECONOMIC&EFFECTIVE METHODS OF DESALTING WATER IRRIGATION TECH & OTHER MEANS OF MAKING ARID LAND AVAIL&ECON USEFUL FOR AGRICULTURAL PURPOSES...SMALL SCALE DEVICES FOR INDIVIDUAL COMMUNITIES&AG HOLDINGS PUBLISH INFO RELATED TO RESEARCH AS WELL AS BUILDING PLANS WRITE FOR PUBLICATIONS LIST MISC REPORT NO M 17 SEND 25 CENTS TO COVER HANDLING SEE ALSO T A LAWAND DIR FIELD OPERATIONS BRACE RESEARCH INST

INTL SOLAR ENGY SOC/UK SECT — PUBLISH LAY + TECHNICAL PERIODICALS
21 ALBEMARLE ST
LONDON — ENGLAND
J K PAGE
CHAIRMAN

A F E D E S
28 RUE DE LA SOURCE
75016 PARIS — FRANCE
PAUL GIRARD
GEN DELEGATE AFEDES

-PHONE-
224 59 35

PERFORM ALL ACTIONS FAVORIZING DEVELOPMENT OF APPLICATIONS OF SOLAR ENERGY...HOLD RESEARCH MEETINGS ABOUT SOLAR FNERGY...INTERNATL CONGRES SUN IN THE SERVICE OF MANKIND JULY 2-6 1973 PARIS

SUBCLASSES — SOLAR SUB SYSTEMS

SISTEMAS DE ENERGIA SOLAR AVENIDA MAGDALA 1151 CONDADO SANTURCE PUERTO RICO DAVID L GUTHRIE	PUBLISH ORIGINAL RESEARCHED INFORMATION PROVIDE INFORMATION SEARCHES TO CLIENTS

-PHONE-
809 7257825
PROVIDE INDIVIDUAL BUILDERS & OTHER INTERESTED PARTIES WITH ACCESS TO SOLAR INFO & CONSULTATION...WRITING SEVERAL BOOKS ON SOLAR ENERGY 'SOLAR ENERGY: A BIBLIOGRAPHY' BY GUTHRIE, DAVID L & RILEY, ROBERT A TO BE PUBLISHED BY OHIO STATE UNIV PUBLICATIONS COMM COLUMBUS OH 43210

------------------------------SUBCLASSES--------------------SOLAR SUB SYSTEMS------------------------

LEHR ASSOC,CONSULTING ENGRS	MEDIUM TEMPERATURE SOLAR COLLECTOR SYSTEMS
ONE PENN PLAZA	HIGH TEMPERATURE SOLAR COLLECTOR SYSTEMS
NEW YORK NY 10001	WIND CONVERSION
VALENTINE A LEHR	INORGANIC SEMI-CONDUCTORS
	PROVIDE INFORMATION SEARCHES TO CLIENTS

-PHONE-
212 9478050

MAKE PRACTICAL ECONOMIC APPLICATION OF NATURAL ENERGY SOURCES TO COMM-ERCIAL RESIDENTIAL & EDUCATIONAL BLDGS...CURRENTLY-PRELIM DESIGN STAGE FOR COLLEGE OF ATLANTIC PROJ BAR HARBOR MAINE/STUDYING FEASIBILITY OF SOLAR H2O HEATERS FOR HOTEL APPLICATION (HILTON INTNL)&FOR RESIDENTIAL COMPLEXES (WITH NYS URBAN DEV CORP)

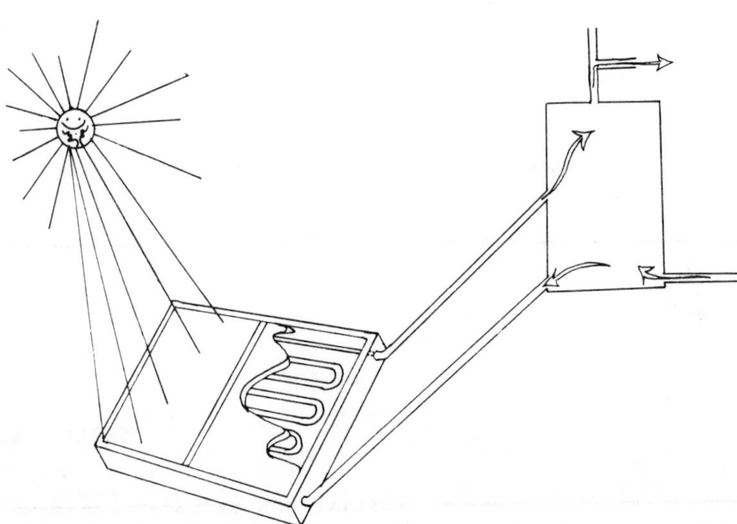

INDEX OF SOLAR/ENERGY CONSERV INFO DISSEMINATION ORG (SOLAR SPACE HEATING) 12/20/74

---------SUBCLASSES---------SOLAR SUB SYSTEMS---------

J K RAMSTETTER
PRODUCT DESIGNER
1520 W ALASKA PLACE
DENVER CO 80223

MEDIUM TEMPERATURE SOLAR COLLECTOR SYSTEMS

CONSULTANT TO RETIRED PEOPLE...CUTTING HEATING COSTS BY 50 PERCENT
WITH SOLAR COLL ON SIDE OF HOME

sun warms
my face
and
makes me
feel good
about
this place

Carolyn

INDEX OF SOLAR/ENERGY CONSERV INFO DISSEMINATION ORG (SOLAR SPACE HEATING & COOLING)

----SUBCLASSES----SOLAR SUB SYSTEMS----

CASSEL SOLAR ENGINEERING — PROVIDE INFORMATION SEARCHES TO CLIENTS
ROUTE 10 BOX 17
ANNAPOLIS MD 21401
DAVID E CASSEL

-PHONE-
301 9740897
ENGINEERING CONSULTANTS-PROMOTE USE OF SOLAR ENGY TO FULFILL ENERGY
REQUIREMENTS OF BLDGS FOR HEATING & COOLING

COGSWELL/HAUSLER ASSOC
P O BOX 2214
CHAPEL HILL NC 27514
EDWARD HOSKINS

-PHONE-
919 9425197
SOLAR APPLICATIONS IN BUILDING DESIGN ARCHITECTURAL CONSIDERATIONS
'SOLAR ENERGY AND THE NATURAL HOUSE',1974
SOLAR APPLICATIONS IN CLIMATILOGICALLY-SENSITIVE BUILDING DESIGN-FOR
FURTHER DETAILS ON PUBLICATION WRITE.

COLORADO STATE UNIVERSITY
PROJECT SUNIF
FT COLLINS CO 80521

-PHONE-
303 4911[illegible]
PACKAGED AUDIO-VISUAL LECTURE SERIES ON SOLAR ENERGY ENGINEERING DIST
IN COLO ONLY TO UNIV & CORPS WITH 3 OR MORE EMPLOYEES REGISTERING

WATT ENGR LTD — SOLAR HEAT,COOL,&ELECTRIC POWER GENERATION; PROVIDE INFORMATION SEARCHES TO CLIENTS — COLLECTOR,MECHANICAL SYSTEMS
8395 BASELINE RD
BOULDER CO 80303
ARTHUR D WATT

-PHONE-
303 4946734
DEV OF SOLAR ENERGY COMPONENTS AND SYSTEMS...CONSULTATION

---SUBCLASSES---SOLAR SUB SYSTEMS---

WORMSER SCIENTIFIC CORP	SOLAR RADIATION MEASUREMENTS
88 FOXWOOD RD	MEDIUM TEMPERATURE SOLAR COLLECTOR SYSTEMS
STAMFORD CT 06903	TOTAL SYSTEMS INTEGRATION
ERIC M WORMSER	SOLAR WATER HEATING
PRESIDENT	SOLAR HEAT,COOL,&ELECTRIC POWER GENERATION
	PROVIDE INFORMATION SEARCHES TO CLIENTS

-PHONE-
203 3221981
203 3226200

PYRAMIDAL OPTICS SOLAR ENERGY COLLECTION SYSTEM...WILL PROVIDE CONSULT SERVICES FOR DESIGN OF THIS SYS ALSO AID IN PLANNING & INSTALLATION OF SYS/LICENSES AVAIL FOR SYS USE

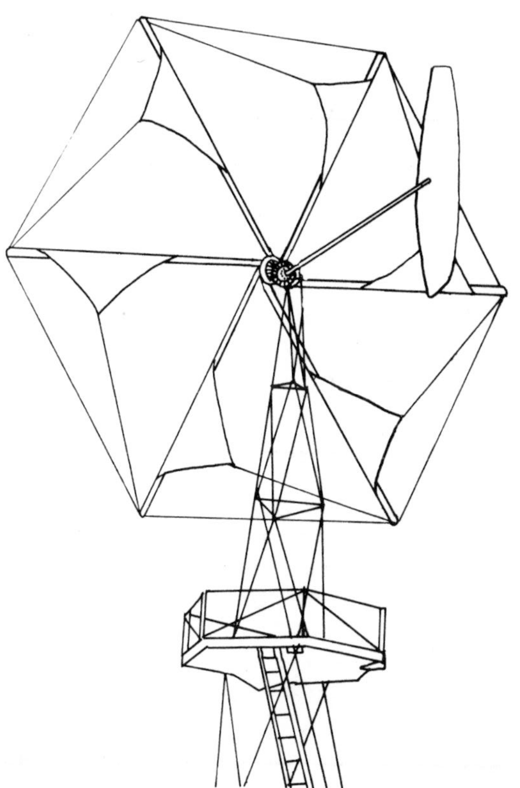

INDEX OF SOLAR/ENERGY CONSERV INFO DISSEMINATION ORG (SOLAR ELECTRIC POWER GENERATION)

----------SUBCLASSES----------SOLAR SUB SYSTEMS----------

UNIVERSITY CITY SCIENCE INST   PROVIDE INFORMATION SEARCHES TO CLIENTS
POWER INFO CENTER
3401 MARKET-RM 2107
PHILADELPHIA   PA   19104

-PHONE-
215 3828683
COLLECTS & PROVIDES INFO ON POWER PRODUCTION BY STATIONARY NON-AUTO-
MOTIVE METHODS.SERVICES AVAIL BY APPROVAL INTERAGENCY ADVANCED PWR GP

"We should think of the sun as our controlled fusion reactor, which is safely installed and operating with essentially an unlimited source of fuel and which has no operating or maintenance requirements."

Dr. Eugene Ralph
Heliotek Division
Textron Industries

	SUBCLASSES	SOLAR SUB SYSTEMS
ARGONNE NATIONAL LABORATORY ACCELERATOR FACILITY 9700 SOUTH CASS AVE ARGONNE IL 60439 JOHN H MARTIN SENIOR PHYSICIST -PHONE- 312 7397711 X4881 312 8529714 BUILDING HEATING BIOMASS	MEDIUM TEMPERATURE SOLAR COLLECTOR SYSTEMS BIO-CONVERSION TOTAL SYSTEMS INTEGRATION SOLAR REFRIGERATION HOMES GREENHOUSES	
COLORADO UNIV OF-BOULDER CTR FOR MGT&TEC PROG BOULDER CO 80302 -PHONE- 303 4434865 OFFER SOLAR HEATING AND COOLING SEMINAR FOR DETAILS CONTACT: DR JAN KREIDER & DR FRANK KREITH	NEW CONSTRUCTION, PASSIVE SYSTEMS NEW CONSTRUCTION, MECHANICAL SYSTEMS PUBLISH LAY + TECHNICAL BOOKS PROVIDE AUDIO & VISUAL SERVICES PROVIDE INFORMATION SEARCHES TO CLIENTS	COLLECTOR,ALSO FUNCTION AS STRUCTURES SKIN COLLECTOR,REFLECTIVE SELECTIVE SURFACES COLLECTOR,ABSORPTIVE SELECTIVE SURFACES COLLECTOR,PASSIVE SYSTEMS COLLECTOR,MECHANICAL SYSTEMS HEAT TRANSFER SYSTEMS, COLLECTOR TO STORAGE STORAGE, SOLID MATERIALS

--------SUBCLASSES--------SOLAR SUB SYSTEMS--------

ILLINOIS, STATE OF
CAPITAL DEVELOP BORD
1030 S LAGRANGE RD
LAGRANGE IL 60525
ROBERT A KEGEL PE

SOLAR SPACE HEATING
PLANNING OFFICES, STATE

-PHONE-
312 3528134
SERVICE AGENCY INVOLVED IN CAPITAL EXPENDITURE FOR CONSTRUCTION...R&D TO FIND IMPROVED WAYS TO SAVE ENGY IN INSTITUTIONAL BLDGS...CURRENTLY CONSIDERING SEVERAL LARGE SCALE R&D PROG SOLAR HEAT UNDER CONSIDERATON SPONSORED'ENERGY CRISIS'REGIONAL CONFERENCE SEPT 19-21 1973

N Y STATE COUNCIL ON ARCHITECT
810 SEVENTH AVE
NEW YORK NY 10019
BARBARA J SCHNEPP

NEW CONSTRUCTION, MECHANICAL SYSTEMS
PUBLISH ORIGINAL RESEARCHED INFORMATION

-PHONE-
212 7657630
CONCERN WITH ALL ENERGY RESEARCH...2 PROJECTS RELATING TO ENERGY CONSV RESULTS FOUND IN 2 DOCUMENTS: AD HOC COMMITTEE ON ENERGY CONSERVATION IN LARGE BLDGS REPORT & AD HOC COMMITTEE ON ENERGY EFFICIENCY IN APP-LIANCES&APPARATUS REPORT

	SUBCLASSES	SOLAR SUB SYSTEMS
ALTERNATIVE SOURCES OF ENERGY RT 2, BOX 90A MILACA MN 56353 DONALD MARINER	WIND CONVERSION SOLAR ENERGY STORAGE TOTAL SYSTEMS INTEGRATION SOLAR COOKING SOLAR WATER HEATING SOLAR SPACE HEATING	COLLECTOR,PASSIVE SYSTEMS COLLECTOR,ALSO FUNCTION AS STRUCTURES SKIN HEAT ENGINE, STERLING ELECTROLYSIS

-PHONE-
612 9836892
ALTERNATIVE SOURCES OF ENERGY IS A MAGAZINE FOR PEOPLE, INTRESTED IN D EVELOPING ALTERNATIVES, DECENTRALIZED TECHNOLOGIES.SUBSCRIPTIONS ARE $5 FOR EACH 6 ISSUES.ASE IS EDIDED AND PUT TOGETHER BY MANY PEOPLE-IT IS A COMMUNICATIONS NETWORK.....BACK ISSUES OF ASE #'S 1-10 AVAILABLE BOUND-$4.00

NEW ALCHEMY INSTITUTE WEST BOX 376 PESCADERO CA 94040 YEDIDA MERRILL	MEDIUM TEMPERATURE SOLAR COLLECTOR SYSTEMS BIO-CONVERSION	STORAGE, METHANE

-PHONE-
805 9690101
NEW ALCHEMY INST HAS MEMBERSHIP PROG AVAIL TO ALL INTRESTED IN ASSIST US ACHIEVE OUR GOALS...MEMBERS RECEIVE OUR PERIODIC PUBLICATIONS WHICH DEAL WITH THEORETIC & PRACTICAL ASPECTS OF NEW WORLD PLANNING...PAST NEWSLETTERS OF INTEREST:'METHANE DIGESTERS BACKGROUND & BIOLOGY USING THE GAS & SLUDGE DESIGNS FOR SIMPLE WORKING MODELS,' NL NO3 SPRING 73, PRICE $3.00

OPEN NORTHWEST INFO NETWORK 608 19TH AVE E SEATTLE WA 98112	PROVIDE INFORMATION SEARCHES TO CLIENTS	

-PHONE-
206 3238506

SUBCLASSES — SOLAR SUB SYSTEMS

OZARK ACCESS CENTER INC
55 SPRING ST BOX 506
EUREKA SPRINGS AR 72632
EDD JEFFORDS

MEDIUM TEMPERATURE SOLAR COLLECTOR SYSTEMS — STORAGE, METHANE
BIO-CONVERSION
WIND CONVERSION
SOLAR SPACE HEATING & COOLING

-PHONE-
501 2539616
501 2539601
RESEARCH SOLAR COLL WIND & WATER GENERATORS FOR PRACTICAL APPLICATION ON HOMESTEADS...CURRENTLY ASSEMBLING METHANE GEN/SOLAR GROWHOLES PUBLISH DATA IN 'OZARK ACCESS CATALOG' SUBSCRIPTIONS $5.00 PER YEAR

ROBERT MOREY ASSOCIATES
ENERGY INFO
PO BOX 98
DANA POINT CA 92629

PUBLISH ENERGY INFO ONCE PER MONTH COVERING LATEST EVENTS IN FIELD OF ENERGY...LISTS COMING MEETINGS ON ALL FORMS ENERGY...PRICE 1 YEAR-12 ISSUES:$30.00

SCHOOL OF LIVING
RT 1 BOX 129
FREELAND MD 21053
MILDRED J LOOMIS

MEDIUM TEMPERATURE SOLAR COLLECTOR SYSTEMS
WIND CONVERSION
TOTAL SYSTEMS INTEGRATION

-PHONE-
301 3575619
301 3296549
PUBLISH THE GREEN REVOLUTION A BI-MONTHLY JOURNAL TO ENCOURAGE PRODUCTIVE HOMESTEADS & DECENTRALIZED INDUSTRY TO INCREASE USE OF ALTERNATIVE POWER & TECHNIQUES...INCLUDES INFO ON SOLAR WIND & H2O POWER TECH...COST $4.00 A YEAR

---SUBCLASSES---SOLAR SUB SYSTEMS---

SOLAR ENERGY DIGEST PO BOX 17776 SAN DIEGO CA 92117 WILLIAM B EDMONDSON CWO-4 EDITOR & PUBLISHER	MEDIUM TEMPERATURE SOLAR COLLECTOR SYSTEMS TOTAL SYSTEMS INTEGRATION ALL SOLAR TECHNOLOGIES SOLAR WATER HEATING SOLAR SPACE HEATING & COOLING PUBLISH ORIGINAL RESEARCHED INFORMATION

-PHONE-
714 2272980

REPORTING ALL IMPORTANT DEVELOPMENTS IN SOLAR ENERGY WORK-WORKING TO ESTABLISH A SOLAR ENERGY RESEARCH CENTER
PLANS AVAILABLE TO BUILD SOLTERRA SOLAR HEAT & COOL HOME
SUBSCRIPTION RATES SOLAR ENERGY DIGEST (PUB 1 PER MO):$27.50 PER YR(US
SEE ALSO:SOLAR EQUIP SALES DIV, SOLAR ENERGY DIGEST-DISTRIBUTE & SELL FLAT PLATE HOT H2O COLLECTORS

SOLAR ENERGY RESEARCH&INFO CTR 1001 CONN AVE NW WASHINGTON DC 20036 LUCILLE ALFIERI ACTING DIRECTOR	ALL SOLAR TECHNOLOGIES PROVIDE INFORMATION SEARCHES TO CLIENTS

A SPECIALTY SERVICE ORGANIZATION DEVOTED EXCLUSIVELY TO ASSISTING PERSONS COMPANIES GOVT ASSOC & OTHER ORGAN.PURPOSE TO PARTICIPATE IN DEVELOPMENT OF SOLAR ENERGY USE IN US & ABROAD PUBLISHES THE FOLLOWING
'SOLAR ENERGY WASHINGTON LETTER'24 ISSUES PUB 1ST & 3RD MONDAYS $75/YR
'SOLAR ENERGY INDUSTRY REPORT'24 ISSUES PUB 2ND & 4TH MONDAYS $75/YR
'SOLAR ENERGY SPECIAL BULLETIN' 4 ISSUES PUB 5TH MONDAYS BONUS
COMBO SUBSCRIPTION $125/YR

SOLAR ENERGY RESEARCH&INFO CTR 1001 CONN AVE NW WASHINGTON DC 20036 RACHEL SYNDER WRITER	ALL SOLAR TECHNOLOGIES PUBLISH ORIGINAL RESEARCHED INFORMATION PROVIDE INFORMATION SEARCHES TO CLIENTS LOBBY TO PROMOTE SOLAR LEGISLATION

-PHONE-
202 2931000

PUBLISHER OF SOLAR ENERGY WASHINGTON LETTER & SOLAR ENERGY INDUSTRY REPORT

----SUBCLASSES----SOLAR SUB SYSTEMS----

AM SOC HEAT REFRIG AIRCON ENGR
PUBLICATION SALES
345 EAST 47TH ST
NEW YORK NY 10017

SOLAR SPACE HEATING & COOLING
PUBLISH ORIGINAL RESEARCHED INFORMATION

-PHONE-
212 7526800
PUBLISH ASHRAE JOURNAL MONTHLY FOR ASHRAE MEMBERS-AUTHORITATIVE JOURNL ON RESEARCH DESIGN DEVELOPMENT&ENGR ALSO REPORTS NEW PROD BOOKS TRENDS &CHANGES RELATED TO INDUSTRY/PUBLISH MANY ADDITIONAL PUB INCLUDING: 'SOLAR-ENGY LOW TEMP ENGR OF: PARTICULARY RELATED TO FLAT PLATE COLL & FEATURING AUTHORATIVE TECH TREATMENTS'. '1974 ASHRAE HANDBOOK APPLICATIONS VOLUME' CH 59 EXCELENT SOLAR INFO...WRITE FOR PUBLICATIONS LIST

----------SUBCLASSES----------SOLAR SUB SYSTEMS----------

ADV SOLAR ENGY TECH NEWSLETTER — ALL SOLAR TECHNOLOGIES
1609 WEST WINDROSE — PUBLISH ORIGINAL RESEARCHED INFORMATION
PHOENIX AZ 85029
CARL M LANGDON
EDITOR-PUBLISHER

-PHONE-
602 9423796
REPORTS TRENDS PRODUCTION APPLICATION CURRENT REVIEWS OF LITERATURE & BASIC RESEARCH...PUB MONTHLY $60/YR IN US CANADA & PUERTO RICO.FOREIGN SUB $65

AMERICAN INST OF ARCHITECTS
1735 NEW YORK AVE NW
WASHINGTON DC 20006

-PHONE-
202 7857300
AIA ANSWERS INQUIRIES PROVIDES REFERENCE SERVICE AND PUBLISHES: AIA JOURNAL (MONTHLY) AIA MEMO (MONTHLY) GOVERNMENT AFFAIRS REVIEW (ALSO MONTHLY)&ARCHITECTS HANDBOOK OF PROFESSIONAL PRACTICES AND OTHERS

ARCHITECTURE PLUS — WIND CONVERSION
1345 SIXTH AVE — SOLAR SPACE HEATING & COOLING
NEW YORK NY 10019 — PUBLISH LAY + TECHNICAL BOOKS
MARGOT VILLECCO
SENIOR EDITOR

-PHONE-
212 4898697
RESCH IN AREAS OF ENGY CONSERVATION AND NATURAL ENGY SYS...AUTHOR OF 'WIND POWER' MAY/JUNE 74 & 'SOLAR POWER' SEPT/OCT 74...EDITOR 'ENERGY CONSERVATION IN BLDG DESIGN' WITH GRANT FROM FORD FOUND...SEE BIBLIO

INDEX OF SOLAR/ENERGY CONSERV INFO DISSEMINATION ORG (PUBLISH LAY + TECHNICAL PERIODICALS)

Organization	State	ZIP	Subclasses	Solar Sub Systems
SOLAR ENERGY SOC OF AMERICA 2780 SEPULA BLVD TORRANCE DAVID SATCHWELL PRESIDENT -PHONE- 213 3263283	CA	90510	PUBLISH LAY + TECHNICAL BOOKS PROVIDE AUDIO & VISUAL SERVICES PROVIDE INFORMATION SEARCHES TO CLIENTS	
CATALYST FOR INFORMATION EXCHANGE SPECIFICALLY DEDICATED TO BRINGING PRACTICAL APPLICATION OF SOLAR ENERGY TO FRUITION...WORK WITH INDIVIDUALS INSTITUTIONS AGENCIES & INDUSTRY TO BRING TOGETHER THOSE WHO HAVE INFORMATION WITH THOSE WHO SEEK IT OFFER PRODUCT & PROGRESS STATUS INFORMATION				
SOLAR ENFRGY SOC OF AMERICA 2780 SEPULA BLVD TORRANCE JUDY STERNBERG * DIRECTOR INFO EXCHNG -PHONE- 213 3263283	CA	90510	PUBLISH LAY + TECHNICAL BOOKS PROVIDE AUDIO & VISUAL SERVICES PROVIDE INFORMATION SEARCHES TO CLIENTS	
SUN PUBLISHING CO P O BOX 4383 ALBUQUERQUE THE SUN SOLAR ENERGY SERIES -PHONE- 505 2556550	NM	87106	SOLAR RADIATION MEASUREMENTS HIGH TEMPERATURE SOLAR COLLECTOR SYSTEMS TOTAL SYSTEMS INTEGRATION ALL SOLAR TECHNOLOGIES PUBLISH LAY + TECHNICAL BOOKS PUBLISH ORIGINAL RESEARCH + REPRINTS	COLLECTOR,ALSO FUNCTION AS STRUCTURES SKIN COLLECTOR,REFLECTIVE SELECTIVE SURFACES COLLECTOR,ABSORPTIVE SELECTIVE SURFACES HEAT TRANSFER SYSTEMS, COLLECTOR TO STORAGE STORAGE, LIQUIDS
REPRINT OF SOLAR ENERGY ARTICLES BY STEVE BAER ORIGINALLY SERIALIZED IN TRIBAL MESSENGER 30 PAGES $3.00 PLUS $.50 POSTAGE...SUBSCRIPTIONS TO 'THE SUN' MONTHLY ALTERNATIVE CULTURE NEWSPAPER $5.00 FOR 8 ISSUES				
SUN PUBLISHING CO P O BOX 4383 ALBUCUERQUE SKIP WATSON * EDITOR -PHONE- 505 2556550	NM	87106	SOLAR RADIATION MEASUREMENTS HIGH TEMPERATURE SOLAR COLLECTOR SYSTEMS TOTAL SYSTEMS INTEGRATION ALL SOLAR TECHNOLOGIES PUBLISH LAY + TECHNICAL BOOKS PUBLISH ORIGINAL RESEARCH + REPRINTS	COLLECTOR,ALSO FUNCTION AS STRUCTURES SKIN COLLECTOR,REFLECTIVE SELECTIVE SURFACES COLLECTOR,ABSORPTIVE SELECTIVE SURFACES HEAT TRANSFER SYSTEMS, COLLECTOR TO STORAGE STORAGE, LIQUIDS

----------SUBCLASSES----------SOLAR SUB SYSTEMS----------

BIOTECHNIC PRESS
P O BOX 26091
ALBUQUERQUE NM 87125

BIO-CONVERSION
SOLAR SPACE HEATING

'LIFE SUPPORT TECHNICS CONF PROCEEDINGS'$1.50 TRANSCRIBED FROM GHOST RANCH CONF FALL 1972 TOPICS INCLUDE:SOLAR ENERGY/WIND ENERGY/LOW ENGY SYSTEMS & MORE
'SOL SHOT POSTER' 50 CENTS-DESIGN FOR CONVECTIVE AIR LOOP ROCK STORAGE SOLAR HEATER
'BIOSPHERE POSTER' 50 CENTS-PROPOSAL FOR FREE FORM HOUSE SOLAR HEATED BY ATTACHED GREENHOUSE
'GROWHOLE POSTER' 50 CENTS-DESIGN FOR AN UNDERGROUND SOLAR HEATED GREENHOUSE FOR GROWING VEGETAGLES IN WINTER
THIS IS A PARTIAL LISTING OF PUBLISHED WORKS

SYNERGY PEOPLES' PAGES
P O BOX AH
STANFORD CA 94305
ALLEN BORNING
TEXT EDITOR

ALTERNATIVE INFO RESOURCE COLLECTIVE.LISTINGS OF ALL ALT INFO ENTERED ON THEIR COMPUTER...THEIR PEOPLES PAGES A COMPUTER PRINT OUT DIRECTORY NOT TO MUCH ON ALT SOURCES OF ENGY OR SOLAR YET...FULL SUBSCRIPTION $5.00 PER YEAR

	SUBCLASSES	SOLAR SUB SYSTEMS
ENVIRONMENTAL INFORMATION CTR 124 EAST 39TH STREET NEW YORK NY 10016 CHRISTOPHER KIMBALL	PUBLISH ORIGINAL RESEARCHED INFORMATION PROVIDE INFORMATION SEARCHES TO CLIENTS HAVE&MAINTAIN ENERGY DATA BANK	

-PHONE-
212 6790810
REPRESENTATIVE OF EIC AN INDEPENDENT RESCH & PUB ORG PROVIDING COMPREHENSIVE INTELLIGENCE REPORTING REFERENCE & RETRIEVAL SERV COVERING ALL ASPECTS OF ENVIRO AFFAIRS...PUBLICATIONS INCLUDE THE ENERGY INDEX &THE ENERGY DIRECTORY & ENVIRONMENT INDEX...SEE BIBLIO SECTION

	SUBCLASSES	SOLAR SUB SYSTEMS
PORTOLA INSTITUTE ENERGY PRIMER 558 SANTA CRUZ AVE MENLO PARK CA 94025 CHARLES E MISSAR EDITOR	TOTAL SYSTEMS INTEGRATION SOLAR HEAT,COOL,&ELECTRIC POWER GENERATION NEW CONSTRUCTION, RESEARCHING CLUSTER HOMES PUBLISH ORIGINAL RESEARCH + REPRINTS PROVIDE INFORMATION SEARCHES TO CLIENTS	COLLECTOR,MULTI-FUNCTION IN DWELLING SYS COLLECTOR,PASSIVE SYSTEMS COLLECTOR,MECHANICAL SYSTEMS STORAGE, LIQUIDS STORAGE, CHEMICAL BATTERIES

-PHONE-
415 3235535
415 3230313
RESEARCHING MATERIAL FOR BOOK ON RENEWABLE SOURCES OF ENERGY TO BE PUBLISHED FALL 1974... BOOK WILL INCLUDE BASIC PRINCIPLES VENDORS AND BOOK REVIEWS IN AREAS OF SOLAR WIND SMALL SCALE WATER METHANE & WOOD ENERGY ARCHITECTURE TRANSPORTATION & SYNERGY WILL ALSO BE EXAMINED AVAILABLE FROM PORTOLA INSTITUTE FOR 4 DOLLARS..THE ENERGY PRIMER

--------SUBCLASSES--------SOLAR SUB SYSTEMS--------

COLORADO, UNIV OF-DENVER	ALL SOLAR TECHNOLOGIES
ENVIRONMTL ACTION-CO	PUBLISH ORIGINAL RESEARCH + REPRINTS
1100 14TH STREET	PUBLISH AND/OR DISTRIBUTE FILMS
DENVER CO 80202	

BEN BILLINGS
ENVIRO ACT REPRINT S
-PHONE-
303 5341602 X11
303 8921117 X471
EARS DISSEMINATES INFO ON ENERGY & ENVIRO ISSUES WE REPRINT CURRENT ARTICLES AS WELL AS OFFER RELEVANT BOOKS & COLOR FILMS CURRENT SOLAR REPRINTS: THE SOLAR RESOURCE 4.95 EA...2 FOR 7.90...WRITE FOR COMPLETE CATALOGUE 25 CENTS

	SUBCLASSES	SOLAR SUB SYSTEMS
BSIC/EFL 3000 SAND HILL RD MENLO PARK CA 94025 JOSHUA A BURNS ASSOCIATE DIRECTOR	RETROFITTING, RESEARCHING NEW CONSTRUCTION, RESEARCHING ACADEMIC BUILDINGS PUBLISH LAY + TECHNICAL PERIODICALS PUBLISH LAY + TECHNICAL BOOKS	

-PHONE-
415 8542300
DEVELOP AND PUBLISH INFORMATION ON SCHOOL DESIGN AND CONSTRUCTION. A MAJOR CURRENT PROJECT IS ON ENERGY CONSERVATION AND IMPROVED BUILDING SYSTEMS.

	SUBCLASSES	SOLAR SUB SYSTEMS
ECOLOGY ACTION EDUCATION INST BOX 3895 MODESTO CA 95352 CLIFFORD C HUMPHREY	SOLAR WATER DISTILLATION SOLAR SPACE HEATING & COOLING	STORAGE, METHANE

-PHONE-
209 5291964
WORKING PROTOTYPES & PLANS FOR HOMECONSTRUCTABLE SYS FOR HEATING,COOL, H2O DISTILLATION,&METHANE GEN SUITED TO CONDITIONS IN CENTRAL VALLEY CALIF/ONE SOLAR H2O COMP/METHANE DIGESTOR FUNCTIONAL LITERATURE AVAIL SPRING 1974

GENERAL RESEARCH CORPORATION   PROVIDE INFORMATION SEARCHES TO CLIENTS
SYSTEMS RESEARCH DIV
1501 WILSON BLVD
ARLINGTON   VA   22209
NANCY C LESTER

-PHONE-
703 5347206
SPECIALISTS IN ENERGY POLICY RESEARCH...SAMPLE OF PAST CONTRACTING AGENCIES: HUDSON INST, NSF, EPA, DEPT OF TRANS, RESOURCES FOR FUTURE.. ...SAMPLES OF POLICY RESEARCH REPORTS,(INQUIRE FOR COMPLETE LIST):
AYRES,R V,'ALTERNATIVE NONPOLLUTING POWER SOURCES,'SAE JOURNAL, VOL 76 NO 12, DEC 1968.
----,'FUTURE OF INNOVATION--TECHNOLOGY IN 21ST CENTURY', PREPARED FOR AMERICAN ASSOC FOR ADVANCEMENT SCI ANNUAL MEET, SYMPOSIUM ON TECH & GROWTH IN RESOURCE LIMITED WORLD, DEC 1971.
CAMBEL,AB,'ENERGY,'SCI JOURNAL, SPECIAL ISSUE ON'FORCASTING FUTURE,'3, 10, 57-62, OCT 1967.
----,'ENERGY R&D & NATIONAL PROGRESS,'STUDY LEADER, US GOVT PRINTING OFF, WASHINGTON DC, 1965.
RENNER,R,'STATUS REPORT: CALIF STEAM BUS PROJECT,' JAN 1972.
----, RU AYRES&K HUMES,'PRIORITIES FOR TECHNOLOGY ASSESSMENT,'PREPARED FOR RANN PROG, NSF, IRT-310-R, FEB 1973.

INFORM   PROVIDE INFORMATION SEARCHES TO CLIENTS
25 BROAD ST
NEW YORK   NY   10004
STEWART W HERMAN

-PHONE-
212 4253550
STUDY THE IMPACTS OF CORPORATIONS ON SOCIETY CURRENTLY STUDYING US COMPANIES INVOLVED IN ALTERNATIVE ENERGY SOURCE RESEARCH RESULTS TO BE PUBLISHED FALL 74

OBELITZ INDUSTRIES INC   MEDIUM TEMPERATURE SOLAR COLLECTOR SYSTEMS
BOX 2788   BIO-CONVERSION
SEAL BEACH   CA   90740   WIND CONVERSION

-PHONE-
213 5985425
ESTABLISH AN OUTLET FOR & MARKET SCIENTIFICALLY SOUND SET OF PLANS & DESIGNS FOR SOLAR & WIND SYS FOR HOMES & FARMS...TO MARKET & FABRICATE ASSOCIATED PRODUCTS, PARTICULARLY SYS FOR HYDROPONIC GARDENING

--------------------SUBCLASSES--------------------SOLAR SUB SYSTEMS--------------------

BUILDING RESEARCH STATION SOLAR RADIATION MEASUREMENTS
ADVISORY SERVICE
GARSTON WATFORD
WD2 7JR UNITED KINGD
*

-PHONE-
092 7 3 766 X12
SAMPLE OF PUBLICATIONS AVAIL INQUIRE FOR COMP LIST: 'DISCOMFORT IN SCHOOLS FROM OVERHEATING IN SUMMER CP 19/70'.'WINDOW DESIGN CRITERIA TO AVOID OVERHEATING BY EXCESSIVE SOLAR HEAT GAINS CP 4/68'.'THE INTERPRETATION OF SOLAR RADIATION MEASUREMENTS FOR BLDG PROB RESEARCH PAPER 73'. SEE ALSO P PERTHERBRIDGE.BLDG RESEARCH STATION

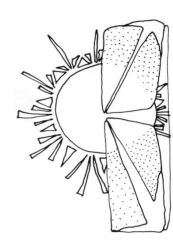

INDEX OF SOLAR/ENERGY CONSERV INFO DISSEMINATION ORG (PUBLISH ORIGINAL RESEARCH + REPRINTS)

12/20/74

SUBCLASSES	SOLAR SUB SYSTEMS
ENVIRONMENTAL EDUCATION GROUP 67 LURLINE AVENUE CANOGO CA 91306 ALLAN TRATNER	OCEAN THERMAL GRADIENT SYSTEMS ALL SOLAR TECHNOLOGIES

----SUBCLASSES----SOLAR SUB SYSTEMS----

CHURCHILL FILMS
CUSTOMER RELATIONS
662 NO ROBERTSON
LOS ANGELES CA 90069
PRISCILLA FLORENCE *

-PHONE-
213 6575110
PLEASE ADDRESS INQUIRY CONCERNING FILMS TO ABOVE

CHURCHILL FILMS
662 NO ROBERTSON
LOS ANGELES CA 90069
GEORGE H MCQUILKIN
PRODUCER-DIRECTOR

-PHONE-
213 6575110
DISTRIBUTE TWO FILMS: ENERGY: NEW SOURCES, 20 MIN COLOR-POT&LIMITS OF
NEW TECHNOLOGIES INCLUDING SOLAR
ENERGY:LESS IS MORE, 18 MIN COLOR-INVESTIGATION OF NEED TO SLOW ENERGY
CONSUMPTION & WAYS TO DO IT

COLORADO UNIV OF-BOULDER
DIV OF CONTINUING ED
BOULDER CO 80302
MARGARET EVANS

-PHONE-
303 4927416
PRODUCING SERIES OF FILMS ON ALTERNATIVE ENERGY INCLUDING SOLAR UNDER
ENVIRO ED GRANT FROM HEW

	SUBCLASSES	SOLAR SUB SYSTEMS
GLEN-KAYE FILMS 100 EAST 21ST STREET BROOKLYN NY 11226 DOUGLAS R KAYE PRODUCER -PHONE- 212 2872929 212 6939618 PRODUCERS OF 'THE AGE OF THE SUN'-A 21 MINUTE FILM ABOUT VARIOUS TECHNOLOGIES OF SOLAR ENERGY UTILIZATION-A GOOD LAYMAN'S FILM WE ARE ALSO ABLE TO PRODUCE OTHER FILMS OR TAPES BY CONTRACTUAL ARRANGEMENTS--INCLUDING EDUCATIONAL, DOCUMENTARY, INDUSTRIAL,&COMMERCIAL FILMS, VIDEO TAPES,& CASSETTES	PROVIDE AUDIO & VISUAL SERVICES	
KNME CHANNEL 5 1130 UNIV BLVD NE ALBUQUERQUE NM FRED MANFREDI PRODUCER -PHONE- 505 2772121 PROD 6 1/2 HR PROGS ON ALTERNATIVE ENGY SOURCES NATL EDUC BROADCST TV	ALL SOLAR TECHNOLOGIES PROVIDE AUDIO & VISUAL SERVICES	
TR PRODUCTIONS SOLAR ENERGY PROJECT 1031 COMMONWEALTH AV BOSTON MA 02215 CHAS SCHWARTZ DIRECTOR -PHONE- 617 7830200 ENCOURAGE WIDER USE OF SOLAR POWER IN USA BY EXPLAINING WHAT CAN DO FOR THEM...STIMULATE CONSUMER DEMAND TO LEAD TO MASS PRODUCTION&LOWER COSTS...PRODUCING A GENERAL AUDIENCE MARKETING TYPE 27 MIN COLOR FILM EXPLORING POTENTIAL OF SOLAR ENERGY & BENEFITS TO CONSUMERS... AVAILABLE JUNE 1975	PUBLISH ORIGINAL RESEARCH + REPRINTS	

SUBCLASSES		SOLAR SUB SYSTEMS
GRAY COMPANY ENTERPRISES SUITE 245 7701 N STEMMONS DALLAS TX 75247 DAN R FOSTER DIR-MEDIA RELATIONS -PHONE- 214 6303841	PUBLISH LAY PERIODICALS	STORAGE, HYDROGEN

CONDUCTS ON-GOING LECTURE&MEDIA PRESENTATIONS RELATIVE TO USE OF HYDROGEN AS ENERGY CARRIER/MATERIAL AVAILABLE: 35 MM COLOR SLIDES, 8X 10 B&W PHOTOS, 2 IN HIGH-BAND COLOR VIDEO (QUAD BROCAST), 1/2 IN COLOR VIDEO-INDUSTRIAL, 16 MM COLOR FILM OF GRAY VAPOR GEN, 8 MM COLOR FILM OF TESTING PROTOTYPE...ALSO SPEAKERS BUREAU...INQUIRE FOR DETAILS

---SUBCLASSES---SOLAR SUB SYSTEMS---

Organization	Subclasses
COLORADO, UNIV OF-DENVER ENVIRONMTL ACTION-CO 1100 14TH STREET DENVER CO 80202 ELIZABETH A KINGMAN SOLAR EXHIBIT PROG -PHONE- 303 5341602 X9 303 8921117 X471	ALL SOLAR TECHNOLOGIES

NSF FUNDED PROJ TO CREATE A MAJOR MUSEUM EXHIBITION TO GENERATE WIDE SPREAD AWARENESS & UNDERSTANDING OF SOLAR ENERGY THIS WILL OPEN AT GATES PLANETARIUM DENVER MUSEUM OF NATURAL HISTORY & THEN TOUR OTHER MUSEUMS & EDUCATIONAL INSTUTIONS THROUGHOUT THE COUNTRY

DENVER MUSEUM OF NATURAL HIST
GATES PLANETARIUM
CITY PARK
DENVER CO 80205
MARK PETERSON

-PHONE-
303 3884201
EXHIBITION OF PREPLANNED DISPLAYS--CURRENTLY PREPARING EXHIBIT SPACE FOR EXHIBIT ON SOLAR ENERGY

Organization	Subclasses
NATIONAL CAPITAL PARKS KLINGLE URBAN ENV CT 1100 OHIO DRIVE S W WASHINGTON DC 20242 NANCY STRADER	ALL SOLAR TECHNOLOGIES GOVERNMENT AGENCIES

-PHONE-
212 2827020
HAVE PREPARED ENVIRONMENTAL EDUCATION ACTIVITIES FOR USE IN PARK PROG SCHOOLS & WORKSHOPS...ALTERNATIVE ENERGY SOURCES PROGRAM FOR CHILDREN & ADULTS, PREPARING CURRICULUM FOR CLASSROOM USE-GRAPHICS DISPLAYS WORKING MODELS & AUDIOVISUAL MATERIALS

--------------------SUBCLASSES--------------------SOLAR SUB SYSTEMS--------------------

UNITED AUTO WORKERS			TOTAL SYSTEMS INTEGRATION
CONSERVATION DEPT			RETROFITTING, INTEREST
UAW FAMILY ED CENTER			PUBLISH LAY PERIODICALS
ONAWAY	MI	49765	PUBLISH ORIGINAL RESEARCHED INFORMATION
PHILIP J PERKINS			LOBBY TO PROMOTE SOLAR LEGISLATION
ENVIRONMENTAL ED			
-PHONE-			
517 7338521			

ENVIRONMENTAL ED TECHNICIAN FOR UAW CONSERVATION DEPT. OEE GRANT FOR
ADULT ED OF UNION MEMBERS. WORK WITH MARGARET ALLEN AS COPROJ TECHNICN

	SUBCLASSES	SOLAR SUB SYSTEMS
ROBERT H BUSHNELL 502 ORD DRIVE BOULDER CO 80303	MEDIUM TEMPERATURE SOLAR COLLECTOR SYSTEMS SOLAR SPACE HEATING	

-PHONE-
303 4947421
PROVIDE DESIGNS FOR HOME & SMALL BUILDING HEATING, DEMONSTRATION DATA/
INITIATE ACTIVITY MARCH 1974

DANA C JENNINGS RT 3 BOX 177 MADISON SD 57042	ALL SOLAR TECHNOLOGIES

-PHONE-
605 2563377
COMPILING GUIDE TO PEOPLE WORKING IN ALT ENERGY FOR GARDEN WAY LABS

ASE INFO SERVICES 1448 E 7TH ST TUCSON AZ 85719 HARRY D BURRIS PERSON	MEDIUM TEMPERATURE SOLAR COLLECTOR SYSTEMS WIND CONVERSION SOLAR COOKING SOLAR WATER HEATING PUBLISH INFORMATIONAL REPRINTS PROVIDE AUDIO SERVICES

-PHONE-
602 8826571
703 2814564
DO INFO SEARCHING ON ALTERNATIVE SOURCES OF ENERGY...WE ARE HAPPY TO SEARCH ON TOPICS RELATED TO DECENTRALIZED APPROACHES...WE ARE COLLECTING AND DISTRIBUTING EDUCATIONAL RESOURCES...WE ARE TRADING SURVIVAL TECHNOLOGY WITH THIRD WORLD COUNTRIES...WIND MEASUREMENT AND RECORDING EQUIPMENT

BECHTEL CORPORATION
ENERGY INFO CENTER
P O BOX 3965
SAN FRANCISCO CA 43201
BARBARA CLEMENTS

-PHONE-
614 2993151
PROVIDES INFORMATION AND DOCUMENT SERVICES ON A FEE BASIS

----SUBCLASSES----SOLAR SUB SYSTEMS----

COLORADO, UNIV OF-DENVER — ALL SOLAR TECHNOLOGIES
ENVIRONMTL ACTION-CO — PUBLISH AND/OR DISTRIBUTE FILMS
1100 14TH STREET — PROVIDE AUDIO & VISUAL SERVICES
DENVER CO 80202
ALBERT NUNEZ JR
SOLAR PROJECT COORD
-PHONE-
303 5341602 X6
303 8921117 X471
COORDINATOR FOR SOLAR HEATING COOLING & ENERGY CONSERVATION CONF MAY 1974-PROCEEDINGS & AUDIO TAPES AVAILABLE/PLANNING 1975 CONF ECONOMIC & POLITICAL CONSIDERATIONS OF SOLAR ENERGY CONVERSION & IMPLEMENTATION

CURTIS&DAVIS ARCHITECTS/PLANER — MEDIUM TEMPERATURE SOLAR COLLECTOR SYSTEMS
126 EAST 38TH STREET — SOLAR SPACE HEATING & COOLING
NEW YORK NY 10016
CARL T GRIMM

-PHONE-
212 6899590
APPLICATION SUB TO NSF TO DEVELOP FURTHER ARCHITECTURAL APPROACHES TO SOLAR ENERGY/TO INTEGRATE SUPPLEMENTARY SOLAR COMFORT CONDITIONING IN TO NEW & EXISTING BUILDINGS

ILLINOIS, UNIVERSITY OF
SMALL HOMES COUNCIL
ONE EAST ST MARYS RD
CHAMPAIGN IL 61820
WAYNE L SHICK

-PHONE-
217 3331801 X457
PROVIDES INFORMATION PRIMARILY ENERGY CONSEVATION...ANSWERS QUESTIONS FEE FOR EXTENSIVE CONSULTATION

```
------------------------------------------------SUBCLASSES-------------------SOLAR SUB SYSTEMS-----------------------

LIBRARY OF CONGRESS                 ALL SOLAR TECHNOLOGIES
SCIENCE POLICY DIV                  SOLAR WATER DISTILLATION
WASHINGTON            DC    20540   SOLAR WATER HEATING
JAMES G MOORE                       SOLAR REFRIGERATION
                                    SOLAR SPACE HEATING & COOLING
                                    SOLAR ELECTRIC POWER GENERATION
-PHONE-
202 4266040
ASSIST MEMBERS OF CONGRESS IN UNDERSTANDING VARIOUS SOLAR APPLICATIONS
THROUGH ASSESSMENTS OF SPECIFIC APPLICATIONS BRIEFINGS BACKGROUND
PAPERS ETC...TRACK SOLAR LITERATURE TO KEEP MEMBERS INFORMED OF RECENT
DEVELOPMENTS...MAINTAIN PERSONAL CONTACT WITH RESEARCHERS BUSINESSES
& AGENCY PERSONNEL INVOLVED IN SOLAR R&D
----------------------------------------------------------------------------------------------------------------------

NATIONAL TECHNICAL INFO SER         WIND CONVERSION
5285 PORT ROYAL ROAD                ALL SOLAR TECHNOLOGIES
SPRINGFIELD           VA    22151   PUBLISH TECHNICAL BOOKS
SALES OFFICE                        PUBLISH ORIGINAL RESEARCHED INFORMATION

-PHONE-
703 3218500
WE ARE THE CENTRAL SOURCE FOR PUBLIC SALE OF GOVT SPONSORED RESEARCH
DEVELOPMENT & ENGINEERING REPORTS & OTHER ANALYSES PREPARED BY FEDERAL
AGENCIES...CURRENT ABSTRACTS OF NTIS DOCUMENTS & OTHER RECORDS OF INT-
EREST PUB WEEKLY...MICROFICHE SERVICE...WRITE FOR FULL LISTING OF SERV
AVAIL
----------------------------------------------------------------------------------------------------------------------

ROARING FORK RESOURCE CENTER        SOLAR HEAT,COOL,&ELECTRIC POWER GENERATION   COLLECTOR,MECHANICAL SYSTEMS
P O BOX 9950                        SOLAR TECHNOLOGY APPLICATION OTHER           COLLECTOR,PASSIVE SYSTEMS
ASPEN                 CO    81611   NEW CONSTRUCTION, PASSIVE SYSTEMS            COLLECTOR,OTHER
GREGORY E FRANTA                    NEW CONSTRUCTION, RESEARCHING                STORAGE, LIQUIDS
                                    PUBLISH ORIGINAL RESEARCH + REPRINTS         STORAGE, METHANE
                                    INCENTIVES FOR SOLAR & ENERGY CONSERVATION
-PHONE-
303 9253481
303 9255125
----------------------------------------------------------------------------------------------------------------------
```

Organization	State	ZIP	SUBCLASSES	SOLAR SUB SYSTEMS
SILVER INSTITUTE INC			MEDIUM TEMPERATURE SOLAR COLLECTOR SYSTEMS	COLLECTOR,REFLECTIVE SELECTIVE SURFACES
COLO REPRESENTATIVE			NON-TERRESTRIAL SOLAR POWER SYSTEMS	COLLECTOR,ABSORPTIVE SELECTIVE SURFACES
1001 CONNECTICUT, NW			WIND CONVERSION	COLLECTOR,PASSIVE SYSTEMS
WASHINGTON	DC	20036	SOLAR ENERGY STORAGE	STORAGE, CHEMICAL BATTERIES
ROBERT R DAVIES			TOTAL SYSTEMS INTEGRATION	STORAGE, FUEL CELLS
LITERATURE CONSULT			HAVE&MAINTAIN ENERGY DATA BANK	STORAGE, HYDROGEN
-PHONE-				
303 4430105 XCOLO				
202 3932285 XDC				
THE SILVER INSTITUTE LETTER				
NEW SILVER TECHNOLOGY - SILVER ABSTRACTS FROM THE CURRENT WORLD LIT				
SOLAR DESIGNS			SOLAR SPACE HEATING & COOLING	
203 HOLLY LANE				
ORINDA	CA	94563		
ROBERT M GRAVEN				
-PHONE-				
415 2543938				
COMPUTATIONAL ANALYSIS OF SOLAR HEAT & COOL EQUIP AND SPECIFICATION OF EQUIPMENT (ENGINEERING & ARCHITECTURE)				
SOUTHERN CALIF, UNIV OF			PUBLISH LAY + TECHNICAL BOOKS	
WESRAC			PUBLISH INFORMATIONAL REPRINTS	
809 WEST 34TH ST			PUBLISH ORIGINAL RESEARCHED INFORMATION	
LOS ANGELES	CA	90007	HAVE&MAINTAIN ENERGY DATA BANK	
HERBERT O ASBURY				
MGR SPECIAL PROJECTS				
-PHONE-				
213 7466132				
213 7463675				
SOLAR ENERGY UTILIZATION COMPEDIUM & UPDATES-BIANNUAL,PLUS WORKSHOP				
INFORMATION SEARCH SERVICES-TECHNICAL,EDUCATIONAL,& MANAGEMENT				
ON-LINE SEARCH SERVICES-TERMINAL ACCESS				
INFORMATION EVALUATION SERVICES				
CURRENT AWARENESS SERVICES (SDI=CAS)				
SOUTHERN CALIF, UNIV OF			PUBLISH LAY + TECHNICAL BOOKS	
WESRAC			PUBLISH INFORMATIONAL REPRINTS	
809 WEST 34TH ST			PUBLISH ORIGINAL RESEARCHED INFORMATION	
LOS ANGLES	CA	90007	HAVE&MAINTAIN ENERGY DATA BANK	
WILLIAM SKINNER *				
MARKETING MGR				
-PHONE-				
213 7466132				
213 7463675				

------SUBCLASSES------SOLAR SUB SYSTEMS------

U S PATENT OFFICE ALL SOLAR TECHNOLOGIES
TECH ASSESMNT&FORCST
WASHINGTON DC 20231
ALFRED C MARMOR
DIRECTOR

-PHONE-
703 5573051
LIST OF SOLAR PATENTS RELATED TO ANY STATISTICAL CAN BE PROVIDED...
STATISTICAL REPORTS & PATENT LISTINGS,WHICH ARE LARGELY COMPUTER GEN,
WILL NORMALLY COST BETWEEN $50.TO $100...REFER TO TECHNOLOGY ASSESMENT
& FORECAST, EARLY WARNING REPORT OF OFFICE OF TECH ASSESMNT & FORCAST
DEC 1973

PUERTO RICO,UNIV OF
CENTRO DE INFO TECNI
RECINTO UNIV MAYAGUZ
MAYAGUEZ PUERTO RICO
ERNESTO Q MARTIN
DIRECTOR
FURNISH TECHNICAL INFO BOTH PRACTICAL & SCIENTIFIC TO PUERTO RICO'S
INDUSTRY & COMMERCE CONCERNING LATEST SCI & TECH DEVELOPMENTS RELATED
TO MFG & MARKETING MADE ON ISLAND

---SUBCLASSES---SOLAR SUB SYSTEMS---

COLORADO, UNIV OF-BOULDER
ATIP
RM 331A UMC
BOULDER CO 80302
MILTON LOEB

MEDIUM TEMPERATURE SOLAR COLLECTOR SYSTEMS
WIND CONVERSION
TOTAL SYSTEMS INTEGRATION
PUBLISH ORIGINAL RESEARCH + REPRINTS
PROVIDE AUDIO & VISUAL SERVICES
PROVIDE INFORMATION SEARCHES TO CLIENTS

-PHONE-
303 4926659
INFO ON TECHNOLOGICALLY FEASIBLE ALTERNATIVES TO PRESENT SYS FOR FOOD PROD, ENGY PROD&CONSUMPTION, WASTE TREATMENT, HOUSING & TRANSPORT
HAVE COLLECTED & ORGANIZED BOOKS, ARTICLES,&TECHNICAL REPORTS ON ABOVE SUBJECTS

COLORADO, UNIV OF-DENVER
ENVIRONMTL ACTION-CO
1100 14TH STREET
DENVER CO 80202
CAROLYN M PESKO
SOLAR DIRECT PROJ DR

ALL SOLAR TECHNOLOGIES
SOLAR SPACE HEATING & COOLING
PUBLISH ORIGINAL RESEARCHED INFORMATION
PROVIDE INFORMATION SEARCHES TO CLIENTS

-PHONE-
303 5341602 X5
303 8921117 X471
HAVE EXTENSIVE SOLAR ENERGY COMPUTERIZED DATA BANK PROGRAMMED TO STORE & ACCESS CURRENT INFO ON ALL ASPECTS OF SOLAR UTILIZATON FROM RESEARCH TO MANUFACTURERS/PUBLISH ALL THESE COMPUTERIZED REPORTS YEARLY/PROVIDE SPECALIZED INFO SEARCHES TO CLIENTS

ENERGY RESEARCH CORP
ENERGY REVIEW
6 EAST VALERIO
SANTA BARBARA CA 93101
RONALD A ZUCKERMAN
EDITOR

PUBLISH LAY PERIODICALS
PUBLISH ORIGINAL RESEARCH + REPRINTS
PROVIDE INFORMATION SEARCHES TO CLIENTS

-PHONE-
805 9631388
HAVE DEVELOPED COMPUTERIZED DATA BANK,PROGRAM TO PROVIDE ACCESS TO UNLIMITED ENERGY SOURCES & ALTERNATIVES...PUBLISH BI-MONTHLY JOURNAL: 'ENERGY REVIEW' WHEREIN SUCH DATA IS ABSTRACTED & INDEXED FOR PUBLICAT

------------------SUBCLASSES------------------SOLAR SUB SYSTEMS------------------

ENVIRONMENTAL ACTION COALITION — PUBLISH LAY PERIODICALS; PROVIDE AUDIO SERVICES
EDUC MATERIALS PROG
235 EAST 49TH ST
NEW YORK NY 10017
JOAN EDWARDS
PROGRAM OFFICER
-PHONE-
212 4869550

HAVE EXTENSIVE LIBRARY OF MATERIALS ON ALL ASPECTS OF ENGY USE,INCLUDE SOLAR ENGY/SPEAKERS BUREAU SPEAKS TO SCHOOLS & COMMUNITY GROUPS/PUBLSH CHILDRENS NEWSLETTER 'ECO-NEWS'/HAVE TEACHING PACKET ON ENGY.....FOR MORE IFO ON ACTIVITIES SEND STAMPED SELF-ADD BUSINESS ENVELOPE TO ABOVE ADDRESS

FRANKLIN INST RESEARCH LABS — ALL SOLAR TECHNOLOGIES; PROVIDE INFORMATION SEARCHES TO CLIENTS
SOLAR APPLC INFO CTR
20TH & RACE STREETS
PHILADELPHIA PA 19103
FRANK WEINSTEIN

-PHONE-
215 4481500

RECENTLY ESTAB INFO CTR OPERATED BY SCI INFO SERVICES DEPT & ENGY SYS LAB AT FIRL...NOW GATHERING INFO...GOAL TO PROVIDE INFO & DATA ON SOLR APPLICATIONS ON INTERNATL SCALE

ILLINOIS, UNIV OF — PROVIDE INFORMATION SEARCHES TO CLIENTS
ADVANCED COMP CENTER
URBANA IL 61801
ROBERT A HERENDEEN

MAIN AREA OF INFO CONCERNS ENERGY CONSERVATION-TOTAL DIRECT & INDIRECT ENERGY COST OF GOODS & SERVICES
PUBLISH REPORTS RESULTING FROM ENERGY RESEARCH DONE AT THE CENTER-NONE DIRECTLY SOLAR RELATED AT PRESENT--WRIE BRUCE HANNON AT ABOVE ADDRESS FOR DETAILS

---SUBCLASSES---SOLAR SUB SYSTEMS---

ROSALYN SWITZEN
SOLR ENGY OMBUDSMAN
BOX 1162
SAN FERNANDO CA 91341

IF YOU SEND SELF ADDRESSED STAMPED ENVELOPE WILL SEND PARTICULARS ON
NATIONAL SOLAR ENERY YEAR 75 IF YOUWANT SET UP CHAPTER IN YOUR CITY

---

sunday
sun
day
a day of rest
official day of quiet
stop what your doing
and listen...
to the sound...

need an official day
so many people
stop all at once
and listen...
stop all the chainsaws
and automobiles
and hammering
and T.V. advertising
and shouting

and listen...
to the sound
of clouds drifting by
alive with birds
of the river of wind
rushing along
to listen to the flowers grow
to feel the sun
sun
day...

Carolyn

	SUBCLASSES	SOLAR SUB SYSTEMS
ARCHITECTS COLLABORATIVE INC 46 BRATTLE ST CAMBRIDGE MA 02138 GERALD L FOSTER -PHONE- 617 8684200	HIGH TEMPERATURE SOLAR COLLECTOR SYSTEMS	
H G MULTITRADE INC P O BOX 01 4781 MIAMI FL 33130 -PHONE- 305 8569603		

RICH CHARTER * MEDIUM TEMPERATURE SOLAR COLLECTOR SYSTEMS
BOX 211 WIND CONVERSION
CAZADERO CA 95421 TOTAL SYSTEMS INTEGRATION
SOLAR SPACE HEATING & COOLING

CREATING PERSONAL FILES ON ALL ALTERNATIVE & SEMICONVENTIONAL ENERGY SOURCES SUITABLE FOR RURAL USES:EXTENSIVE WORK ON PELTON WHEEL/IMPLUSE TURBINE TO HARNESS WATER POWER--I CAN RESPOND TO LIMITED NUMBER OF SERIOUS INQUIRIES CONCERNING UTILIZATION-WOULD LIKE TO RECEIVE LIT FROM OTHERS TO UPDATE MY FILES

L M CUDDY
722 CRESTFOREST RD
KNOXVILLE TN 37919

INTEREST FORMING CO TO DISTRIBUTE SOLAR PRODUCTS

SARAH K DUBIN-VAUGHN SOLAR SPACE HEATING & COOLING
ONE SEASIDE PL
E NORWALK CT 06855

-PHONE-
203 8534953

BULLOCK ENGINEERING & DEV CO
70 WEST 6TH AVE
DENVER CO 80222
HARRY C BULLOCK
PRESIDENT

-PHONE-
303 8926416
303 8926483

----------SUBCLASSES----------SOLAR SUB SYSTEMS----------

ENERGY MANAGEMENT CO
8900 MELROSE AVE
LOS ANGELES CA 90069

-PHONE-
213 2746881

LEROY S TROYER & ASSOC
112 1/2 LINCOLNWAY E
MISHAWAKA IN 46544
LEROY S TROYER

-PHONE-
219 2599976
TERRESTERIAL APPLICATIONS OF SOLAR ENERGY

	SUBCLASSES	SOLAR SUB SYSTEMS
MARTIN MARIETTA AEROSPACE MAIL I661 PO BOX 179 DENVER CO 80201 DAVID H BACHMAN SR FACILITIES ENGR -PHONE- 303 7945211 X4051 303 7945211 X3427	MEDIUM TEMPERATURE SOLAR COLLECTOR SYSTEMS SOLAR SPACE HEATING & COOLING INCENTIVES FOR SOLAR & ENERGY CONSERVATION	COLLECTOR,MECHANICAL SYSTEMS HEAT TRANSFER SYSTEMS, COLLECTOR TO STORAGE HEAT PUMPS STORAGE, PHASE CHANGE MATERIALS STORAGE, LIQUIDS STORAGE, GAS

UNIVERSITY OF COLORADO STUDIES,VOL.29,NO.4,APR 55,"THE PERFORMANCE OF A THIN LIQUID FILM TYPE OF FLAT PLATE SOLAR HEAT COLLECTOR.

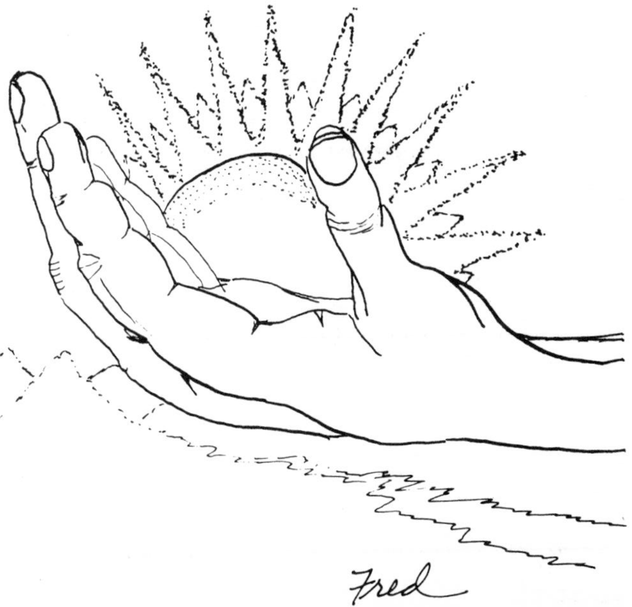

--------SUBCLASSES--------SOLAR SUB SYSTEMS--------

BOARD OF COUNTY COMMISSIONERS
BROWARD COUNTY
CO COURTHOUSE RM 270
FT LAUDERDALE FL 33301
THOMAS HUGHES

The sun explodes, reflects, suspends
and penetrates this Earth every
waking/sleeping momentary gleam.

Offices creak; lightbulbs signal
the whispy, rattlely spray of
spent-mineral compound span.

On and through solar glare, photosynthetic warmth and green;
green plants outside the venetian-wink
of tiny office-world.

Phone-hooks heated by solar strays
buzz, sing melted sirens; guard
and rub-off mans call.

Sunray lands and strays.
Collected and luminous.
Unbounded by its confinement.

The sun is a joy today...
Greg

	SUBCLASSES	SOLAR SUB SYSTEMS
ALBRIGHT COLLEGE DEPT OF PHYSICS READING PA 19604 DR THURMAN R KREMSER	MEDIUM TEMPERATURE SOLAR COLLECTOR SYSTEMS	
NEVADA,STATE OF HIGHWAY DEPARTMENT 1263 SO STEWART ST CARSON CITY NV 89701 JAMES O HUBBARD ARCHITECT	WIND CONVERSION SOLAR ENERGY STORAGE RETROFITTING, PASSIVE SYSTEMS NEW CONSTRUCTION, PASSIVE SYSTEMS SINGLE FAMILY, DEMONSTRATION GARAGES/GAS STATIONS	COLLECTOR,PASSIVE SYSTEMS COLLECTOR,ALSO FUNCTION AS STRUCTURES SKIN HEAT TRANSFER SYSTEMS, COLLECTOR TO STORAGE HEAT RECOVERY, STORAGE TO UTILIZATION STORAGE, SOLID MATERIALS STORAGE, PHASE CHANGE MATERIALS STORAGE, LIQUIDS
PELCASP SOLAR DESIGNERS 240 LANGDON ST MADISON WI 53703 MICHAEL B PELLETT PRES	WIND CONVERSION SOLAR ENERGY STORAGE SOLAR ELECTRIC POWER GENERATION RETROFITTING, PASSIVE SYSTEMS RETROFITTING, PRACTICING	COLLECTOR,ABSORPTIVE SELECTIVE SURFACES
SAMUEL PAUL ARCHITECT, OFF OF 107-40 QUEENS BLVD FOREST HILLS NY 11375 SAMUEL PAUL		

ALBRIGHT COLLEGE
-PHONE-
215 9212381 X266
MAIN INTEREST IS UTILIZATION OF SOLAR ENERGY FOR RESIDENTIAL HEATING & LIGHTING/AT PRESENT NO ACTIVE RESEARCH BEING DONE---NOT IN POSITION TO RESPOND TO INQUIRIES AT THIS TIME

NEVADA,STATE OF
-PHONE-
702 8827485
RESPONSIBILITY FOR THE DESIGN, CONSTRUCTION AND MAINTENANCE OF APPROX THREE HUNDRED HIGHWAY OWNED BUILDINGS THROUGHOUT THE STATE OF NEVADA

PELCASP SOLAR DESIGNERS
-PHONE-
608 2519967
DESIGNING AND BUILDING CUSTOM RETROFITTED SYSTEMS FOR EXISTING HOMES

SAMUEL PAUL ARCHITECT, OFF OF
-PHONE-
212 2615100

	SUBCLASSES	SOLAR SUB SYSTEMS
DEETER HOMES 246 PALM AVE AUBURN CA 95603 DAN DEETER -PHONE- 916 8858448 CURRENTLY BUILDER OF APPROX 500 HOMES YEARLY IN CALIF/PLIMINARY CONTAT WITH UNIV OF CALIF AT BERKELEY TO BUILD TEST HOUSE/I WOULD LIKE TO BLD SEVERAL EXPERIMENTAL HOMES IN FOOTHILLS AREA NEAR SACRAMENTO		
ENVIRONMENTAL SYS FOR PEOPLE 213 SUN 2030 E SPDWY TUCSON AZ 85719 ROBERT G WASON CONSULTING ENGR -PHONE- 602 7955437		
FEUER CORPORATION 5401 MCCONNELL AVE LOS ANGELES CA 90066 STANLEY FEUER ASHRAE PRESIDENT -PHONE- 213 3904046 213 4562894 INTEREST IN RETROFITTING & NEW CONSTRUCTION MECHANICAL SYS	RETROFITTING, MECHANICAL SYSTEMS NEW CONSTRUCTION, MECHANICAL SYSTEMS APARTMENTS ESTATES HIGH RISE MULTI-STRUCTURE,PROPOSED 500 THOUS-1 MILL SQ FT,PROPOSED	
FEUER CORPORATION ENGINEERING 5401 MCCONNELL AVE LOS ANGELES CA 90066 RICHARD D MILLER ASHRAE ASME CHIEF ENGINEER -PHONE- 213 3904046 213 8217209 INTEREST IN RETROFITTING & NEW CONSTRUCTION MECHANICAL SYS SPEAKER AT SOLAR COOLING WORKSHOP SPONSERED BY JPL-NSF AT SEMIANNUAL '74 ASHRAE MEETING	APARTMENTS CONDOMINIUMS HIGH RISE MULTI-STRUCTURE,PROPOSED HOTEL/MOTEL/DORM BUILDINGS 500 THOUS-1 MILL SQ FT,PROPOSED INCENTIVES FOR SOLAR & ENERGY CONSERVATION	COLLECTOR,ALSO FUNCTION AS STRUCTURES SKIN HEAT TRANSFER SYSTEMS, COLLECTOR TO STORAGE HEAT PUMPS

	SUBCLASSES	SOLAR SUB SYSTEMS
LENNOX INDUSTRIES INC HEATING EQUIP ENGR PO BOX 250 MARSHALLTOWN IA 50158 GARRY O HANSON MANAGER -PHONE- 515 7544011		
PROF ENG CONSULTANTS MECHANICAL 1440 E ENGLISH WICHITA KS 67211 ERVIN E HYSOM CONSULTING MECH ENG -PHONE- 316 2622691 ALSO APPLICATIONS FOR INDUSTRIAL BLDGS OFFICE/BANK BLDGS RELIGIOUS/ PUBLIC/AMUSEMENT BLDGS	MEDIUM TEMPERATURE SOLAR COLLECTOR SYSTEMS HIGH TEMPERATURE SOLAR COLLECTOR SYSTEMS NEW CONSTRUCTION, MECHANICAL SYSTEMS HOMES HEALTH/HOSPITAL BUILDINGS OFFICE/BANK BUILDINGS	HEAT TRANSFER SYSTEMS, COLLECTOR TO STORAGE COLLECTOR,ALSO FUNCTION AS STRUCTURES SKIN COLLECTOR,MECHANICAL SYSTEMS HEAT PUMPS STORAGE, LIQUIDS
RUDKIN-WILEY CORPORATION POLARCH 760 HONEYSPOT STRATFORD CT 06497 SAM HILL MGR SALES&MARKETING -PHONE- 202 3755966 MFG POLYARCH HOUSING SYS: FIBERGLASS SEGMENTS MODULAR HOUSING...HEAT LOSS .07 U FACTOR...INTEREST IN INCORPORATING SOLAR HEAT IN SYS	NEW CONSTRUCTION, TECHNOLOGIES HOMES	

	SUBCLASSES	SOLAR SUB SYSTEMS
GENERAL ELECTRIC COMPANY TPD MOUNTAIN VIEW ROAD LYNCHBURG VA 24502 OTWARD M MUELLER ELECTRONICS ENGINEER -PHONE- 804 8467311 X2765 804 2371161	OCEAN THERMAL GRADIENT SYSTEMS INORGANIC SEMI-CONDUCTORS ALL SOLAR TECHNOLOGIES HOMES SINGLE FAMILY, DEVELOPMENT PUBLISH TECHNICAL PERIODICALS	HEAT TRANSFER SYSTEMS, COLLECTOR TO STORAGE COLLECTOR,REFLECTIVE SELECTIVE SURFACES COLLECTOR,ABSORPTIVE SELECTIVE SURFACES HEAT PUMPS STORAGE, GAS ELECTROLYSIS
PERRY-DEAN-STEWART ARCHITECTS 955 PARK SQ BLDG BOSTON MA 02116 CHARLES F ROGERS PARTNER -PHONE- 617 4829160 APPLICATIONS FOR HEALTH/HOSPITAL BLDGS HOMES ACADEMIC BLDGS INDUSTRIAL BLDGS OFFICE/BANK BLDGS	TOTAL SYSTEMS INTEGRATION SOLAR SPACE HEATING & COOLING RETROFITTING, INTEREST NEW CONSTRUCTION, INTEREST PROVIDE INFORMATION SEARCHES TO CLIENTS HAVE&MAINTAIN ENERGY DATA BANK	
WARNER BURNS TOAN LUNDE 724 FIFTH AVE NEW YORK NY 10019 JEAN DUFFY -PHONE- 212 7578900	WIND CONVERSION SOLAR ENERGY STORAGE SOLAR WATER HEATING RETROFITTING, TECHNOLOGIES	

	SUBCLASSES	SOLAR SUB SYSTEMS
PACIFIC POWER & LIGHT CUST TECH SVCS 920 SW 6TH AVE PORTLAND OR 97204 LAWTHAN M AUSTIN JR TECH SVCS DIR -PHONE- 503 2267411 X1054 503 6469894	SOLAR WATER HEATING SOLAR SPACE HEATING NEW CONSTRUCTION, MECHANICAL SYSTEMS HOMES MOBILE HOMES	HEAT PUMPS COLLECTOR,ALSO FUNCTION AS STRUCTURES SKIN COLLECTOR,PASSIVE SYSTEMS COLLECTOR,MECHANICAL SYSTEMS

ALSO: OFFICE BANK BLDGS UP TO 50 THOUSAND SQFT & 50-500 THOUSAND SQFT

	SUBCLASSES	SOLAR SUB SYSTEMS
LOS ALAMOS SCIENTIFIC LAB ENGINEERING ENG-9 MS 636 LOS ALAMOS NM 87544 RICHARD A HEMPHILL ENGINEER -PHONE- 505 6676261 505 6721965	MEDIUM TEMPERATURE SOLAR COLLECTOR SYSTEMS HIGH TEMPERATURE SOLAR COLLECTOR SYSTEMS TOTAL SYSTEMS INTEGRATION SOLAR SPACE HEATING & COOLING RETROFITTING, MECHANICAL SYSTEMS LOW RISE MULTI-STRUCTURE,PROPOSED	COLLECTOR,MECHANICAL SYSTEMS HEAT ENGINE, RANKINE STORAGE, MECHANICAL/PUMPED LIQUIDS TURBINES

-----SUBCLASSES-----SOLAR SUB SYSTEMS-----

MASSACHUSETTS, UNIVERSITY OF PHYSICS DEPT HASBROUCK LABORATORY AMHERST MA 01002 ROBERT A JOHNSON RESEARCH ASSOCIATE -PHONE- 413 5450346	MEDIUM TEMPERATURE SOLAR COLLECTOR SYSTEMS SOLAR SPACE HEATING CONDOMINIUMS 6-14 FAMILY, PROPOSED PUBLISH ORIGINAL RESEARCHED INFORMATION

			SUBCLASSES	SOLAR SUB SYSTEMS
CUTTER LABS			SOLAR ENERGY STORAGE	COLLECTOR,REFLECTIVE SELECTIVE SURFACES
808 PARKER			SOLAR WATER HEATING	HEAT TRANSFER SYSTEMS, COLLECTOR TO STORAGE
BERKELEY	CA	94710	SOLAR ELECTRIC POWER GENERATION	
EDGAR M FRIED, PE			APARTMENTS	
STAFF ENG				
-PHONE-				
415 8410123 X278				
415 2842858				
HOME HTG SYS				

SUBCLASSES	SOLAR SUB SYSTEMS
NEW CONSTRUCTION, MECHANICAL SYSTEMS	

U S GOVT GSA
CONSTRUCTION MGMT
BLDG#41 DFC
DENVER CO 80225
JAMES H OTTMER
MECHANICAL ENGINEER
-PHONE-
303 2342641
DESIGN AND CONSTRUCTION OF FEDERAL BUILDING

NEW CONSTRUCTION, MECHANICAL SYSTEMS

U S GOVT GSA
CONSTRUCTION MGMT
BLDG#41 DFC
DENVER CO 80225
CARL SWENSON *
CONST ENGR SUPERVISR
-PHONE-
303 2342641

---SUBCLASSES---SOLAR SUB SYSTEMS---

DENVER PLANNING OFFICE 1445 CLEVELAND PLACE DENVER CO 80202 ALAN L CANTER DIRECTOR OF PLANNING	ZONING ORDINANCES FOR SOLAR UTILIZATION BUILDING CODES FOR SOLAR UTILIZATION

-PHONE-
303 2972736
303 2972737
ALSO INTEREST: ENERGY CONSV FINANCIAL&POLITICAL CONSIDERATIONS

DENVER PLANNING OFFICE 1445 CLEVELAND PLACE DENVER CO 80202 WAYLAND WALKER * CITY PLANNER	ZONING ORDINANCES FOR SOLAR UTILIZATION BUILDING CODES FOR SOLAR UTILIZATION

-PHONE-
303 2972736
ALSO INTEREST: ENERGY CONSV FINANCIAL&POLITICAL CONSIDERATIONS

Manufacturer			Subclasses	Solar Sub Systems
ALCOA ALUMINUM 1501 ALCOA BLDG PITTSBURGH WILLIAM S LEWIS MARKET DEV MANAGER	PA	15219	WIND CONVERSION INORGANIC SEMI-CONDUCTORS	COLLECTOR,ABSORPTIVE SELECTIVE SURFACES

-PHONE-
412 5532748
PRODUCE ALUM MILL PRODUCTS...RAW GALLIUM(PHOTOVOLTAIC APPL)...HAVE EXPERIMENTAL SELECTIVE SURFACE FOR LAB TESTING OR PILOT PROJ...COMPOSITE BORON GRAPHITE FILAMENT IN EPOXY USED FOR WINDMILL BLADES

Manufacturer			Subclasses	Solar Sub Systems
ALCOA ALUMINUM 1501 ALCOA BLDG PITTSBURGH JOHN E WRIGHT * SUPERVISOR PUB REL	PA	15219	WIND CONVERSION INORGANIC SEMI-CONDUCTORS	COLLECTOR,ABSORPTIVE SELECTIVE SURFACES

-PHONE-
412 5534751

Manufacturer			Subclasses	Solar Sub Systems
BROWN MANUFACTURING CO P O BOX 14546 OKLAHOMA CITY RUSSELL BROWN PRESIDENT	OK	73114	RETROFITTING, MECHANICAL SYSTEMS NEW CONSTRUCTION, MECHANICAL SYSTEMS CONDOMINIUMS HEALTH/HOSPITAL BUILDINGS HOTEL/MOTEL/DORM BUILDINGS OFFICE/BANK BUILDINGS	

-PHONE-
405 7511323
MFG ENVIRO LOUVERS...FROM COOL TOWER SCREENS TO SUN LOUVERS WITH MECH CONTROLS

Manufacturer			Subclasses	Solar Sub Systems
DEKO LABS BOX 12841 UNIV STATN GAINESVILLE DONALD F DEKOLD	FL	32604	SOLAR WATER HEATING SOLAR SPACE HEATING	

-PHONE-
904 3726009
MFG TC CONTROLLER/HAS 2 TEMP SENSORS TO MANAGE CIRCULATING PUMP IN A SOLAR WATER HEATING SYS

-----SUBCLASSES-----SOLAR SUB SYSTEMS-----

KALWALL CORPORATION
KAL-LITE DIVISION
P O BOX 237
MANCHESTER NH 03105
RONALD K HERMSDORF

-PHONE-
603 6273861
MFG TWO FIBERGLASS SHEETS:SUN-LITE UV LIFE OF 7YRS & LONG LIFE SUN-LITE UV LIFE OF 20YRS FOR USE AS COVER PLATES FOR SOLAR COLLECTORS-HIGH LIGHT TRANSMITTANCE, RESISTANCE TO BREAKAGE, LIGHT WEIGHT EASY TO HANDLE...FOR MORE INFO WRITE

OLIN CORPORATION,BRASS GROUP
ROLL-BOND PRODUCTS
EAST ALTON IL 62024
JOHN I BARTON
MARKETING MANAGER

-PHONE-
618 2582000
618 2582443
MANUFACTURE ROLL-BOND PRODUCTS...PLATE-TYPE HEAT EXCHANGER PANELS

OPTICAL COATING LAB INC
P O BOX 1599
SANTA ROSA CA 95403
JOSEPH H APFEL
DIRECTOR RESEARCH

Subclasses: HIGH TEMPERATURE SOLAR COLLECTOR SYSTEMS; INORGANIC SEMI-CONDUCTORS

Solar sub systems: COLLECTOR,ABSORPTIVE SELECTIVE SURFACES; COLLECTOR,REFLECTIVE SELECTIVE SURFACES

-PHONE-
707 5456440
OCLI PRODUCES- SOLAR CELL COVERS FOR PHOTOVOLTAIC CONVERTERS AND IS PREPARED TO PROVIDE ANTIREFLECTION COATINGS & SELECTIVE COATINGS FOR SOLAR COLLECTORS...COATINGS ON GLASS PLASTIC & METAL SURFACES
SEE ALSO: DANFORTH JOSLYN OPTICAL COATING LAB

	SUBCLASSES	SOLAR SUB SYSTEMS
OPTICAL COATING LABORATORY INC 2789 GIFFEN AVENUE SANTA ROSA CA 95403 DANFORTH JOSLYN	HIGH TEMPERATURE SOLAR COLLECTOR SYSTEMS INORGANIC SEMI-CONDUCTORS	COLLECTOR,ABSORPTIVE SELECTIVE SURFACES COLLECTOR,REFLECTIVE SELECTIVE SURFACES

-PHONE-
707 5456440
PUBLISHED: 'THIN FILMS AND SOLAR ENERGY,'OPTICAL COATING LAB...SEE
SOLAR TECH BIBLIO FOR ANNOTATION...SEE ALSO: JOSEPH APFEL O C L I

	SUBCLASSES	SOLAR SUB SYSTEMS
PAUL MUELLER CO P O BOX 828 SPRINGFIELD MO 65801	SOLAR SPACE HEATING & COOLING	HEAT TRANSFER SYSTEMS, COLLECTOR TO STORAGE

-PHONE-
417 8652831
MANUFACTURE 'TEMP-PLATE' STAINLESS STEEL HEAT TRANSFER SURFACE

	SUBCLASSES	SOLAR SUB SYSTEMS
REYNOLDS METALS EXTRUSION DIV 6601 W BROAD ST RICHMOND VA 23230 ADAM J DRESCHER	SOLAR SPACE HEATING & COOLING	

-PHONE-
804 2822331 X2197
EXTRUDED ALUM FLAT PLATE ABSOBER PANEL...PRODUCE EXTRUSIONS 6 IN WIDE
& UP TO 20 FT LONG...ALSO SUPPLY TUBING SHEET & PLATE...PRODUCT DEV &
SALES RESCH DIV OPEN FOR CONSULATION

	SUBCLASSES	SOLAR SUB SYSTEMS
3 M CO MARKETING DEPT 3 M CENTER ST PAUL MN 55101 JOHN J MUELLER MARKETING COORD		COLLECTOR,ABSORPTIVE SELECTIVE SURFACES

-PHONE-
612 7337201
PRODUCE NEXTEL VELVET COATING IN VARIETY OF VEHICLES...3M TECH SERVICE
PROVIDES CONSULTING SERVICES...WILLAPPLY COATINGS TO PROTOTYPES FOR NO
CHARGE ON LIMIT BASIS...ALSO MFG SCOTCHCAL ALUM POLYESTER FILMS WITH
HIGH REFLECT SURF...ALSO SCOTCH TINT GLASS FILM

	SUBCLASSES	SOLAR SUB SYSTEMS
LIBBEY OWENS FORD CO 811 MADISON AVE TOLEDO OH 43695 DAVID R GROVE SALES MGR	MEDIUM TEMPERATURE SOLAR COLLECTOR SYSTEMS	COLLECTOR,REFLECTIVE SELECTIVE SURFACES COLLECTOR,ABSORPTIVE SELECTIVE SURFACES

-PHONE-
419 2425781
MANUFACTURE MEDIUM TO LARGE SIZED FLAT &/OR CURVED GLASS PRODUCTS COATED WITH FILMS WHICH HAVE SELECTIVE TRANSMISSION & REFLECTION CHAR AS SPECIFIED BY USER FOR USE IN MEDIUM OR HIGH TEMP CONCENTRATORS OR COLLECTORS

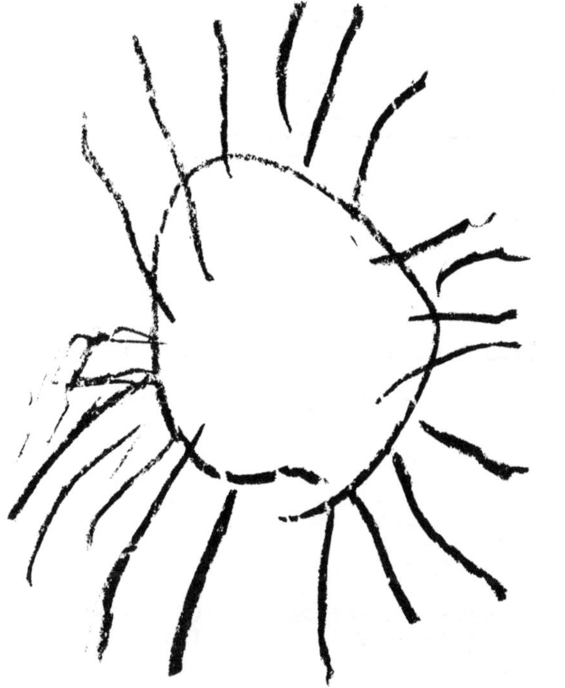

SENSENICH CORPORATION
PO BOX 1168
LANCASTER PA 17604
A C WEDGE
EXECUTIVE VP

MFG WOODEN PROPELLERS-INTERESTED INWORKING WITH WINDPOWER PLANT INNOVA
TORS & GENERATOR MFGS

WINDWORKS — SOLAR ELECTRIC POWER GENERATION
BOX 329 RTE 3
MUKWONAGO WI 53149
HANS MEYER

DESIGNED SIMPLE RIG SUITABLE FOR HOME BUILD...HAVE AVAIL EXPANDED
PAPER BLOCKS FOR BLADES AVAIL FOR $10.00 A SET

	SUBCLASSES	SOLAR SUB SYSTEMS
ENERGY DEVELOPMENT ASSOC 1100 W WHITCOMB AVE MADISON HTS MI 48071 NEVILLE MAPHAM GENERAL MANAGER	SOLAR POWERED TRANSPORTATION	STORAGE, CHEMICAL BATTERIES

-PHONE-
313 5664000
313 5839434
LONG LIFE, LOW COST ZINC-CHLORIDE STORAGE BATTERY:2000 TO 5000 CYCLES
AT $10 PER KWH & 70 PERCENT EFF APPLICATION TO PRACTICAL ELECTRIC CARS

With self-less wonder
sunlight fuses life
to float home on the ozone.
Greg

	SUBCLASSES	SOLAR SUB SYSTEMS
DUPONT E I DE NEMOURS CO DEVELOPMENT DEPT 1007 MARKET ST WILMINGTON DE 19898	MEDIUM TEMPERATURE SOLAR COLLECTOR SYSTEMS HIGH TEMPERATURE SOLAR COLLECTOR SYSTEMS SOLAR SPACE HEATING & COOLING	

PRODUCES 'MYLAR'(POLYETHYLENE TERA PHTHALATE) LUCITE(POLYMETHYL METHA-
CRYLATE) AND TEDLAR...PLASTICS SUITARLE FOR USE ON COLLECTORS

SELECT RAD SURFACES PLATING CO P O BOX 8364 SAN JOSE CA 95155 RICHARD T AVALON PRES	SOLAR SPACE COOLING SOLAR REFRIGERATION SOLAR ELECTRIC POWER GENERATION	COLLECTOR,ABSORPTIVE SELECTIVE SURFACES

-PHONE-
408 2959040
MANUF & MARKET SELECTIVE RADIATION SURFACES WITH APPLICATION TO REFRIG
AIR COND & ELECTRICAL GEN EQUIP WE PLATE SURFACES TO YOUR SPECS

------SUBCLASSES------SOLAR SUB SYSTEMS------

NALLE PLASTICS INC
203 COLORADO STREET
AUSTIN TX 78701
GEORGE S NALLE JR

-PHONE-
512 4776168
SEA WATER DESALINIZATION WE MAKE PLASTIC NET TO SUPPORT FILM & FILTER
MEDIA

	SUBCLASSES	SOLAR SUB SYSTEMS
ISOTHERMICS P O BOX 86 AUGUSTA NJ 07822 -PHONE- 201 3833500 SECONDARY HEAT RECOVERY DEVICES	RETROFITTING, MECHANICAL SYSTEMS NEW CONSTRUCTION, MECHANICAL SYSTEMS HEALTH/HOSPITAL BUILDINGS INDUSTRIAL BUILDINGS OFFICE/BANK BUILDINGS COMMERCIAL OTHER	HEAT RECOVERY, STORAGE TO UTILIZATION STORAGE, LIQUIDS
MEGATHERM CORP TAUNTON AVE EAST PROVIDENCE RI 02914 ROBERT H STEVENSON -PHONE- 401 4383800 THERMAL STORAGE USING H2O & CHEMICAL SOLUTION INSTALLED IN REDE OLD FORGE BLDG SOLAR PROJECT		STORAGE, LIQUIDS HEAT RECOVERY, STORAGE TO UTILIZATION
RHO SIGMA UNLIMITED 5108 MELVIN AVE TARZANA CA 91356 ROBERT J SCHLESINGER P E -PHONE- 213 3424376 MFG DIFFERENTIAL THERMOSTAT FOR CONTROL OF SOLAR HEATING &/OR HOT H2O SYS	MEDIUM TEMPERATURE SOLAR COLLECTOR SYSTEMS SOLAR WATER HEATING	

			SUBCLASSES	SOLAR SUB SYSTEMS
ARKLA INDUST INC 509 MARSHALL ST SHREVEPORT PHILIP ANDERSON	LA	71151	MEDIUM TEMPERATURE SOLAR COLLECTOR SYSTEMS	

-PHONE-
318 4251271
3 1/2, 15, 25 & 100 TON LIBR COOLING UNITS OPERABLE WITH ENERGY SOURCE TEMP LESS THAN 200F...PROTOTYPE 3 TON UNIT INSTALLED IN NSF COLORADO ST UNIV DEMO HOME

			SUBCLASSES	SOLAR SUB SYSTEMS
NATIONAL SUN CONTROL PO BOX 330205 MIAMI JACK TRENT OWNER	FL	33133	SOLAR RADIATION MEASUREMENTS CONDOMINIUMS ACADEMIC BUILDINGS HEALTH/HOSPITAL BUILDINGS	COLLECTOR,REFLECTIVE SELECTIVE SURFACES

-PHONE-
305 4447494
WE WHOLESALE A METALIZED MYLAR FILM TO DEALERS: FILM IS BONDED TO EXISTING GLASS TO CONTROL SOLAR HEAT FADE & GLARE COOLING COSTS MAY BE DRASTICALLY REDUCED WHEN REFLECTIVE FILM IS APPLIED TO WINDOWS THAT ALLOW SUNLIGHT TO COME INTO A BUILDING

			SUBCLASSES	SOLAR SUB SYSTEMS
NIAGARA BLOWER CO 405 LEXINGTON AVE NEW YORK ERWIN LODWIG	NY	10017	SOLAR REFRIGERATION	

-PHONE-
212 6976151
DEHUMIDIFICATION OF INCOMING AIR THROUGH USE OF INLET AIR SCRUBBING WITH HYGROL...CONCEPT HAS BEEN IN WIDE USE FOR COOLING IN LARGE SCALE INDUSTRIAL PROCESSES...ADDITIONAL DEV IN PROGRESS TO ADAPT TO SMALLER SCALE RESIDENTIAL COOL...PROTOTYPE UNIT TO BE USED IN NEW HAMP HOUSE

INDEX OF SOLAR COMPONENTS MANUFACTURER (SOLAR SPACE COOLING)

-SUBCLASSES- -SOLAR SUB SYSTEMS-

SOLAR-X CONTROL PRODUCTS CORP 217 CALF ST NEWTON MA 02158	RETROFITTING, PASSIVE SYSTEMS RETROFITTING, PRACTICING NEW CONSTRUCTION, PASSIVE SYSTEMS NEW CONSTRUCTION, PRACTICING

-PHONE-
617 2448686
MAIN DIST FOR SOLAR-X-THIN SUN SCREEN FILMS...APPLY TO WINDOW REDUCE
SOLAR HEAT GAIN

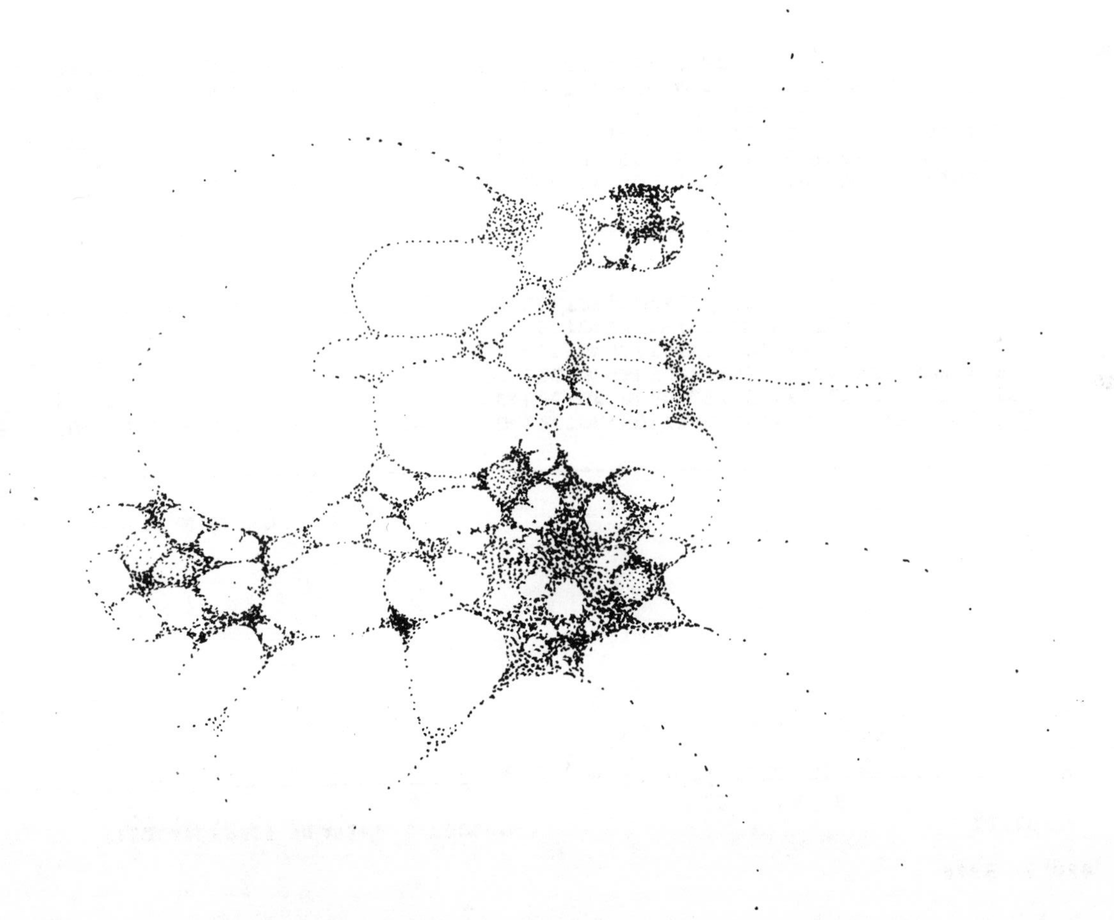

			SUBCLASSES	SOLAR SUB SYSTEMS
AMELCO WINDOW CORP 77 RT 17 BOX 333 HASBROUCK HTS R C CURRENT JR -PHONE- 212 2440610	NJ	07604		
MFG DUAL GLAZED THERMAL BREAK HORIZONTALLY PIVOTED VENETIAN BLIND WINDOWS/CONSERVE ENERGY				
ASG INDUSTRIES, INC BOX 929 KINGSPORT JAMES S HERBERT * DIR CUST TECH SERV -PHONE- 615 2450211 X325	TN	37662	MEDIUM TEMPERATURE SOLAR COLLECTOR SYSTEMS SOLAR WATER HEATING SOLAR SPACE HEATING SOLAR SPACE COOLING RETROFITTING, MECHANICAL SYSTEMS NEW CONSTRUCTION, MECHANICAL SYSTEMS	COLLECTOR,MULTI-FUNCTION IN DWELLING SYS COLLECTOR,REFLECTIVE SELECTIVE SURFACES COLLECTOR,ABSORPTIVE SELECTIVE SURFACES COLLECTOR,PASSIVE SYSTEMS COLLECTOR,MECHANICAL SYSTEMS COLLECTOR, EVACUATED
ASG INDUSTRIES,INC 1B BELMAR ROAD CRANBURY CLARENCE W CLARKSON DIR LTG PROD TECH SV -PHONE- 609 6550058 609 6550057	NJ	08512	MEDIUM TEMPERATURE SOLAR COLLECTOR SYSTEMS SOLAR WATER HEATING SOLAR SPACE HEATING SOLAR SPACE COOLING RETROFITTING, MECHANICAL SYSTEMS NEW CONSTRUCTION, MECHANICAL SYSTEMS	COLLECTOR,MULTI-FUNCTION IN DWELLING SYS COLLECTOR,REFLECTIVE SELECTIVE SURFACES COLLECTOR,ABSORPTIVE SELECTIVE SURFACES COLLECTOR,PASSIVE SYSTEMS COLLECTOR,MECHANICAL SYSTEMS COLLECTOR, EVACUATED
NEWS BULLETIN ON"GLASS FOR SOLAR ENERGY COLLECTORS" ADDRESS INQUIRY TO J S HERBERT ASG KINGSPORT TN				
ROHM AND HAAS PLASTICS DEPT INDEPENDENCE MALL W PHILADELPHIA ROBERT F RITLER TECH COORD -PHONE- 215 5926799	PA	19105	MEDIUM TEMPERATURE SOLAR COLLECTOR SYSTEMS SOLAR COOKING SOLAR WATER DISTILLATION SOLAR WATER HEATING	
MFG... "PLEXIGLAS"(POLYMETHYL METHACRYLATE...A TYPE OF PLASTIC SUIT-ABLE FOR COLLECTOR COVERS				

----------SUBCLASSES----------SOLAR SUB SYSTEMS----------

	SOLAR SUB SYSTEMS
LOW IMPACT TECHNOLOGY LTD 73 MOLESWORTH ST CORNWALL ENGLAND ANDREW MACKILLOP DIRECTOR	HEAT ENGINE, STERLING

STERLING HEAT PUMPS WITH SOLAR IMPUT FOR HEATING AND COOLING...PHONE
WADEBRIDGE 2996

---SUBCLASSES---SOLAR SUB SYSTEMS---

SOLAR SUNSTILL INC SOLAR WATER DISTILLATION
SETAUKET NY 11733
CHAD J RASEMAN

-PHONE-
516 9414078
MFG COATINGS FOR PLASTIC & GLASS THAT CONTROL CONDENSATION & LIGHT

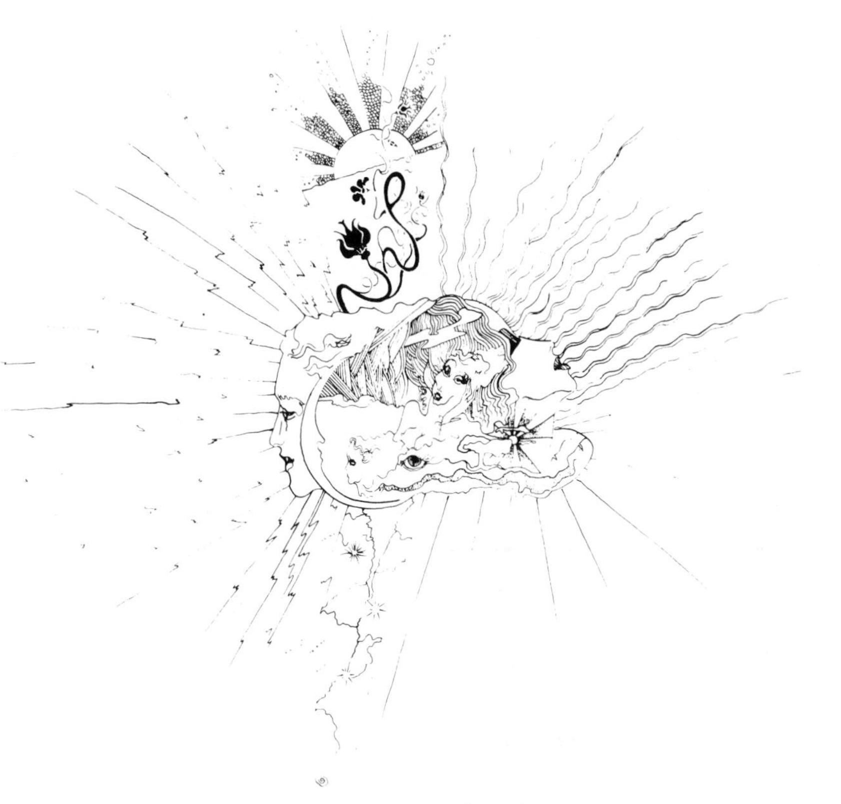

John

SUBCLASSES — SOLAR SUB SYSTEMS

B F SALES ENGR CO
4770 FOX ST NO 9
DENVER CO 80216
GLENN SELCH

REPRESENT PAUL MUELLER CO WHO MFG EXPANDED STEEL FLAT PLATE LIQUID TYPE COLL PANELS-TEMPLATE HEAT TRANSFER PANELS ALSO CAPABLE OF FABRICATION COMPLETE COLL ASSEMBLIES

	SUBCLASSES — SOLAR SUB SYSTEMS
AMERICAN ENERGY ALTERNATVS INC P O BOX 905 BOULDER CO 80302 J P SAYLER	MEDIUM TEMPERATURE SOLAR COLLECTOR SYSTEMS INORGANIC SEMI-CONDUCTORS SOLAR WATER HEATING SOLAR ELECTRIC POWER GENERATION

-PHONE-
303 4471865
WIND TURBINE SHROUDED UNIT 15.25 FT DIAMETER GIVES 10 KW/DAY IN 11 MPH
WINDS/PHOTOVOLTAICS 6-12 VOLT APPLICATIONS/SOLAR HOT WATER HEATERS

PENNWALT CORPORATION AUTOMATIC POWER DIV P O BOX 18738 HOUSTON TX 77023 ROBERT J DODGE	INORGANIC SEMI-CONDUCTORS SOLAR ELECTRIC POWER GENERATION

-PHONE-
713 2285208
DISTRIBUTE WINDMILLS FROM 24 TO 4100 WATTS MADE BY AEROWATT IN FRANCE
ALSO ASSEMBLE PHOTOVOLTAIC ARRAYS FOR MARINE NAVIGATION

WALLACE AND TIERMAN LTD 925 WARDEN AVE SCARBOROUGH ONTARIO CANADA	SOLAR ELECTRIC POWER GENERATION

DISTRIBUTE WINDMILLS FROM 1/2KW TO 5KW MADE BY AEROWATT IN FRANCE

----SUBCLASSES----SOLAR SUB SYSTEMS----

ZURN INDUSTRIES INC
5533 PERRY HWY
ERIE PA 16509
HARRY LINDSAY

-PHONE-
814 8644041
DISTRIBUTOR FOR SOLAREX PRODUCTS

```
------------------------------------------------SUBCLASSES-------------------SOLAR SUB SYSTEMS-------------------------

SOLAR ENERGY DIGEST                      MEDIUM TEMPERATURE SOLAR COLLECTOR SYSTEMS
P O BOX 17776                            SOLAR SPACE HEATING
SAN DIEGO                 CA    92117    PUBLISH LAY PERIODICALS
SOLAR EQUIP SALES DIVISION               PUBLISH ORIGINAL RESEARCHED INFORMATION

-PHONE-
714 2772980
DISTRIBUTE SOLAPAK FLAT PLAT HOT WATER HEATERS MADE BY BEASLEY OF AUST
AVAIL SINGLE & DOUBLE GLAZED 1 SOLAPAK YIELDS 10 GAL PER DAY
WRITE FOR BROCHURE
SEE ALSO:WILLIAM EDMONDSON,SOLAR ENERGY DIGEST FOR SUBSCRIPTION INFO
------------------------------------------------------------------------------------------------------------------------

T-K SOLAR DISTRIBUTORS
5 TH FLOOR
9 EAST 16TH STREET
NEW YORK                  NY    10003
THOMAS P KAY
PRESIDENT
-PHONE-
212 9894422
MARKETING ESTAB SOLAR PRODUCTS SUCH AS H2O HEATERS...PERSONS WITH
RELATED PRODUCTS CONTACT US...PUBLIC INQUIRY INVITED
------------------------------------------------------------------------------------------------------------------------

OBC INTERNATIONAL MARKETING              MEDIUM TEMPERATURE SOLAR COLLECTOR SYSTEMS
P O BOX 205
GLEN IRIS VICTORIA        AUSTRALIA


-PHONE-
203 063
208 321
SALES AGENTS FOR BRAEMAR ENGINEERING SOLAR HOT WATER HEATERS
------------------------------------------------------------------------------------------------------------------------
```

	SUBCLASSES	SOLAR SUB SYSTEMS
COMFORT TRAIN AIR COND CO 4060 S KALAMATH ENGLEWOOD CO 80110 WILLIAM BROWNING -PHONE- 303 7894497 DEALER OF SOLORON SOLAR SYS	MEDIUM TEMPERATURE SOLAR COLLECTOR SYSTEMS SOLAR WATER HEATING	COLLECTOR,MECHANICAL SYSTEMS
VANTAGE HEATING&AIR COND INC 5611 KENDALL CT ARVADA CO 80002 RICHARD PENCE -PHONE- 303 4204410 DEALER SOLORON SOLAR SYS...300 MILE LIMIT ON DISTRIBUTION	MEDIUM TEMPERATURE SOLAR COLLECTOR SYSTEMS SOLAR WATER HEATING	COLLECTOR,MECHANICAL SYSTEMS

	SUBCLASSES	SOLAR SUB SYSTEMS
WEATHER ENGINEERING & MFG INC 4703 N PARK DR COLORADO SPRINGS CO 80907 GEORGE R PEACORE PRESIDENT	WIND CONVERSION RETROFITTING, MECHANICAL SYSTEMS NEW CONSTRUCTION, MECHANICAL SYSTEMS OFFICE/BANK BUILDINGS UP TO 50 THOUSAND SQ FT,PROPOSED	HEAT RECOVERY, STORAGE TO UTILIZATION COLLECTOR,MECHANICAL SYSTEMS

-PHONE-
303 5988228

WEATHER ENGINEERING&MFG INC IS A DESIGN CONSTRUCT MECH CONTRACTOR SPECIALIZING IN TOTAL ENGINEERING DESIGN & INSTALLATION OF ENERGY CONSERVATION SYSTEMS UTILIZING SOLAR COLLECTION STORAGE&WIND GENERATORS THIS FIRM IS STAFFED BY ENGINEERS WITH BROAD BACKGROUNDS IN HVAC SYS & BACKED BY SHOP & MFG FACILITIES & A COMPLETE SERVICE DEPT

	SUBCLASSES	SOLAR SUB SYSTEMS
WEATHER ENGINEERING & MFG INC 4703 N PARK DR COLORADO SPRINGS CO 80907 LANE A PINNOW MGR ENGR & SALES	SOLAR SPACE HEATING & COOLING RETROFITTING, MECHANICAL SYSTEMS NEW CONSTRUCTION, MECHANICAL SYSTEMS OFFICE/BANK BUILDINGS UP TO 50 THOUSAND SQ FT,PROPOSED	COLLECTOR,MECHANICAL SYSTEMS STORAGE, MECHANICAL/PUMPED LIQUIDS

-PHONE-
303 5988228

	SUBCLASSES	SOLAR SUB SYSTEMS
ACORN GLAS-TINT INTERNATIONAL 1123 W CENTURY BLVD LOS ANGELES CA 90044 CHRIS T DUNKLE PRESIDENT	HOTEL/MOTEL/DORM BUILDINGS INDUSTRIAL BUILDINGS MERCANTILE BUILDINGS REFRIGERATION STORAGE BUILDINGS RELIGIOUS/PUBLIC/AMUSEMENT BUILDINGS WAREHOUSE BUILDINGS	

-PHONE-
213 7561471
GLASS TINTING PROCESS-REDUCES FADING GLARE,HEAT-INSULATES GLASS
ENERGY CONSERVATION

SUBCLASSES — SOLAR SUB SYSTEMS

EPPLEY LABORATORY INC
12 SHEFFIELD AVE
NEWPORT RI 02840
GEORGE L KIRK
PRESIDENT

-PHONE-
4 0184710 X20
EPPLY CONTINUES TO DEVELOP, IMPROVE & MFG SOLAR RAD INSTRUMENTATION,
BOTH FOR TOTAL MEASUREMENTS & DIFF SPECTRAL BANDS---SINCE 1920'S

IBM
ELECTRONICS SYSTEMS
HUNTSVILLE AL 35804
LOLLAR MANDT

SUNFALL MONITOR AUTOMATIC SOLAR ENGY RECORDING SYS IBM NO 732000485
DESIGN ACCOMPLISHED THROUGH:MARSHALL SPACE FLIGHT CTR NASA WHO IS EVAL
PERFORMANCE

"In one second the sun radiates more energy than man has used since the beginning of civilization..."

National Geographic 1965

	SUBCLASSES	SOLAR SUB SYSTEMS
DOME EAST CORPORATION 325 DUFFY AVENUE HICKSVILLE NY 11801 DAVID A ROBINSON DESIGN ENGINEER	SOLAR WATER HEATING GREENHOUSES	

-PHONE-
516 9380545
DOME EAST IS A DESIGNER, MANUFACTTURER, AND INSTALLER OF ENERGY CONSERVING WOOD DOME HOMES. THESE HOMES ARE WELL SEALED,HEAVILY INSULATED, & OWNER-ERECTABLE. WE ALSO MANUFACTURE SOLAR-HEATABLE POOL COVERS, GREEN HOUSES, AS WELL AS SOLAR WINDOWS & HOT H2O HEATERS,AVAIL FALL 1974...WRITE..CALL..OR..STOP..BY.

	SUBCLASSES	SOLAR SUB SYSTEMS
GAYDARDT INDUSTRIES INC 8542 EDGEWORTH DR CAPITAL HEIGHTS MD 20028 GEORGE R GAYDOS	SOLAR SPACE HEATING	COLLECTOR,MECHANICAL SYSTEMS

-PHONE-
301 3500614
MFG & INSTALL SOLAR SYS...COLL COPPER TUBING ON RUBY MESH(MADE OF 3 NATURAL MINERALS) STYROFOAM INSULATION...PRICE $3-4000 FOR 4 TO 5 UNIT 2 BY 3 FT...HAVE INSTALLED SYS IN 9HOMES

	SUBCLASSES	SOLAR SUB SYSTEMS
HELIO-DYNAMICS INC 518 SOUTH VAN NESS A LOS ANGELES CA 90020 TRUMAN D TEMPLE PRES	SOLAR RADIATION MEASUREMENTS SOLAR WATER HEATING SOLAR SPACE COOLING	

-PHONE-
213 3849853
SOLAR HEAT PANELS SINGLE GLAZED FOR LOW TEMP APP-SWIMMING POOLS.SOLAR DOMESTIC HOT H2O SYS WITH STORAGE. SOLAR AIRCOND CHILLERS&AIR HANDLING UNITS WITH CHILLER INTEGRAL WITH SOLAR ABSORBERS 3TONS & 5TONS COOLING & DEHUMIDIFY. SOLID STATE WIDE BAND PYRHELIOMETER DIRECT READ REMOTE RECORDING SINGLE OR DUAL CHANNEL FOR MULTIPLE LISTINGS

			SUBCLASSES	SOLAR SUB SYSTEMS
HELIOTEC INC 33 EDINBORO ST BOSTON	MA	02160	SOLAR WATER HEATING SOLAR SPACE HEATING & COOLING	COLLECTOR,ALSO FUNCTION AS STRUCTURES SKIN COLLECTOR,PASSIVE SYSTEMS COLLECTOR,MECHANICAL SYSTEMS

-PHONE-
617 4821245
SOLAR MEMBRANE-TRANSPARENT INSULAT MAT WITH MAX SOLAR TRANSMISSION
VIRTUALLY OPAQUE TO LONG IRRADIATION

			SUBCLASSES	SOLAR SUB SYSTEMS
NORTHRUP INC 2208 CANTON ST DALLAS LYNN L NORTHRUP	TX	75201		

-PHONE-
214 7415631
MANUF COLLECTORS TESTING PROTOTYPE

			SUBCLASSES	SOLAR SUB SYSTEMS
PPG INDUSTRIES INC NEW PROD DEVELOPMENT ONE GATEWAY CENTER PITTSBURGH R R LEWCHUK PROJECT MGR	PA	15222	PUBLISH ORIGINAL RESEARCHED INFORMATION SOLAR WATER HEATING SOLAR SPACE HEATING	

-PHONE-
412 4342645
CURRENTLY SUPP GLASS & GLASS PROD FOR SOLAR COLLECTORS OR COLLECT SYS
COST ESTIMATES FABRICATION & INSTALLATION OF COLLECTORS MAYBE ARRANGED
THROUGH PPG ARCHITECTURAL REP IN YOUR AREA...LITERATURE ON APPLICATION
OF GLASS & GLASS PROD AVAIL UPON REQUEST
MFG BASELINE SOLAR COLLECTOR A FLAT PLATE DOUBLE GLAZED LIQUID TYPE
COLL AT PRESENT PPG IS A COMPONENT SUPPLIER TO SOLAR INDUSTRY

			SUBCLASSES	SOLAR SUB SYSTEMS
RAYPAK INC 31111 AGOURA RD THOUSAND OAKS ALLEN BONIFACE	CA	91360	SOLAR WATER HEATING SOLAR SPACE HEATING	

-PHONE-
213 8891500
H2O COOLED COLL DOUBLE GLAZE AL FLAT PLATES 84X24 IN...PRICE COMPLETE
$15.50 PER SQ FT PLUS INSTALL...ALSO MFG SWIMMING POOL HEATERS 44 INS
BY 8 FT & 10 FT...ALL COLL AL SHEETS WITH COPPER TUBES &FLAT BK SURF

SOLAR SYS INC 323 COUNTRY CLUB DR REHOBOTH BEACH DANE DE REIMER	DE	19971	SOLAR SPACE HEATING	COLLECTOR,MECHANICAL SYSTEMS

-PHONE-
302 2272323

SOLERG ASSOCIATES PO BOX 90691 LOS ANGELES EGAN J RATTIN	CA	90009		

-PHONE-
213 8364200
IMPORT MEDIUM TEMP CONCENTRATORS

SOLORON 4850 OLIVE ST DENVER JOHN BAYLESS PRES & DIRECTOR	CO	80022	SOLAR WATER HEATING SOLAR SPACE HEATING & COOLING	COLLECTOR,MECHANICAL SYSTEMS

-PHONE-
303 2892288
PROVIDE COMPLETE AIR OR LIQUID TYPE SOLAR SYS WILL INSTALL..ENGINEERED
SYS FOR RESIDENTIAL COMM AGRICULTURAL APPLICATIONS...GOG LOF VP

			SUBCLASSES	SOLAR SUB SYSTEMS
STEELCRAFT CORP ENVIRONMENTAL DESIGN PO BOX 12408 MEMPHIS GARY FORD V P -PHONE- 901 3242151 901 4581885	TN	38112	WIND CONVERSION SOLAR AGRICULTURAL DRYING SOLAR WATER HEATING SOLAR SPACE HEATING & COOLING HOMES	COLLECTOR,ABSORPTIVE SELECTIVE SURFACES COLLECTOR,ALSO FUNCTION AS STRUCTURES SKIN COLLECTOR,MULTI-FUNCTION IN DWELLING SYS COLLECTOR,PASSIVE SYSTEMS COLLECTOR,MECHANICAL SYSTEMS HEAT TRANSFER SYSTEMS, COLLECTOR TO STORAGE HEAT RECOVERY, STORAGE TO UTILIZATION

MFG & MKT FLAT PLATE COLLECTORS (SELECTIVE SURFACE) WIND GENERATORS & GEODESIC DOMES EXPERIMENTAL RESEARCH RESIDENCE & LAB SOLAR ENERGY SYS ENGINEERING

STOLLE CORP 1501 MICHIGAN ST SIDNEY EDWARD G BECK JR	OH	45365	SOLAR WATER HEATING SOLAR SPACE HEATING	

MFG FLATE PLATE LIQUID TYPE SOLAR COLLECTOR PANELS...COPPER TUBES BOND TO ALUMINUM SHEET INQUIRE FOR DETAILS

SUNWORKS INC 669 BOSTON POST ROAD GUILFORD EVERETT M BARBER JR PRESIDENT -PHONE- 203 4536191	CT	06437	WIND CONVERSION SOLAR ENERGY STORAGE TOTAL SYSTEMS INTEGRATION RETROFITTING, PRACTICING NEW CONSTRUCTION, PRACTICING HOMES	COLLECTOR,MULTI-FUNCTION IN DWELLING SYS COLLECTOR,ALSO FUNCTION AS STRUCTURES SKIN COLLECTOR,MECHANICAL SYSTEMS HEAT TRANSFER SYSTEMS, COLLECTOR TO STORAGE STORAGE, OTHER

MFG FLAT PLATE COPPER HOT H2O TYPE SOLAR HEAT COLLECTORS.AVAIL FLUSH-MOUNTED TO FUNCTION ALSO AS ROOF OR WALL AND SURFACE-MOUNTED FREESTAND MODEL...WRITE FOR DETAILS...ALSO OFFERING ENGINEERING SERVICES FOR LOW GRADE THERMAL APPLICATIONS OF SOLAR ENERGY

TRANTER INC SPECIAL PROJECTS 735 EAST HAZEL ST LANSING ROBERT S BOWLAND ASST TO VP -PHONE- 517 3728410	MI	48909	SOLAR ENERGY STORAGE	

MFG SOLAR FLAT PLATE LIQUID TYPE COLL WITH 90 PERCENT INTERNAL WETTED SURFACE FOR MAX EFF...ALSO SUPPLY PLATECOIL FOR USE IN HEAT STORAGE TANKS

----------SUBCLASSES----------SOLAR SUB SYSTEMS----------

COLLECTOR,MECHANICAL SYSTEMS

UNITSPAN ARCH SYS INC
9419 MASON AVE
CHATSWORTH CA 91311
AL OTTUM

-PHONE-
213 9981131
ALUM FLAT PLATE COLL MODULES

SOLAR ENERGY INST OF TURKEY PUBLISH ORIGINAL RESEARCHED INFORMATION
POBOX 37 BAKANLIKLAR
ANKARA TURKEY
MUAMMER CETINCELIK
PRESIDENT

-PHONE-
234 255378
RESEARCH & MFG TERRESTRIAL THERMAL COLLECTOR SYS ALSO TECHNICAL PUB
ON SOLAR ENERGY ACTIVITIES

	SUBCLASSES	SOLAR SUB SYSTEMS
TOKIWA OPTIC&ELECT MANUFACTORY 2-9 1CHOME KOYAMADAI SHINAGAWA-KU TOKYO JAPAN HIROSHI MIZOJIRI PRESIDENT -PHONE- 03 7126212 03 7149947	SOLAR SPACE HEATING & COOLING	

TRIAL PRODUCTION OF PARABOLAS...IMPROVEMENT OF OUR SOLAR HEATING & COOLING APPARATUS

---SUBCLASSES---SOLAR SUB SYSTEMS---

DYNA TECHNOLOGY INC — SOLAR ELECTRIC POWER GENERATION
P O BOX 3263
SIOUX CITY IA 51102

MFG 200 WATT 12 VOLT 'WINDCHARGER'

GARDEN WAY LABORATORIES — BIO-CONVERSION
P O BOX 66
CHARLOTTE VT 05445
DR DOUGLAS C TAFF

-PHONE-
802 8626501
DEVELOPMENT OF HOME SIZE WIND ENGY CONVERSION-NOW IN MARKETING
PHASE
RESEARCH BEING CARRIED ON-PHOTOSYNTHESIS,ORGANIC GARDENING&COMPOSTING
& EFFICIENCIES INVOLVED

SENCENBAUGH WIND ELECTRIC — SOLAR ELECTRIC POWER GENERATION
P O BOX 11174
PALO ALTO CA 94306

SHAW PUMP COMPANY
9660 EAST RUSH ST
SOUTH EL MONTE CA 91733

-PHONE-
213 4431784
WINDMILL FOR PUMPING WATER

SUBCLASSES — SOLAR SUB SYSTEMS

SOLAR WIND — PUBLISH LAY BOOKS
PO BOX 7
EAST HOLDEN ME 04429
HENRY M CLEWS
PRES

-PHONE-
207 8435168

DISSEMINATE INFO ON WIND ELECT SYS...32 PAGE BOOKLET 'ELECTRIC POWER FROM THE WIND' AVAIL $2.00....ALSO $12.00 SET PLANS FOR 500WATT WIND GENERATOR CAN BE BUILT FOR $400...AGENTS FOR DUNLITE & CENTURY OF AUSTRALIA ELEKTRO GMBH OF SWITZERLAND DYNA TECHNOLOGY & SENCENBAUGH WIND ELECT OF USA...COMPLETE WIND ELECTSYS FROM 50 TO 12000 WATTS AS WELL AS INDIVIDUAL COMPONENTS...WRITE FOR PRICES

SIDNEY WILLIAMS CO PTY LTD
WM PARADE P O BOX 22
DULWICH HILL NSW AUSTRALIA

SELL WATER PUMPING WINDMILLS FROM 8TO 30 FT IN DIAMETER-FAN MILL TYPES

AEROWATT — SOLAR ELECTRIC POWER GENERATION
37 RUE CHANZY
PARIS 75 FRANCE

PRODUCE WINDMILLS FROM 1/2KW TO 5KW...HAVE DISTRIBUTERS IN US AND CANADA...SEE PENNWALT CORP& WALLACE&TIERNAN

ENAG S A — SOLAR ELECTRIC POWER GENERATION
DE PONT L'ABBE 29S
QUIMPER FRANCE

MFG WINDMILLS FROM 250W TO 2KW

SUBCLASSES — SOLAR SUB SYSTEMS

MASCHINEFABRIK LUBING BENING
2847 BARNSTORF PO110
GERMANY

-PHONE-
054 42625
MAKE SMALL WINDMILLS FOR WATER PUMPING AND ELECTRICITY CONVERSION

DOMENICO SPERANDO — SOLAR ELECTRIC POWER GENERATION
VIA CIMAROSA 13-21
FOLLONICA (GR) ITALY

AGER AEROGENERATORS IN 12 24 AND 48VOLTS FROM 300W TO 1KW

DUNLITE ELECTRICAL CO LTD — SOLAR ELECTRIC POWER GENERATION
FROME ST
ADELAIDE SOUTH AUSTRA

MFG WIND GENERATORS FROM 1/2 TO 2 KW

WALTER SCHOENBALL — SOLAR ELECTRIC POWER GENERATION
45 CH M DUBOULE
CH-1211
GENEVA 19 SWITZERLAND

GERMANY ROTOR DIA 36FT 5BLADES ON EACH CONTRAROTATING PROPELLER 70KW
NOAH DOUBLE ROTOR WINDPOWER PLANT PROTOTYPE INSTALLED ON WEST COAST
ELECT OUTPUT/PRODUCTION & DISTRIBUTION RIGHTS OFFERED TO AMERICAN MFGS
PROTOTYPE UNITS OF 30KW PLANT CAN BE PROVIDED ON SPECIAL ORDER FOR
PILOT PROJECTS PLEASE WRITE FOR DETAILS

SUBCLASSES — SOLAR SUB SYSTEMS

INST FOR FARM ELECTRIFICATION
1U VESHNIAKOVSKI
DOM 2 MOSCOW USSR

SOLAR ELECTRIC POWER GENERATION

NUMEROUS WINDMILLS FROM 1/2KW TO 25KW

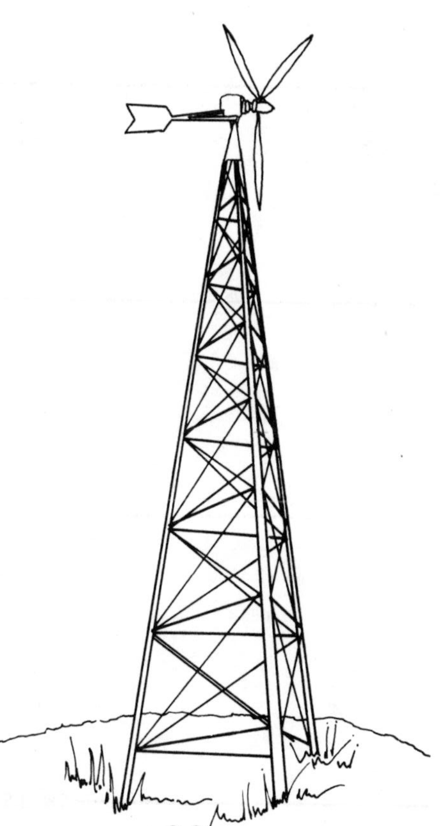

SUBCLASSES — SOLAR SUB SYSTEMS

ENERGY CONVERSION DEVICES INC
TROY MI 48084
STANFORD R OVSHINSKY

-PHONE-
313 5497300

INTERNATIONAL RECTIFIER PUBLISH LAY BOOKS
SEMICONDUCTOR DIV
233 KANSAS ST
EL SEGUNDO CA 90245

-PHONE-
213 6786281
MARKET: SOLAR CELLS & PHOTOCELLS
PUBLISH:'SOLAR CELL EXPERIMENTERS HANDBOOK,'EXPERIMENTERS GUIDE TO SOLAR CELLS INCLUDES SIMPLE DEMOS & PRACTICAL CIRCUITS-HB99-$.50 'SOLAR CELLS & PHOTOCELLS,'DISCUSSION & PRACTICAL CONSIDERATIONS & ILLUSTRATED APPLICATION DATA-HB-30-$2.00

SOLAR ENERGY SYSTEMS INC
70 S CHAPEL ST
NEWARK DE 19711
DR KARL W BOER
-PHONE-
302 7310990
MANUF LOW-COST,LONG LIFE, CADIUM SULFIDE SOLAR CELLS DEV BY UNIV OF DE WILL MANUF SOLAR ENERGY SYS TO SUPPLY SUPP ELECT POWER,HEATING & COOLING OF RES,BUSINESS&INDUSTRY...SHELL OIL CO MINORITY STOCK HOLDER PROVID 3MILL FINANCIAL SUPPORT

SOLAR POWER CORP
186 FORBES ROAD
BRAINTREE MA 02184
ROBERT W WILLIS
VICE PRESIDENT

-PHONE-
617 8486877
SELL LOW COST PHOTOVOLTAIC SYSTEMS FOR TERRESTRIAL USE

---SUBCLASSES---SOLAR SUB SYSTEMS---

SOLAREX CORP — SOLAR ELECTRIC POWER GENERATION
1335 PICCARD DR
ROCKVILLE MD 20850
PETER F VARADI
EXEC V P

-PHONE-
301 9480202
MFG ROUND SILICON SOLAR CELLS FOR REMOTE LOCATIONS

SPECTROLAB DIV OF TEXTRON — SOLAR ELECTRIC POWER GENERATION
SOLAR POWER SYSTEMS
12484 GLADSTONE AVE
SYLMAR CA 91342
DICK DONNELLY

-PHONE-
213 3654611
MFG & MARKET COMPLETE LINE TERRESTRIAL PHOTOVOLTAIC SOLAR POWER SUPPLY
FOR 15 YRS HAVE DESIGN&MFG SOLAR CONVERSION SYS...ALSO MFG SOLAR
SIMULATION SYS SOLAR INSTRUMENTATION & SOLAR SENSORS

SPECTROLAB DIV OF TEXTRON — SOLAR ELECTRIC POWER GENERATION
SOLAR POWER SYSTEMS
12484 GLADSTONE AVE
SYLMAR CA 91342
JERRY W RAVIN *
MARKETING MGR
-PHONE-
213 3654611

ENERGEX CORP			SOLAR RADIATION MEASUREMENTS
418 TROPICANA			WIND CONVERSION
LAS VEGAS	NV	89109	SOLAR WATER DISTILLATION
ALFRED JENKINS			SOLAR WATER HEATING
PRESIDENT			SOLAR SPACE HEATING & COOLING
			HOMES

-PHONE-
702 7362994
PRODUCE & MARKET SOLAR ENGY PRODUCTS, SYSTEMS-OFF-THE-SHELF DELIVERY OF SOLAR COLLECTORS, HOT H2O HEATERS AND MANY MORE....OVER 150 COMMER SOLAR ENGY INSTALLED SINCE 1970....PLEASE WRITE FOR BROCHURE

----------SUBCLASSES----------SOLAR SUB SYSTEMS----------

SOLAR PRODUCTS CORP
3500 EAST 1ST AVE
DENVER CO 80206

HIGH TEMPERATURE SOLAR COLLECTOR SYSTEMS

MANUF SUNFLOWER SOLAR POWERED COOK STOVE REACHES 450F IN 10 MIN.WEIGHT
2 LBS.CORR PAPER AL FOIL PLASTIC.$5.95

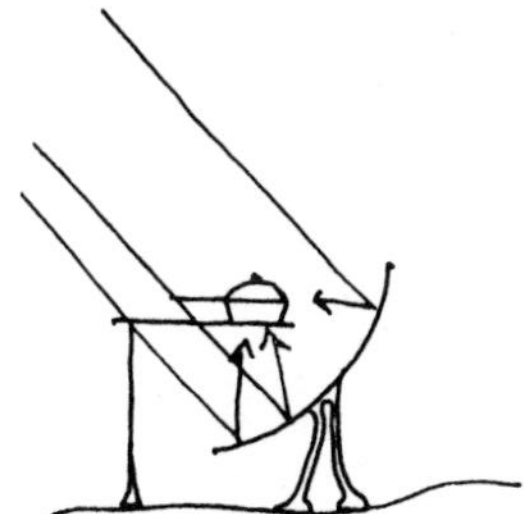

	SUBCLASSES	SOLAR SUB SYSTEMS
BEUTELS SOLAR HEATER INC 1527 N MIAMI AVE MIAMI FL 33136 -PHONE- 305 3711426	MEDIUM TEMPERATURE SOLAR COLLECTOR SYSTEMS	
COLEMAN SOLAR SERVICE 8900 NW 34 AVE RD MIAMI FL 33147 JAMES J GAY -PHONE- 305 6914126		COLLECTOR,MECHANICAL SYSTEMS

MAINTENANCE & REPAIR WORK ON EXISTING SOLAR H2O SYS...MAKES NO NEW IN-STALLATIONS BUT WOULD IF DEMAND GREAT...MOST PRESENT WORK ON HOMES IN CORAL GABLES AREA BUILT IN 20S & 30S

	SUBCLASSES	SOLAR SUB SYSTEMS
D&J SHEET METAL CO 10555 NW 7TH AVE MIAMI FL 33150 JAKE STICHER -PHONE- 305 7577033		COLLECTOR,MECHANICAL SYSTEMS
DAYLIN INC SUN SOURCE 9606 SANTA MONICA BEVERLY HILLS CA 90035 DAVID L COLLINS GENERAL MANAGER -PHONE- 213 8783211	MEDIUM TEMPERATURE SOLAR COLLECTOR SYSTEMS SOLAR SPACE HEATING & COOLING UP TO 50 THOUSAND SQ FT,PROPOSED PUBLISH LAY PERIODICALS PUBLISH INFORMATIONAL REPRINTS LOBBY TO PROMOTE SOLAR LEGISLATION	COLLECTOR,ABSORPTIVE SELECTIVE SURFACES TURBINES

HIGH PRODUCTION ADVANCED TECHNOLOGY COLLECTOR PANELS
SOLAR WATER HEATER SYSTEMS MANUFACTURING
INDUSTRIAL SYSTEM DESIGN AND PROJECT COORDINATION
DO IT YOURSELF SOLAR SYSTEMS
SPECIAL PURPOSE LOW INCOME HOUSING HOT WATER AND AREA HEATING SYSTEMS
APPLICATIONS: MOBILE HOMES PROPOSED SINGLE FAMILY PROPOSED 2-5 FAMILY CLUSTER HOMES PROPOSED HEALTH/HOSPITAL/INDUSTRIAL BLDGS UP TO 50 THOUS SQFT PROPOSED

------SUBCLASSES------SOLAR SUB SYSTEMS------

FAFCO INC
2860 SPRING ST
REDWOOD CITY CA 94063
RALPH SCHNEIDER
MARKETING MGR

MEDIUM TEMPERATURE SOLAR COLLECTOR SYSTEMS

-PHONE-
415 3646772
MANUFACTURE BLACK PLASTIC LIQUID TYPE SOLAR WATER HEATERS MARKET AS INEXPENSIVE SWIMMING POOL HEATERS...RESEARCHING APPLICATIONS FOR HOT WATER HEATING, SPACE HEAT & COOL, HYDROPONICS,& AQUACULTURE WRITE FOR BROCHURE...SEE:TOM BOYD,'POOL-HEATING PANELS MAKE SUN WORK FOR YOU,'POPULAR SCIENCE,MAY 1973.

FRED RICE PRODUCTIONS INC
6313 PEACH AVE
VAN NUYS CA 91401
FRED RICE
PRESIDENT

INORGANIC SEMI-CONDUCTORS
ALL SOLAR TECHNOLOGIES
SOLAR COOKING
SOLAR WATER DISTILLATION
SOLAR SPACE HEATING & COOLING
MOBILE HOMES

-PHONE-
213 7863860
MANUF & MARKET HIGH SPEED CYLINDRICAL SOLAR WATER HEATER SYSTEM WRITE FOR BROCHURE
SOLAR/SONIC HOME:COMPLETELY SOLAR LOCATION: 48780 EISENHOWER DR, LA QUINTA,CA 92253/P O BOX 644/PHONE:714-564-4823
LIMITED PRODUCTION/PROTOTYPES-SOLAR HOMES FACTORY BUILT&CUSTOM
SOLAR/SONIC MOBILE HOME:SOLAR ASSISTED
LIMITED PRODUCTION ON SOLAR CLOCKS,RADIOS,SECURITY SYS,SOLAR FLASH-LIGHTS, SWIMMING POOL HEATER,SOLAR OUTDOOR AIRCONDITIONING

FUN & FROLIC INC
P O BOX 277
MADISON HEIGHTS MI 48071
EDWARD J KONOPKA
PRESIDENT

MANUFACTURE BLACK PLASTIC LIQUID TYPE SOLAR WATER HEATERS AS INEXPENSE SWIMMING POOL HEATERS.....ALSO COULD APPLY TO DOMESTIC HOT WATER HEATERS

	SUBCLASSES	SOLAR SUB SYSTEMS
HITACHI 437 MADISON AVE NEW YORK NY 10022 -PHONE- 212 8384804 MFG HI HEATER...CYLINDER COLL OF POLYETHYLENE.BOX CARBON STEEL.COVER SHEET POLYCARBONATE.FOAM PLASTIC INSULATION		
JOHNSON DIVERSIFIED INC 2340 QUEEN ANN ST MERRITT ISLAND FL 32952 STAN JOHNSON -PHONE- 305 4525545		
LOF BROS SOLAR APPLIANCES P O BOX 10594 DENVER CO 80210 L LOF -PHONE- 303 7446852 303 7582585 MFG & DISTRIBUTE CUSTOM LIGHTWEIGHT POOL COVERS TO SOLAR HEAT SWIMMING POOLS IN SUMMER..CHEMICALLY TREATEDFILM WITH SELECTIVE COAT...GUARANTE 18-24 MOS 27 CENTS/SQFT WITH 8 MAY DAYS DENVER HEAT POOL FROM AMBIENT TEMP TO 82F	MEDIUM TEMPERATURE SOLAR COLLECTOR SYSTEMS	COLLECTOR,ABSORPTIVE SELECTIVE SURFACES
SOL-THERM CORP 7 WEST 14TH STREET NEW YORK NY 10011 ITAMAR SITTENFELD VICE PRESIDENT -PHONE- 212 6914632 HOT H2O SYS WITH COLLECTORS & SUPPLEMENTAL IMMERSION HEATER & REQ PIPING & CONTROLS & STORAGE TANK...RETAIL $695. ADDITIONAL COLL $245. 14 TO 16 WKS DELIVERY TIME	MEDIUM TEMPERATURE SOLAR COLLECTOR SYSTEMS	

Manufacturer	State	ZIP	Subclasses	Solar Sub Systems
SOLAR ENERGY CO DEERWOOD DR MERRIMACK ROGER PAPINEAU PRESIDENT	NH	03054		

-PHONE-
603 4245168
SWIMMING POOL SOLAR HEATERS

Manufacturer	State	ZIP	Subclasses	Solar Sub Systems
SUHAY ENTERPRISES 1505 E WINDSOR RD GLENDALE FRANK L SUHAY	CA	91204	MEDIUM TEMPERATURE SOLAR COLLECTOR SYSTEMS SOLAR SPACE HEATING	

-PHONE-
213 2469352
PORTABLE HOT H2O HEATER FOR PICNIC OR CAMPING WEIGHT 1 LB PROD 4 1/2 LBS HOT H2O...SWIMMING POOL HEATER...EDUCATIONAL TOY VERSION...I WISH TO LICENSE THE MFG & DISTRIBUTION RIGHTS TO PRODUCTS SO I MAY RETURN TO RESEARCH...HOLDS SOLAR PATENTS

Manufacturer	State	ZIP	Subclasses	Solar Sub Systems
SUNDU COMPANY 3319 KEYES LANE ANAHEIM ANTHONY MEAGHER	CA	92804	MEDIUM TEMPERATURE SOLAR COLLECTOR SYSTEMS	

-PHONE-
714 8282873
MANUF PLASTIC FLAT PLATE HOT WATER TYPE COLLECTORS TO HEAT SWIMMING POOLS

Manufacturer	State	ZIP	Subclasses	Solar Sub Systems
SUNWATER CO 1112 PIONEER WAY EL CAJON HORACE MCCRACKEN	CA	92020	MEDIUM TEMPERATURE SOLAR COLLECTOR SYSTEMS WIND CONVERSION SOLAR WATER DISTILLATION SOLAR SPACE HEATING & COOLING HOMES	COLLECTOR,MECHANICAL SYSTEMS COLLECTOR,ALSO FUNCTION AS STRUCTURES SKIN COLLECTOR,MULTI-FUNCTION IN DWELLING SYS COLLECTOR,PASSIVE SYSTEMS STORAGE, SOLID MATERIALS STORAGE, MECHANICAL/PUMPED LIQUIDS STORAGE, LIQUIDS

-PHONE-
714 2830454
SOLAR STILLS FROM COMPLETELY AUTOMATIC SYS TO DO IT YOURSELF...1,1 1/2 & 2 GAL/DAY YIELD SIZES COVER MOST RESIDENTIAL NEEDS...ALSO H2O HEATRS & SWIMMING POOL HEATERS

SUBCLASSES — SOLAR SUB SYSTEMS

BEASLEY INDUSTRIES PTY LTD — MEDIUM TEMPERATURE SOLAR COLLECTOR SYSTEMS
BOLTON AVENUE
DEVON PARK, S A AUSTRALIA
E L BEASLEY
SALES MANAGER

-PHONE-
082 464871
AUSTRALIAS LARGEST PRODUCER OF SOLAR WATER HEATERS WITH EXPORTS TO MANY COUNTRIES...MARKET A RANGE OF DOMESTIC SOLAR WATER HEATERS ALSO OTHER SOLAR ENERGY APPLIANCES INCLUDING FLAT PLATE COLLECTORS

BRAEMAR ENGINEERING PTY LTD — MEDIUM TEMPERATURE SOLAR COLLECTOR SYSTEMS
BOX 63 PO
BROOKLY PARK, S A AUSTRALIA

-PHONE-
56 6644

BRAEMAR ENGR PTY LTD — MEDIUM TEMPERATURE SOLAR COLLECTOR SYSTEMS
P O BOX 58
GEEBUNG AUSTRALIA

SOLAR KITS INCLUDE ABSORBER PANELS STORAGE TANK CONNECTING PIECES & TUBING...FOB 1 PANNEL $74...OFFICES THROUGHOUT AUSTRALIA

SMALLS SOLA HEETA CO
10 GOONGARRIE ST
BAYSWATER AUSTRALIA

THERMAX ELECTRIC WATER HEATERS
P O BOX 173
HAMILTON CENTRAL AUSTRALIA

---SUBCLASSES---SOLAR SUB SYSTEMS---

TURBON ENGINEERING PTY LTD — MEDIUM TEMPERATURE SOLAR COLLECTOR SYSTEMS
BIRUDI STREET
COORPAROO AUSTRALIA

SOLAR HEAT LTD
99 MIDDLETON HALL RD
KINGS NORTON BIRMGHM ENGLAND
A A BLANCO
DIRECTOR

MFG SUNTRAP SOLAR HOT H2O HEATERS

METALCO — MEDIUM TEMPERATURE SOLAR COLLECTOR SYSTEMS
NICOSIA CYPRUS GREECE
MARIUS IONNIDES
PROD MGR

ASKELON METAL PRODUCTS LTD — MEDIUM TEMPERATURE SOLAR COLLECTOR SYSTEMS
TEL AVIV ISRAEL

MFG SOLAR HOT WATER HEATERS FOR MIROMIT LTD TEL-AVIV

MIROMIT LTD
44 MONTEFORT ST
TEL AVIV ISRAEL

-PHONE-
611481
SOLAR HOT WATER HEATERS FOR RESIDENTIAL & COMMERCIAL USES SPECIAL
SELECTIVE BLACK COATING

----SUBCLASSES----SOLAR SUB SYSTEMS----

TERRUM SUN&ELECT BOILER INDUST
14 HAYOTZER HOLDN
TEL AVIV ISRAEL 61148

SEKISUI KAYAKU KOGIYO CO
2 KINUGASA MACHI
OSAKA JAPAN

RADISOL S A
31 BOULEVARD DANTON
CASABLANCO MOROCCO

	SUBCLASSES	SOLAR SUB SYSTEMS
ENERGY SYSTEMS INC 634 CREST DR EL CAJON CA 92021 TERRANCE R CASTER PRESIDENT	6-14 FAMILY, DEVELOPMENT	COLLECTOR,MECHANICAL SYSTEMS

-PHONE-
714 4471000
DEV & INSTALL SOLAR HEAT PACKAGES IN RESIDENTIAL DEVEL.PLAN TO MARKET NATIONWIDE END OF 1974...FIRST 7 HOUSES HAVE ROOF AIR COLL FOR HEAT... ROCK BED STORAGE...VARIOUS SYS WILL BE TRIED FOR OPTIMAL DESIGN

	SUBCLASSES	SOLAR SUB SYSTEMS
HELTER-SKELTER ENGINEERING BOX 479 FOREST KNOLLS CA 94933 WILLIAM WARD-SWANN	MEDIUM TEMPERATURE SOLAR COLLECTOR SYSTEMS WIND CONVERSION	

-PHONE-
415 4572986
INEXPENSIVE FLAT PLATE COLLECTORS FOR THE HOBBIEST-AND HEAT STORAGE FOR RESIDENTIAL HEATING
WINDMILLS FOR LOW TO MEDIUM POWER GENERATION 0.5-5 KW-NO MODELS TO DATE

	SUBCLASSES	SOLAR SUB SYSTEMS
KALWALL CORPORATION P O BOX 237 MANCHESTER NH 03105 KEITH HARRISON DIVISION MGR	MEDIUM TEMPERATURE SOLAR COLLECTOR SYSTEMS SOLAR WATER HEATING HOMES GREENHOUSES INDUSTRIAL BUILDINGS COMMERCIAL OTHER	COLLECTOR,ALSO FUNCTION AS STRUCTURES SKIN COLLECTOR,MULTI-FUNCTION IN DWELLING SYS COLLECTOR,OTHER STORAGE, SOLID MATERIALS

-PHONE-
603 6273861
603 6277887
HAVE 4 PRODUCTS SOLAR RELATED:SUN-LITE FIBERGLASS SHEETING FOR GLAZING COLLECTORS/DOUBLE GLAZED SOLAR GLAZING ASSEMBLIES/WALL PANELS WHICH CAN BE MODIFIED TO COLLECTORS/&SOLAR COLLECTORS BOTH AIR & WATER TYPES FLAT PLATE

	SUBCLASSES	SOLAR SUB SYSTEMS
SOLAR INC 4706 FAIRVIEW AVE BOISE ID 83704 JAMES H BALLANTYNE JR		
ZOMEWORKS CORPORATION BOX 712 ALBUQUERQUE NM 87103 STEVE BAER PRESIDENT	MEDIUM TEMPERATURE SOLAR COLLECTOR SYSTEMS SOLAR WATER HEATING HOMES GREENHOUSES PROVIDE INFORMATION SEARCHES TO CLIENTS	COLLECTOR,ALSO FUNCTION AS STRUCTURES SKIN COLLECTOR,MULTI-FUNCTION IN DWELLING SYS COLLECTOR,PASSIVE SYSTEMS STORAGE, LIQUIDS

-PHONE-
505 2425354
HOT H2O HEATER WE ARE MARKETING/WE DO DESIGN & CONSULTATION WORK
HAVE DEVELOPED SKY LID WATER HEATER&BEAD WALL WRITE FOR BROCHURE/MANY
PUBLISHED WORKS

	SUBCLASSES	SOLAR SUB SYSTEMS
HOFLAR INDUSTRIES 5511 128TH ST SURREY B C CANADA ERICH W HOFFMANN	MEDIUM TEMPERATURE SOLAR COLLECTOR SYSTEMS	

-PHONE-
604 5962665
DESIGN & CONSTRUCTION OF SOLAR HEATED HOUSES & SWIMMING POOLS...EXPER-
IMENTAL HOUSE IN OPERATION

"Right now, a home heating system can be designed—and we have reliable systems—that can compete with the cost of electricity in most regions of the country."
L. Herwig, N.S.F.

	SUBCLASSES	SOLAR SUB SYSTEMS
A A I CORP WEAPONS&AERO SYS DIV PO BOX 6767 BALTIMORE MD 21204 IRWIN R BARR	MEDIUM TEMPERATURE SOLAR COLLECTOR SYSTEMS WIND CONVERSION SOLAR SPACE HEATING PUBLISH ORIGINAL RESEARCHED INFORMATION	COLLECTOR,MECHANICAL SYSTEMS STORAGE, LIQUIDS

-PHONE-
301 6661400 X232

DESIGN BUILDING & TESTING COMPLETE SOLAR HEATING & COOLING SYS FIRST MAJOR PROJ RETROFIT 1/3 OF TIMONIUM ELEMENTARY SCH 5000 SQFT COLL 15000 GAL WATER STORAGE...WE DESIGN & MANUF HEATING & COOLING SYS TO MEET YOUR COMMERCIAL OR RESIDENTIAL NEEDS

	SUBCLASSES	SOLAR SUB SYSTEMS
REVERE COPPER & BRASS INC BLDG PROD DEPT P O BOX 151 ROME NY 13440 WILLIAM J HEIDRICH	MEDIUM TEMPERATURE SOLAR COLLECTOR SYSTEMS SOLAR WATER HEATING	COLLECTOR,MECHANICAL SYSTEMS COLLECTOR,ALSO FUNCTION AS STRUCTURES SKIN COLLECTOR,ABSORPTIVE SELECTIVE SURFACES STORAGE, LIQUIDS

-PHONE-
315 3382022

MANUF & MARKET FLAT PLATE LIQUID TYPE COLLECTORS FOR APPLICATION TO BLDG HEAT&COOL HOT H2O & SWIMMING POOLS COLLECTOR IS COPPER WITH TUBES STANDARD 2X8 FT PANELS OTHER SIZES AVAIL ON REQUEST

	SUBCLASSES	SOLAR SUB SYSTEMS
SOLAR ENERGY RESEARCH CORP RT 4 BOX 26B LONGMONT CO 80501 JAMES B WIEGAND PRESIDENT	TOTAL SYSTEMS INTEGRATION SOLAR AGRICULTURAL DRYING SOLAR WATER DISTILLATION SOLAR WATER HEATING PROVIDE INFORMATION SEARCHES TO CLIENTS PROVIDE VENTURE CAPITOL	COLLECTOR,MECHANICAL SYSTEMS COLLECTOR,ALSO FUNCTION AS STRUCTURES SKIN COLLECTOR,PASSIVE SYSTEMS HEAT TRANSFER SYSTEMS, COLLECTOR TO STORAGE HEAT RECOVERY, STORAGE TO UTILIZATION STORAGE, MECHANICAL/PUMPED LIQUIDS STORAGE, LIQUIDS

-PHONE-
303 7724522

AVAILABLE NOW AT 4DOLLARS/SQFT COMPLETE MECH COL STORAGE UTILIZATION SYS NOW OPERATING 2000 SQFT MECH HEAT SYS IN COMMERCIAL TOMATO GREEN HOUSE WILL MARKET SOLID STATE CONTROL LATE 1974 LESS THAN 100 DOLLARS INSTALLING 4FTX12FT ROOF MODULE. WILL MANUFACTURE MID 1974 CONSULTING AVAIL ON INDUSTRIAL SPACE OR PROCESS HEATING AND WASTE UTILIZATION SYS LOW COST

Manufacturer			Subclasses	Solar Sub Systems
SOLARSYSTEMS INC 1515 WSW LOOP 323 TYLER JOHN L DECKER PRESIDENT	TX	75701	HIGH TEMPERATURE SOLAR COLLECTOR SYSTEMS SOLAR WATER HEATING RETROFITTING, MECHANICAL SYSTEMS RETROFITTING, PRACTICING NEW CONSTRUCTION, MECHANICAL SYSTEMS NEW CONSTRUCTION, PRACTICING	COLLECTOR,ALSO FUNCTION AS STRUCTURES SKIN COLLECTOR,MECHANICAL SYSTEMS HEAT TRANSFER SYSTEMS, COLLECTOR TO STORAGE STORAGE, MECHANICAL/PUMPED LIQUIDS

-PHONE-
214 5920945
214 5975537
MANUFACTURERS OF HIGH EFFICIENCY FLAT PLATE SOLAR ENERGY COLLECTORS...
PRODUCE TEMPERATURES CAPABLE OF OPERATING ABSORPTION TYPE AIRCONDITION
SYS 230 TO 260 DEGREE WATER AT COST COMPETITIVE WITH OTHER ENERGY FORM
& CHEAPER THAN MOST IN MANY AREAS...COLLECTORS CAN BE DELIVERED WITHIN
60 TO 120 DAYS DEPENDING ON QUANT

Manufacturer			Subclasses	Solar Sub Systems
SOLARSYSTEMS INC 1515 WSW LOOP 323 TYLER CONE RICE * V P SALES	TX	75701	HIGH TEMPERATURE SOLAR COLLECTOR SYSTEMS SOLAR WATER HEATING RETROFITTING, MECHANICAL SYSTEMS RETROFITTING, PRACTICING NEW CONSTRUCTION, MECHANICAL SYSTEMS NEW CONSTRUCTION, PRACTICING	COLLECTOR,ALSO FUNCTION AS STRUCTURES SKIN COLLECTOR,MECHANICAL SYSTEMS HEAT TRANSFER SYSTEMS, COLLECTOR TO STORAGE STORAGE, MECHANICAL/PUMPED LIQUIDS

-PHONE-
214 5920945
214 5975534

Manufacturer			Subclasses	Solar Sub Systems
SUNEARTH CONSTRUCTION CO INC BOX 99 MILFORD SQUARE	PA	18935	MEDIUM TEMPERATURE SOLAR COLLECTOR SYSTEMS SOLAR WATER HEATING	COLLECTOR,MECHANICAL SYSTEMS

-PHONE-
215 5368555
MFG FLAT PLATE LIQUID TYPE COLL FOR HEAT COOL & HOT H2O...SHIPPED
READY TO INSTALL WITH ALL REQ HARDWARE TAPES & CAUKS...$7.50 /SQFT

Manufacturer			Subclasses	Solar Sub Systems
THOMASON SOLAR HOMES INC 6802 WALKER MILL SE WASHINGTON DR HARRY E THOMASON	DC	20027	HOMES PUBLISH ORIGINAL RESEARCHED INFORMATION PROVIDE INFORMATION SEARCHES TO CLIENTS	

-PHONE-
301 3364042
301 3365329
HOUSE PLANS & DISCRIPTIVE LITERATURE AVAIL THROUGH EDMUND SCI CO

---SUBCLASSES---SOLAR SUB SYSTEMS---

CHEMALLOY-ELECTRONICS CORP 7947 TOWERS-PO DR 10 SANTEE CA 92071 MR FRIEDMAN -PHONE- 714 4485715 MFG FUEL CELLS	STORAGE, FUEL CELLS

"If 1% of the solar energy falling on the Sahara Desert were converted to electrical power it would supply all the worlds needs for electrical power for the year 2000..."

Van W. Bearinger
Honeywell

----SUBCLASSES----SOLAR SUB SYSTEMS----

PHYSICAL INDUSTRIES CORP	TOTAL SYSTEMS INTEGRATION
SOLAR DIVISION	RETROFITTING, PRACTICING
P O BOX 357	NEW CONSTRUCTION, PRACTICING
LAKESIDE CA 92040	SINGLE FAMILY, EXPERIMENT
JOHN H HEDGER	PUBLISH LAY + TECHNICAL BOOKS
PRESIDENT	PUPLISH ORIGINAL RESEARCH + REPRINTS
-PHONE-	
714 5611266	

'THE SLEEPING GIANT-SOLAR ENERGY' TEACH SOLAR ENGY BASICS 4 NIGHTS/ WK HIGH SCH. MFG SOLAR LIQUID HEATING PANELS SWIMPOOL HEATERS HOUSE HEAT & COOL HEAT ENGINE & HEAT STORAGE TANKS MFG 16 FT SOLAR POWERED BOAT IN FUTURE. EXP WITH AIRCOND USING OUR HEAT ENGINE. MFG TOYS STEAM GEN STEAM ENGINES EXP WITH HEAT STORAGE EXP WITH SOLAR POWERED AUTO-HEAT ENGINE & HEAT STORAGE

			SUBCLASSES	SOLAR SUB SYSTEMS
ALBUQUERQUE WESTERN INDUS INC			MEDIUM TEMPERATURE SOLAR COLLECTOR SYSTEMS	
ALBUCUERQUE	NM	87107	SOLAR WATER HEATING	
			SOLAR SPACE HEATING & COOLING	

-PHONE-
505 3447224
MANUFACTURE SUN KING MODULAR SOLAR HEATED HOMES:1,2,3,&4 BEDROOMS
RANGE FROM 750 SQFT TO 1500 SQFT...FLAT PLATE WATER TYPE COLLECTORS
36 HOURS STORAGE

			SUBCLASSES	SOLAR SUB SYSTEMS
BOB SCHMITT HOMES INC			SOLAR SPACE HEATING & COOLING	HEAT PUMPS
13079 FALLING WATER				
STRONGSVILLE	OH	44136		
D J DIRKSE				

-PHONE-
216 2386915
HOME BUILDER- THE SUN HOUSE...HOUSE HEATED BY SUN ATRIUM AT HOUSE CORE
NO OTHER WINDOWS TO CUT ENERGY COSTS NO OTHER OUTSIDE WINDOWS
HOUSE WELL INSULATED EMPLOYS HEAT PUMP SYS
WRITE FOR BROCHURE

			SUBCLASSES	SOLAR SUB SYSTEMS
SOLAR HOMES INC			MEDIUM TEMPERATURE SOLAR COLLECTOR SYSTEMS	
2 NARRAGANSETT AVE			SOLAR SPACE HEATING	
JAMESTOWN	RI	02835	NEW CONSTRUCTION, PRACTICING	
STENCER DICKENSON				

-PHONE-
401 4231025
BUILDER OF SOLAR HEATED HOMES

---SUBCLASSES---SOLAR SUB SYSTEMS---

LAB SCIENCES INC
PO BOX 1236
BOCA RATON FA 33432
JOSEPH A CARISEO
PRESIDENT

-PHONE-
305 3951917
MFG DEMO APPARATUS...SOLAR PANEL UNIT $2195. THERMAL RAD TEST & RESCH
UNIT $1495. SOLAR ENGY TEST & RESCH UNIT

	SUBCLASSES—SOLAR SUB SYSTEMS
EDUCATIONAL MATERIALS&EQUIP CO	MEDIUM TEMPERATURE SOLAR COLLECTOR SYSTEMS
BOX 17	HIGH TEMPERATURE SOLAR COLLECTOR SYSTEMS
PELHAM NY 10803	INORGANIC SEMI-CONDUCTORS
THOMAS J MCMAHON	SOLAR WATER DISTILLATION
	SOLAR WATER HEATING

-PHONE-
914 5761121
HIGH QUALITY EQUIP & A V AIDS FOR ENTIRE EDUCATIONAL SPECTRUM...MFG & DISTRIBUTE SOLAR ENGY SCIENCE EXP OUTFIT $89.95 INCLUDES FULL SPECTRUM SOLAR EXPERIMENTS. WRITE FOR DETAILS

--------SUBCLASSES--------SOLAR SUB SYSTEMS--------

EDMUND SCIENTIFIC CO			HIGH TEMPERATURE SOLAR COLLECTOR SYSTEMS
ADV & MKT			WIND CONVERSION
101 E GLOUCESTER PK			INORGANIC SEMI-CONDUCTORS
BARRINGTON	NJ	08007	SOLAR COOKING
JACK SCHARFF			SOLAR WATER HEATING
DIR MKT & ADV			PUBLISH LAY + TECHNICAL PERIODICALS
-PHONE-			
609-547-3488			

PUBLISHER OF:SOLAR HOUSE PLANS, DR H THOMASON AUTHOR SK.9440.
H THOMASON,'SOLAR HOUSES & SOLAR HOUSE MODELS'NO 9069,$1.00
H THOMASON,'SOLAR HOUSE HEATING & AIR-CONDITIONING SYS:COMPARISONS & LIMITATIONS'NO 9463,$5.00
SELL VARIED SOLAR RELATED PRODUCTS SUCH AS:SOLAR FURNACE, FRESNEL LENS RADIOMETER, 12 VOLT PANEL OF SILICON SOLAR CELLS, SOLAR MOTORS, SINGLE SOLAR CELLS, REFLECTORS, HEAT PIPES, SOLAR DEMONSTRATORS...SEND FOR CATALOG 742 FREE...OVER 4000 UNIQUE ITEMS OF INTEREST TO ALL

	SUBCLASSES	SOLAR SUB SYSTEMS
BIG OUTDOORS PEOPLE P O BOX 936 MINNEAPOLIS MN 55440 DENNIS O JOHNSON	SOLAR WATER HEATING SOLAR SPACE HEATING & COOLING	

-PHONE-
612 3315344
612 3315430
MARKET & MANUF GEODESIC DOMES & OTHER ALTERNATIVE BUILD SYS
CURRENTLY RESEARCHING SOLAR UTILIZATION IN RELATION TO DOMES

	SUBCLASSES	SOLAR SUB SYSTEMS
CERTAIN-TEED PRODUCTS CORP P O BOX 860 VALLEY FORGE PA 19482 PHILIP W BACON SOLAR ENGY COORD		STORAGE, OTHER

-PHONE-
215 6875000 X7989
MANUF SOLAR COLLECTORS, STORAGE TANKS, & PLASTIC PIPE SYS- CURRENTLY
RESEARCHING MATERIALS & MARKET ANALYSIS

	SUBCLASSES	SOLAR SUB SYSTEMS
CORNING GLASS WORKS LIGHTING PROD DIV CORNING NY 14830 DONALD A URQUHART * MANAGER SPECIAL PROJ	HIGH TEMPERATURE SOLAR COLLECTOR SYSTEMS SOLAR WATER HEATING SOLAR SPACE HEATING & COOLING SOLAR ELECTRIC POWER GENERATION SOLAR HEAT,COOL,&ELECTRIC POWER GENERATION	COLLECTOR,ABSORPTIVE SELECTIVE SURFACES COLLECTOR,REFLECTIVE SELECTIVE SURFACES

-PHONE-
607 9747306
HIGH PERFORMANCE TUBULAR COLLECTORS VACUUM INSULATED CONFIGURATIONS,
GLASS COMPONENTS, CERAMIC PARTS...NO PUBLIC INQUIRY PLEASE

	SUBCLASSES	SOLAR SUB SYSTEMS
GRUMMAN AEROSPACE CORPORATION ENERGY SYSTEMS DEPT PLANT 30 BETHPAGE NY 11714 DR ROBERT MADLEY MGR RESEARCH&DEVEL -PHONE- 516 5751573	OCEAN THERMAL GRADIENT SYSTEMS NON-TERRESTRIAL SOLAR POWER SYSTEMS WIND CONVERSION	STORAGE, HYDROGEN

OBJECTIVE: DEVELOP & MARKET SOLAR ENERGY COMPONENTS/SYSTEMS.....FLAT PLATE & TUBULAR COLLECTORS
STATUS: DEVELOPING SOLAR COLLECTORS&SMALL WIND GENERATORS, MARKETING ENERGY CONSV SERVICES TO OUTSIDE ORGANIZATIONS
FOR INQUIRY SEE: JOHN MOCKOVCIAK,JR,GRUMMAN AEROSPACE

GRUMMAN AEROSPACE CORPORATION PLANT 30 BETHPAGE NY 11714 JOHN MOCKOVCIAK, JR * MGR ENERGY SYS DEPT -PHONE- 516 5753785	OCEAN THERMAL GRADIENT SYSTEMS NON-TERRESTRIAL SOLAR POWER SYSTEMS WIND CONVERSION	STORAGE, HYDROGEN

FOR INFO ON PRODUCTS INQUIRE TO ABOVE

LIBBEY OWENS FORD CO 811 MADISON AVE TOLEDO OH 43695 Y K PEI DR -PHONE- 419 2473731	SOLAR SPACE HEATING RETROFITTING, MECHANICAL SYSTEMS	COLLECTOR,MECHANICAL SYSTEMS COLLECTOR, EVACUATED

DEVELOPING FLAT PLATES WITH TUBULAR GLASS ABSORBER UNITS...GLASS PIPES COVERED BY AN OUTER VACUUM JACKET

UNIVERSAL OIL PRODUCTS CO WOLVERINE TUBE DIV P O BOX 2202 DECATUR AL 35601 JOHN K THORNE MGR PRODUCT DEVELOP -PHONE- 205 3531310 X290		COLLECTOR,ABSORPTIVE SELECTIVE SURFACES

APPLICATION OF SEAMLESS METAL TUBULAR PRODUCTS TO SOLAR THERMAL COLL INCLUDING HEAT PIPES SUB-ASSEMBLIES & SELECTIVE COATINGS ON TUBULAR COMPONENTS

--SUBCLASSES--------------------SOLAR SUB SYSTEMS--------------------------

UNIVERSAL OIL PRODUCTS COMPANY   SOLAR WATER HEATING
10 UOP PLAZA
DES PLAINES   IL   60016
WILLIAM C HOLT JR

-PHONE-
312 3913330
SOLAR HOT H2O HEATING...PROTOTYPE SOLAR THERMAL COLL DEVELOPEMENT &
CONSTRUCTION FOR IN HOUSE APPLICATION

----------SUBCLASSES----------SOLAR SUB SYSTEMS----------

CARSON ASTRONOMICAL INSTRUMENT
P O BOX 5566
VALENCIA CA 91355
CURTIS L OLSON
DIR OF MARKETING

-PHONE-
805 2551234
DEVELOP & MANUFACTURE LOW COST SOLAR TRACKING SYSTEM FOR HIGH TEMP
CONCENTRATORS-CONCEPTUAL STAGES

RAM PRODUCTS INC
BOX 340
STURGIS MI 49091
PAUL SORKORAM

-PHONE-
616 6519351
HIGH TEMP CONCENTRATORS-ACRYLIC TROUGHS & CONCENTRATORS

SHELDAHL INC — COLLECTOR,REFLECTIVE SELECTIVE SURFACES
CORP R & D
HWY 3
NORTHFIELD MN 55057
DONALD E ANDERSON
VICE PRES
-PHONE-
507 6455633
REFLECTIVE TAUT MEMERANE PLANAR MIRRORS FOR SOLAR CONCENTRATION

YUP COMPANY LTD
1-10-3 JUJONAKAHARA
KITA-KU TOKYO JAPAN
YOSHIO YATABE

-PHONE-
03 9008987
GIANT FRESNEL LENSES UP TO 50 METERS IN DIAMETER MANUFACTURE NOW BEING
PLANNED AVAIL IN 3YRS

SUBCLASSES	SOLAR SUB SYSTEMS
ENVIRONMENTAL ENERGIES INC 11350 SCHAFFER ST DETROIT MI 48227	SOLAR ELECTRIC POWER GENERATION

MFG WIND ELECT EQUIP

superman wind
speeding through space
you blow my tent up
like a big balloon
then deflate it... and
you're gone again.

superman wind
that bends the trees
and makes the pines
to sing
that makes the
ground squirrels run
and hide
won't you take me for
a ride?

Carolyn

------SUBCLASSES------SOLAR SUB SYSTEMS------

ION PHYSICS CORPORATION
SOUTH BEDFORD ST
BURLINGTON MA 01803
ROBERT BERMAN
MANAGER

-PHONE-
617 2721313 X462
PHOTOVOLTAIC SYS: PROCESS DEVELOP FOR POLYCRYSTALLINE SOLAR MATERIAL & DEVICE FAB...INTEGRAL COVERSLIPPING WITH BOROSILICATE & QUARTZ ON SINGLE CELLS OR MODULES AS SERVICE UP TO 6MILS OF 7070 GLASS. COVER-SLIPPING SERVICE AVAIL UP TO 6MILS BOROSILICATE GLASS & 2MILS MAXIMUM FUSED SILICA FOR DEVELOPMENT OF THIN POLYCRYSTALLINE SOLAR CELLS HAVE DEVEL ION IMPLANTED SOLAR CELLS & IMPLANTATION EQUIP FOR DEVICE STUDIES AIR FORCE CONTRACTS

MONSANTO RESEARCH CORP
1515 NICHOLAS ROAD
DAYTON OH 45407
FRANK WINSLOW

-PHONE-
513 2683411
DEVELOP ENERGY CONVERSION MATERIALS & DEVICES...HAVE DEVELOPED HI TEMP SOLAR CELL MATERIALS & THERMOELECTRICS

TEXAS INSTRUMENTS INC MEDIUM TEMPERATURE SOLAR COLLECTOR SYSTEMS
GOVT PROD MS10-10
ATTLEBORO MA 02703
WILLIAM C PAYNTON
MANAGER

-PHONE-
617 2222800 X6654
617 2222800 X484
PROVIDE COLL SYS COMPONENTS & TOTAL SYS AS REC...SOLAR CELLS USED ON SATELLITES...RESEARCHING NEW MFG TECHNIQUES TO REDUCE COMPONENT COSTS STATUS REPORTS TO NSF

----------SUBCLASSES----------SOLAR SUB SYSTEMS----------

ATLANTIC RESEARCH CORPORATION	WIND CONVERSION
5390 CHEROKEE AVENUE	SOLAR ENERGY STORAGE
ALEXANDRIA VA 22314	TOTAL SYSTEMS INTEGRATION
GEORGE D SUMMERS	SOLAR SPACE HEATING & COOLING
DIRECTOR	RETROFITTING, INTEREST
	NEW CONSTRUCTION, INTEREST

-PHONE-
703 3543400 X315
703 3543400 X312

MULTIDISCIPLINARY R&D COMPANY INVOLVED IN DIVERSE ANALYTICAL & HARDWARE PROGRAMS HAVE INTEREST IN SOLAR ENERGY ANALYSES TECHNOLOGY & HARDWARE

------------SUBCLASSES------------SOLAR SUB SYSTEMS------------

PORTABLE ALUMINUM IRRIG CO INC	BIO-CONVERSION
PO BOX 878	SOLAR FURNACE
VISTA CA 92083	SOLAR HEAT,COOL,&ELECTRIC POWER GENERATION
WILL C KINNEY	

-PHONE-
714 7242163
SOLAR H2O PURIFIERS FOR HOME OFFICE FACTORY.FINISHED PRODUCTION MODLES
1 GAL TO 2700 GAL DAILY PROD
SOLAR 2000F FURNACE DEVELOP STAGE TO HEAT LIGHT STRUCTURES
ALSO RESEARCH SOLAR RELATION OF PLANT CONVERSION AEROBIC NITROGEN TO
PLANT PROTEIN

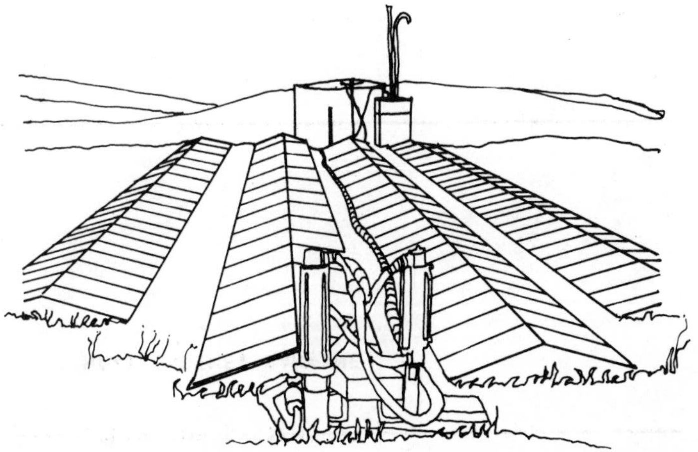

	SUBCLASSES	SOLAR SUB SYSTEMS
THERMO ELECTRON CORPORATION 85 FIRST AVENUE WALTHAM MA 02154 JERRY P DAVIS -PHONE- 674 8908700 X266 DEVELOPMENT OF ORGANIC RANKINE CYCLE SYS...WORKING FLUIDS SELECTED & SYS DESIGNED		HEAT ENGINE, RANKINE

	SUBCLASSES	SOLAR SUB SYSTEMS
COBALT ENGINEERING CO 164 PENINGTN-HBOURTN PENNINGTON NJ 08534 BRAD OWEN,PE	MEDIUM TEMPERATURE SOLAR COLLECTOR SYSTEMS HIGH TEMPERATURE SOLAR COLLECTOR SYSTEMS	

-PHONE-
609 7371376
609 8871414 X360
CONCENTRATORS,STORAGE,BLDG SYS-DEVELOPING SOLAR HEATER FOR RESIDENTIAL
HEATING/PROTOTYPE COMPLETE

IRVIN INDUSTRIES INC STRUCTURES GROUP 555 S BROADWAY LEXINGTON KY 40508 LLOYD RAIN MANAGER-PROD DEVEL	MEDIUM TEMPERATURE SOLAR COLLECTOR SYSTEMS RETROFITTING, PASSIVE SYSTEMS RELIGIOUS/PUBLIC/AMUSEMENT BUILDINGS WAREHOUSE BUILDINGS COMMERCIAL OTHER UP TO 50 THOUSAND SQFT,COMPLETED/DATA AVAIL	COLLECTOR,ALSO FUNCTION AS STRUCTURES SKIN COLLECTOR,PASSIVE SYSTEMS

-PHONE-
606 2546406
MFG & DIST FOR SINGLE & DBL WALLED AIR SUPPORTED STRUCTURES...SIZE
RANGE 5000 - 55000 SQFT LEASING AVAILABLE

NORTHROP RESEARCH&TECH CENTER 3401 WEST BROADWAY HAWTHORNE CA 90250 WALTER E CRANDALL	HOMES	HEAT TRANSFER SYSTEMS, COLLECTOR TO STORAGE HEAT RECOVERY, STORAGE TO UTILIZATION

-PHONE-
213 6754611 X1526
RESEARCHING SOLAR CONVERTERS SOLAR SYS DESIGN...PROTOTYPE HOME SOLAR
SYS BEING CONSTRUCTED

SERVICE UNLIMITED INC SECOND & WALNUT STS WILMINGTON DE 19801 RICHARD W WOLF		HEAT PUMPS

-PHONE-
302 6551568
DEV OF MATERIALS & MFG TECHNIQUES TO PRODUCE&MARKET BOTH COLLECTION
SURFACES & HIGH ENERGY HEAT PUMPS...PRESENTLY INSTALLING RESIDENTIAL
UNIT 810 SQFT COLL...LIMITED SUPPLY WILL BE FURNISHED AS AVAIL

---SUBCLASSES---SOLAR SUB SYSTEMS---

ARKLA INDUSTRIES INC
BOX 751
LITTLE ROCK AR 72203
ROBERT J DESTICHE
VP SALES

-PHONE-
501 3726241
PRODUCE AND SELL HEAT OPERATED ABSORPTION COOLING EQUIPMENT

CONTOUR CO INC — SOLAR WATER HEATING
9300 WHITMORE ST
EL MONTE CA 91731
DAN MAROVSH

-PHONE-
212 2834101
IN RESEARCH & DEV STAGES

DYNATHERM CORP — MEDIUM TEMPERATURE SOLAR COLLECTOR SYSTEMS
MARBLE CT OFF IND LN
COCKEYSVILLE MD 21030
J A STREB
V.P.MARKETING

-PHONE-
301 6669151
DEV MEDIUM TEMP HEAT PIPED SOLAR ENGY COLL ELEMENTS FOR CENTRAL POWER
STATION...DEV SOLAR SPACE HEAT&COOL SYS USING MODULAR COMPONENTS...
HEAT PIPES FOR MED TEMP COLL BEING TESTED..PLEMIN STUDIES COMP ON SOLR
SPACE HEAT LOW COST MODULES WITH WIDE SCALE APPLICABILITY

EDWARDS ENGINEERING CORP — MEDIUM TEMPERATURE SOLAR COLLECTOR SYSTEMS
101 ALEXANDER AVE
POMPTON NJ 07444
EDWARD BOGUCZ *

-PHONE-
201 8352808
ADDRESS INQUIRY TO CONCERNING NEW PRODUCTS/SEE: RAY EDWARDS FOR PROD
DETAILS

----SUBCLASSES----SOLAR SUB SYSTEMS----

EDWARDS ENGINEERING CORP
101 ALEXANDER AVE
POMPTON NJ 07444
RAY C EDWARDS
PRESIDENT

MEDIUM TEMPERATURE SOLAR COLLECTOR SYSTEMS

-PHONE-
202 8352808 X49
BUILDING A COMPLETE SOLAR HEATING & COOLING SYS USING FLAT PLATE TYPE
COLLECT...BE AVAILABLE JULY 1974...ADDRESS PUB INQ TO: EDWARD BOGUCZ

ITEK CORPORATION
OPTICAL SYSTEMS DIV
10 MAGUIRE ROAD
LEXINGTON MA 02173
GARY NELSON

MEDIUM TEMPERATURE SOLAR COLLECTOR SYSTEMS
SOLAR WATER HEATING

-PHONE-
617 2763411
AS OF NOV 73 TESTING PROTOTYPES: SOLAR WATER & SPACE CONDITIONING COLL
SEVERAL CONCEPTS UNDER TEST

JOHNSON SERVICE COMPANY
RESEARCH DIVISION
507 E MICHIGAN ST
MILWAUKEE WI 53201
S R BUCHANAN
ASST DIR OF RESEARCH

RESIDENTIAL OTHER
COMMERCIAL OTHER

-PHONE-
414 2769200 X677
CONTROL SYS FOR HOME & COMMERCIAL BLDGS...KEEPING ABREAST OF SOLAR DEV
TO DETERMINE SYS CONTROL NEEDS

ROCKWELL INTERNATIONAL
SPACE DIVISION
12214 LAKEWOOD BLVD
DOWNEY CA 90241
CHARLES L GOULD SF26

NON-TERRESTRIAL SOLAR POWER SYSTEMS
SOLAR ELECTRIC POWER GENERATION
COMMERCIAL OTHER

-PHONE-
213 5943921
DEVELOP PRACTICAL TECH FOR SOLAR HEATING & COOLING & MARKET...COMPANY
FUNDED PRELIM R&D WORK ONLY
ALSO R&D SOLAR TERRESTERIAL POWER PLANTS

	SUBCLASSES	SOLAR SUB SYSTEMS
SKY THERM PROCESS& ENGINEERING 945 WILSHIRE BLVD LOS ANGELES CA 90017 HAROLD R HAY	MEDIUM TEMPERATURE SOLAR COLLECTOR SYSTEMS SOLAR WATER DISTILLATION SOLAR WATER HEATING PROVIDE AUDIO & VISUAL SERVICES	COLLECTOR,ALSO FUNCTION AS STRUCTURES SKIN COLLECTOR,MULTI-FUNCTION IN DWELLING SYS COLLECTOR,PASSIVE SYSTEMS STORAGE, LIQUIDS

-PHONE-
213 6247261 X306
DEVELOPMENT OF SELECTIVE THERMAL BALANCE NATURAL AIRCOND (HEAT & NIGHT SKY COOLING) ON HORIZONTAL ROOFS & VERTICAL WALLS/ DEV SOLAR STILLS & SOLAR WATER HEATERS MODULAR & INTEGRATED WITH ROOFTOP SELECTIVE THERML BAL NAT AIRCOND/TEST HOUSE COMP IN ATASCADERO/LECTURES AVAIL

SOLAR HOME SYSTEMS ROOM 202 38518 OAKHILL CIRCLE WILLOUGHEY OH 44094 JOSEPH BARBISH	MEDIUM TEMPERATURE SOLAR COLLECTOR SYSTEMS HIGH TEMPERATURE SOLAR COLLECTOR SYSTEMS RESIDENTIAL OTHER	COLLECTOR, EVACUATED

-PHONE-
216 9511119
DEVELOPING SYS TO SUPPLY 100 PERCENT HEAT&COOL RESIDENT DWELLINGS AT FEASIBLE COST.....DESIGNED & TESTED VACUMATED FLAT PLATE COLL...DESIGN STORAGE SYS USING SYNTHETIC FLUID...HAVE BLDG DESIGNED FOR SYS BUILDING STARTING DATE:JULY 1974

SUNDSTRAND HYDRAULICS 2210 HARRISON AVE ROCKFORD IL 61101 JAMES H MEYER SR PROJ ENGR	TOTAL SYSTEMS INTEGRATION ALL SOLAR TECHNOLOGIES SOLAR WATER DISTILLATION SOLAR ELECTRIC POWER GENERATION	HEAT RECOVERY, STORAGE TO UTILIZATION COLLECTOR,MECHANICAL SYSTEMS HEAT TRANSFER SYSTEMS, COLLECTOR TO STORAGE HEAT PUMPS STORAGE, MECHANICAL/PUMPED LIQUIDS STORAGE, MECHANICAL/COMPRESSED GAS TURBINES

-PHONE-
815 2267246
MFG: AIR COMFORT CONTROLS HEAT PUMPS HEAT EXCHANGERS HOT WATER CIRCULATORS FUEL PUMPS...WILL MFG TOTAL ENERGY PACKAGES

SUNDSTRAND HYDRAULICS 2210 HARRISON AVE ROCKFORD IL 61101 JAMES F NELSON * PRODUCT MANAGER	TOTAL SYSTEMS INTEGRATION ALL SOLAR TECHNOLOGIES SOLAR WATER DISTILLATION SOLAR ELECTRIC POWER GENERATION	HEAT RECOVERY, STORAGE TO UTILIZATION COLLECTOR,MECHANICAL SYSTEMS HEAT TRANSFER SYSTEMS, COLLECTOR TO STORAGE HEAT PUMPS STORAGE, MECHANICAL/PUMPED LIQUIDS STORAGE, MECHANICAL/COMPRESSED GAS TURBINES

-PHONE-
815 2267212

			SUBCLASSES	SOLAR SUB SYSTEMS
TRW SYSTEMS BLDG 42 ONE SPACE PARK REDONDO BEACH JACK M CHERNE MGR SOLAR ENGY PROG -PHONE- 213 5352871	CA	90278	MEDIUM TEMPERATURE SOLAR COLLECTOR SYSTEMS SOLAR AGRICULTURAL DRYING RETROFITTING, MECHANICAL SYSTEMS NEW CONSTRUCTION, MECHANICAL SYSTEMS ACADEMIC BUILDINGS	COLLECTOR,MECHANICAL SYSTEMS

WORKING ON COLLECTOR DESIGN...CROP DRYING PROGRAM...PLAN TO DO PHASE I NSF SOLAR HEATING DEMO PROGRAM

TRW SYSTEMS ONE SPACE PARK REDONDO BEACH KEN MORITZ * -PHONE- 213 5362253	CA	90278	SOLAR ENGINES PROVIDE INFORMATION SEARCHES TO CLIENTS	

	SUBCLASSES	SOLAR SUB SYSTEMS
SUNSTRAND AVIATION RESEARCH DEPT 4747 HARRISON AVE ROCKFORD IL 61101 R G MOKADAM -PHONE- 815 2266780 USE OF SOLAR ENERGY COLLECTED AT MODERATE HIGH TEMP FOR HEAT COOL & ELECT GEN...DEVELOP ORGANIC RANKINE CYCLE SYS FOR SOLAR PART OF TOTAL ENERGY PACKAGE SEE ALSO: SUNDSTRAN HYDRAULICS		HEAT ENGINE, RANKINE
WESTINGHOUSE ELECTRIC CORP MFG DEVEL. LAB R & D CENTER PITTSBURGH PA 15235 FREDERICK J MICHEL MGR INT'L SERVICES -PHONE- 412 2563548 412 2564724	MEDIUM TEMPERATURE SOLAR COLLECTOR SYSTEMS HIGH TEMPERATURE SOLAR COLLECTOR SYSTEMS SOLAR ENERGY STORAGE TOTAL SYSTEMS INTEGRATION SOLAR ELECTRIC POWER GENERATION RETROFITTING, TECHNOLOGIES	COLLECTOR,MULTI-FUNCTION IN DWELLING SYS COLLECTOR,REFLECTIVE SELECTIVE SURFACES COLLECTOR,ABSORPTIVE SELECTIVE SURFACES COLLECTOR,PASSIVE SYSTEMS HEAT PUMPS STORAGE, SOLID MATERIALS STORAGE, PHASE CHANGE MATERIALS

	SUBCLASSES	SOLAR SUB SYSTEMS
STANFIELD AIR SYSTEMS 110 TARA WAY ATHENS GA 30601 LYNN M STANFIELD PRESIDENT -PHONE- 404 5494767 404 5434202	NEW CONSTRUCTION, INTEREST HOMES	

SUBCLASSES	SOLAR SUB SYSTEMS
U S HOME CORP RESOURCE SAVING HOME 1437 S BELCHER RD CLEARWATER FL 33516 ALAN C BOMSTEIN DIRECTOR -PHONE- 813 5310441	SOLAR WATER HEATING SOLAR SPACE HEATING & COOLING

DEVELOPING PRODUCTION-TYPE HOME TO UTILIZE UP TO 50 PERCENT LESS ENGY CONSUME 1/3 LESS H2O...FIRST HOUSES IN FLA WILL USE 2 SOLAR SYS- HOT H2O & POOL HEATER...CONSTRUCTION START MAY COMPLETE DEC 1974

Company			SUBCLASSES	SOLAR SUB SYSTEMS
COLSPAN ENVIRONMENTAL SYS INC P O BOX 3467 BOULDER ALAN M FRANK -PHONE- 303 4494411	CO	80303		COLLECTOR,ABSORPTIVE SELECTIVE SURFACES
GENERAL DYNAMICS CORP CONVAIR AEROSPACE P O BOX 748 FORT WORTH DR RALPH O DOUGHTY GROUP ENGINEER -PHONE- 817 7324811 X3113 817 7324811 X3292	TX	76101	HIGH TEMPERATURE SOLAR COLLECTOR SYSTEMS SOLAR ENERGY STORAGE SOLAR WATER HEATING SOLAR HEAT,COOL,&ELECTRIC POWER GENERATION RETROFITTING, TECHNOLOGIES RETROFITTING, MECHANICAL SYSTEMS	COLLECTOR,ALSO FUNCTION AS STRUCTURES SKIN COLLECTOR,ABSORPTIVE SELECTIVE SURFACES COLLECTOR,MECHANICAL SYSTEMS COLLECTOR, EVACUATED HEAT TRANSFER SYSTEMS, COLLECTOR TO STORAGE
GENERAL DYNAMICS CORP CONVAIR AEROSPACE P O BOX 748 FT WORTH DONALD W GOODWIN DESIGN SPECIALIST -PHONE- 817 7324811 X3113 817 7324811 X3292	TX	76101	HIGH TEMPERATURE SOLAR COLLECTOR SYSTEMS SOLAR ENERGY STORAGE SOLAR WATER HEATING SOLAR HEAT,COOL,&ELECTRIC POWER GENERATION RETROFITTING, TECHNOLOGIES RETROFITTING, MECHANICAL SYSTEMS	COLLECTOR,ALSO FUNCTION AS STRUCTURES SKIN COLLECTOR,ABSORPTIVE SELECTIVE SURFACES COLLECTOR,MECHANICAL SYSTEMS COLLECTOR, EVACUATED HEAT TRANSFER SYSTEMS, COLLECTOR TO STORAGE
GENERAL MOTORS CORP HARRISON RADIATOR UPPER MOUNTAIN ROAD LOCKPORT VERNON L ERIKSEN DIRECTOR OF R AND D -PHONE- 716 4393362	NY	14094	SOLAR WATER HEATING SOLAR SPACE HEATING SOLAR SPACE COOLING SOLAR SPACE HEATING & COOLING	HEAT TRANSFER SYSTEMS, COLLECTOR TO STORAGE HEAT RECOVERY, STORAGE TO UTILIZATION

COLSPAN ENVIRONMENTAL SYS INC: DEVELOPING PROCESS FOR LOW COST MASS PRODUCTION OF SELECTIVE ABSORBER SURFACES-HAVE EQUIPMENT&CAN PRODUCE SMALL SAMPLES OF MATERIAL&PROTYPES

GENERAL MOTORS CORP: DESIGN, ANALYSIS AND FABRICATION OF SOLAR COLLECTORS HEAT EXCHANGERS FOR SOLAR ENERGY SYSTEMS

			SUBCLASSES	SOLAR SUB SYSTEMS
SEA SOLAR POWER INC			TOTAL SYSTEMS INTEGRATION	TURBINES
1615 HILLOCK LANE			SOLAR WATER HEATING	
YORK	PA	17403	SOLAR SPACE COOLING	
JAMES H ANDERSON *			SOLAR SPACE HEATING & COOLING	
VICE PRES			SOLAR ELECTRIC POWER GENERATION	
			SOLAR HEAT,COOL,&ELECTRIC POWER GENERATION	

-PHONE-
717 8438594
ADDRESS INQUIRY TO ABOVE FOR DESC SEE: J HILBERT ANDERSON

			SUBCLASSES	SOLAR SUB SYSTEMS
SEA SOLAR POWER INC			TOTAL SYSTEMS INTEGRATION	TURBINES
1615 HILLOCK LANE			SOLAR WATER HEATING	
YORK	PA	17403	SOLAR SPACE COOLING	
J HILBERT ANDERSON			SOLAR SPACE HEATING & COOLING	
PRESIDENT			SOLAR ELECTRIC POWER GENERATION	
			SOLAR HEAT,COOL,&ELECTRIC POWER GENERATION	

-PHONE-
717 8438594
SEA SOLAR POWER IS ENGAGED IN DESIGN & DEVELOPMENT OF PLANTS TO PRODUCE POWER, FRESH H2O, HYDROGEN, METHANOL, FOOD & OTHER MINERALS FROM TROPICAL OCEAN WATERS-IN CONJUNCTION WITH THIS DEVELOPING LOW TEMP VAPOR TURBINES FOR USE IN DIRECT SOLAR HEATED POWER PLANTS & LOW TEMP WASTE HEAT PLANTS

ZEPHER WIND DYNAMO — SOLAR ELECTRIC POWER GENERATION
P O BOX 241
BRUNSWICK MA 04011
ALLEN LISHNESS

-PHONE-
207 7256534
PROVIDE NEW WINDMILL DESIGN

------------SUBCLASSES------------SOLAR SUB SYSTEMS------------

RICHARD R SMITH
5325 LUDLOW DRIVE
CAMP SPRINGS MD 20031

-PHONE-
301 4235324
PHOTOVOLTAIC SYS-DEVELOPING COMBINATION SOLAR CELL & ENERGY STORAGE
SYS WITH MINIMUM MASS & MAX ENERGY COLLECTION & STORAGE

--------SUBCLASSES--------SOLAR SUB SYSTEMS--------

SUNPOWER
RT 4 BOX 275
ATHENS OH 45701
WILLIAM T BEALE
PRESIDENT

HEAT ENGINE, STERLING

-PHONE-
614 5936934
DEVELOPING PRACTICAL STERLING ENGINE COLLECTOR SYS WITH PUMPED H2O OR
ELECTRIC OUTPUT...PROTOTYPES IN TEST...SMALL DEMO MODELS IN PRODUCTION

----------SUBCLASSES----------SOLAR SUB SYSTEMS----------

JOHN R HENSLEY MEDIUM TEMPERATURE SOLAR COLLECTOR SYSTEMS
2238 MOFFETT DRIVE
FORT COLLINS CO 80521

-PHONE-
303 4931688
MATERIALS RESEARCH...PROTOTYPE TO BE TESTED IN OWN HOUSE HOT WATER SYS
FLAT PLATE TYPE COLLECTOR

J C LOGAN
GRANT RD
NEWMARKET NH 03857

-PHONE-
603 6595774
LOW TECH HOT H2O HEATER USING RECYCLED MATERIALS & STORAGE SYS...EARLY
STAGES OF DEVELOPMENT

JOSEPH A NORDONE MEDIUM TEMPERATURE SOLAR COLLECTOR SYSTEMS
70 N W 94TH ST
MIAMI SHORES FL 33150

-PHONE-
305 7572194
FLAT PLATE TYPE HOT H2O HEATER

			SUBCLASSES	SOLAR SUB SYSTEMS
EDWIN K SWANSON 5615 S JOLLY ROGER TEMPE	AZ	85283	HOMES RESIDENTIAL OTHER	
ECO-SOLUTIONS,INC. P O BOX 4117 BOULDER ROBERT HESS * PRESIDENT	CO	80302	MEDIUM TEMPERATURE SOLAR COLLECTOR SYSTEMS SOLAR COOKING SOLAR WATER HEATING SINGLE FAMILY, PROPOSED UP TO 50 THOUSAND SQ FT,PROPOSED	COLLECTOR,ALSO FUNCTION AS STRUCTURES SKIN COLLECTOR,REFLECTIVE SELECTIVE SURFACES COLLECTOR,ABSORPTIVE SELECTIVE SURFACES HEAT RECOVERY, STORAGE TO UTILIZATION STORAGE, LIQUIDS
ECO-SOLUTIONS,INC. PO BOX 4117 BOULDER DAVID A SCHOEN EXECUTIVE SECRETARY	CO	80302	MEDIUM TEMPERATURE SOLAR COLLECTOR SYSTEMS SOLAR COOKING SOLAR WATER HEATING SINGLE FAMILY, PROPOSED	COLLECTOR,ALSO FUNCTION AS STRUCTURES SKIN COLLECTOR,REFLECTIVE SELECTIVE SURFACES COLLECTOR,ABSORPTIVE SELECTIVE SURFACES HEAT RECOVERY, STORAGE TO UTILIZATION STORAGE, LIQUIDS
R&D CONSULTANTS 8713 BASELINE ROAD LAFAYETTE ROGER D DENNETT	CO	80026	MEDIUM TEMPERATURE SOLAR COLLECTOR SYSTEMS RETROFITTING, TECHNOLOGIES HOMES	

Notes to the entries above, in order:

EDWIN K SWANSON
-PHONE-
602 8383727
FLAT PLATE COLL SYS FOR SPACE HEAT...

ECO-SOLUTIONS,INC. (ROBERT HESS)
-PHONE-
303 4436798

ECO-SOLUTIONS,INC. (DAVID A SCHOEN)
-PHONE-
303 4436798

R&D CONSULTANTS
-PHONE-
303 4946131
DEVELOPING LOW COST DO-IT-YOURSELF STYLE HOUSE HEATERS FOR EXISTING
HOMES & STRUCTURES
BROCHURES LITERATURE & DO-IT-YOURSELF MANUAL AVAIL SEPT 1974

	SUBCLASSES	SOLAR SUB SYSTEMS
SUNSHINE ENERGY CORP ROUTE 25 BROOKFIELD CENTER CT 06805 BRUCE GRIFFEN	SOLAR WATER HEATING	COLLECTOR,MECHANICAL SYSTEMS

-PHONE-
203 7753369
DO-IT YOURSELF COLL ASSEMBLED OR INKIT FORM

			SUBCLASSES	SOLAR SUB SYSTEMS
JONH R FEEMSTER * 20089 PIERCE RD SARATOGA	CA	95070	MEDIUM TEMPERATURE SOLAR COLLECTOR SYSTEMS	

-PHONE-
408 7394880 X2733
DEVELOPING RELIABLE,TECHNICALLY SOUND,&COST EFFECT HEATING & COOLING
SYS...PROTYPES BEING TESTED

MEEK'S 6520 E 25 PLACE TULSA IVAN R MEEK PRES	OK	74129	SINGLE FAMILY, PROPOSED WIND CONVERSION SOLAR ENERGY STORAGE NEW CONSTRUCTION, MECHANICAL SYSTEMS RESIDENTIAL OTHER	COLLECTOR,MECHANICAL SYSTEMS STORAGE, HYDROGEN

-PHONE-
918 8356081
918 8356638

--------SUBCLASSES--------SOLAR SUB SYSTEMS--------

JAMES A POTTER
12 GREEN HOUSE BLVD
WEST HARTFORD CT 06110

-PHONE-
203 5611637
DEVELOP MFG & MARKET HOME-SITE SYS COMPONENTS & EQUIP

			SUBCLASSES	SOLAR SUB SYSTEMS
SOLAR POWER SUPPLY RT 3 BOX A10 EVERGREEN DOUG DONNER * V P MARKETING -PHONE- 303 6744734	CO	80439	MEDIUM TEMPERATURE SOLAR COLLECTOR SYSTEMS HIGH TEMPERATURE SOLAR COLLECTOR SYSTEMS TOTAL SYSTEMS INTEGRATION SINGLE FAMILY, DEVELOPMENT GREENHOUSES PROVIDE AUDIO & VISUAL SERVICES	COLLECTOR,PASSIVE SYSTEMS COLLECTOR,REFLECTIVE SELECTIVE SURFACES COLLECTOR,ABSORPTIVE SELECTIVE SURFACES HEAT RECOVERY, STORAGE TO UTILIZATION STORAGE, PHASE CHANGE MATERIALS STORAGE, LIQUIDS TURBINES
SOLAR POWER SUPPLY RT 3 BOX A10 EVERGREEN MALCOM LILLYWHITE V P RESEARCH&DEVELOP -PHONE- 303 6744734	CO	80439	MEDIUM TEMPERATURE SOLAR COLLECTOR SYSTEMS HIGH TEMPERATURE SOLAR COLLECTOR SYSTEMS TOTAL SYSTEMS INTEGRATION SINGLE FAMILY, DEVELOPMENT GREENHOUSES PROVIDE AUDIO & VISUAL SERVICES	COLLECTOR,PASSIVE SYSTEMS COLLECTOR,REFLECTIVE SELECTIVE SURFACES COLLECTOR,ABSORPTIVE SELECTIVE SURFACES HEAT RECOVERY, STORAGE TO UTILIZATION STORAGE, PHASE CHANGE MATERIALS STORAGE, LIQUIDS TURBINES

A NATURAL ENERGY SYSTEMS CONSULTING FIRM ENGAGED IN DESIGN & CONSTRUCT OF ECONOMICAL ALTERNATIVE SOURCES OF POWER/COMPRISED OF TEAM OF SPECIALISTS INVOLVED IN NETWORK OF ACTIVITIES COVERING BROAD SPECTRUM OF PROF FIELDS/WE SERVE INDIVIDUALS ORGANIZATIONS & COMM ACCOUNTS/ALSO SYS INTEGRATION SOLAR METHANE & WIND

--------SUBCLASSES--------SOLAR SUB SYSTEMS--------

JACK HEDGER
PO BOX 357
LAKESIDE CA 92040

MEDIUM TEMPERATURE SOLAR COLLECTOR SYSTEMS

SOLAR POWERED BOAT WILL MARKET SOLAR TOY&FULL SIZED PLEASURE BOATS
DESCRIPTION: CATAMARAN HULLS WITH CANOPY OF FLAT PLATE COLLECTORS-CON-VERT METHYLENE CHLORIDE OR OTHER LOW BOILING LIQUIDS TO GAS WITH SUFF PRESSURE TO DRIVE PRIME MOVER EXAUSTING TO WATER-COOLED CONDENSER MOUNTED ON HULLS

---

gliding through the universe
on this
our spaceship
orbiting
our central power plant
without you...
sun
our spaceship
would be
an
empty
cinder ship.

Carolyn

			SUBCLASSES	SOLAR SUB SYSTEMS
NEW AGE DOMES EVENING STAR RANCH RR 4 BOX 90 GOLDEN STEVEN COFFEL	CO	80422	MEDIUM TEMPERATURE SOLAR COLLECTOR SYSTEMS SOLAR SPACE HEATING	

-PHONE-
303 5825719
SELF SUFFICIENT ENVIRONMENTS...SOLAR HEATED BUILDING CONST PLANTS/ PEOPLE...HOPE TO MARKET GREENHOUSE IN 1974 THAT WILL REQUIRE NO AUX ENERGY

			SUBCLASSES	SOLAR SUB SYSTEMS
SOLTEC PO BOX 6844 DENVER RICHARD S SPEED PRESIDENT	CO	80206	MEDIUM TEMPERATURE SOLAR COLLECTOR SYSTEMS BIO-CONVERSION TOTAL SYSTEMS INTEGRATION SOLAR SPACE HEATING	

-PHONE-
303 3338869
DESIGN TEST & MARKET INTEGRATED LIFE SUPPORT SYSTEMS CURRENT RESEARCH & DEVELOPMENT OF SOLAR HOT AIR FLAT PLATE SPACE HEATERS & SOLAR HEATED GREENHOUSES PROTOTYPES UNDER EVAL

---SUBCLASSES---SOLAR SUB SYSTEMS---

CENTER FOR ENVIRONMENT & MAN
275 WINDSOR STREET
HARTFORD CT 06120
RAYMOND J DOLAN

WIND CONVERSION
SOLAR SPACE HEATING & COOLING

-PHONE-
203 5494400 X348

ANALYZE SHORT TERM REGIONAL & LOCAL VARIATIONS OF INSOLATION, WIND, & SPACE HEATING/COOLING DEMAND. TO DEFINE LOCAL LIMITATIONS ON WIND ENGY EXTRACTION. STUDY & DEVELOP SOLAR POWERED DESICCANT SYS FOR DEHUMIDIFY

BUILDING RESCH STATION
GARSTON WATFORD
WD2 7JR UNITED KINGD
P PERTHERBRIDGE

PUBLISH ORIGINAL RESEARCHED INFORMATION

-PHONE-
092 7 3 740 X40

THERMAL RESPONSE TO SOLAR GAINS IN BLDGS: DEVELOPMENT OF THEORIES FOR CALCULATION OF INDOOR TEMP & COOLING WADS IN BLDGS SUBJECT TO ENERGY GAINS INCLUDING SOLAR GAINS THROUGH WINDOWS WALLS &/OR ROOF...ANALYSIS OF SOLAR RAD IRRAD & DRY BULB TEMP TO PROD DESIGN VALUES DEVELOPING ADMITTANCE PROCEDURE FOR MANUAL & COMPUTER CALCULATIONS OF INDOOR TEMP & COOLING LOADS SEE ALSO ADVISORY SERVICE BUILDING RESEARCH STATION

--------SUBCLASSES--------SOLAR SUB SYSTEMS--------

TECHNION-ISRAEL INST OF TECH	SOLAR REFRIGERATION
MECHANICAL ENGINEERG	SOLAR SPACE HEATING & COOLING
TECHNION CITY	SOLAR ELECTRIC POWER GENERATION
HAIFA ISRAEL	SOLAR HEAT,COOL,&ELECTRIC POWER GENERATION
ALEXANDER SOLAN	
ASSOCIATE PROFESSOR	
RESEARCH IN ENERGY CONSERVATION	

	SUBCLASSES	SOLAR SUB SYSTEMS
CALIF DEPT PARKS & RECREATION SPECIAL STUDIES SECT 1416 9TH ST SACRAMENTO CA 95811 DR ROBERT B DEERING	RETROFITTING, RESOURCES RETROFITTING, TECHNOLOGIES RETROFITTING, PASSIVE SYSTEMS NEW CONSTRUCTION, RESOURCES NEW CONSTRUCTION, TECHNOLOGIES NEW CONSTRUCTION, PASSIVE SYSTEMS	

-PHONE-
916 4453130

PRODUCE MORE LIVABLE ENVIRO IN HOT CLIMATE ENVIRON THROUGH REDUCTION OF HEAT USING LANDSCAPE MATERIALS: PLANTS & ENVIRONMENTS
PUBLICATION:'TEMPERATURE CONTROL FOR HOUSES,'JOURNAL OF HOME ECONOMICS VOL 50, NO 3, MARCH 1958, PP175-184
'THE EFFECT OF PLANT MATERIAL UPON MICROCLIMATE OF HOUSE & GARDEN, NATIONAL HORTICULTURE MAG, JULY 1954, PP 162-167.

MASSACHUSETTS, UNIV OF DEPT CHEM ENGR AMHERST MA 01002 W LEIGH SHORT PROF	SOLAR TECHNOLOGY APPLICATION OTHER NEW CONSTRUCTION, TECHNOLOGIES	STORAGE, METHANE

-PHONE-
413 5451588
413 5452507

ANEROBIC CONVERSION DOMESTIC WASTE AND CELLULOSE TO METHANE...WORK IN PROGRESS

---SUBCLASSES---SOLAR SUB SYSTEMS---

OHIO STATE UNIV
ELECT ENGR LAB
2015 NEIL RM300 CL
COLUMBUS OH 43210
ARTHUR E MIDDLETON DIRECTOR
ELECT MAT&DEVICE LAB
-PHONE-
614 4226210
614 4869561

NON-TERRESTRIAL SOLAR POWER SYSTEMS

EFFECTS OF POSITIVE ION IMPLANTATION INTO ANTIREFLECTION COATING OF SILICON SOLAR CELLS...RESEARCH ON PREPARATION & PROPERTIES OF SEMI-CONDUCTOR COMPOUNDS AS ALUMINUM ANTIMONIDE

---SUBCLASSES---SOLAR SUB SYSTEMS---

LYALL J LICHTY
TUBAC AZ 85640

-PHONE-
602 3982439
ENERGY CONVERSION

John

--------SUBCLASSES--------SOLAR SUB SYSTEMS--------

CALIF, UNIV OF-DAVIS
DEPT ENVIRO HORTCULT
DAVIS CA 95616
JONATHAN HAMMOND

RETROFITTING, TECHNOLOGIES
RETROFITTING, PASSIVE SYSTEMS
NEW CONSTRUCTION, TECHNOLOGIES
NEW CONSTRUCTION, PASSIVE SYSTEMS

-PHONE-
916 7523697

PLANNING&DESIGN OF CLIMACTLY ADAPTED BLDGS BY MANIPULATING COLOR, ORIENTATION, HEAT SINKS / DEVELOP AMENDMENTS TO BLDG CODE&ZONING ORD IN DAVIS GOAL: ENGY CONSV FOR HEAT&COOL BY DESIGN FOR DAVIS CLIMATE RESPECT TO ORIENTATION,INSULATION,GLASS EXPOSURE,OVERHANGS

CALIF, UNIV OF-DAVIS
AGR ENGR DEPT
DAVIS CA 95616
DR L W NEUBAUER

RETROFITTING, RESOURCES
RETROFITTING, TECHNOLOGIES
RETROFITTING, PASSIVE SYSTEMS
NEW CONSTRUCTION, RESOURCES
NEW CONSTRUCTION, TECHNOLOGIES
NEW CONSTRUCTION, PASSIVE SYSTEMS

-PHONE-
916 7520102

PLANNING CLIMACTLY ADAPTED BLDGS THROUGH ORIENTATION,SHAPE,INSULATION, GLASS EXPOSURE,OVER HANGS / DATA TAKEN ON MANY HOUSES&APARTMENTS SUMER 1973/CONSTRUCTION OF SOLARANGER /PUBLISHED:'SUMMER HEAT CONTROL FOR SMALL HOUSES'TRANSACTIONS OF ASAE VOL 12 NO 1 PP 102-105 1959.'SHADING DEVICES TO LIMIT SOLAR HEAT GAIN BUT INCREASE COLD SKY RADIATION' TRANSACT ASAE VOL 8 NO 4 PP 471-475 1965.'THERMAL EFFECTIVENESS OF SHAPE-1' SOLAR ENERGY VOL X NO 3 PP141-149 JULY 1966.'SHAPES & ORIENT-ATIONS OF HOUSES FOR NATURAL COOLING' TRANSACT ASAE VOL 15 PP 126-128 1972.'EFFECT OF SHAPE OF BLDG ON INTERIOR AIR TEMP' TRANSACT ASAE VOL 11 NO 4 PP 537-539 1968.

GEORGIA INST OF TECHNOLOGY
SCHOOL OF ARCH
ATLANTA GA 30332
FRANK J CLARKE

NEW CONSTRUCTION, RESEARCHING
HOMES

-PHONE-
404 8944885

DEVELOP ENERGY CONSV HOMES FOR ALANTA GA AREA USING SUPP SOLAR HEAT & ABSORPTION COOL SYS...GOAL TO DEVEL COMPETITIVELY PRICE HOUSE WITH LOW MONTHLY UTILITY COSTS...SPONSORED BY FULTON FEDERAL SAVINGS & LOAN ASSOC ALANTA

-----SUBCLASSES-----SOLAR SUB SYSTEMS-----

WM PATTERSON COLLEGE
SECONDARY EDUCATION
300 POMPTON RD
WAYNE NJ 07470
ZWEIG JONAS
ASSOCIATE PROFESSOR
-PHONE-
201 8812119

RETROFITTING, RESOURCES
PUBLISH ORIGINAL RESEARCH + REPRINTS
PROVIDE AUDIO & VISUAL SERVICES

MEDIUM TEMP COLL SYS/BIO-CONVERSION/WIND CONVERSION/SOLAR ENGY STORAGE
SOLAR COOKING/MOBILE HOMES,SINGLE FAMILY,2-5 FAMILY,6-14 FAMILY,&15FAM
COMPLETED DATA AVAIL/
SUMMER WORKSHOP 'ECOLOGY ENERGY CONSERVATION AND THE ENERGY CRISIS'

---

passing under power lines
be aligned
electricity in our mind

dwell in a cocoon...
wound with power lines
blanket ourselves
with more power lines
to slumber by

align the brain
feel so fine
in my mind

crave it feel it electric flesh
electricity...
our brain runs on it

power lines
the spaceship's brain
we...
but a synapse
in the brain...

Carolyn

------SUBCLASSES------SOLAR SUB SYSTEMS------

CONNECTICUT, UNIV OF
ENVIRONMENTAL ECON.
BOX U-21 UNIV. OF CT
STORRS CT 06268
CARLOS D STERN
ASSISTANT PROFESSOR
-PHONE-
203 4862740

RETROFITTING, PASSIVE SYSTEMS
RETROFITTING, MECHANICAL SYSTEMS
RETROFITTING, RESEARCHING
RETROFITTING, PRACTICING
PUBLISH ORIGINAL RESEARCHED INFORMATION
PUBLISH ORIGINAL RESEARCH + REPRINTS

RESEARCH PROJECT INTO THE POTENTIAL FOR ENERGY CONSERVATION AT LARGE PUBLIC FACILITIES. NO PUBLICATIONS AS YET. INTEREST IN INCENTIVES FOR ENERGY CONSERVATION

CONNECTICUT, UNIVERSITY OF
CHEMICAL ENGR
STORRS CT 06268
DR MICHAEL HOWARD *
ASST PROF
-PHONE-
203 4864020

RETROFITTING, PASSIVE SYSTEMS
RETROFITTING, MECHANICAL SYSTEMS
RETROFITTING, RESEARCHING
RETROFITTING, PRACTICING
PUBLISH ORIGINAL RESEARCHED INFORMATION
PUBLISH ORIGINAL RESEARCH + REPRINTS

FORD MOTOR CO
SCIENTIFIC RES STAFF
PO BOX 2053
DEARBORN MI 48121
ROY L GEALER
PRINCIPAL STAFF ENGR
-PHONE-
313 3231219

RETROFITTING, INTEREST
RETROFITTING, RESEARCHING
NEW CONSTRUCTION, TECHNOLOGIES
NEW CONSTRUCTION, INTEREST
NEW CONSTRUCTION, RESEARCHING

IN HOUSE INVOLVEMENT IN ENERGY CONSERVATION OPPORTUNITIES IN MANUF PROCESSES, & IN FUEL & ENERGY CONSERVATION AND RECOVERY MEASURES

----SUBCLASSES----SOLAR SUB SYSTEMS----

DR METIN LOKMANHEKIM
9190 RED BRANCH RD
COLUMBIA MD 21045

SOLAR RADIATION MEASUREMENTS
SOLAR SPACE HEATING & COOLING
NEW CONSTRUCTION, MECHANICAL SYSTEMS

HEAT ENGINE, RANKINE

-PHONE-
301 7307800

CURRENT NSF PROJECT STUDYING COOLING OF BLDGS WITH RANKINE CYCLE TO BE COMP JUNE 74 PAST WORK: WROTE COMPUTER PROG TO CALCULATE SOLAR RADIATN INTENSITY SIMULATE SOLAR COLLECTORS CALCULATE ENERGY UTILIZATION OF BLDGS & SHADOW CALCULATIONS

--------SUBCLASSES--------SOLAR SUB SYSTEMS--------

OAK RIDGE NATIONAL LABORATORY
PO BOX X
OAK RIDGE TN 37830
R CARLSMITH

RETROFITTING, RESOURCES
NEW CONSTRUCTION, RESOURCES

FIND & COMMUNICATE STRATEGIES FOR ENERGY CONSERVATION...ENERGY USE PATTERNS & OPPORTUNITIES FOR ENERGY CONSV IN RESIDENTIAL & TRANSPORT

-----SUBCLASSES----- -----SOLAR SUB SYSTEMS-----

COLORADO UNIV OF-BOULDER
CIVIL & ENVIR ENGR
BOULDER CO 80302
DAVID FENG

RETROFITTING, PASSIVE SYSTEMS

COLLECTOR,PASSIVE SYSTEMS

-PHONE-
303 4927112

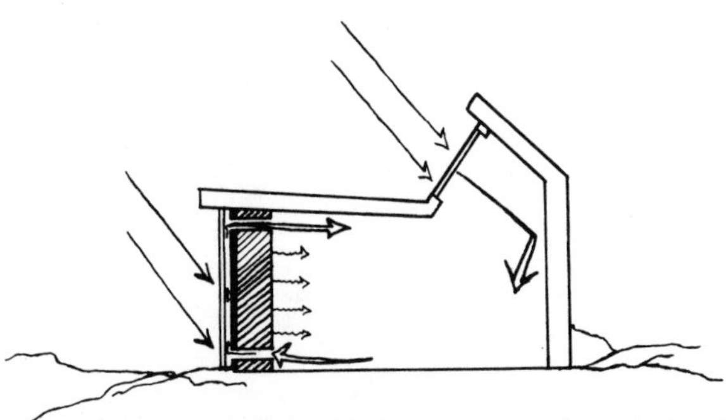

---SUBCLASSES---SOLAR SUB SYSTEMS---

JOHNS-MANVILLE CORP
GROWTH PLAN&DEVELOP
GREENWOOD PLAZA
DENVER CO 80217
JACK D VERSCHOOR
MANAGER-PLANNING
-PHONE-
303 7701000 X2879
303 9865757
RES & COMM BUILDING INSULATION

RETROFITTING, TECHNOLOGIES
RETROFITTING, MECHANICAL SYSTEMS
RETROFITTING, RESEARCHING
NEW CONSTRUCTION, TECHNOLOGIES
NEW CONSTRUCTION, INTEREST
NEW CONSTRUCTION, RESEARCHING

PGA ENGINEERS INC
10 BROADWAY
ST LOUIS MO 63102
WILLIAM EVERS JR
VICE PRESIDENT

SOLAR SPACE HEATING & COOLING

-PHONE-
314 2317318
SPECIALIZE IN ENERGY CONSV...PAST WORK:SOLAR ENGY STUDIES FOR MO STATE COMM INVOLVEMENT IN DEVELOP COMPUTER ENGY ANALYSIS PROG...SOLAR ENERGY DESIGN OF NEW SCHOOL IN MISSOURI

	SUBCLASSES	SOLAR SUB SYSTEMS
OWENS-CORNING FIBERGLAS CORP	SOLAR ENERGY STORAGE	COLLECTOR,MULTI-FUNCTION IN DWELLING SYS
TECHNICAL CENTER	TOTAL SYSTEMS INTEGRATION	COLLECTOR,ALSO FUNCTION AS STRUCTURES SKIN
PO BOX 415	SOLAR SPACE HEATING & COOLING	COLLECTOR,MECHANICAL SYSTEMS
GRANVILLE OH 43023	RETROFITTING, MECHANICAL SYSTEMS	COLLECTOR,OTHER
DR GEORGE W GRIMM	RETROFITTING, RESEARCHING	HEAT TRANSFER SYSTEMS, COLLECTOR TO STORAGE
MGR CORP RESEARCH	NEW CONSTRUCTION, RESEARCHING	STORAGE, SOLID MATERIALS
-PHONE-		
614 5870610		
ENERGY UTILIZATION & CONSERVATION MATERIALS R&D		

			SUBCLASSES	SOLAR SUB SYSTEMS
BORG WARNER CORP. RESEARCH CENTER ALGONQUIN&WOLF RDS DES PLAINES NEAL F CARDNER RESEARCH PHYSICIST -PHONE- 312 8273131 X266 THERMOELECTRIC RESEARCH	IL	60018		
WISCONSIN, UNIV OF MARINE STUDIES CENTE 1225 WEST DAYTON ST MADISON THOMAS W SMITH RESEARCH SPECIALIST -PHONE- 608 2634578 LOW COST HOUSING DESIGN FOR ENERGY CONSERVATION AND STUDY OF POSSIBLE SOLAR ENERGY APPLICATIONS TO LOW COST HOUSING	WI	53706	SOLAR RADIATION MEASUREMENTS SOLAR SPACE HEATING & COOLING CLUSTER HOMES HOMES SINGLE FAMILY, PROPOSED UP TO 50 THOUSAND SQ FT,PROPOSED	STORAGE, SOLID MATERIALS COLLECTOR,ALSO FUNCTION AS STRUCTURES SKIN HEAT TRANSFER SYSTEMS, COLLECTOR TO STORAGE HEAT RECOVERY, STORAGE TO UTILIZATION STORAGE, PHASE CHANGE MATERIALS STORAGE, OTHER

----------SUBCLASSES----------SOLAR SUB SYSTEMS----------

JOHNS-MANVILLE
RES AND DEV
PO BOX 5108
DENVER CO 80217
JOHN O COLLINS, JR
MGR GOVT RES LIASON
-PHONE-
303 7701000 X404

TOTAL SYSTEMS INTEGRATION
SOLAR WATER HEATING
SOLAR SPACE HEATING & COOLING
RETROFITTING, TECHNOLOGIES
RETROFITTING, RESEARCHING

COLLECTOR,MULTI-FUNCTION IN DWELLING SYS

PERTINENT PRODUCTS INCLUDING THERMAL INSULATIONS FOR COLLECTORS&STRUCTURES, ROOF DECK INSULATION/SUPPORT PRODUCTS FOR COLLECTORS & ASBESTOS CEMENT ROOF SYS FOR INTEGRATION OF COLLECTOR WITH ROOF. RESEARCH CAPABILITIES INCLUDE HEAT TRANSFER STUDIES & EVALUATIONS & SYS DEVELOP

JOHNS-MANVILLE CORP
PRODUCT INFO CENTER
GREENWOOD PLAZA
DENVER CO 80217
ANN PADOVANI *
CURRENT PRODUCTS
-PHONE-
303 7701000

TOTAL SYSTEMS INTEGRATION
SOLAR WATER HEATING
SOLAR SPACE HEATING & COOLING
RETROFITTING, TECHNOLOGIES
RETROFITTING, RESEARCHING

COLLECTOR,MULTI-FUNCTION IN DWELLING SYS

	SUBCLASSES	SOLAR SUB SYSTEMS
NAHB RESEARCH FOUNDATION INC	SOLAR SPACE HEATING & COOLING	COLLECTOR,ALSO FUNCTION AS STRUCTURES SKIN
PO BOX 1627	RETROFITTING, RESEARCHING	COLLECTOR,MULTI-FUNCTION IN DWELLING SYS
ROCKVILLE MD 20850	NEW CONSTRUCTION, TECHNOLOGIES	COLLECTOR,OTHER
RALPH J JOHNSON	NEW CONSTRUCTION, RESEARCHING	HEAT RECOVERY, STORAGE TO UTILIZATION
STAFF VICE PRESIDENT	APARTMENTS	STORAGE, OTHER
	HOMES	
-PHONE-		
301 7624200		
INSULATION MANUAL-HOMES & APARTMENTS		

--------SUBCLASSES--------SOLAR SUB SYSTEMS--------

RESOURCES FOR THE FUTURE INC
ENERGY RESEARCH&TECH
1755 MASS AVE NW
WASHINGTON DC 20036
SAM H SCHURR

NEW CONSTRUCTION, TECHNOLOGIES
RETROFITTING, TECHNOLOGIES

RESEARCH & DEVELOPMENT TASKS IN SOCIAL SCIENCE ASPECTS OF ENERGY SYS
ENERGY SYSTEMS MODELLING,ENGY CONSV TECHNIQUES,ENVIRONMENTAL & CONSV
ISSUES-TRADEOFFS,MATINENCE OF RELIABLE ENERGY SOURCE,R&D DEV OF ENGY
FINAL REPORT TO NSF SEPT 1973

FOR ADDITIONAL FINANCIAL INSTITUTIONS SEE:
INDEX OF SOLAR PROJECTS SECTION 2 PROJECT FUNDING SOURCES

---SUBCLASSES---SOLAR SUB SYSTEMS---

RAND CORP
1700 MAIN ST
SANTA MONICA CA 90406
D N MORRIS

-PHONE-
213 3930411

EVALUATION OF MEASURES TO CONSERVE ENERGY...RESEARCH DESIGNED TO STRENGTHEN ANALYTICAL BASIS FOR DEFINING ROLE OF GOVT POLICY IN ENERGY CONSERVATION...NSF GRANT

"We could burn up spaceship Earth itself to provide energy, but that would give us very little future."

Fuller, Operating Manual for Spaceship Earth

--------SUBCLASSES--------SOLAR SUB SYSTEMS--------

COLORADO UNIV OF-BOULDER
GEOGRAPHY DEPT
BOULDER CO 80302
R G BARRY

-PHONE-
303 4926387

MICHIGAN STATE UNIV
211 AG ENGINEERING
EAST LANSING MI 48824
ERNEST H KIDDER

MEDIUM TEMPERATURE SOLAR COLLECTOR SYSTEMS
SOLAR ENERGY STORAGE
SOLAR AGRICULTURAL DRYING

-PHONE-
517 3554720
COLLECTION OF DATA ON INCOMMING SOLAR ENGY MEASURED BY EPPLEY PYRO-HEIOMETER.PAST WORK: SOLAR DRYING OF FARM CROPS & HEATING FARM BLDGS SOLAR AIR HEATERS HOT ROCK STORAGE

NASA GODDARD SPACE FLIGHT CTR
CODE 912
GREENBELT MD 20770
DR MATTHEW P THEKAEKARA

-PHONE-
301 9825034
DETERMINATION OF EXTRATERRIAL SOLAR SPECTRUM...COMPUTATION OF SPECTRUM AT GROUND LEVEL...VARIATIONS & SOLAR CONSTANT...AS RESULT OF THIS RESEARCH WE ESTABLISHED VALUES WHICH ARE THE STANDARD FOR ASTM & NASA SAMPLE PUB:'STANDARD VALUES FOR THE SOLAR CONSTANT & ITS SPECTRAL COM-PONENTS,'NATURE PHYSICAL SCIENCE, VOL 229, JAN 4 1974, PP 6-9. 'EVALUATING THE LIGHT FROM THE SUN,' OPTICAL SPECTRA, VOL 6, NO 3, MARCH 1972, PP 32-35.

----SUBCLASSES----SOLAR SUB SYSTEMS----

YALE UNIV
FORESTRY&ENVIRO STDY
360 PROSPECT ST
NEW HAVEN CT 06511
WILLIAM E REIFSNYDER

-PHONE-
203 4360020

RESEARCH ON DISTRIBUTION OF SOLAR RADIATION IN FOREST STANDS & URBAN COMPLEXES.PUBLISHED:'SOLAR-RADIATION ATTENUATION THROUGH THE LOWEST 100 METERS OF AN URBAN ATMOSPHERE,'JOURNAL OF GEOPHYSICAL RESEARCH, VOL 77 NO 33, NOV 20'72,PP6499-6507...'SPATIAL & TEMPORAL DISTRIBUTION OF SOLAR RADIATION BENEATH FOREST CANOPIES,'AGRICULTURAL METEOROLOGY,9 (1971/1972),PP 21-37.

INSTITUT ROYAL METEOROLOGIQUE
RADIOMETRY SECTION
3 AVENUE CIRCULAIRE
B 1180 BRUSSELS BELGIUM
R DOGNIAUX

-PHONE-
02 746788

RECORDS & MEASUREMENTS OF COMPONENTS OF RADIATION&ILLUMINATION&THEIR ANALYSIS FOR STUDIES OF PHYSICAL PROPERTIES OF THE ATMOSPHERE&APPLICAT TO CLIMATOLOGICAL DISTRIBUTION WORLD WIDE...DEVELOPMENT OF ABSOLUTE INSTRUMENTS FOR RADIOMETRY...APPLICATION OF SOLAR ENERGY & DAYLIGHT IN GREENHOUSES

ALBERTA, UNIV OF
DEPT OF MECH ENGR
EDMONTON 7 ALBERTA CANADA
GERALD W SADLER

-PHONE-
403 4323450

MEASURE SOLAR RAD AT EDMONTON&APPLY RESULTS TO HEAT & AIRCOND PROBLEMS NORMAL INCIDENCE HORIZONTAL&DIFFUSE MEASUREMENTS TAKEN ON CONTINUOUS BASIS DURNING SUNNY PERIODS...CLEAR SKY DATA IS MAINLY LISTED BUT TRANSMITTANCE DURNING PARTIAL CLOUD DAYS MEASURED FILTER MEASUREMENTS TAKEN & ATMOSPHERIC TURBIDITY STUDIED MEASURED EXTINCTION COEFFICIENT & APPARENT SOLAR CONSTANT REPORTED

-SUBCLASSES- -SOLAR SUB SYSTEMS-

ST JEROME UNIVERSITY
DEPT D'HELIOPHYSIQUE
13013 MARSEILLE FRANCE
GEORGES PERI
FACULTY OF SCIENCE

-PHONE-
93 98 0901

HIGH TEMPERATURE SOLAR COLLECTOR SYSTEMS

RESEARCH ON VARIOUS POSSIBILITES IN IMPROVING THE EFFICIENCY IN THE THERMAL BALANCE OF SURFACE EXPOSED TO SOLAR RADIATION & OTHER ENVIRONMENTAL CONDITIONS...PROCESSES FOR STORAGE OF ENERGY ESP THERMAL FORMS
PAST WORK: SOLAR COLLECTORS FOCUSING DEVICES SOLAR BOILERS UP TO 200KW ANTIRADIANT&ANTICONVECTIVE STRUCTURES SELECTIVE SURFACES ETC
NO PUBLIC INQUIRY PLEASE

Yon

--------SUBCLASSES--------SOLAR SUB SYSTEMS--------

EDWARD F ELLIS *
6126 VRAIN ST
ARVADA CO 80003

HIGH TEMPERATURE SOLAR COLLECTOR SYSTEMS
WIND CONVERSION

-PHONE-
303 5717561

ROBERT F GIRVAN
STANHOPE IA 50246

BIO-CONVERSION
GREENHOUSES

-PHONE-
515 8263287
FLAT PLATE COLLECTORS-PLAN TO DEVELOP A LIVING UNIT, PERHAPS A MANU-
FACTURING UNIT USING NATURAL ENERGY SOURCES...PREPARING TO BUILD A
GREENHOUSE ON SOUTH SIDE OF HOUSE & PASS WASH WATER THROUGH HEAT
EXCHANGER

DAVID SEALANDER
ROUTE 4 BOX 371
IDAHO FALLS ID 83401

HIGH TEMPERATURE SOLAR COLLECTOR SYSTEMS
BIO-CONVERSION
WIND CONVERSION
INORGANIC SEMI-CONDUCTORS

AUBURN UNIVERSITY
DEPT OF INDUST ENGR
AUBURN AL 36830
CHARLES R WHITE

WIND CONVERSION

-PHONE-
205 8264340

---SUBCLASSES---SOLAR SUB SYSTEMS---

BATTELLE COLUMBUS LABS
THERM&MECH ENGY SYS
505 KING AVENUE
COLUMBUS OH 43201
SHERWOOD G TALBERT
SR MECH ENGR
-PHONE-
614 2993151
SYSTEMS ANALYSIS AND ECONOMIC STUDY OF PHOTOCHEMICAL SOLAR ENGY SYS

CALIF, UNIV OF--L A — HEAT TRANSFER SYSTEMS, COLLECTOR TO STORAGE; COLLECTOR,ABSORPTIVE SELECTIVE SURFACES
SCHOOL OF ENGR
ENGY & KINETICS DEPT
LOS ANGELES CA 90024
D K EDWARDS

-PHONE-
213 8255313
HEAT TRANSF IN HONEYCOMB STRUCTURES-INTERCELL LEAKAGE, INTRACELL
CONVECTION & RADIATION, ANGLE OF TILT,NSF SPONSORED
INVESTIGATION SELECTIVE SURFACES
PUBLICATIONS:'PERFORMANCE CHARACTERISTICS OF RECTANGULAR HONEYCOMB
SOLAR-THERMAL CONVERTERS,'SOLAR ENERGY, VOL 13, 1971, PP 193-221.

CALIF,UNIV OF--BERKELEY — WIND CONVERSION — COLLECTOR,ABSORPTIVE SELECTIVE SURFACES
MATERIALS SCI DEPT
HEARST MINING BLDG
BERKELEY CA 94720
MARSHAL F MERRIAM

-PHONE-
415 6423815
IMPROVED MATERIALS FOR SELECTIVE SURFACES/DESIGN & EVALUATION OF LOW
COST COLLECTORS-RESEARCH MOSTLY PROPOSALS
TEACH COURSES AT UNIV ON SOLAR ENERGY

	SUBCLASSES	SOLAR SUB SYSTEMS
COLUMBIA UNIVERSITY MECHANICAL ENGR 214 MUDD NEW YORK NY 10027 PROF ROBERT TAUSSIG -PHONE- 212 2802961	SOLAR ENERGY STORAGE SOLAR SPACE HEATING & COOLING	

R&D OF THERMAL COLL SYS OPTIMIZED FOR OVERNIGHT ENERGY STORAGE FOR USE IN NE USA WITH SMALL AUX POWER SUPPLY FOR HOME HEAT & COOL...PRORAM JUST STARTED-SOME COMPUTER SIMULATION OF ENERGY USE IN BUILDINGS TO RESULT IN SYS OPTIMIZATION PROG

	SUBCLASSES	SOLAR SUB SYSTEMS
CON EDISON CO OF NY INC RESEARCH & DEVELOP 4 IRVING PLACE NEW YORK NY 10003 DR ROBERT A BELL DIRECTOR -PHONE- 212 4603882	HIGH TEMPERATURE SOLAR COLLECTOR SYSTEMS SOLAR ELECTRIC POWER GENERATION	

INSTALLATION & TEST SOLAR ENGY SYSTEMS IN BUILDING IN METRO AREA TO DETERMINE ENERGY CONVERSION & ELECTRIC LOAD LEVELING EFFECT--PRIMARY STUDIES TO SELECT MOST PROMISING SYSTEMS FOR DETAILED EXAMINATION

	SUBCLASSES	SOLAR SUB SYSTEMS
CORNING GLASS WORKS PHYSICAL RESEARCH CORNING NY 14830 UGUR ORTABASI SENIOR PHYSIST -PHONE- 607 9743370	SOLAR WATER HEATING SOLAR SPACE HEATING SOLAR SPACE COOLING SOLAR REFRIGERATION SOLAR SPACE HEATING & COOLING	COLLECTOR,ALSO FUNCTION AS STRUCTURES SKIN COLLECTOR,MULTI-FUNCTION IN DWELLING SYS COLLECTOR,REFLECTIVE SELECTIVE SURFACES COLLECTOR,ABSORPTIVE SELECTIVE SURFACES COLLECTOR,PASSIVE SYSTEMS COLLECTOR,OTHER

DEVELOP MEDIUM TEMP COLL SYSTEMS

	SUBCLASSES	SOLAR SUB SYSTEMS
DAYTON UNIV OF DEPT OF MECH ENGR COLLEGE PK AV DAYTON OH 45469 JOHN E MINARDI -PHONE- 513 2290123	WIND CONVERSION SOLAR SPACE HEATING & COOLING	COLLECTOR,MECHANICAL SYSTEMS

'BLACK' LIQUID FLAT PLATE COLLECTOR

SUBCLASSES — SOLAR SUB SYSTEMS

GEORGIA INST OF TECHNOLOGY — HIGH TEMPERATURE SOLAR COLLECTOR SYSTEMS
ENGINEERING EXP STAT
ATLANTA GA 30332
J F KINNEY

-PHONE-
404 8943661
DESIGN LOW COST PROTOTYPE FLAT PLATE COLL FOR CHEAP MASS PROD...ALSO
DEVELOPING CONCENTRATING COLL WITH CONCENTRATE FACTOR OF 2 TO 3

JOHN SIMKINS SCHOOL — WIND CONVERSION
MAIN ST
SOUTH YARMOUTH MA 02664
JOHN SILVER

-PHONE-
617 3982412
FLAT PLATE COLL&WIND ENERGY...HAVE CONST SIMPLE FLAT PLATE...RESEARCH
& TEACHING SOLAR

LAMA FOUNDATION — WIND CONVERSION; SOLAR SPACE HEATING; GREENHOUSES
PO BOX 444
SAN CRISTOBAL NM 87564
HANS VON BREIEN JR
COORDINATOR

OBJECTIVE:INSTALLATION & USE OF PRACTICAL APPLICATIONS FOR OUR OWN
ENERGY USE...STATUS: FEASIBILITY STUDY

MINNESOTA STATE HEALTH DEPT — HIGH TEMPERATURE SOLAR COLLECTOR SYSTEMS
ENVIRO HEALTH DIV
5900 PENN AVE SOUTH
MINNEAPOLIS MN 55419
PAUL T PANAGOS
PUBLIC HEALTH ENGR
-PHONE-
614 9229225
SOLAR ENERGY CONVERSION & UTILIZATION

	SUBCLASSES	SOLAR SUB SYSTEMS
MISSISSIPPI, UNIVERSITY OF CHEMICAL ENGINEERING UNIVERSITY MS 38677 FRANK A ANDERSON -PHONE- 601 2327023	WIND CONVERSION	

ENGINEERING INSTRUCTION...BUILT & OPERATED FINNED PIPE COLLECTOR ABOUT 15 YRS AGO

	SUBCLASSES	SOLAR SUB SYSTEMS
NASA G C MARSHALL SPACE FLT CT PUBLIC AFFAIRS MARSHAL SPACE F C AL 35812 JOSEPH M JONES DIRECTOR	SOLAR SPACE HEATING & COOLING	COLLECTOR,ABSORPTIVE SELECTIVE SURFACES

HAVE DEVELOPED SPECIAL ABSORPTIVE COATING FOR ALUMINUM PANELS FLAT PLATE COLL...ABSORBS 93 PERCENT TOTAL SOLAR HEAT RERAD ONLY 6 PERCENT INFRARED HEAT...LICENSING INFO AVAIL FROM OFFICE OF CHIEF COUNSEL... TECH INFO AVAIL TECH UTILIZATION OFFICE MSC

	SUBCLASSES	SOLAR SUB SYSTEMS
NASA LEWIS RESCH CTR POWER APPLICATIONS MAIL DROP 500 201 CLEVELAND OH 44142 ROBERT RAGSDALE ASST TO CHIEF -PHONE- 216 4334000 X6943	HIGH TEMPERATURE SOLAR COLLECTOR SYSTEMS SOLAR SPACE HEATING & COOLING	COLLECTOR,ABSORPTIVE SELECTIVE SURFACES COLLECTOR,OTHER

RESPON SOLAR ENRY PROGRAMS AT CTR...COLL TESTING LAB MODELING OF SYS COATINGS RESCH EVAL SOLAR THERMAL ELECT POWER PLANT

	SUBCLASSES	SOLAR SUB SYSTEMS
NATIONAL SCIENCE FOUNDATION ADVNCED ENGY RES&TEC WASHINGTON DC 20550 GEORGE M KAPLAN PROG MGR SOLAR THERM -PHONE- 202 6327364	HIGH TEMPERATURE SOLAR COLLECTOR SYSTEMS GOVERNMENT AGENCIES PROVIDING GRANTS	

	SUBCLASSES	SOLAR SUB SYSTEMS
NEW ALCHEMY INSTITUTE WEST BOX 376 PESCADERO CA 94040 RICHARD MERRILL DIRECTOR -PHONE- 805 9690101 COMPARISON OF FLAT PLATE COLL DESIGNS MADE FROM READILY AVAIL MATERIAL (METAL FIBERGLASS)...ALSO METHANE SYS & ALGAE CONVERSION SYS	BIO-CONVERSION PUBLISH LAY PERIODICALS	STORAGE, METHANE
NORTHERN ARIZ UNIV MECH ENGR NAU BOX 15600 FLAGSTAFF AZ 86001 WILLIAM W DAVIS CHAIRMAN -PHONE- 602 5233698 EXPERIMENTAL INVESTIGATION EFFICIENCY OF DIFFUSE SOLAR ENERGY COLL... RESCH USING CINDER BLOCK FOR ENGY STORAGE...RESCH AIR COOLED COLL		STORAGE, SOLID MATERIALS COLLECTOR,MECHANICAL SYSTEMS
NUCLEAR TECHNOLOGY P O BOX 1 AMSTON CN 06231 FRANK KERNOZEK -PHONE- 203 5372387 RESCH POLYCARBOXCYLIC ACID AS ALUM CORROSION INHIBITOR		
SOLAR ENERGY DEVELOPMENT INC 1457 ALAMEDA AVE LAKEWOOD OH 44107 NICHOLAS S MACRON -PHONE- 216 2213500 RESEARCH & DEV LOW PRICED FLAT PLATE VACUUM COLLECTOR		COLLECTOR, EVACUATED

---SUBCLASSES---SOLAR SUB SYSTEMS---

STANFORD RESEARCH INSTITUTE
333 RAVENSWOOD AVE
MENLO PARK CA 94025
RICHARD L GOEN

-PHONE-
415 3266200 X3198
ASSESSMENT OF POTENTIAL APPLICABILITY OF NEW ENGY SYS TO MILITARY INSTALLATIONS INCLUDING SOLAR...ASSESSMENT OF TOTAL ENERGY SYS FOR DEFENSE DEPT

STANFORD UNIVERSITY — HIGH TEMPERATURE SOLAR COLLECTOR SYSTEMS
DEPT OF MECH ENGR
STANFORD CA 94305
JOEL H FERZEGER

-PHONE-
415 3212300 X3615
SOLAR COLLECTORS FOR MED TO HIGH TEMP APPLICATION

TEXAS A&M UNIV
CHEMICAL ENGRG DEPT
COLLEGE STATION TX 77843
RICHARD R DAVISON

-PHONE-
713 8453361
FLAT PLATE COLL STORAGE..SOLAR HEATER DEVELOPMENT
GOPFFARTH, DAVIDSON, HARRIS,&BAIRD,'PERFORMANCE CORRELATION OF HORIZONTAL PLASTIC SOLAR WATER HEATERS,'SOLAR ENERGY, VOL 12, PERGAMON PRES 1968, PP 183-196.

TEXAS A&M UNIV — SOLAR WATER DISTILLATION
CHEMICAL ENGR DEPT
COLLEGE STATION TX 77843
W B HARRIS

-PHONE-
713 8453361
DEVELOP LOW COST HORIZONTAL COLL FOR LGE SCALE USE & LONG TERM (6 MO) STORAGE...COLL COST $1.00 SQFT 50 PERCENT EFF AT 150F SEPT SUN ANGLE 10 PERCENT HEAT LOSS OVER 6 MO STORAGE

	SUBCLASSES	SOLAR SUB SYSTEMS
TEXAS INSTRUMENTS 34 FOREST ST ATTLEBORO MA 02703 TUEVO SANTALA -PHONE- 217 2222800 X6150 RESCH UNDER NSF GRANT	SOLAR WATER HEATING SOLAR SPACE HEATING & COOLING	COLLECTOR,MECHANICAL SYSTEMS
TEXAS INSTRUMENTS INC MS10-16 ATTLEBORO MA 02703 UNTO SAVOLAINEN -PHONE- 617 2222800 X272 PLANNING STAGES		COLLECTOR,ABSORPTIVE SELECTIVE SURFACES
UTAH, UNIV OF MECH ENGR DEPT SALT LAKE CITY UT 84112 ROBERT F BOEHM -PHONE- 801 5816441 DEVELOP MORE EFF HIGHER TEMP FLAT PLATE TYPE COLLECTORS...HEATING SUPP WITH HEAT PUMPS...COOLING WITH ABSORPTION DEVICES	SOLAR SPACE HEATING & COOLING	HEAT PUMPS
WHIRLPOOL CORP R&E MONTE ROAD BENTON HARBOR MI 49022 DR EUGENE P WHITLOW -PHONE- 616 9265341 SMALL RESEARCH PROG AIMED AT KEEPING UP WITH DEVELOPMENTS...LITERATURE SURVEY IN PROGRESS...NO EXP WORK YET		

---SUBCLASSES---SOLAR SUB SYSTEMS---

WILLIAMS COLLEGE
PHYSICS DEPT
WILLIAMSTOWN MA 01267
JAY SHELTON

-PHONE-
413 5972123

C I R S O		SOLAR WATER DISTILLATION
DIV OF MECH ENGR		SOLAR WATER HEATING
P O BOX 26		SOLAR SPACE HEATING
HIGHETT, VICTORIA	AUSTRALIA	SOLAR SPACE COOLING
ROBERT V DUNKLE		PUBLISH INFORMATIONAL REPRINTS
CHIEF RESEARCH SCI		PUBLISH ORIGINAL RESEARCHED INFORMATION

-PHONE-
950333

THERMAL COLLECTOR SYS R&D MAINLY LOW TEMP APPLICATIONS...INQUIRE FOR PUBLICATIONS LIST OF AVAILABLE ARTICLES & CIRCULARS ON SOLAR ENERGY 'CIRSO DIV OF MECH ENGR CIRCULAR NO2 (1964) PRINCIPLES OF DESIGN CONST & INSTALLATION OF SOLAR WATER HEATERS' A GOOD MANUAL ON HARDWARE REQU-IREMENTS

C I R S O		SOLAR WATER DISTILLATION
DIV OF MECH ENGR		SOLAR WATER HEATING
P O BOX 26		SOLAR SPACE HEATING
HIGHETT, VICTORIA	AUSTRALIA	PUBLISH ORIGINAL RESEARCHED INFORMATION

F G HOGG *
CHIEF
-PHONE-
03 950333

EFFECTIVE&EFFICIENT METHODS&EQUIP FOR UTILIZATION OF SOLAR ENERGY BY INDUSTRY & INDIVIDUALS IN AREAS WHERE ECONOMICALLY JUSTIFIED &/OR CAN RESULT IN SIGNIFICANT SAVINGS IN FOSSIL FUELS...SOLAR H2O HEAT SUCCESS COMMERCIALLY&WIDELY USED...SOLAR STILL DEVELOP FAR ADVANCED...SOLAR AIR HEATERS IN LIMITED USE

	SUBCLASSES	SOLAR SUB SYSTEMS
MELBOURNE, UNIV OF MECH ENGR DEPT MELBOURNE PARKVILLE, VICTORIA AUSTRALIA W W CHARTRES	SOLAR SPACE HEATING	HEAT TRANSFER SYSTEMS, COLLECTOR TO STORAGE HEAT RECOVERY, STORAGE TO UTILIZATION

-PHONE-
345 1844 X6744
RESEARCH INTO UTILIZATION OF NATURAL ENERGY SOURCES&FUNDAMENTAL ASPECT OF COMBINED MODE HEAT TRANSFER PROCESSES...CONTINUING PROGRAM OF SOLAR ENERGY AND BASIC HEAT TRANSFER WORK PAST WORK: INVESTIGATION OF FREE CONVECTIVE FORCED CONVECTIVE & RADIATIVE HEAT TRANSFER PROCESSES IN SOLAR EQUIP PRIMARILY SOLAR AIR HEATERS MANY PUBLISHED TECH PAPERS

	SUBCLASSES
QUEENSLAND, UNIV OF DEPT OF MECH ENGR ST LUCIA, QUEENSLAND AUSTRALIA NORMAN R SHERIDAN	INORGANIC SEMI-CONDUCTORS SOLAR SPACE COOLING SOLAR SPACE HEATING & COOLING

-PHONE-
072 700111
INVESTIGATION INTO METHODS OF HEAT & COOL LOW-COST BLDGS LOW ENERGY USAGE PRIME INTEREST ALSO CONCENTRATING PHOTOVOLTAIC SYS PAST WORK ANALYSIS OF SOLAR AIR COND SYS USEING FLAT PLATE COLL&LITHIUM BROMIDE WATER ABSORPTION REFRIG PROTOTYPE HOUSE BUILT & TESTED MANY PUBLISHED WORKS

	SUBCLASSES
WILLIAM F KYRYLUK 1578 WEST 71ST AVE VANCOUVER 14 B C CANADA	SOLAR WATER DISTILLATION SOLAR WATER HEATING

-PHONE-
2638315
TO COME UP WITH GOOD SOLAR STILL & WATER HEATER...HOLDER OF US PATENT 3514942

SUBCLASSES — SOLAR SUB SYSTEMS

UNIV TECNICA FED SANTA MARIA — SOLAR WATER DISTILLATION; SOLAR WATER HEATING
LAB DE ENERGIA SOLAR
CASILLA 110 V
VALPARAISO CHILE
JULIO HIRSCHMANN R

-PHONE-
61268

RESEARCH OF POSSIBLE SOLAR APPLICATIONS IN NORTH CHILE...CURENT WORK ON SOLAR DESTILLATION & SOLAR WATER HEATING OF SOLAR ENERGY LAB... WORKING IN AGREEMENT WITH STATE ORGANISATIONS FOR INDUSTRIAL PRODUCTN PUBLISHED:'THEORY & EXPERIENCE WITH SOLAR STILLS IN CHILE'SOLAR ENERGY 1973 VOL 14 PP 405-413

GEORGE KASABOV
23 STRATFORD VILLAS
LONDON N W 1 ENGLAND

ARCHITECTURAL DESIGN OF BUILDINGS USING AMBIENT ENERGY PROJECTS IN PROGRESS IN FRANCE & ENGLAND

PILKINGTON BROTHERS LTD
R&D LABS
LATHOM ORMSHIRE
LANCASHIRE ENGLAND
A PULFORD
HEAD NEW VENTURES

USE OF GLASS & GLASS FIBRES IN ENGY CONS AND COLL DEVICES...MKT RESCH & BUSINESS DEV OF SOLAR ENERGY

UNIVERSITY OF TECHNOLOGY — SOLAR SPACE HEATING
LOUGHBOROUGH
LEIES ENGLAND
D J CROOME

-PHONE-
LOU 63171

TO DESIGN AN EFFICIENT ECONOMICAL SOLAR COLLECTOR FOR BLDGS

	SUBCLASSES	SOLAR SUB SYSTEMS
KAPUR SOLAR FARMS BIJWASAN-NAJAFGARH R P O KAPAS HERA NEW DELHI 110037 INDIA J C KAPUR	SOLAR WATER HEATING SOLAR REFRIGERATION SOLAR SPACE HEATING & COOLING	

-PHONE-
391747
CARRYING ON WORK IN HOME HEATING COOLING STORAGE SYS ALSO SOLAR WATER HEATERS...NUMBER OF PROTOTYPES & FULL MODEL PROTOTYPE PLANTS UNDER INVESTIGATION/ PAST WORK ON SOLAR REFRIGERATION

ROORKEE, UNIV OF
DEPT OF MECH ENGR
ROORKEE 247667 INDIA
RAJENDRA PRAKASH

-PHONE-
289
FLAT PLATE COLLECTORS UTILIZATION OF SOLAR ENERGY

	SUBCLASSES	SOLAR SUB SYSTEMS
NATIONAL PHYSICAL LAB ISRAEL HEBREW UNIV CAMPUS DANZIGER BLDG A JERUSALEM ISRAEL DR HARRY TABOR	INORGANIC SEMI-CONDUCTORS SOLAR PONDS SOLAR COOKING SOLAR WATER HEATING SOLAR SPACE HEATING & COOLING	

-PHONE-
30211 X475
LOW TEMP SYS-H2O HEATING HOUSE HEAT&COOL RENEWED ACTIVITY DUE TO ENGY CRISIS PAST WORK: SOLAR COLL SOLAR POWER UNITS SOLAR COOKERS CADMIUM SULFIDE SOLAR CELLS SOLAR PONDS LOW TEMP TURBINE IN KW RANGE SEE ALSO DR H TABOR SCIENTIFIC RESEARCH FOUNDATION

	SUBCLASSES	SOLAR SUB SYSTEMS
SCIENTIFIC RESEARCH FOUNDATION P O B 3745 JERUSALEM ISRAEL DR H TABOR	HIGH TEMPERATURE SOLAR COLLECTOR SYSTEMS	

-PHONE-
02 534515
EXTENSION OF WORK STARTED IN THE NATIONAL PHYSICAL LABORATORY ISRAEL
SEE ALSO DR HARRY TABOR NTL PHY LAB ISRAEL

----------SUBCLASSES----------SOLAR SUB SYSTEMS----------

PUERTO RICO, UNIV OF
GEN ENGR DEPT
MAYAGUEZ PUERTO RICO
DR MANUEL RODRIGUEZ PERAZZA

-PHONE-
809 8324040 X209
809 8324040 X455
TO DESIGN A FLAT PLATE COLL TO YIELD OUTPUT TEMP OVER 200C AT EFF AROUND 70 PERCENT FOR UNDER 2 DOLLARS/SQFT MATERIALS COST EXP WITH HONEYCOMB FLAT PLATES

CHAMBER OF MINES OF S AFRICA SOLAR COOKING
RESEARCH LABS
PO BOX 61809
JOHANNESBURG SOUTH AFRICA
AUSTIN WHILLIER

-PHONE-
JOH 31161
FLAT PLATE SOLAR COOKER FOR COOKING BY BOILING PERSONAL INTEREST ONLY ORGANIZATION WILL NOT RESPOND TO PUBLIC INQUIRY

PHILIPS FORSCHUNGSLABORATORIUM INORGANIC SEMI-CONDUCTORS
AACHEN GMBH
D-51 AACHEN
WEISSHAUSSTRASSE WEST GERMANY
DR HORST HORSTER

-PHONE-
024 1 62071
DEVELOPMENT OF SOLAR COLLECTORS FOR LOW TEMP OPERATION CURRENTLY ASSEMBLING EXISTING COMPONENTS TO LAB MODLES & INVESTIGATION OF THESE MODLES PAST WORK ON CONDUCTIVE COATINGS FOR SILICON SOLAR CELLS

----SUBCLASSES----SOLAR SUB SYSTEMS----

PAUL DIEGES *
1051 DAVIS RD
PERRIS CA 92370

MEDIUM TEMPERATURE SOLAR COLLECTOR SYSTEMS
WIND CONVERSION
SOLAR WATER HEATING

TURBINES

-PHONE-
714 6572822
CREATE A WHIRLWIND EFFECT THAT WILL SUSTAIN & DRAW THROUGH A TURBINE. CONST STARTED...HAVE CONST 3FT LONG 1FT WIDE COLL WITH 4 FLAT MIRRORS TO CONCENTRATE SUN ON BLACK TUBE INSULATE WITH FIBERGLASS, 2LAYERS GLASS-HEATS H2O

LARRY DOBSON
RT 1 BOX 459
CLINTON WA 98236

WIND CONVERSION
SOLAR SPACE HEATING & COOLING

-PHONE-
204 3211361
DEV CENTRIFUGAL FIBERGLASS-RESIN PARABOLIC REFLECTOR MOLDING TECHNIQUE SUTABLE FOR LOWTECH/DESIGNS FOR SOLAR HEATED HOUSE/SIMPLE RIGID BIMET-ALIC SPRING FOR GREENHOUSE VENTS...THIS IS NOT MY FULL-TIME OCCUPATION BUT I WOULD LIKE TO SHARE IDEAS WITH OTHERS

AMERICAN SCIENCE&ENGINEER INC
995 MASS AVE
CAMBRIDGE MA 02139
RYSZARD GAJEWSKI
DIRECTOR OF RESCH

MEDIUM TEMPERATURE SOLAR COLLECTOR SYSTEMS
SOLAR WATER HEATING
SOLAR SPACE HEATING & COOLING

COLLECTOR,OTHER

-PHONE-
617 8681600
SOLAR PLASMAS & DESIGN & EVAL COLL...PAPER PRESENTED ISES 74 MEETING CYLINDRICAL BLACKBODY SOLAR COLL

--------SUBCLASSES--------SOLAR SUB SYSTEMS--------

ARGONNE NATIONAL LABORATORY
9700 SOUTH CASS AVE
ARGONNE IL 60439
PROF ROLAND WINSTON

SOLAR ELECTRIC POWER GENERATION
SOLAR FURNACE

COLLECTOR:UPSIDE-DOWN CONE WITH CURVING PARABOLIC SIDES WITH SMALL HOLE AT BOTTOM-COLLECTS DIFFUSE & DIRECT LIGHT
HAS DEV COLLECTOR TO CONCENTRATE LIGHT 10 TIMES CAPABLE OF ELECT GEN
LARGE CONES COULD PROVIDE THOUSANDS DEGREES FOR 2-3 HOURS/DAY SOLAR FURNACE.....GRANTS FROM NSF & AEC TO RESEARCH

ARIZONA, UNIV OF
OPTICAL SCIENCES CTR
TUCSON AZ 85721
ADEN B MEINEL

COLLECTOR,ABSORPTIVE SELECTIVE SURFACES

-PHONE-
602 7493322
602 8843138

RESCH ON APPLICATION & ECONOMICS OF SOLAR...MOST NOTABLE RESCH ON SOLAR FARMS FOR SOUTHWEST REGION & TROUGH CONCENTRATING COLL ALSO WORK WITH SELECTIVE SURFACES

ARIZONA, UNIV OF
OPTICAL SCIENCES CTR
TUCSON AZ 85721
MARJORIE P MEINEL

COLLECTOR,REFLECTIVE SELECTIVE SURFACES

-PHONE-
602 7493322
602 8843138

SEE ADEN MEINEL FOR DESC OF WORK

----SUBCLASSES---- ----SOLAR SUB SYSTEMS----

ARIZONA, UNIVERSITY OF
ENVIRONMENT RES LAB
TUCSON INT'L AIRPORT
TUCSON AZ 85706
DR JOHN F PECK

SOLAR WATER DISTILLATION
SOLAR REFRIGERATION
SOLAR SPACE HEATING & COOLING

COLLECTOR,REFLECTIVE SELECTIVE SURFACES
COLLECTOR, EVACUATED

-PHONE-
602 8842931
DESIGN & TESTING OF 300F COLLECTORS FOR AUGMENTATION OF STEAM SOURCES FOR ABSORPTION REFRIG & STEAM/HOT WATER HEATING-INITIAL STAGES COMPLETED PROJECTS: HOUSE HEAT&COOL BY SOLAR/SOLAR DISTILL SEA WATER BLISS,RAYMOND W,'THE PERFORMANCE OFEXPERIMENTAL SYS USING SOLAR ENERGY FOR HEATING & NIGHT RADIATION FOR COOLING A BUILDING,' UN CONFERENCE ON NEW SOURCES OF ENERGY III.C.2 (1961).

CALIF, UNIV OF-RICHMOND FLD ST
RICHMOND FIELD STA
1301 SOUTH 46TH ST
RICHMOND CA 94704
DON O HORNING

-PHONE-
415 2356000 X211
DEVELOP HIGH TEMP COLL SYS FOR PRODUCTION OF STABLE NITROGEN COMPOUNDS FOR AGRICULTURE/ALSO FOR METALURGICAL RESEARCH

COLORADO UNIV OF-BOULDER
CIRES
BOULDER CO 80302
HARTMUT SPETZER

MEDIUM TEMPERATURE SOLAR COLLECTOR SYSTEMS

-PHONE-
303 4928028
TROUGH COLL USING ALUM FOIL...40 PERCENT EFFICIENCY

COLORADO UNIV OF-BOULDER
AEROSPACE ENG SCI
BOULDER CO 80302
MAHINDER UBEROI

COLLECTOR,REFLECTIVE SELECTIVE SURFACES

-PHONE-
303 4926612

	SUBCLASSES	SOLAR SUB SYSTEMS
GROUP SEVEN CHEMICALS CO 1210 DONA DRIVE CORPUS CHRISTI TX 78407 GEORGE E ULVILD, FAIC OWNER	SOLAR WATER DISTILLATION	STORAGE, PHASE CHANGE MATERIALS STORAGE, HYDROGEN

-PHONE-
512 8886191
512 8835809
HIGH TEMP CONCENTRATORS, CONVERSION WITH RADIOMETERS, CHEMICALS FOR HE
AT-SINK STORAGE OF SOLAR HEAT, SOLAR WARMING OF HIGH FREEZING LIQUIDS,
IE.CAUSTIC-SODA&LOW TEMP THERMAL DECOMP OF H2O...STUDIES ON SCALE
MODEL
PAST RESEARCH: INCREASE EFFICIENCY OF SOLAR SALT PROD IN NE MEXICO

INT'L RESEARCH&TECH CORP 1501 WILSON BLVD ARLINGTON VA 22209 THEODORE B TAYLOR	BIO-CONVERSION	

-PHONE-
703 5245834
ASCESS ALTERNATIVES FOR USE OF SOLAR ENGY FOR FUELS OR ELECTRICITY
PROSPECTS FOR PROD PLANT FUELS IN GREENHOUSES
TAYLOR & HUMPSTONE,'THE RESTORATION OF THE EARTH,'NEW YORK, HARPER &
ROW, 1973.

NAVAL RESEARCH LAB UPPER AIR PHYSICS BR CODE 7120 WASHINGTON DC 20375 TALBOT CHUBB	SOLAR FURNACE SOLAR ELECTRIC POWER GENERATION	

-PHONE-
202 7673580
GAS DISSOCIATION SOLAR THERMAL POWER SYSTEM...RESCH THAT CAN BE EMPLO
YED IN DESTRUCTIVE DISTILL PROCESS SUCH AS OIL SHALE RECOVERY

	SUBCLASSES	SOLAR SUB SYSTEMS
PENN, UNIVERSITY OF N CTR ENGY MGT&POWER TOWNE BUILDING PHILADELPHIA PA 19104 JESSE C DENTON * -PHONE- 215 5945122	SOLAR SPACE HEATING & COOLING SOLAR ELECTRIC POWER GENERATION	
PENN, UNIVERSITY OF N CTR ENGY MGT&POWER TOWNE BUILDING PHILADELPHIA PA 19104 GEORGE L SCHRENK -PHONE- 215 5945121 ADVANCED RESEARCH & EDUCATION (PHD PROGRAM) IN ENERGY MANAGEMENT & POWER...I HAVE DONE EXTENSIVE WORK ON SOLAR CONCENTRATORS	SOLAR SPACE HEATING & COOLING SOLAR ELECTRIC POWER GENERATION	
PENNSYLVANIA STATE UNIV MATERIAL SCIENCE DPT 320 M I BLDG UNIVERSITY PARK PA 16802 HOWARD B PALMER PROF OF FUEL SCIENCE -PHONE- 814 8656512 SOLAR FARMS UTILIZING LOW PRESSURE CLOSED CYCLE GAS TURBINES PAPER IN PROC 8TH IECEC	SOLAR ENERGY STORAGE ALL SOLAR TECHNOLOGIES SOLAR ELECTRIC POWER GENERATION SOLAR TECHNOLOGY APPLICATION OTHER	COLLECTOR,ABSORPTIVE SELECTIVE SURFACES COLLECTOR,REFLECTIVE SELECTIVE SURFACES HEAT TRANSFER SYSTEMS, COLLECTOR TO STORAGE HEAT RECOVERY, STORAGE TO UTILIZATION STORAGE, MECHANICAL/PUMPED LIQUIDS STORAGE, GAS TURBINES
SANDIA LABORATORIES DIVISION 8184 LIVERMORE CA 94550 THOMAS D BRUMLEVE STAFF MEMBER -PHONE- 415 4552941 HIGH TEMPERATURE SOLAR ENERGY SYSTEM (POINT FOCUS CENTRAL TOWER) SYS ANALYSIS DESIGN EXPERIMENTATION & DEVELOPMENT...THERMAL ENERGY STORAGE INVESTIGATIONS OF INEXPENSIVE FLAT PLATE COLLECTOR TECHNIQUE	MEDIUM TEMPERATURE SOLAR COLLECTOR SYSTEMS SOLAR ENERGY STORAGE SOLAR FURNACE SOLAR SPACE HEATING & COOLING SOLAR ELECTRIC POWER GENERATION	COLLECTOR,ALSO FUNCTION AS STRUCTURES SKIN COLLECTOR,REFLECTIVE SELECTIVE SURFACES HEAT TRANSFER SYSTEMS, COLLECTOR TO STORAGE STORAGE, LIQUIDS

	SUBCLASSES	SOLAR SUB SYSTEMS
SANDIA LABORATORIES P O BOX 969 LIVERMORE CA 94550 CLIFFORD T YOKOMIZO TECHNICAL STAFF	MEDIUM TEMPERATURE SOLAR COLLECTOR SYSTEMS SOLAR ENERGY STORAGE SOLAR SPACE HEATING & COOLING	COLLECTOR,REFLECTIVE SELECTIVE SURFACES COLLECTOR,ABSORPTIVE SELECTIVE SURFACES STORAGE, SOLID MATERIALS STORAGE, PHASE CHANGE MATERIALS STORAGE, LIQUIDS

-PHONE-
415 4557011 X2668
DETAILED THERMAL ANALYSIS OF FLAT PLATE COLL...MIRROR MODULE DESIGN
FOR POINT FOCUS SOLAR COLLECTION SYSTEM THERMAL STORAGE

	SUBCLASSES	SOLAR SUB SYSTEMS
WESTINGHOUSE RESEARCH LAB CHURCHILL BORO PITTSBURGH PA 15235 ROGER W WARREN MANAGER	SOLAR ELECTRIC POWER GENERATION SOLAR HEAT,COOL,&ELECTRIC POWER GENERATION	COLLECTOR,ABSORPTIVE SELECTIVE SURFACES COLLECTOR,OTHER

-PHONE-
412 2565187
HIGH TEMPERATURE COLLECTOR DESIGN, SELECTIVE COATINGS

NEW SOUTH WALES, UNIV OF
SCHOOL OF PHYSICS
BOX 1
KENSINGTON, N S W AUSTRALIA
J GIUTRONICH

-PHONE-
663 0351
NO PRESENT WORK PAST WORK: BUILT & EVALUATED 12 FT PARABOLIC MIRROR
TYPE SOLAR FURNACE

ELECTROTECHNICAL LAB
ENGY SYSTEM SECTION
TANASHI
TOYKO JAPAN
TATSU TANI

--SUBCLASSES--SOLAR SUB SYSTEMS--

SOPHIA UNIVERSITY
DEPT OF PHYSICS
7 KIOI-CHO CHIYODA-K
TOKYO JAPAN
ISAO OSHIDA
FACULTY OF SCI & TEC
-PHONE-
03 2659211 X567
USE OF PLASTIC FRESNEL LENSES AS SOLAR COLLECTORS STUDYING EFF OF 40CM
PLASTIC LENS

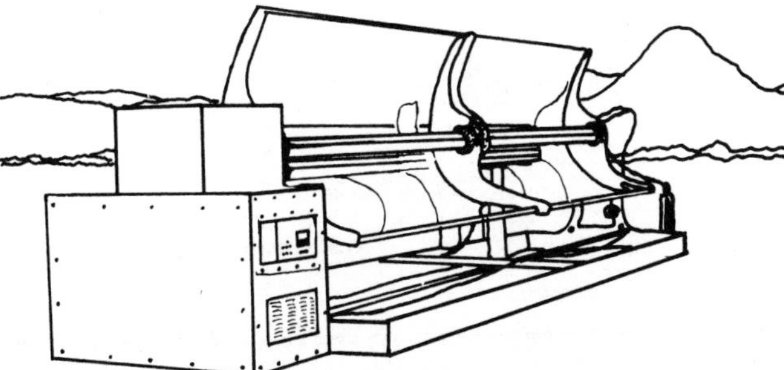

	SUBCLASSES	SOLAR SUB SYSTEMS
CARNEGIE MELLON UNIV 6123 SCIENCE HALL PITTSBURG PA 15213 CLARENCE ZENER		STORAGE, HYDROGEN

-PHONE-
412 6212600 X229
SOLAR POWER SYS BASED ON THERMODYNAMIC CYCLES OF HEAT ENGINES USEING NATURAL TEMP GRAD OF OCEAN/PLANTS COULD PROD ELECTRICITY OR HYDROGEN & PRODUCE VOLUME FRESH H2O/THIS PROJECT WILL DEV COMPUTER BASED ANALYTIC MODELS FOR TECH & ECONOMIC ANALYSIS NSF CONTRACT GI 39114
PUB:'SOLAR SEA POWER,'PHYSICS TODAY, VOL 26, NO 1, JAN 1973, PP 48-53.

	SUBCLASSES	SOLAR SUB SYSTEMS
ESCHER TECHNOLOGY ASSOC P O BOX 189 ST JOHNS MI 48879 WILLIAM ESCHER	HIGH TEMPERATURE SOLAR COLLECTOR SYSTEMS SOLAR ENERGY STORAGE SOLAR WATER DISTILLATION	STORAGE, HYDROGEN HEAT ENGINE, RANKINE

-PHONE-
517 2246726
SOLAR TO HYDROGEN ENERGY CONVERSION VARIOUS METHODS: LGE CENTRAL STAT-ION-OCEAN BASING APP,THERMAL,OTG ETC.....MANY TECHNICAL PUBLICATIONS ON SUBJECT AVAIL THROUGH ESCHER TEC ASSOC...PLAN TO ENTER LABORATORY RESEARCH IN COLLAB WITH UNIV GROUPS

	SUBCLASSES	SOLAR SUB SYSTEMS
HAWAII, UNIV OF HI INST OF GEOPHYSIC 2525 CORREA ROAD HONOLULU HI 96822 MARTIN VITOUSEK	MEDIUM TEMPERATURE SOLAR COLLECTOR SYSTEMS HIGH TEMPERATURE SOLAR COLLECTOR SYSTEMS WIND CONVERSION SOLAR ELECTRIC POWER GENERATION	STORAGE, HYDROGEN

-PHONE-
808 9487668
TO ASSESS VARIOUS NATURAL SOURCES OF ENERGY&THEIR POTENTIAL FOR SMALL TROPICAL ISLAND COMMUNITIES...TO DEVELOP POWER SYS USEING HYDROGEN & DEVELOP STORAGE SYS USEING PRIMARILY HYDROGEN GAS
RESEARCH IN ITIAL STAGE

	SUBCLASSES	SOLAR SUB SYSTEMS
JOHNS HOPKINS UNIV APPLIED PHYSICS LAB 8621 GEORGIA AVE SILVER SPRINGS MD 20910 W H AVERY	SOLAR ELECTRIC POWER GENERATION	

-PHONE-
301 9537100 X2411
DEVELOPMENT OF 100 MW SOLAR SEA POWER PLANT IN 5 YEARS LEADING TO SUBSEQUENT EXTENSIVE USE OF SUCH PLANTS AS MAJOR POWER SOURCE FOR NATION...PRELIM PERFORMANCE & COST EST COMPLETE.PROPOSAL SUBMITTED TO AEC FOR DEVELOPMENT PROJ

MASSACHUSETTS, UNIV OF CIVIL ENGINEERING AMHERST MA 01002 W E HERONEMUS	SOLAR ELECTRIC POWER GENERATION	STORAGE, HYDROGEN

CONCEPT & FEASIBILITY STUDIES FOR POWER GEN FROM OCEAN TEMP GRAD TO DETERMINE ECONOMIC & TECH FEASIBILITIES

NATIONAL SCIENCE FOUNDATION ADVNCED ENGY RES&TEC WASHINGTON DC 20550 ROBERT COHEN PROG MGR OCEAN THERM	GOVERNMENT AGENCIES PROVIDING GRANTS INCENTIVES FOR ENERGY CONSERVATION	HEAT ENGINE, RANKINE COLLECTOR,OTHER HEAT TRANSFER SYSTEMS, COLLECTOR TO STORAGE STORAGE, OTHER TURBINES

-PHONE-
202 6327364
ALSO CONCERNED WITH GEOPHYSICAL DATA REQUIREMENTS FOR SOLAR ENERGY APPLICATIONS

UNITED ENGRS&CONSTRUCTORS INC 1401 ARCH PHILADELPHIA PA	SOLAR ELECTRIC POWER GENERATION PROVIDE INFORMATION SEARCHES TO CLIENTS	

PERFORMING ENGINEERING & FEASIBILITY STUDY OF SEA SOLAR POWERS CONCEPT OF ELECT GEN THROUGH OCEAN TEMP GRADIENTS

--------------------SUBCLASSES--------------------SOLAR SUB SYSTEMS--------------------

AGRO-CITY PLANNING CONSULTANTS
3410 GARROTT RD
HOUSTON TX 77006
GEOFFREY STANFORD

-PHONE-
713 5297266
PHOTOSYNTHESIS-ENERGY FARMING: FORESTRY ON CONTINUOUS SHORT-ROTATION HARVESTING TO SUPPLY CONSISTENT QUALITY OF FUEL INDEFINITELY/ 10 MILES SQUARE WILL MAINTAIN 1000 MW ELECTRIC STEAM GENERATOR
FIELD TRIALS IN PROCESS

BROOKHAVEN NATIONAL LAB — SOLAR ELECTRIC POWER GENERATION
PUBLIC RELATIONS
40 BROOKHAVEN AVE
UPTON NY 11973
CARL R THIEN *
INFORMATION OFFICER
-PHONE-
516 3452345

CALIF, UNIV OF-BERKELEY
BIOCHEM DEPT
BERKELEY CA 94720
J B NEILANDS

-PHONE-
415 6427460
DEVELOP MODEL CATALYSTS IMITATING THOSE USED IN PHOTOSYNTHESIS/USE OF IRON-CONTAINING ENZYMES IN FUEL CELLS/+USE OF ENZYMES OR SYNZYMES TO CONSERVE ENGY FOR INDUSTRIAL PURPOSES
INTERESTED IN CORRES WITH OTHER BIOLOGICAL SCIENTISTS IN SIMILAR AREAS

CALIF, UNIV OF-BERKELEY — STORAGE, METHANE
SCH OF PUBLIC HEALTH
BERKELEY CA 94720
W J OSWALD

SOLAR ENERGY FIXATION & CONVERSION WITH ALGAE-BACTERIA SYS TO PRODUCE METHANE FOR USE AS COMBUSTIBLE GAS OR AS ELECTRICITY

CALIF, UNIV OF-LA JOLLA
CHEM DEPT
LA JOLLA CA 92037
JOHN BENEMAN

STORAGE, HYDROGEN

-PHONE-
714 4532000 X1639
BASIC RESEARCH-PHOTOSYNTHESIS & BLUE-GREEN ALGAE TO DETERMINE FEASIBIL OF HYDROGEN PRODUCTION FROM WATER & SUNLIGHT BY THESE SYS 'HYDROGEN EVOLUTION BY CHLOROPLAST-HYDROGENESE SYSTEMS' PROCEEDINGS OF NAT ACADEMY OF SCIENCES,AUG 1973

CHARLES F KETTERING RESRCH LAB
150 E SO COLLEGE ST
YELLOW SPRINGS OH 45387
DR DONALD L KEISTER

-PHONE-
513 7677271
USE OF BIOLOGICAL PHOTOCHEMICAL SYS IN ENERGY CONVERSION...WE ARE NOW TRYING TO DETERMINE IF OUR WORK SINCE 1927 IN PHOTOSYNTHESIS CAN BE APPLIED TO THE PROBLEM

CHICAGO, UNIV OF
BIOLOGY DEPT
5630 S INGLESIDE AVE
CHICAGO IL 60637
RONALD G ALDERFER

-PHONE-
312 7532666
TO UNDERSTAND MECHANISMS OF ENVIRO REGULATION OF PHOTOSYNTHESIS IN PLANT CANOPIES/ HAVE WORKING MODELS EXPLAIN ASPECTS OF PROBLEMS PRESENT RESEARCH-PHYSIOLOGICAL&BIOCHEMICAL CHAR PLANT TISSUE TO DETERM BIOLOGIC POT FOR PHOTOSYNTHESIS/PUBLISHED:'ENGY EXCHANGE:PLANT CANOPIE ECOLOGY 52/5/:855-861/&/'INTERACTION SOLAR RAD WITH PLANT SYS',SOLAR ENERGY,15:77-82.

--------SUBCLASSES--------SOLAR SUB SYSTEMS--------

DELAWARE, UNIV OF
CHEMISTRY DEPT
NEWARK DE 19711
MARJORIE P KRAUS

-PHONE-
302 7382458
215 2748114
PHOTOSYNTHESIS, ALGAL CULTURE, LAGOONING OF ORGANIC WASTES & HARVEST
OF SUITABLE ALGAL PRODUCT/STUDIES ON FACTORS INFLUENCING ALGAL PRODUCT

ILLINOIS, UNIV OF
SCHOOL OF ENGINEERNG
URBANA IL 61801
J T PFEFFER

PROVIDE ECONOMIC ANALYSIS OF SOLAR TECH — STORAGE, METHANE

CONVERSION REFUSE & WASTE H2O TO METHANE GAS ON LAB SCALE...MATH SYS
DEVELOP TO SIMULATE SYSTEM...ANALYSIS OF ECONOMIC FEASIBILITY OF SYS

KANSAS STATE UNIV
CHEM ENGR DEPT
MANHATTAN KS 66506
ROBERT H SHIPMAN

-PHONE-
913 5325584
PHOTOSYNTHESIS-RESEARCH INTO PRODUCTION OF SINGLE-CELL PROTEINS (SCP)
FROM PHOTOSYNTHETIC BACTERIA

MICHIGAN, UNIVERSITY OF
DEPT OF ELECT ENGR
ANN ARBOR MI 48104
DALE M GRIMES

ALL SOLAR TECHNOLOGIES

-PHONE-
313 7649546
ATTEMPTING TO MODEL CHLOROPLAST ANTENNAS.HOPFFULLY SUCH A MODELING &
UNDERSTANDING WOULD PERMIT CONSTRUCT OF SIMILAR & CONTROLLABLE ONES IN
LAB...ALSO TEACH SENIOR SEMINAR IN ENVIRO SCI FNGR ON SUBJ SOLAR ENGY
STUDENTS HAVE PROJ

NASA LEWIS RESEARCH CTR
PHYSICAL SCIENCE DIV
21000 BROOKPARK ROAD
CLEVELAND OH 44135
YIH-YUN HSU
MAIL STOP 301-1
-PHONE-
216 4334000 X205
PHOTOSYNTHESIS OF CLEAN FUELS...OBJECT TO PRODUCE PORTABLE FUELS FROM BIOMASS...IDENTIFYING POTENTIAL CROPS, EXPERIMENTING ON VARIOUS CONVERSION PROCESSES&SYS ANALYSIS

NATIONAL SCIENCE FOUNDATION — GOVERNMENT AGENCIES PROVIDING GRANTS
ADVNCED ENGY RES&TEC
WASHINGTON DC 20550
RICHARD BOGAN
PROG MGR BIOCONVERSN

-PHONE-
202 6327364

NAVAL UNDERSEAS CENTER
SAN DIEGO CA 92132
HOWARD WILCOX

-PHONE-
714 2257957
MARINE FARM TO GROW GIANT KELP FOR BIOCONV FOOD PLASTIC & FERTILIZER

SMITHSONIAN RADIATION BIO LAB — SOLAR RADIATION MEASUREMENTS
12441 PARKLAWN DR
ROCKVILLE MD 20852
WALTER SHROPSHIRE JR

-PHONE-
301 4432333
REGULATION OF BIOLOGICAL PROCESSES BY SUNLIGHT&MEASUREMENT OF SPECTRAL DISTRIBUTION OF SOLAR RAD IMPORTANT FOR BIOLOGICAL PHOTOMORPHOGENESIS PHOTOTROPISM SEED GERMINATION & PIGMENT SYNTHESIS
MANY PUBLISHED WORKS SAMPLE:CRAKER,ABELES,&SHROPSHIRE,'LIGHT INDUCED ETHYLENE PRODUCTION IN SORGHUM,'PLANT PHYSIOLOGY 51:1082-1083, 1973.

---SUBCLASSES---SOLAR SUB SYSTEMS---

STANFORD RESEARCH INSTITUTE — SOLAR AGRICULTURAL DRYING
FOOD&PLANT SCIENCES — PUBLISH ORIGINAL RESEARCHED INFORMATION
333 RAVENSWOOD AVE
MENLO PARK CA 94025
ROBERT E INMAN
PLANT BIOLOGIST
-PHONE-
415 3266200 X2155
408 2558853
EFFECTIVE UTILIZATION OF SOLAR ENERGY TO PRODUCE CLEAN FUEL

TEXAS, UNIV OF
ZOOLOGY DEPT
AUSTIN TX 78712
JACK M YERS

-PHONE-
512 4711686
RESCH MECHANISMS INVOLVED IN ELECTRON TRANSPORT IN PHOTOSYNTHESIS
PUB: WANG & MYERS,'ENERGY TRANSFER BETWEEN PHOTOSYNTHETIC UNITS,' PHOTOCHEM PHOTOBIOL, 17:321-32, (1973)...PAST WORK ON MASS CULTURE OF ALGAE SCIENTIFIC PROBLEMS

WOODS HOLE OCEANOGRAPHIC INST — OCEAN THERMAL GRADIENT SYSTEMS
CHEMISTRY DEPT
WOODS HOLE MA 02543
OLIVER C ZAFIRIOU

-PHONE-
617 5481400 X302
FUNDAMENTAL RESEARCH IN LIGHT ABSORPTION & PHOTOCHEMISTRY OF SEA WATER
MANY STUDIES ON BIOLOGICAL PRODUCTIVITY OF SEA WATER & ITS UTILIZATION BY MAN

---SUBCLASSES--- ---SOLAR SUB SYSTEMS---

BIOMASS ENERGY INST INC — PUBLISH ORIGINAL RESEARCHED INFORMATION — STORAGE, METHANE

BIOMASS ENERGY INST INC
PO BOX 129 P STA C
WINNIPEG MANITOBA CANADA
JOHN MORRISS

-PHONE-
204 2840472

RESEARCH & DEV USE OF RENEWABLE BIOLOGICAL MATERIAL FOR ENERGY PURPOSE OPERATING METHANE PRODUCING ANIMAL WASTE RECYCLER GLENEA MANITOBA... ISOLATE & DEVELOP USEFUL ALGAL STRAINS...PUBLISHED:PROCEEDINGS INTERNATIONAL BIOMAS ENERGY CONFERENCE AVAILABLE FOR 10 DOLL ADDRESS ABOVE

"He had been eight years upon a project for extracting sunbeams out of cucumbers, which were to be put in phials hermetically sealed, and let out to warm the air in raw inclement summers."

Gullivers Travels (1727)
Part III, Chap V., Voyage to Laputa

-----SUBCLASSES-----SOLAR SUB SYSTEMS-----

FRED E DAVISON
HIGHWOOD MT 59450

SOLAR ELECTRIC POWER GENERATION

-PHONE-
406 7332013
FEASIBILITY ECONOMIC STRUCTURE STUDY FOR 25000KW & UP GENERATING PLANT
WIND POWER MACHINE COMPUTER ANALYSIS OF 3 PHASE SYS-NSF SPONSORER
RESEARCH FOR INFO INQUIRE TO: DR R POWE MONTANA STATE UNIV MECH ENGR
BOZEMAN MT

GREGORY P HENKE
2190 GARFIELD
EUGENE OR 97405

MEDIUM TEMPERATURE SOLAR COLLECTOR SYSTEMS

-PHONE-
503 3448065
APPLICATION OF WIND ENGY&SOLAR COLLECTORS-ENERGY TRANSFER

WILLIAM SMITH
P O BOX 281
JAMESTOWN RI 02835

SOLAR SPACE HEATING
SOLAR ELECTRIC POWER GENERATION

WORKING ON A 12 FT DIAMETER 12KW WIND GENERATOR...ALSO COLLABORATING
ON CONSTRUCTION OF SOLAR HEATED HOUSE

DAN WEITZEL
CONSULTANT
3040 ASH AVE
BOULDER CO 80303

MEDIUM TEMPERATURE SOLAR COLLECTOR SYSTEMS
HIGH TEMPERATURE SOLAR COLLECTOR SYSTEMS

STUDY & PROTOTYPES NOT READY FOR PUBLIC INQUIRY YET

--------------------SUBCLASSES--------------------SOLAR SUB SYSTEMS--------------------

ALASKA,UNIV OF
GEOPHYSICAL INSTITUT
FAIRBANKS AK 99701
DR KEITH MATHER
DIRECTOR

-PHONE-
907 4797607
907 4797558
DETERMINE WIND POTENTIAL OF ALASKA FOR NATIONAL ENERGY NEEDS/ALSO SEEK IMPLEMENTATION OF WIND POWER IN ALASKA COMMUNITIES/PROPOSALS UNDER REVIEW

AVCO EVERETT RESEARCH LAB INC
2385 REVERE BEACH PK
EVERETT MA 02149
RAYMOND JANNEY *
ASSIST TO CHAIRMAN

-PHONE-
617 3893000

AVCO EVERETT RESEARCH LAB INC
2385 REVERE BEACH PK
EVERETT MA 02149
ARTHUR KANTROWITZ
CHAIRMAN

-PHONE-
617 3893000

BROOKHAVEN NATIONAL LABORATORY SOLAR ELECTRIC POWER GENERATION
BIOLOGY DEPT
50 BELL AVENUE
UPTON NY 11973
JOHN M OLSEN
BIOPHYSICIST
-PHONE-
516 3453382
A SOLAR ENERGY CONVERTER MODELED AFTER A PHOTOSYNTHETIC ECOSYSTEM, IN PREPARATION

SUBCLASSES — SOLAR SUB SYSTEMS

FLORIDA, UNIV OF
DEPT MECH ENGR
GAINESVILLE FL 32601
DR HERBERT SCHAEPER

SOLAR AGRICULTURAL DRYING
SOLAR POWERED TRANSPORTATION

-PHONE-
904 3920809

UTILIZATION OF WIND & SOLAR ENGY IN CARIBBEAN & CENTRAL AMERICA FOR AGRICULT, FISHING,&OTHER INDUSTRIES/SOLAR HORTICULTURAL DRYERS-EASTERN JAMAICA/SOLAR FISH DRYERS-LAKE IZABEL GUATEMALA/WIND ELECTRIC POWER ST THOMAS PARISH,JAMAICA...ALSO WORK ON SOLAR ELECT TRANSPORT/PUBLISH PAPERS INSTITUTE OF ELECTRICAL&ELECTRONICS ENGRS INC,APRIL 30,MAY 1-2 1973,LOUISVILLE KY:'UNIVERSITY OF FLORIDA SOLAR ELECTRIC CAR'

SEE: DR E A FARBER,FLORIDA,UNIV OF

KANSAS,UNIV OF
DEPT OF GEOGRAPHY
LAWRENCE KS 66044
STEVE BLAKE

TOTAL SYSTEMS INTEGRATION

-PHONE-
913 8644276

TO EVOLVE DESIGN FOR WIND GEN FROM LOW COST REDDILY AVAIL MATERIALS WHICH CAN COMPETE SUCCESSFULLY WITH RURAL ELECT RATES...DESIGN TO USE SALVAGE ITEMS & SAILWING BLADE...PROTOTYPE BUILT TESTING BEGUN APRIL 1974...GOAL WORKING SELF-SUFF HOMESTEADS

MONTANA STATE UNIVESITY
SCHOOL OF ENGINEERG
BOZEMAN MT 59715
R E POWE

SOLAR ELECTRIC POWER GENERATION

TECHNICAL FEASIBILITY STUDY OF A WIND ENERGY CONVERSION SYSTEM BASED ON TRACKED VEHICLE-AIRFOIL CONCEPT

	SUBCLASSES	SOLAR SUB SYSTEMS
NASA LEWIS RESEARCH CTR	SOLAR ELECTRIC POWER GENERATION	
PLUM BROOK STATION		
21000 BROOKPARK ROAD		
CLEVELAND OH 44135		
RICHARD L PUTHOFF		

-PHONE-
216 4334000
100 KW WIND TURBINE GENERATOR CONSISTS ROTOR TURBINE TRANSMISSION SHAFT ALTENATOR AND TOWER...TWO BLADES 125 FT DIAM...40RPM 100 KW AT GENERATOR IN 18 MPH VELOCITY

NATIONAL SCIENCE FOUNDATION	GOVERNMENT AGENCIES PROVIDING GRANTS	
ADVNCED ENGY RES&TEC		
WASHINGTON DC 20550		
LOUIS DIVONE		
PROG MGR WIND ENGY		

-PHONE-
202 6327364

OKLAHOMA STATE UNIVERSITY	SOLAR ENERGY STORAGE	ELECTROLYSIS
ELECTRICAL ENGG	TOTAL SYSTEMS INTEGRATION	HEAT ENGINE, RANKINE
ENGINEERING SOUTH	SOLAR ELECTRIC POWER GENERATION	STORAGE, FUEL CELLS
STILLWATER OK 74074	SOLAR TECHNOLOGY APPLICATION OTHER	STORAGE, GAS
WILLIAM L HUGHES	PUBLISH ORIGINAL RESEARCHED INFORMATION	STORAGE, OTHER
HEAD SCHOOL OF E E	PUBLISH ORIGINAL RESEARCH + REPRINTS	TURBINES

-PHONE-
405 3726211 X7581
ENGAGED IN ENERGY CONVERSION & ENERGY STORAGE RESEARCH SINCE 1961 MORE THAN 20 PUBLISHED TECHNICAL PAPERS MORE THAN 20 REPORTS 7 PATENTS & SEVERAL MASTERS & DOCTORS THESES ...INTERESTS HIGH PRESSURE ELECTRO-LYSIS APHODID BURNER SOLAR & WIND ENERGY SYS&FIELD MODULATED GENERATOR SYS

	SUBCLASSES	SOLAR SUB SYSTEMS
OKLAHOMA STATE UNIVERSITY ELECTRICAL ENGG ENGINEERING SOUTH STILLWATER OK 74074 RAMACHANDRA G RAMAKUMAR ASSOC PROF ELEC ENGG -PHONE- 405 3726211 X6476	SOLAR ENERGY STORAGE TOTAL SYSTEMS INTEGRATION SOLAR ELECTRIC POWER GENERATION SOLAR TECHNOLOGY APPLICATION OTHER PUBLISH ORIGINAL RESEARCHED INFORMATION PUBLISH ORIGINAL RESEARCH + REPRINTS	STORAGE, OTHER STORAGE, FUEL CELLS STORAGE, GAS ELECTROLYSIS TURBINES

MORE THAN 20 PUBLISHED TECHNICAL PAPERS IN THE AREAS OF ENERGY CONVERSION & ENERGY STORAGE-INVOLVED SINCE 1967 IN SOLAR & WIND ENERGY SYS RESEARCH...3 PATENTS...PRIMARY INTERESTS ARE IN FIELD MODULATED POWER GENERATION & SOLAR ENGY SYS

WEST TEXAS STATE UNIV
BOX 248
CANYON TX 79016
DR VAUGHN NELSON

-PHONE-
806 6563904

PERFORM ANALYSIS OF WIND & ATMOSPHERIC DATA FROM NATL WEATHER STATIONS IN SOUTHERN HIGH PLAINS..SET UP WEATHER DATA INSTRUMENTATION SITE NEAR AMARILLO TX TO CORRELATE WIND & ATMO DATA WITH WEATHER BUREAU RESULTS

	SUBCLASSES
NATIONAL RESEARCH COUNCIL OTTAWA CANADA PETER SOUTH ASSOC RESCH OFFICER	SOLAR ELECTRIC POWER GENERATION

VERTICAL AXIS WIND TURBINE

-----SUBCLASSES-----SOLAR SUB SYSTEMS-----

JON T ELIASON *
6501 ROBINHOOD LANE
HUNTSVILLE AL 35806

-PHONE-
205 8374782
PAST RES: INVESTIGATE & RECOMMEND TO NASA-MSFC APPROPRIATE SOLAR ENGY
APPLICATIONS-SUPPORT CONTRACT THROUGH SPERRY RANN SUPPORT SERV
NO CURRENT ACTIVE EFFORT

FIDEL ORONA
3417 MEMPHIS DR
EL PASO TX 79902

SOLAR WATER HEATING
SOLAR ELECTRIC POWER GENERATION
SOLAR POWERED TRANSPORTATION

-PHONE-
915 7793877
SILICON SOLAR CELL APPLICATIONS: 6TH GEN CLEAN POWER PLANT-APPLYING
HIGH PRESSURE ELECTROLYSIS CHAMBERS WITH 3 HOT H2O HEATERS BEFORE
ENTERING COMBUSTION CHAMBER...IN PROCESS OF PURCH MATERIALS TO BUILD
ELECT CAR WITH PHOTOVOLTAIC CELLS
PRESENTED 1ST GEN OF CLEAN POWER PLANT AT ISES MEETING OCT 1973

YVON F PIERRE
8709 SOUTHERN PINES
VIENNA VA 22180

-PHONE-
703 9388206
STUDYING CONVERSION METHODS:UNDER GRAD VA POLYTECH INST-BS JUNE 74

SUBCLASSES — SOLAR SUB SYSTEMS

AMERICAN CYANAMID CO
CHEM RESEARCH DIV
1937 WEST MAIN ST
STAMFORD CT 06904
DR GOTTFRIED HAACKE

MEDIUM TEMPERATURE SOLAR COLLECTOR SYSTEMS — COLLECTOR,REFLECTIVE SELECTIVE SURFACES

-PHONE-
203 3487331 X329

DEVELOPMENT OF TRANSPARENT ELECTRICALLY CONDUCTIVE HEAT REFLECTING COATING
RESEARCH PHASE-QUARTERLY PROGRESS REPORTS TO NSF/RANN

ARIZONA, UNIV OF
OPTICAL SCIENCE CTR
TUCSON AZ 85721
BERNHARD O SERAPHIN

HIGH TEMPERATURE SOLAR COLLECTOR SYSTEMS — COLLECTOR,ABSORPTIVE SELECTIVE SURFACES; COLLECTOR,REFLECTIVE SELECTIVE SURFACES

-PHONE-
602 8842263

NEW APPROACH TO A SELECTIVE SOLAR ENERGY CONVERTER USED TO TRANSFORM SOLAR RAD TO HIGH TEMP HEAT / A TWO-LAYERED CONSTRUCTION: TOP LAYER SEMICONDUCTOR MATERIAL, BOTTOM LAYER METAL FILM WITH HIGH REFLECTANCE. USE OF CHEMICAL VAPOR DEPOSITION TECHNIQUES TO APPLY SEMICOND MATERIAL FOR OPTICAL STRUCTURES

BOSTON COLLEGE
PHYSICS DEPT
140 COMMONWEALTH
CHESTNUT HILL MA 02167
P H FANG

SOLAR ELECTRIC POWER GENERATION

LOW COST POLYCRYSTALINE SILICON PHOTOVOLTAIC CELLS FOR LGE SOLAR POWER SYSTEMS -1ST YR OF STUDY:3 GROWTH TEC USED TO PROD POLYCR SIL 10 MICON THICK/CRYSTALS UNIFORM /ELECTRICAL CONDUCTIVITY 20-100 OMEGA CM CURRENTLY CHARACTERIZ CRYSTAL FILMS & JUNCTION INFO TECH WILL BE DEVEL

--------SUBCLASSES--------SOLAR SUB SYSTEMS--------

BOSTON UNIVERSITY
CHEMISTRY DEPT
685 COMMONWEALTH AVE
BOSTON MA 02215
N N LICHTIN
CHAIRMAN
-PHONE-
617 3532493

IDENTIFY & CHARACTERIZE INORGAN PHOTOREDOX SYSTEMS USE IN SOLAR PHOTO-GALVANIC CELLS OR IN PHOTOFORMATION OF FUELS/FUNDAMENTAL RESEARCH BY CHEM DEPT/APPLIED RES PERFORMED AT EXXON RESEARCH & ENGR CO,CORP RES ENERGY CONVERSION UNIT
OVER ALL GOAL PHOTOGALVANIC CELL WITH 5 PERCENT ENGINEER EFF

BROWN UNIVERSITY
ENGINEERING DEPT
UNIV HALL BOX 1867
PROVIDENCE RI 02912
J J LOFERSKI

SOLAR ELECTRIC POWER GENERATION

OBJEC:INVESTIGATION OF THIN FILM SOLAR CELLS BASED ON CU2S & TERNARY COMPOUNDS OF TYPE CUINS2 & CUINSE2 FOR LARGE SCALE/LOW COST/TERRESTRIL SOLAR UTILIZATION
'SOME PROBLEMS ASSOC WITH LGE SCALE PRODUCTION ELECTRIC POWER FROM SOL ENGY VIA PHOTOVOLTAIC EFFECT' ASME PUBLICATION,V Y,NOV 26-30 1972.

CALIF INSTITUT OF TECHNOLOGY
JET PROPULSION LAB
4800 OAK GROVE DRIVE
PASADENA CA 91103
PAUL A BERMAN

-PHONE-
213 3542816

R&D SPACE RELATED PHOTOVOLTAIC SOLAR POWER SYS & ADVANCED CONCEPTS OF LIGHT WEIGHT CELLS & ARRAYS IMPROVED EFFICIENCY RELIABILITY RADIATION RESISTANCE...I AM PRIMARILY CONCERNED WITH ADVANCED CONCEPT ASPECTS CANNOT SPEAK FOR ORGANIZATION...PUBLISHED: JPL TECHNICAL REPT 32-1558 'CONSIDERATIONS WITH RESPECT TO DESIGN OF SOLAR PHOTOVOLTAIC POWER SYS FOR TERRESTRIAL APPLICATIONS'...JPL TECHNICAL REPT 32-1573 'PHOTOVOLT SOLAR ARRAY TECHNOLOGY REQUIRED FOR THREE WIDE-SCALE GENERATING SYS FOR TERRESTRIAL APPLICATIONS:ROOFTOP SOLAR FARM & SATELLITE'.

----------SUBCLASSES----------SOLAR SUB SYSTEMS----------

CALIF, UNIV OF-BERKELEY
LAWRENCE BERKELEY LB
BERKELEY CA 94720
MICHAEL WAHIG
ENERGY&ENVIRO PROG

-PHONE-
415 8432740 X5787
415 8432740 X5001
DEVELOPMENT OF TECHNOLOGY TO POINT WHERE ECONOMIC FEASIBILITY ACHIEVD
STUDIES OF SURFACE STRUCTURE & ELECTRONIC PROPERTIES OF POLYCRYSTALLIN
PHOTOVOLTAIC MATERIALS & DEVICES

CARNEGIE MELLON UNIVERSITY
ELECTRICAL ENGR DEPT
PITTSBURG PA 15213
PROF A G MILNES

-PHONE-
412 6212600
GALLIUM ARSENIDE SOLAR CELLS FOR HIGH TEMP SPACE APPLICATIONS/HETERO-
FACE CELLS OF PALGAAS-PGAAS-NGAAS HAVE BEEN FABRICATED & PGAAS-NGAAS
CELLS STUDIED
TECHNICAL RESULTS PUBLISHED

CATHOLIC UNIVERSITY OF AMERICA
DEPT OF ELECT ENGR
4 ST ON MICH AVE N E
WASHINGTON DC 20017
CHARLES F PULVARI

-PHONE-
202 6355193
202 5262400
DEVELOP TEMP DEPENDENT DIELECTRIC THIN FILMS & UTILIZE THESE FOR SOLAR
ENGY CONVERSION

COLORADO UNIV OF-BOULDER SOLAR RADIATION MEASUREMENTS
ELECT ENGR
BOULDER CO 80302
FRANK BARNES

-PHONE-
303 4927327

--------SUBCLASSES--------SOLAR SUB SYSTEMS--------

COMSTAT LABORATORIES
APPLIED SCI DIV
P O BOX 115
CLARKSBURG MD 20734
DR EDMUND S RITTNER
DIRECTOR
-PHONE-
301 4284541
DEVELOP LIGHTWEIGHT & HIGHLY EFFICIENT SOLAR ENGY CONVERSION&STORAGE SYS SPACE APPLICATION/POSSIBLE TERRESTRIAL APP BEING CONSIDERED MAJOR ADVANCES MADE IN ENGY CONVERSION WITH COMSTAT VIOLET SOLAR CELL & NICKEL-HYDROGEN RECHARGEABLE BATTERY--HAVE USED SOLAR TO POWER COMM-UNICATIONS SATELLITES FROM EARLY BIRD THROUGH INTELSTAT IV

DOW CORNING
SEMICONDUCTOR PROD
12334 GEDDES RD
HEMLOCK MI 48626
LEON D CROSSMAN
MGR SOLID STATE RES
-PHONE-
517 6425201 X344

Subclasses: NON-TERRESTRIAL SOLAR POWER SYSTEMS; SOLAR ENERGY STORAGE; SOLAR ELECTRIC POWER GENERATION

DOW CORNING CORP
RESEARCH DEPT
S SAGINAW RD
MIDLAND MI 48640
H A CLARK

-PHONE-
517 6368676
PRODUCTION FLAT-STRIP SINGLE CRYSTAL SILICON/DEVELOPING PROCESS FOR CHEAPER SOLAR CELL GRADE SILICON/NSF SUBCONTRACT ON CRYSTAL GROWTH ALSO RESEARCH CONST MATERIALS FOR THERMAL COLLECTORS-RESIN PAINTS (SEALANTS) UNDER EXPOSURE TESTS

Subclasses: MEDIUM TEMPERATURE SOLAR COLLECTOR SYSTEMS
Solar sub systems: COLLECTOR,ABSORPTIVE SELECTIVE SURFACES

EXXON RESCH&ENGR CO
SOLAR ENGY CONV UNIT
LINDEN NJ 07036
J A ECKERT
PROJECT MANAGER

-PHONE-
201 4740100
PHOTOGALVANIC CELLS BASED ON IRON THIONINE

------SUBCLASSES------SOLAR SUB SYSTEMS------

FLORIDA, UNIV OF
ELECTRIC ENGR DEPT
GAINESVILLE FL 32601
ROBERT L BAILEY

SOLAR ELECTRIC POWER GENERATION
SOLAR ENGINES

-PHONE-
904 3920916
904 3920921

ELECTROMAGNETIC WAVE ENERGY CONVERTER-RESEARCH UTILIZING WAVE-LIKE PROPERTIES OF RADIATION INTERACTING WITH ABSORBER-CONVERTER ELEMENTS ROUGH SURFACE TEXTURE & YIELD D-C OUTPUT FOR FURTHER INFO SEE:
R L BAILEY,'PROPOSED NEW CONCEPT FOR A SOLAR ENGY CONVERTER,'JOURNAL OF ENGINEERING FOR POWER, APRIL 1974, PP7377
ALSO RESEARCHING: SOLAR ELECTRIC MOTORS, PROTYPE MOTORS BEING TESTED
SEE: DR E A FARBER, FLORIDA,UNIV OF

HARVARD UNIVERSITY
GRADUATE SCHOOL
CAMBRIDGE STATION
CAMBRIDGE MA 02138
B CHALMERS

LOW COST FABRICATION OF SILICON SOLAR CELLS: GROWTH OF SILICON SINGLE CRYSTAL RIBBONS TO BE USED IN CONTINUOUS PRODUCTION OF SILICON CELLS

I B M
T WATSON RESERCH CTR
PO BOX 218
YORKTOWN HEIGHTS NY 10598
ROBERT W KEYES

-PHONE-
914 9453000

DEVELOP HIGH EFF GALLIUM ARSENIDE PHOTOVOLTAIC SOLAR CELLS
WOODALL & HOVEL,'HIGH-EFFICIENCY GA1-X ALX AS-GAAS SOLAR CELLS,'APPL PHYS LETT, VOL 21, NO 8, OCT 1972.
TECHNICAL DISCUSSION OF ABOVE RESEARCH

--------SUBCLASSES--------SOLAR SUB SYSTEMS--------

INNOTECH CORP
181 MAIN ST
NORWALK CN 06851
RAY PENNOYER
V P

-PHONE-
203 8462041

LOUISIANA TECH UNIVERSITY — BIO-CONVERSION; WIND CONVERSION — STORAGE, FUEL CELLS
ELECTRICAL ENGINEER
PO BOX 4967,TECH STA
RUSTON LA 71270
DEAN L SMITH
ASSISTANT PROFESSOR
-PHONE-
318 2572262
PHOTOCELLS, METHANE PRODUCTION FROM ORGANIC WASTES, FUEL CELLS, AEOLIAN POWER

MITRE CORPORATION, THE
SOLAR ENERGY LAB
WESTGATE RESEARCH PK
MCLEAN VA 22101
JAMES S BURTON
DIRECTOR
-PHONE-
703 8933500

MITRE CORPORATION, THE — ALL SOLAR TECHNOLOGIES
SYSTEMS DEVELOP DIV
WESTGATE RESEARCH PK
MCLEAN VA 22101
FRANK R ELDRIDGE

-PHONE-
703 5368103
R&D&SYSTEMS ANALYSIS OF SOLAR ENERGY SYSTEMS...BUILDING A 1KW (PEAK) PHOTOVOLTAIC SYS FOR DEMO PURPOSES USE HYDROGEN AS ENERGY STORAGE MEDIA PROVIDE AC OR DC ELECT OR HYDROGEN FUEL GAS SEE ALSO:RICHARD GREELEY MITRE...PUBLISHED REPORTS AVAIL THROUGH NATL TECH INFO SERV

--SUBCLASSES--SOLAR SUB SYSTEMS--

NASA LANGLEY RESEARCH CTR
MAIL STOP 231A
HAMPTON VA 23665
DR EDMUND J CONWAY

-PHONE-
804 8273127
DEVELOPMENT OF HIGH EFFICIENCY GAAS SOLAR CELLS...RESEARCH IN PROGRESS DETERMINE EFFECT OF GAAS SURFACE PROPERTIES ON EFFICIENCY...ION IMPLA-NTED GAAS CELLS BEING STUDIED...GAAS DIFFUSED SOLAR CELLS BEING CONST

NASA LEWIS RESEARCH CTR
POWER ELECTRONICS BR
21000 BROOKPARK RD
CLEVELAND OH 44135
THOMAS J RILEY

-PHONE-
216 4334000 X288
EXAMINING THE EDGE ILLUMINATED MULTI JUNCTION SOLAR CELL AS IT RELATES TO USE UNDER CONCENTRATED ENERGY...PRELIMINARY DATA UP TO 2000 SUNS

NASA LEWIS RESEARCH CTR — HIGH TEMPERATURE SOLAR COLLECTOR SYSTEMS
M S 54-4
21000 BROOKPARK ROAD
CLEVELAND OH 44135
BERNARD L SATER

-PHONE-
216 4334000 X220
DEVEL TERRESTRIAL PHOTOVOLTAIC (& COMBINED THERMAL SYS) USING SOLAR CONCENTRATORS IN CONJ WITH THE EDGE ILLUMINATED MULTIJUNCT SOLAR CELL SATER, ET AL 'A NEW SOLAR CELL FOR HIGH INTENSITY APPLICATIONS,' INTL SOLAR ENERGY CONF, OCT 3-4 73, CLEVELAND OHIO

NATIONAL SCIENCE FOUNDATION — GOVERNMENT AGENCIES PROVIDING GRANTS
ADVNCED ENGY RES&TEC
WASHINGTON DC 20550
H RICHARD BLIEDEN
PROG MGR PHOTOVOLTAC

-PHONE-
202 6327364

SUBCLASSES — SOLAR SUB SYSTEMS

NOTRE DAME, UNIV OF
ELECT ENGR DEPT
NOTRE DAME IN 46556
WILLIAM B BERRY

-PHONE-
219 2837193
RESEARCH ON ELECTRONIC CONVERSION DEVICES...PUBLISHED:'ROLE OF DEFECTS IN DETERMINING ELECTRICAL PROPERTIES OF CDS THIN FILMS, I GRAIN BOUNDARIES & SURFACES,& II STACKING FAULTS,'JOURNAL OF APPLIED PHYSICS, VOL 43, NO 8, AUGUST 1972 PP 3516-3521 & 3522-3527.

PHILCO-FORD CORP
3939 FABIAN WAY
PALO ALTO CA 94303
HOWARD E POLLARD
SR ENG SPECIALIST

SOLAR RADIATION MEASUREMENTS
NON-TERRESTRIAL SOLAR POWER SYSTEMS

-PHONE-
415 3264350 X6329
RESPONSIBLE FOR ANALYSIS DESIGN & FABRICATION OF SOLAR CELL ARRAYS OF PHILCO FORD SATELLITE PROGRAMS...CONDUCT RESEARCH NECESSARY TO SUPPORT ARRAY DESIGN & PERFORMANCE ASSESSMENT INCLUDING ENVIRONMENTAL DEGRADATION OF PHOTOVOLTAIC DEVICES...RECENT REPORT AFAPL-TR-72-44 REAL TIME EFFECTS OF SPACE & NUCLEAR ENVIRONMENT ON LITHIUM-DOPED CELLS

RCA CORPORATION
D SARNOFF RESCH CTR
PRINCETON NJ 08540
PAUL RAPPAPORT
DIRECTOR M&P LAB

-PHONE-
609 4522700
OBJECTIVE TO CONTRIBUTE TO TECH & UNDERSTANDING OF LOW COST LARGE AREA SOLAR CELLS...RESEARCH OVER PAST 20 YEARS...NEW VMJ CELL DEVELOPED SEVERAL NEW IDEAS ARE BEING INVESTIGATED FOR LOW COST CELLS

SUBCLASSES — SOLAR SUB SYSTEMS

RUTGERS UNIVERSITY
ELECTRIC ENGR DEPT
NEW BRUNSWICK NJ 08903
WAYNE ANDERSON

-PHONE-
201 9323871
FABRICATION OF SCHOTTKY BARRIER PHOTOVOLTAIC SOLAR CELLS...USING SCHOTTKY BARRIER DIODE PRINCIPLES DEVELOP MORE EFF CHEAPER PHOTOVOLTAC

SOUTHERN METHODIST UNIVERSITY
INSTITUTE OF TECHN
DALLAS TX 75275
TING L CHU
PROFESSOR

-PHONE-
214 6923014
214 6923017
DEVELOPMENT OF LOW COST THIN FILM POLYCRYSTALLINE SILICON SOLAR CELLS SUITABLE FOR LARGE SCALE TERRESTRIAL UTILIZATION...SILICON FILMS HAVE BEEN DEPOSITED ON STEEL SUBSTRATES & RESISTIVITY & CONDUCTIVITY TYPE OF DEPOSIT HAVE BEEN CONTROLLED REASONABLY WELL

STANFORD UNIVERSITY
NEWS SERVICE
STANFORD CA 94305
ROBERT W BEYERS *
DIR NEWS SERVICE

SOLAR HEAT,COOL,&ELECTRIC POWER GENERATION
PROVIDE AUDIO SERVICES
PROVIDE AUDIO & VISUAL SERVICES

-PHONE-
415 3212300 X2558

--------SUBCLASSES--------SOLAR SUB SYSTEMS--------

STANFORD UNIVERSITY
MATERIALS SCIENCE
STANFORD CA 94305
RICHARD H BUBE
PROF MAT SCI AND EE

SOLAR ELECTRIC POWER GENERATION
PUBLISH ORIGINAL RESEARCHED INFORMATION
PUBLISH ORIGINAL RESEARCH + REPRINTS

-PHONE-
415 3212300 X2535
415 3215796

PHOTOVOLTAIC HETEROJUNCTION OF II-VI COMPOUNDS FOR SOLAR CELLS

J APPL PHYS 41 1694 1970
APPL PHYS LETT 19 459 1971
J ELECTROCHEM SOC 119 936 1972
7 47 1968 8 1 1970 9 118 1972
J APPL PHYS 41 3731 1970
J APPL PHYS 43 2839 1972
IEEE PHOTOVOLTAIC SPECIALISTS CONF
10 85 1973

TYCO LABORATORIES INC
SILICON DEPARTMENT
16 HICKORY DRIVE
WALTHAM MA 02154
DAVID N JEWETT
MANAGER
-PHONE-
617 8902400

SILICON RIBBON GROWTH DEMONSTRATED...PHYSICAL & ELECTRICAL QUALITY IMPROVED

TYCO LABORATORIES INC
16 HICKORY DRIVE
WALTHAM MA 02154
A I MLAVSKY

-PHONE-
617 8902400

R &D OF TYCO-DEVELOPED EFG PROCESS TO PRODUCE LOW COST SILICON RIBBON SOLAR CELLS...BASIC FEASIBILITY OF TECHNIQUE ESTAB...CELLS HAVE BEEN MADE FROM EFG SILICON RIBBON

SUBCLASSES — SOLAR SUB SYSTEMS

US AFCRL-PHF — ORGANIC SEMI-CONDUCTORS
HANSCOM FIELD
BEDFORD MA 01730
NICHOLAS F YANNONI

-PHONE-
617 8612265
IMPROVE EFFICIENCY & STABILITY OF SILICON SOLAR CELLS...
PILOT PLANT FOR SILICON CELL FABRICATION & TEST HAS BEEN SET UP...
ORGANIC PHOTOVOLTAIC SOLAR CELLS RESEARCH ON CONDUCTIVITY PROBLEMS OF
ORGANIC COMPOUNDS HAS BEEN STUDIED OVER 5 YEARS

VERMONT, UNIV OF — WIND CONVERSION — STORAGE, HYDROGEN
ELECT ENGR DEPT
BURLINGTON VT 05401
LLOYD M LAMBERT

-PHONE-
802 6563332
OPTO-MECHANICAL ENERGY CONVERTER-USES CONVENTIONAL SOLAR CELLS OUTPUT
AMPLIFIED BY WIND POWER HYDROELECTRIC OCEAN WAVE MOTION ETC..PURPOSE
TO INCREASE ATTRACTIVNESS OF SOLAR CELLS AS ELECT ENGY CONVERTER
SOLAR-CATALYTIC HYDROGEN GENERATOR-THERMALLY DECOMPOSE H2O BY SOLAR AS
HEAT SOURCE..CONCEPTUAL STAGE

WESTINGHOUSE RESEARCH LABS
BEULAH RD
PITTSBURG PA 15235
R K RIEL *

-PHONE-
412 2563614

WESTINGHOUSE RESEARCH LABS
SOLID STATE RES
BEULAH RD
PITTSBURGH PA 15235
F A SHIRLAND

-PHONE-
412 2563613
R & D VERY LOW COST THIN FILM SOLAR CELLS FOR LARGE SCALE HARVESTING
OF SOLAR ENERGY

--------SUBCLASSES--------SOLAR SUB SYSTEMS--------

FLINDERS UNIV OF S AUSTRALIA PUBLISH ORIGINAL RESEARCHED INFORMATION COLLECTOR,ABSORPTIVE SELECTIVE SURFACES
BEDFORD PARK
SOUTH AUSTRALIA AUSTRALIA
J O'M BOCKRIS

-PHONE-
76 0511
TO DEVELOP CHEAP SILICON TIME DEPENDENCE OF THE PROPERTIES OF SELECTVE
COATINGS & CHEAP ELECTROLYSERS

JOHN B AYER
P ENG
2013 DEERHURST CRT
OTTAWA ONT K1J842 CANADA

DESIGN & DEVELOPMENT OF SILICON SOLAR CELL POWER SYS FOR SATELLITES
(COMMUNICATION) THESIS: DIRECT CONVERSION OF SOLAR TO ELECTRICAL ENGY

READING UNIV OF
CHEMISTRY
READING ENGLAND
D BRYCE-SMITH

PHOTOELECTROCHEMICAL ENERGY CONVERSION

THE ROYAL INSTITUTION
LONDON ENGLAND
M D ARCHER

PHOTOELECTROCHEMICAL ENGY CONVERSION

----------SUBCLASSES----------SOLAR SUB SYSTEMS----------

UK ATOMIC ENERGY AUTHORITY
REACTOR GROUP
RISLEY
WARRINGTON LANCASHIR ENGLAND
N KIRKER
TECH OPERATIONS DIR
-PHONE-
WAR 31244 X2290
R &D PROVING MECHANISMS FOR USE IN APPLICATIONS SATELLITES EG SOLAR
PANEL DRIVE MECH

C N E S
91 BRETIGNY BOX 04
FRANCE
WOLFGANG PALZ

-PHONE-
4909220
DEVELOPMENT OF THIN FILM SOLAR CELLS

CNTRL SALT&MARNE CHEM RES INST — SOLAR WATER DISTILLATION
BHAVNAGAR (GUJARAT) INDIA — SOLAR ELECTRIC POWER GENERATION
DR R L DATTA

-PHONE-
3960
SELECTIVE FILMS SOLAR CELLS-DEVELOPMENT OF SOLAR GADGETS FOR: ICE
PRODUCTION KITS FOR LIGHTING SMALL POWER GEN PRODUCING DISTILLED H2O

GOVT OF INDIA YOJANA BHAVAN — SOLAR WATER DISTILLATION
POWER & ENERGY SECT — SOLAR WATER HEATING
PARLIAMENT STREET — SOLAR POWERED TRANSPORTATION
NEW DELHI 1 INDIA — PLANNING OFFICES, FEDERAL
MANAS KUMAR CHATTERJEE
DIR ENGY PLANN COMM
-PHONE-
381109
OUR SECTOR RESPONSIBLE TO ENSURE ENERGY SUPPLY AT MINIMUM COST CONSIS-
TENT WITH GROWTH IN OTHER SECTORS OF ECONOMY...STUDIES BEING CARRIED
OUT TO ASCERTAIN ECONOMICS OF SOLAR CELLS FUEL CELLS & HIGH ENERGY
DENSITY BATTERIES AS POWER SOURCES IN ISOLATED RURAL AREAS AS WELL AS
ELECTRIC VEHICLE PROPULSION

---SUBCLASSES---SOLAR SUB SYSTEMS---

JADAVPUR UNIVERSITY
ELEC & TELECOMM ENGR
CALCUTTA 700032 INDIA
DR J S CHATTERJEE *
PROF IN CHARGE

JADAVPUR UNIVERSITY
ELEC & TELECOMM ENGR
CALCUTTA 700032 INDIA
MANISH KR MUKHERJEE
PROFESSOR

GOAL TO FABRICATE LOW COST SOLAR CELLS PUBLISHED: 'SOLAR BATTERIES & AGRO-ECONOMY OF DEVELOPING NATIONS' J. SCI. INDUST. RESEARCH, VOL 28, PP 81-5, 1969.

KALYANI, UNIV OF
DEPT OF PHYSICS
KALYANI NADIA
WEST BENGAL INDIA
HIRANMAY SAHA

PHOTOVOLTAIC SYS-DEVELOPMENT OF CHEAP & EFFICIENT SOLAR CELLS SUITABLE FOR TERRESTRIAL APPLICATIONS PUBLISHED:'SECONDARY IONISATION AND ITS POSSIBLE BEARINGS ON THE PERFORMANCE OF A SOLAR CELL' SOLID STATE ELECTRONICS VOL 15, 1972, P 1389

KALYANI, UNIV OF
DEPT OF PHYSICS
KALYANI NADIA
WEST BENGAL INDIA
PROF D C SARKAR *
HEAD

ISRAEL INSTITUTE OF TECHNOLOGY
TECHNION
HAIFA ISRAEL
M S ERLICKI

-PHONE-
04 235151
APPLICATION OF SOLAR CELLS IN AUTONOMOUS SYS

---SUBCLASSES---SOLAR SUB SYSTEMS---

ISRAEL INSTITUTE OF TECHNOLOGY
TECHNION
HAIFA ISRAEL
S GAVRIL

-PHONE-
04 235151

ISRAEL INSTITUTE OF TECHNOLOGY
TECHNION
HAIFA ISRAEL
D SCHIEBER

-PHONE-
04 235151

	SUBCLASSES	SOLAR SUB SYSTEMS
ALLIED CHEMICAL CORP MATERIAL RESERCH CTR P O BOX 1021R MORRISTOWN NJ 07960 A J NOZIK		STORAGE, HYDROGEN
AUST INVENT & DEVELOPMENTS PLD EVELOPMENTS P O BOX 227 FILLMORE UT 84631 GORDON D GRIFFIN MANAGING DIRECTOR -PHONE- 801 7436817	MEDIUM TEMPERATURE SOLAR COLLECTOR SYSTEMS SOLAR WATER HEATING SOLAR SPACE HEATING RETROFITTING, PASSIVE SYSTEMS NEW CONSTRUCTION, PASSIVE SYSTEMS GREENHOUSES	STORAGE, SOLID MATERIALS COLLECTOR,REFLECTIVE SELECTIVE SURFACES COLLECTOR,ABSORPTIVE SELECTIVE SURFACES HEAT TRANSFER SYSTEMS, COLLECTOR TO STORAGE HEAT RECOVERY, STORAGE TO UTILIZATION STORAGE, LIQUIDS
BATTELLE COLUMBUS LABS 505 KING AVENUE COLUMBUS OH 43201 RICHARD A NATHAN PRINCIPAL CHEMIST -PHONE- 614 2993151 X2049	NON-TERRESTRIAL SOLAR POWER SYSTEMS TOTAL SYSTEMS INTEGRATION ALL SOLAR TECHNOLOGIES SOLAR WATER HEATING SOLAR SPACE HEATING & COOLING SOLAR HEAT,COOL,&ELECTRIC POWER GENERATION	

ALLIED CHEMICAL CORP: RESEARCH INITIATED FOR DIRECT USE OF SOLAR RADIATION FOR PRODUCTION OF HYDROGEN BY PHOTOLYSIS OF H2O & OTHER ENERGY CONVERSION SYSTEMS

AUST INVENT & DEVELOPMENTS PLD: PROJECT INVOLVES DESIGN & TESTING OF REMOTE SOLAR STORAGE HEATER PRIME AIM REQUIREMENTS:1)EASY MASS PRODUCTION 2)MINIMUM ONSITE INSTALATION 3)COST EFFECTIVE 4)VERY LOW UPKEEP 5)MODULAR CONSTRUCTION ON UNBREAK-ABLE LONG LIFE 6)PRICE 10-15 PERCENT OF HEATED BUILDING 7)SUPPLY 70-95 PERCENT HEATING DEMAND 8)STORAGE 10-15 DAYS 9)ASTHETICALLY ACCEPTABLE 10)SUITABLE ALL AREAS HEATING NEEDS 11)MINIMUM SPACE REQUIREMENT WE ARE DEVELOPERS WITH INTEREST IN APPLICATIONS TO RESIDENTIAL COMM-ERCIAL&GREENHOUSE STRUCTURES

BATTELLE COLUMBUS LABS: SEE ALSO: JAMES EIBLING, & ROBERT SCHWERZEL,BATTELLE COLUMBUS LABS

			SUBCLASSES	SOLAR SUB SYSTEMS
DELAWARE, UNIV OF			MEDIUM TEMPERATURE SOLAR COLLECTOR SYSTEMS	STORAGE, PHASE CHANGE MATERIALS
ENERGY CONVERSION			SOLAR SPACE HEATING	
NEWARK	DE	19711	SOLAR SPACE COOLING	
MARIA TELKES			RETROFITTING, TECHNOLOGIES	
ADJUNCT PROFESSOR			HOMES	
			SINGLE FAMILY, EXPERIMENT	

-PHONE-
302 7388481
ACTIVE: SOLAR-THERMAL STORAGE WITH PHASE-CHANGE MATERIALS FOR HEATING & COOLING. DESIGNED HEATING SYSTEM OF SOLAR HOMES BUILT AT DOVER, MASS PRINCETON,N J & THERMAL STORAGE OF UNIV OF DELAWARE BUILDING: SOLAR 1. EXPERIMENTAL WORK: SOLAR COLLECTORS SELECTIVE BLACKS. BUILT SOLAR STILLS & SOLAR COOKING OVENS...LIST OF PUBLICATIONS AVAIL.
SEE ALSO: DR KARL W BOER, DELAWARE UNIV OF,& L R CRITTENDON,DELAWARE UNIV OF

FORD MOTOR COMPANY
SCIENTIFIC RES STAFF
ADVANCED TECH APP/RA
DEARBORN MI 48121
J V PETROCELLI

ELECTROLYTE & ELECTRODES FOR SODIUM-SULFUR BATTERY FOR IMPROVED ENERGY TECHNOLOGY/FABRICATION OF CERAMIC ELECTROLYTE FROM POWERS & DESIGN OF SULFUR ELECTRODE

OKLAHOMA STATE UNIVERSITY			WIND CONVERSION	ELECTROLYSIS
ELECTRICAL ENGG			TOTAL SYSTEMS INTEGRATION	STORAGE, FUEL CELLS
ENGINEERING SOUTH			SOLAR ELECTRIC POWER GENERATION	STORAGE, GAS
STILLWATER	OK	74074	SOLAR TECHNOLOGY APPLICATION OTHER	STORAGE, OTHER
H JACK ALLISON			PUBLISH ORIGINAL RESEARCHED INFORMATION	TURBINES
ASSOC PROF ELEC ENGG			PUBLISH ORIGINAL RESEARCH + REPRINTS	

-PHONE-
405 3726211 X6476
MORE THAN 20 PUBLISHED TECHNICAL PAPERS IN AREAS OF ENERGY CONVERSION & ENERGY STORAGE...INVOLVED SINCE 1961 IN SOLAR & WIND ENGY RESEARCH 5 PATENTS...PRIMARY INTERESTS ELECTROLYSIS SYS FIELD MODULATED GENERAT SYS & SOLAR ENGY SYS

	SUBCLASSES	SOLAR SUB SYSTEMS
STONEWALLEN ENGINEERING RT 2 BOX 381A DELTA CO 81416 R TODD MAC DOUGALL ENGINEER	HIGH TEMPERATURE SOLAR COLLECTOR SYSTEMS BIO-CONVERSION INORGANIC SEMI-CONDUCTORS TOTAL SYSTEMS INTEGRATION SOLAR WATER HEATING	STORAGE, FUEL CELLS HEAT ENGINE, STERLING ELECTROLYSIS

PAST WORK: FAST RECOVERY HOT H2O HEATER CONSISTING OF 2 CONCENTRATING TUBES...PRESENT WORK: DEVELOPING SOLAR PUMP FOR FARM IRRIGATION..NO WORKING PROTOTYPES YET

Here comes the sun,
Source of life, breath is warm
Do you need me, as I need you?

I look up to you, as guide for me,
you're so free, dependably,
heat is the color of life, light.

So I worship you at dawn, at dusk
But daytime passes, nary a nod.
Time of my life and the lemon escapes.

Til I need you again, to feed me
my candles burn low, slow
to shoot up again, as green graceful stem.

Here comes the sun.

Fred

----SUBCLASSES---- ----SOLAR SUB SYSTEMS----

			SUBCLASSES	SOLAR SUB SYSTEMS
RODNEY F KUHARICH			WIND CONVERSION	COLLECTOR,MULTI-FUNCTION IN DWELLING SYS
327 E. PLATTE			ALL SOLAR TECHNOLOGIES	COLLECTOR,ALSO FUNCTION AS STRUCTURES SKIN
COLORADO SPRINGS	CO	80902	SOLAR SPACE HEATING	STORAGE, LIQUIDS
			ACADEMIC BUILDINGS	
			50-500 THOUSAND SQ FT,PROPOSED	
			PROVIDE INFORMATION SEARCHES TO CLIENTS	

-PHONE-
303 4711111
303 6337011
SOLAR HOME HEATING CONSULTANT, CURRENT PROJECTS: SYS DESIGN FOR WRIGHT INGRAHAM INSTITUTE PERMANENT STRUCTURE UTILIZING SUN WIND & WASTE; CONSTRUCTION OF MONITORING & MEASURING STATION FOR SOLAR&WIND ENERGY; SOLAR HEATING FEASIBILITY STUDY;RESEARCH ADM PHOENIX CORP COLORADO SPRINGS CO

CAMERON B LAIRD
R R 3
RANSSELAER IN 47906

-PHONE-
317 4932456
219 8663628
TOTAL ENGY SYS-INCORPORATION OF SOLAR ENGY ELEMENTS INTO ECONOMICALLY FEASIBLE RESIDENCES...PROCESSING & DIRECTING OF ENERGY CONVERSION-ENVIRONMENTAL CONCERN INFO

THOMAS WATSON			SOLAR SPACE HEATING & COOLING
1144 CANYON RD			SOLAR ELECTRIC POWER GENERATION
SANTA FE	NM	87501	RETROFITTING, PRACTICING

-PHONE-
505 9881800
DESIGN OF OPTIMUM SELF-CONTAINED HOUSING UNITS LOW COST UNDER $10.PER SQFT...HAVE DESIGNED & LIVED IN TWO RETROFIT SYS OF LOW COST BOTH WORKED USED WASTE HEAT FROM ATTACHED GREENHOUSE...NOW DEVELOP PORTABLE FACTORY TO BUILD LOW COST RESIDENCES & RELATED STRUCTURES AT SITE... I AM PLEASED TO SHARE IDEAS PLEASE CALL

	SUBCLASSES	SOLAR SUB SYSTEMS
ALABAMA,UNIV OF/IN HUNTSVILLE	WIND CONVERSION	
CTR ENVIRONMENT STUD	SOLAR WATER DISTILLATION	
P O BOX 1247	SOLAR WATER HEATING	
HUNTSVILLE AL 35807	SOLAR SPACE HEATING & COOLING	
D L CHRISTENSEN		

-PHONE-
205 8956257
DEVELOP SELF-SUFFICIENT HOUSING & COMMUNITIES USING SOLAR & WIND ENGY&
RECYCLING SYSTEMS:COMPONENT & SYSTEMS DESIGN COMPLETE/MODELS COMPLETE
BREADBOARD TEST UNITS UNDER CONST PHASE I TESTING UNDER WAY
PHASE IV SCHEDULED FOR SUMMER 1975

	SUBCLASSES	SOLAR SUB SYSTEMS
BALL BROTHERS RESEARCH CORP	SOLAR RADIATION MEASUREMENTS	HEAT TRANSFER SYSTEMS, COLLECTOR TO STORAGE
AEROSPACE DIV	MEDIUM TEMPERATURE SOLAR COLLECTOR SYSTEMS	COLLECTOR,REFLECTIVE SELECTIVE SURFACES
P O BOX 1062	HIGH TEMPERATURE SOLAR COLLECTOR SYSTEMS	HEAT RECOVERY, STORAGE TO UTILIZATION
BOULDER CO 80302	WIND CONVERSION	STORAGE, PHASE CHANGE MATERIALS
M W FRANK	SOLAR ENERGY STORAGE	STORAGE, LIQUIDS
PROG MGR	SOLAR HEAT,COOL,&ELECTRIC POWER GENERATION	STORAGE, GAS
-PHONE-		ELECTROLYSIS
303 4414627		

PUBLISH INFORMATIONAL REPRINTS

	SUBCLASSES	SOLAR SUB SYSTEMS
ENGINEERS COLLABORATIVE LTD	SOLAR HEAT,COOL,&ELECTRIC POWER GENERATION	
101 E ONTARIO		
CHICAGO IL 60611		
L A BIHLER		
EXEC V P		

-PHONE-
312 6449600
INTEGRATED SYS DESIGN TO ENHANCE FEASIBILITY OF SOLAR POWERED BLDGS

	SUBCLASSES	SOLAR SUB SYSTEMS
NEW ALCHEMY INSTITUTE EAST	BIO-CONVERSION	
P O BOX 432	WIND CONVERSION	
WOODS HOLE MA 02543	PUBLISH LAY PERIODICALS	
JOHN TODD	PUBLISH ORIGINAL RESEARCHED INFORMATION	
PRES		

-PHONE-
617 5632655
PROJECT THE ARK INTEGRATED SYS USING SOLAR WIND & AQUICULTURE...PROTO-
TYPE MINI ARK COMP TAKING DATA...FOR $25/YR SUSTAINING MEMBERS RECEIVE
QUARTERLY JOURNAL & CALENDAR

	SUBCLASSES	SOLAR SUB SYSTEMS
NEW LIFE ENVIRO DESIGNS INST P O BOX 648 KALAMAZOO MI 49005 RICHARD TILMANN	MEDIUM TEMPERATURE SOLAR COLLECTOR SYSTEMS SOLAR SPACE HEATING GREENHOUSES	

-PHONE-
616 3452948
INTEGRATION OF LOW-COST SOLAR SYS INTO COMPREHENSIVE DEVELOPMENT STRATEGY FOR DECENTRALIZED COMMUNITIES...HAVE DEVELOPED THIN FILM FLAT PLATE COLLECTOR FOR DUEL HEATING-GREENHOUSE USE...IMPROVED DESIGN IN PROGRESS

NEW LIFE FARM DRURY MO 65638 TED LANDERS FARMER	MEDIUM TEMPERATURE SOLAR COLLECTOR SYSTEMS WIND CONVERSION RETROFITTING, PRACTICING HOMES SINGLE FAMILY, UNDER CONSTRUCTION	COLLECTOR,MULTI-FUNCTION IN DWELLING SYS COLLECTOR,ALSO FUNCTION AS STRUCTURES SKIN HEAT RECOVERY, STORAGE TO UTILIZATION STORAGE, LIQUIDS

-PHONE-
417 2612553
WE ARE INTEGRATING WIND,GRAVITY,SOLAR & PYRAMID ENERGIES TO ACHIEVE A TECHNOLOGICALLY LOW LEVEL, EFFICIENT SYSTEM OF HIGH QUALITY (SENSITIVE CRYSTALLIZATION ANALYSIS) FOOD PRODUCTION. WE WILL BE OPERATED BY & DISTRIBUTING CURRENT NOT PAST ENERGIES. WE WILL BE VERY COMPETITIVE IN THE MARKETPLACE DUE TO LOW COST OF PRODUCTION & HIGH YIELDS & QUALITY

NORTH MOUNTAIN COMMUNITY RR 2 BOX 207 LEXINGTON VA 24450 WYNN SOLOMON	WIND CONVERSION SOLAR WATER DISTILLATION SOLAR WATER HEATING SOLAR SPACE HEATING & COOLING SOLAR ELECTRIC POWER GENERATION	STORAGE, METHANE STORAGE, LIQUIDS STORAGE, FUEL CELLS TURBINES STORAGE, HYDROGEN

-PHONE-
703 4637095
COMMUNITY WITH PROJECTED POP OF 50 PEOPLE WISH TO LIVE IN HARMONY WITH EARTH SELFSUFF..CONSTRUCTING SOLAR H2O & SPACE HEATER WITH STORAGE FOR WASHHOUSE.PLANNED: H2O PREHEATER FOR MAIN HOUSE WINDMILL FOR ELECTRICY H2O WHEEL FOR ELECT SOLAR POND HEAT FOR NEW DORM...FURTHER ALT ENGY SOURCES BEING PLANNED

	SUBCLASSES	SOLAR SUB SYSTEMS
RHODE ISLAND, UNIV OF MECHANICAL ENGR DEPT KINGSTON RI 02881 DR RICHARD C LESSMANN	MEDIUM TEMPERATURE SOLAR COLLECTOR SYSTEMS WIND CONVERSION	

STUDY & BUILD INTEGRATED ENERGY SYS IN NEW ENGLAND USEING FLAT PLATES & WIND MILLS

SANDIA LABORATORIES DIVISION 8184 LIVERMORE CA 94550 ALAN C SKINROOD SUPERVISOR	HIGH TEMPERATURE SOLAR COLLECTOR SYSTEMS SOLAR ENERGY STORAGE SOLAR SPACE HEATING & COOLING SOLAR ELECTRIC POWER GENERATION SOLAR TECHNOLOGY APPLICATION OTHER PUBLISH ORIGINAL RESEARCHED INFORMATION	COLLECTOR,OTHER COLLECTOR,REFLECTIVE SELECTIVE SURFACES COLLECTOR,MECHANICAL SYSTEMS STORAGE, SOLID MATERIALS STORAGE, PHASE CHANGE MATERIALS STORAGE, LIQUIDS

-PHONE-
415 4552501

RESEARCH & DEVELOPMENT ON SOLAR ENERGY COLLECTION & UTILIZATION METH-ODS EMPHASIS ON HIGH TEMP APPLICATIONS

SANDIA LABS
ORGANIZATION 5717
ALBUQUERQUE NM 87110
RAYMOND HARRISON

-PHONE-
505 2648669

DETERMINE FEASABILITY OF SOLAR COMMUNITY-ANALYTICAL & EXPLORATORY STAGE/MODEL CONSTRUCTION STARTING

SANDIA LABS DIV 4736 BOX 5800 ALBUQUERQUE NM 87115 R P STROMBERG	MEDIUM TEMPERATURE SOLAR COLLECTOR SYSTEMS HIGH TEMPERATURE SOLAR COLLECTOR SYSTEMS SOLAR SPACE HEATING	

-PHONE-
505 2648170

R&D: SOLAR THERMAL ENERGY SYS CURRENTLY RESEARCHING SOLAR TOTAL ENERGY CONCEPT

-----SUBCLASSES-----SOLAR SUB SYSTEMS-----

US NAVAL WEAPONS CENTER
CHINA LAKE CA 93555
RICHARD D ULRICH
CODE 45701

-PHONE-
714 9397384
DEVELOP TOTAL ENERGY COMMUNITY

Organization	Subclasses	Solar Sub Systems
YOUNGSTOWN STATE UNIV ELECT ENGR DEPT YOUNGSTOWN OH 44503 CHARLES K ALEXANDER, JR, PHD DIR SOLAR ENGY TSK G	ALL SOLAR TECHNOLOGIES SOLAR HEAT,COOL,&ELECTRIC POWER GENERATION RETROFITTING, TECHNOLOGIES NEW CONSTRUCTION, RESEARCHING PUBLISH LAY + TECHNICAL PERIODICALS	COLLECTOR,MULTI-FUNCTION IN DWELLING SYS COLLECTOR,MECHANICAL SYSTEMS COLLECTOR,OTHER STORAGE, GAS STORAGE, OTHER STEAM ENGINES TURBINES

-PHONE-
216 7461851 X425
216 7592698
SOLAR ENERGY ENGR AREA OF SPECIALIZATION IN DEPT OF ELECT ENGR LEADING TO MASTER OF SCIENCE IN ENGR...SPECIALIZED COURSES IN SOLAR ENGR ON UNDERGRAD LEVEL...PROGRAMS DEVELOPED WITH DR ROST, DIR,SETG, ACTIVE IN SE R&D, SYS EMPHASIS CONTRIBUTING EDITOR TO TERRASTAR

Organization	Subclasses	Solar Sub Systems
YOUNGSTOWN STATE UNIV ELECT ENGR DEPT YOUNGSTOWN OH 44503 DR D F ROST DIR SOLAR ENGY TSK G	ALL SOLAR TECHNOLOGIES SOLAR HEAT,COOL,&ELECTRIC POWER GENERATION RETROFITTING, TECHNOLOGIES NEW CONSTRUCTION, RESEARCHING PUBLISH LAY + TECHNICAL BOOKS	COLLECTOR,MULTI-FUNCTION IN DWELLING SYS COLLECTOR,PASSIVE SYSTEMS HEAT TRANSFER SYSTEMS, COLLECTOR TO STORAGE HEAT RECOVERY, STORAGE TO UTILIZATION STORAGE, SOLID MATERIALS STORAGE, LIQUIDS ELECTROLYSIS

-PHONE-
216 7461851 X425
216 5336128
SOLAR ENERGY ENGR AREA OF SPECIALIZATION IN DEPT OF ELECT ENGR LEADING TO MASTER OF SCIENCE IN ENGR...SPECIALIZED COURSES IN SOLAR ENGR ON UNDERGRAD LEVEL...PROGRAMS DEVELOPED WITH DR ALEXANDER, DIR, SETG, ACTIVE IN SE R&D, SYS EMPHASIS TO STUDY HARDWARE-HUMAN FACTORS INTERACTION

		SUBCLASSES	SOLAR SUB SYSTEMS
MANITOBA 410- 352 DONALD ST WINNIPEG MANITOBA JOHN M GLASS CO ENVIRONMENTAL DES -PHONE- 204 9424735	CANADA	SOLAR SPACE HEATING NEW CONSTRUCTION, PASSIVE SYSTEMS NEW CONSTRUCTION, MECHANICAL SYSTEMS 6-14 FAMILY, PROPOSED UP TO 50 THOUSAND SQ FT,PROPOSED MORTGAGE INSTITUTIONS	
CCR-EURATOM SCIENTIF DIRECTORATE ISPRA (VARESE) JOACHIM GRETZ -PHONE- 003 9332780 X131 TERRESTRIAL THERMAL COLL FOR HABATAT BIOLOGICAL CONV H2 PRODUCTION PROTOTYPE BEING DEVELOPED	ITALY	MEDIUM TEMPERATURE SOLAR COLLECTOR SYSTEMS BIO-CONVERSION SOLAR ENERGY STORAGE	STORAGE, HYDROGEN

Aspens gently
turning
set me yearning
thoughts of far away places.
and faces

of highlands without roads
and paths five toes wide
leading me home
gently
home

Carolyn

----SUBCLASSES----SOLAR SUB SYSTEMS----

KJERSTI A COCHRAN *
1326 PERALTA AVE
BERKELEY CA 94702

CREATE INNOVATIVE WAYS OF COLLECTING & CONVERTING SOLAR ENGY IN ARCH-
ITECTURE OF LIVING & WORKING ENVIRO WITH AWARENESS OF TOTAL ECOLOGY OF
HUMAN NEED

LEON P GAUCHER *
CONSULTANT
R D 1
FISHKILL NY 12524

-PHONE-
914 6312465
ENERGY OVERVIEW AND IMPORTANT ROLE SOLAR WILL PLAY IN OUR ENERGY
FUTURE PROJECTIONS
SEE:'ENERGY IN PERSPECTIVE,'CHEMICAL TECHNOLOGY,VOL 1,MARCH 1971,PP153
-158.
'THE SOLAR ERA,'MECHANICAL ENGR,AUGUST 1972,PP9-12.

MARTIN WEBER
67-29-218 STREET
BAYSIDE NY 11364

-PHONE-
212 2249352

AEROSPACE CORP
ADVANCED TECH APPLIC
P O BOX 92957
LOS ANGELES CA 90009
DR A B GREENBERG

1)DEV METHOD TO ASSESS ROLE OF SOLAR ENGY IN NAT ENGY NEEDS 2)DEV METH
FOR COMPAR ANALYSIS FOR COMPETING SOLAR THERMAL CONV SYS 3)APPLY METH
TO ACCESS POT VARIOUS SOLAR SYS FORUSE IN SOUTH CALIF

----SUBCLASSES----SOLAR SUP SYSTEMS----

ARCHITECTS COLLABORATIVE INC
46 BRATTLE STREET
CAMBRIDGE MA 02138
SARAH P HARKNESS

MEDIUM TEMPERATURE SOLAR COLLECTOR SYSTEMS
WIND CONVERSION
INORGANIC SEMI-CONDUCTORS

-PHONE-
617 8684200 X333
TO INTEGRATE ARCHITECTURAL DESIGN WITH USE OF SOLAR ENERGY DESIGN
COURSES TAUGHT: STUDIO HELIOTECHNICA BOSTON ARCH CTR BY BROOKS CALVIN
BUILT FORM AS AFFECTED BY USE OF SOLAR ENERGY HARVARD GRAD SCH DESIGN
BY S HARKNESS...CONDUCTED SOLAR ENERGY WORKSHOP HARVARD GRAD SCH DESGN
FEB 20 1974

ARTHUR D LITTLE, INC.
3601 S WADSWORTH
DENVER CO 80235
JOHN K REYNOLDS
CONSULTANT

GREENHOUSES
HEALTH/HOSPITAL BUILDINGS

-PHONE-
303 9854505
303 9854741
PUBLIC AFFAIRS PROJECT SECTION INTEREST IN SOLAR FINANCIAL LEGAL
POLITICAL ENVIRONMENTAL SOCIAL & PSYCHOLOGICAL CONSIDERATIONS

BATTELLE COLUMBUS LABS
UNCONVENTIONALENERGY
505 KING AVENUE
COLUMBUS OH 43201
JAMES A EIBLING
PROGRAM MANAGER
-PHONE-
614 2993151
SEE ALSO: RICHARD NATHAN & ROBERT SCHWERZEL, BATTELLE COLUMBUS LABS

	SUBCLASSES	SOLAR SUB SYSTEMS
BATTELLE COLUMBUS LABS	BIO-CONVERSION	
ORGANIC CHEMISTRY	INORGANIC SEMI-CONDUCTORS	
505 KING AVENUE	ORGANIC SEMI-CONDUCTORS	
COLUMBUS OH 43201	SOLAR ENERGY STORAGE	
DR ROBERT E SCHWERZEL	TOTAL SYSTEMS INTEGRATION	
RESEARCH CHEMIST	SOLAR SPACE HEATING & COOLING	
-PHONE-		
614 2993151 X3457		
614 2993151 X3154		

WE ARE ENGAGED IN VARIETY OF PROGRAMS IN THE GENERAL AREA OF PHOTO-CHEMICAL SOLAR ENGY CONVERSION & STORAGE THESE INCLUDE APPLICATIONS BOTH TO FUEL PRODUCTION & TO BUILDING HEATING & COOLING
SEE ALSO: RICHARD NATHAN & JAMES EIBLING, BATTELLE COLUMBUS LABS

	SUBCLASSES	SOLAR SUB SYSTEMS
CONTINENTAL CAN CO INC	MEDIUM TEMPERATURE SOLAR COLLECTOR SYSTEMS	COLLECTOR,ABSORPTIVE SELECTIVE SURFACES
PROCUREMENT PLANNING	WIND CONVERSION	STORAGE, FUEL CELLS
633 THIRD AVE	INORGANIC SEMI-CONDUCTORS	
NEW YORK NY 10017		
R W DIEHL		
CORP DIRECTOR		
-PHONE-		
212 5517851		

STUDY POSSIBILITY OF USING SOLAR ENERGY AS ALTERNATIVE TO FOSSIL FUEL

	SUBCLASSES	SOLAR SUB SYSTEMS
CONTINENTAL CAN COMPANY INC	MEDIUM TEMPERATURE SOLAR COLLECTOR SYSTEMS	COLLECTOR,ABSORPTIVE SELECTIVE SURFACES
CORP R&D DEPT	WIND CONVERSION	STORAGE, FUEL CELLS
7622 S RACINE AVE	INORGANIC SEMI-CONDUCTORS	
CHICAGO IL 60620		
DR J G BUCK		
-PHONE-		
312 8463800		

STUDY POSSIBILITY OF USEING SOLAR ENERGY AS ALTERNATIVE TO FOSSIL FUEL

	SUBCLASSES	SOLAR SUB SYSTEMS
COSANTI FOUNDATION	SOLAR WATER HEATING	COLLECTOR,PASSIVE SYSTEMS
6433 DOUBLETREE ROAD	SOLAR SPACE HEATING	COLLECTOR,ALSO FUNCTION AS STRUCTURES SKIN
SCOTTSDALE AZ 85253	NEW CONSTRUCTION, TECHNOLOGIES	COLLECTOR,MULTI-FUNCTION IN DWELLING SYS
PAOLO SOLERI	1MILLION SQ FT + UP,UNDER CONSTRUCTION	HEAT TRANSFER SYSTEMS, COLLECTOR TO STORAGE
PRESIDENT	PUBLISH LAY + TECHNICAL BOOKS	HEAT RECOVERY, STORAGE TO UTILIZATION
	PUBLISH ORIGINAL RESEARCH + REPRINTS	STORAGE, SOLID MATERIALS
-PHONE-		STORAGE, LICUIDS
602 9486145		

SOLERI,'ARCOLOGY:THE CITY IN THE IMAGE OF MAN,'MIT PRESS,CAMBRIDGE, MASS, 1969.
'CONSANTI FOUNDATION,' ARCHITECTURE PLUS, MARCH 1973, PP 13-19.

---SUBCLASSES---SOLAR SUB SYSTEMS---

FEDERAL ENGY OFFICE REGION II
26 FEDERAL PLAZA
NEW YORK NY 10017
KENNETH A PORTER

-PHONE-
212 2641041
FEDERAL ENGY PROGRAMS-REGION II:NY,NJ,PR,VI...SOLAR INCLUDED IN FY1975
BUDGET

FLORIDA, UNIV OF — HIGH TEMPERATURE SOLAR COLLECTOR SYSTEMS; SOLAR WATER DISTILLATION
126 MECH ENGR BLDG
GAINESVILLE FL 32611
JOHN C REED
PROF EMERITUS

-PHONE-
904 3920805
TEACHING UNDER-GRAD&GRAD COURSES RESEARCH IN MANY FIELDS SEE: DR E A
FARBER, FLORIDA, UNIV OF

HARVARD UNIV
GRADUATE SCH DESIGN
DEPT OF ARCHITECTURE
CAMBRIDGE MA 02138
DAVID LORD
ASST PROF
-PHONE-
617 4952570
SYSTEMS APPROACH TO ARCHITECTURAL INTEGRATION OF SOLAR ENGY COLLECTION
SYSTEMS

JOHNS HOPKINS UNIV — MEDIUM TEMPERATURE SOLAR COLLECTOR SYSTEMS; HIGH TEMPERATURE SOLAR COLLECTOR SYSTEMS; OCEAN THERMAL GRADIENT SYSTEMS; SOLAR COOKING
APPLIED PHYSICS LAB
8621 GEORGIA AVE
SILVER SPRING MD 20910
CHARLES J SWET

-PHONE-
301 9537100
EXPLORATORY DEV SOLAR PUMPS & SOLAR COOKERS...DESIGN STUDIES ON SOLAR
COLLECTORS FOCUSSING & FLAT

SUBCLASSES—SOLAR SUB SYSTEMS

MIDWEST RESEARCH INST
ENGR SCIENCES DIV
425 VOULKER BLVD
KANSAS CITY MO 64110
MICHAEL C NOLAND

-PHONE-
816 5610202
CONTRACTED BY OFF OF TECH ASSES TO DO TECH ASSES OF SOLAR ENGY UTIL

MITRE CORPORATION, THE
WESTGATE RESEARCH PK
MCLEAN VA 22101
RICHARD S GREELEY

INORGANIC SEMI-CONDUCTORS
PUBLISH ORIGINAL RESEARCHED INFORMATION

-PHONE-
703 8933500
ASSISTING NSF IN FORMULATING & EVALUATING 5YR SOLAR ENGY RESEARCH PROG SYS ANALYSIS OF ON GOING & PLANNED RESEARCH RELATIVE TO PROBLEMS OF DEVELOP TECHNICALLY& ECONOMICALLY VIABLE SOLAR ENGY SYS HAS BEEN COMP SYSTEMS ANALYSIS OF SOLAR ENERGY SYSTEMS SEE ALSO:FRANK ELDRIDGE MITRE PUBLISHED REPORTS AVAIL THROUGH NATL TECH INFO SERV

NASA
CODE R-2
WASHINGTON DC 20546
PUBLIC AFFAIRS OFFICER *

PROVIDE ECONOMIC ANALYSIS OF SOLAR TECH

NASA
CODE RPP
WASHINGTON DC 20546
ERNST M COHN

PROVIDE ECONOMIC ANALYSIS OF SOLAR TECH

-PHONE-
202 7552400
TO EXPLORE TECH & ECONOMIC POSSIBILITIES OF TERRESTRIAL SOLAR ENERGY USES...RESEARCH MANAGEMENT & COORDINATION

---SUBCLASSES---SOLAR SUB SYSTEMS---

NATIONAL SCIENCE FOUNDATION GOVERNMENT AGENCIES PROVIDING GRANTS
ADVNCED ENGY RES&TEC
WASHINGTON DC 20550
DONALD A BEATTIE
ACTING DIVISION DIR

-PHONE-
202 6325726

NATIONAL SCIENCE FOUNDATION GOVERNMENT AGENCIES PROVIDING GRANTS
ADVNCED ENGY RES&TEC
WASHINGTON DC 20550
LLOYD O HERWIG *
DIRECTOR

-PHONE-
202 6327230

NATIONAL SCIENCE FOUNDATION GOVERNMENT AGENCIES PROVIDING GRANTS
ADVNCED ENGY RES&TEC
WASHINGTON DC 20550
GEORGE JAMES
INFORMATION PROG MGR

-PHONE-
202 6327398

RENSSELAER POLYTECHNIC INST NEW CONSTRUCTION, RESOURCES
ARCHITECT RESERCH CT
TROY NY 12181
WALTER KRONER

-PHONE-
518 2706461
IMPACT OF SOLAR TECHNOLOGY ON PLANNING & ARCHITECTURE
ALSO ADVISING STATE LEGISLATURE ON ENERGY CONSERVATION

----SUBCLASSES----SOLAR SUB SYSTEMS----

UNITED NATIONS — PUBLISH LAY + TECHNICAL PERIODICALS
OFFICE FOR SCI &TECH
NEW YORK NY 10017
GERTRAND H CHATEL
SCI APPL SEC RM3146E

-PHONE-
212 7541234 X2591
INTERNATIONAL COOPERATION IN RESEARCH DEVELOPMENT IN SOLAR ENERGY

UNITED NATIONS — PUBLISH LAY + TECHNICAL PERIODICALS
RESOURCES& TRANSPORT
NEW YORK NY 10017
THE SECTION HEAD
ENERGY SECTION ESA

-PHONE-
212 7541234 X4339
ASSEMBLY OF AVAIL TECHNICAL INFO ON USES OF SOLAR ENERGY...ECONOMIC & TECHNICAL EVAL OF EXISTING INSTALLATIONS & CRITICAL REVIEW OF NEW PROPOSALS...SOLAR RELAT TO PRESENT & FUTURE WORLD ENGY REQ PROCEEDINGS OF UN CONFERENCE ON NEW SOURCES OF ENERGY, SOLAR WIND & GEOCHEMICAL. ROME 21-31 AUGUST 1961 VOL 1-7 (SOLAR VOLS 4 5&6 UN PUBL SALES NOS 63.I.38, 63.I.39,&63.I.40).

WILLIAM M BROBECK & ASSOCIATES — SOLAR ENERGY STORAGE; PROVIDE INFORMATION SEARCHES TO CLIENTS — HEAT ENGINE, RANKINE
1011 GILMAN STREET
BERKELEY CA 94710
WARREN W EUKEL

-PHONE-
415 5248664
TO DEVELOP HARDWARE THAT CONVERTS SOLAR ENERGY TO PRESENT EQUIPMENT NEEDS

---SUBCLASSES---SOLAR SUB SYSTEMS---

ADELAIDE, UNIV OF
ARCH & TOWN PLANNING
ADELAIDE, S A AUSTRALIA
J D KENDRICK
SR LECTURER BLDG SCI

SOLAR RADIATION MEASUREMENTS

-PHONE-
223 4333 X2444

INTEGRATION OF SOLAR ENERGY & COLLECTOR SYS IN ARCH & TOWN PLANNING LONG TERM PROJECT: APPLICATION OF AVAILABLE DATA ON SOLAR RADIATION MEASUREMENT DEVELOPMENT OF ARCH DESIGN AIDS...COLLECTION & TESTING OF AVAIL DATA

UNIVERSIDADE FED DE PARAIBA
ESCOLA DE ENGENHARIA
58.000 JOAO PESSOA
PARAIBA BRASIL
CLEANTHO DA CAMARA TORRES
LAB DE ENERGIA SOLAR

HIGH TEMPERATURE SOLAR COLLECTOR SYSTEMS
SOLAR AGRICULTURAL DRYING
SOLAR COOKING
SOLAR WATER DISTILLATION
SOLAR SPACE HEATING & COOLING
SOLAR FURNACE

STAFF: CLEANTHO DA CAMARA TORRES SOLAR STILL&COOKER/ ANTONIO MARIA AMAZONAS MAC DOWELL SOLAR MEASUREMENTS/ JULIO GOLDFARB SOLAR FURNACES/ EMERSON FREITAS JAGUARIDE SOLAR STILL/ ROGERIO PINHEIRO KLUPPEL SOLAR MEASUREMENTS&HEATING/ PAULO MARTINS DE AEREU DRYING FRUITS/ ANTONIO SOUTO COUTINHO COOLING & AIR COND/ ZENONAS STASEVSKAS SOLAR FURNACES & STEEL WORK

CENTRAL RESEARCH ORGANIZATION
PHYSICS & ENGR RESCH
KANBE
YANKIN P O RANGOON BURMA
U MAUNG MAUNG

MEDIUM TEMPERATURE SOLAR COLLECTOR SYSTEMS
WIND CONVERSION
SOLAR AGRICULTURAL DRYING
SOLAR COOKING
SOLAR SPACE HEATING
SOLAR REFRIGERATION

-PHONE-
50544

HAVE DEVELOPED DOMESTIC SOLAR COOKERS SOLAR CIGARETTE LIGHTERS HOT H2O HEATERS SOLAR STILLS FOOD DRYING FRUIT DRYING FISH DRYING SOLAR REFRIG

---SUBCLASSES---SOLAR SUB SYSTEMS---

MCGILL UNIV MACDONALD COLLEGE	WIND CONVERSION
BRACE RESEARCH INST	SOLAR COOKING
ST ANN DE BELEVUE800	SOLAR WATER DISTILLATION
QUEBEC H9X 3MI CANADA	SOLAR WATER HEATING
T A LAWAND *	GREENHOUSES
DIR FIELD OPERATIONS	PUBLISH ORIGINAL RESEARCHED INFORMATION
-PHONE-	
514 4576580 X341	

RESEARCH PRINCIPALLY IN FIELDS OF WATER SUPPLY FOR SMALL COMM IN ARID AREAS ESP DESALINIZATION USE OF SOLAR&WIND TO PROVIDE ENERGY NEEDS FOR POOR RURAL COMM FOR COOKING CROP DEHYDRATION WATER HEATING COOLING ETC ADAPTATION OF GREENHOUSES TO ARID AREAS GREENHOUSES FED WITH SALINE WATER REDESIGN OF GREENHOUSES & REGULAR HABITATION FOR COLDER AREAS TO REDUCE HEATING LOADS HAVE BUILT: SOLAR STILLS & DEV DESIGNS FOR PREFAB MFG OF SAME/ WINDMILL FOR PUMPING DEEP WELLS&IMPOUNDED SURFACE H2O/ LUBING WIND GENERATOR/ SMALL SCALE SOLAR CABINET DRYERS/TECH MANUAL ON SOLAR AG DRYERS BEING PREPARED/SOLAR COOKERS...SEE ALSO:RON ALWARD BRACE RESEARCH INST (INFO DISSEM GP)

ELECTRICAL RESEARCH ASSOC
CLEEVE RD
LEATHERHEAD SURRER ENGLAND

"...Although this enormous energy is dispersed over an immense area, its power is so great that the amount falling on less than one half of one percent of the land area of the U.S. is more than enough to meet the nations total energy needs to the year 2000..."

Hammond "Solar Energy-Largest Resource"
Energy Crisis in America Congress Quarterly

SUBCLASSES—SOLAR SUB SYSTEMS

PURDUE UNIV
DEPT OF AG ENGR
W LAFAYETTE IN 47907
BRIAN HORSFIELD

-PHONE-
317 7492971
DETERMINE FEASIBILITY OF USING SOLAR ENGY TO DRY ANIMAL WASTES TO FAC-
ILITATE STORAGE HANDLING & CROP APPLICATION...COMPUTER STIMULATION STU
DY COMPLETE INDICATES CONCEPT FEASIBLE
'DRYING ANIMAL WASTES WITH SOLAR ENGY & EXHAUST VENTILATION AIR,'
PRESENTED 1973 ANNUAL MEET AM SOC OF AG ENGRS,UNIV KY,LEXINGTON KY,
JUNE 17-20,1973

----------SUBCLASSES----------SOLAR SUB SYSTEMS----------

	SUBCLASSES	SOLAR SUB SYSTEMS
DESERT SUNSHINE EXPOSURE TESTS	SOLAR RADIATION MEASUREMENTS	COLLECTOR,REFLECTIVE SELECTIVE SURFACES
BOX185 BLACKCAN STGE	MEDIUM TEMPERATURE SOLAR COLLECTOR SYSTEMS	COLLECTOR,ABSORPTIVE SELECTIVE SURFACES
PHOENIX AZ 85020	ALL SOLAR TECHNOLOGIES	COLLECTOR,MECHANICAL SYSTEMS
GENE A ZERLAUT	SOLAR ENGINES	
PRESIDENT	SOLAR HEAT,COOL,&ELECTRIC POWER GENERATION	
	HAVE&MAINTAIN ENERGY DATA BANK	

-PHONE-
602 4657525
602 4657521

DESERT SUNSHINE EXPOSURE TESTS IS A 25 YR OLD SUN-WEATHERING STATION LOCATED IN CENTRAL ARIZONA DESERT CURRENTLY TESTING CLIENTS PAINTS PLASTICS FABRICS GLASS AND SOLAR MATERIALS SUCH AS GLAZINGS AND TRANSPARANCIES REFLECTORS (MIRRORS) BLACK HIGH A/E RECEIVERS ANTIREFLECTION COATINGS SOLAR CELL MATERIALS AND ARRAYS AND SOLAR DEVICES/SYSTEMS

"Shadow and sun — so too our
lives are made —
Here learn how great the sun,
how small the shade!
Richard Le Gallienne, For Sundials

---SUBCLASSES--- ---SOLAR SUB SYSTEMS---

OHIO ST UNIV
DEPT OF PHYSICS
COLUMBUS OH 43210
ARI RAHL

-PHONE-
614 4226446

MCGILL UNIVERSITY — SOLAR ENERGY STORAGE
CIVIL ENGR DEPT
MONTREAL P Q CANADA
NICOLAS CHEPURNIY

-PHONE-
514 8437821
CONSTRUCTION OF EFFECTIVE SOLAR POND FOR STORING SOLAR ENERGY...EXPER-
IMENTAL & THEORETICAL WORK UNDERWAY SEE MCGILL UNIV MACDONALD COLLEGE
BRACE RESEARCH INST FOR PUB LIST

NEGEV, UNIV OF THE — BIO-CONVERSION, WIND CONVERSION
R & D AUTHORITY
P O B 1025
BEER SHEVA ISRAEL
J SCHECHTER
DIRECTOR
-PHONE-
0573382
LIQUID SOLAR POND GAS SOLAR POND WIND ENERGY DESIGN & OPERATION OF
SHROUDS PHOTOSYNTHESIS LARGE SCALE PROD OF ALGAE THERMAL DESIGN OF
RESIDENCES

---SUBCLASSES---SOLAR SUB SYSTEMS---

PHILIP WERLEIN
7927 ST CHARLES AVE
NEW ORLEANS LA 70118

-PHONE-
504 8610544
AIMING AT SIMPLEST DEVICE POSSIBLE TO FOCUS 110F SUN ENERGY ON THIN LAYER OF SPROUTED WHEAT DOUGH...HOPEFULLY COOKER WILL TRACK SUN ITSELF

VOLUNTEERS IN INTL TECH ASSIST
3706 RHODE ISLAND AV
MT RANIER MD 20822
THOMAS H FOX

SOLAR AGRICULTURAL DRYING
SOLAR WATER HEATING
SOLAR REFRIGERATION

-PHONE-
301 2777000
SOLAR APPLICATIONS IN DEVELOPING AREAS WATER HEATING COOKING DRYING REFRIGERATION...HAVE DEVELOPED & EVALUATED SOLAR COOKERS FOR ARID REGIONS

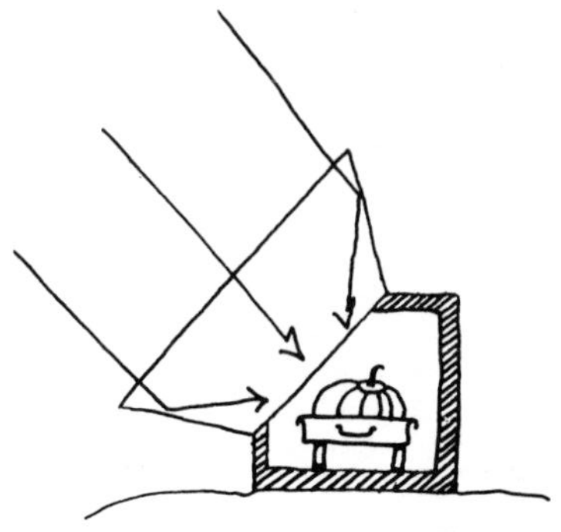

---SUBCLASSES---SOLAR SUB SYSTEMS---

GEORGIA INST OF TECHNOLOGY HIGH TEMPERATURE SOLAR COLLECTOR SYSTEMS
ENGINEERING EXP STAT SOLAR ELECTRIC POWER GENERATION
ATLANTA GA 30332
STEVE H BOMAR

-PHONE-
404 8943661
FURNACE DESCRIPTION: REFLECTOR FOCAL LENGTH 59 FT, 130 FT HIGH& 175 FT WIDE--9500 MIRRORS 17.7 IN X 17.7 IN---63 OTHER MIRRORS (HELIOSTATS) IN 8 TIERS TRACK SUN &REFLECT RAYS IN PARALLEL BEAMS TO PARABOLA--- HELIOSTATS 24.6 FT X 19.7 FT...SOLAR INTENSITY 1000 WATTS PER SQ METER IMAGE OF SUN AT FOCAL POINT 6.61 IN ABOUT 270 KW CONCENTRATED AT FOCUS
LOCATION: PYRENEES AT ODEILLO-FONT ROMEU,FRANCE (ALTITUDE 5900FT) 20 MILES EAST ANDORRA...SUN SHINES AS MANY AS 180 DAYS PER YEAR
COMP DATE: FULL POWER OPERATION IN 1974 AFTER 12 YEARS CONSTRUCTION
COST: $2,000,000.
RESEARCH: TESTING RADAR TRANSMISSION PROPERTIES OF MATERIALS EXPOSED TO HIGH RADIANT HEAT FLUX ENVIRO...2 PROG COMP FOR US ARMY
SEE ALSO: JESSE D WALTON, GEORGIA INST OF TECH, ENGINEERING EXP STAT

MAGRUDER MIDDLE SCHOOL WIND CONVERSION
TORRANCE CA 90510
ALLEN GOLDSMITH

TEACHING SOLAR ENERGY TO 8TH GRADE SCIENCE CLASS
CLASS PROJECTS:HAVE BUILT 32 SQFT SOLAR FURNACE TO PRODUCE 3000-4500F TEMPERATURES-PLAN TO INSTALL STEAM GENERATOR TO PRODUCE ELECT
ALSO EXPERIMENTING WITH WINDMILLS

GOVT INDUST RESCH INST NAGOYA SOLAR SPACE HEATING & COOLING
SOLAR RESEARCH LAB
1 HIRATE-MACHI
KITA-KU NAGOYA JAPAN
TETSUO NOGUCHI

-PHONE-
052 9112111
HIGH TEMP PROPERTIES MEASUREMENTS OF REFRACTORY OXIDES MATERIALS PLUS TEMP & EMISSIVITY MEASUREMENTS BY USE OF SOLAR FURNACE/SELECTIVE SURFACE & HEAT STORAGE MATERIALS FOR THERMAL CONVERSION/ SPACE HEATING & COOLING

--------SUBCLASSES--------SOLAR SUB SYSTEMS--------

CALIF, UNIV OF-RICHMOND FLD ST — MEDIUM TEMPERATURE SOLAR COLLECTOR SYSTEMS
177 RICHMOND FLD STA
1301 SOUTH 46TH ST
RICHMOND CA 94704
EVERETT D HOWE

-PHONE-
415 6426352
415 5254990 X237
PRODUCTION OF LARGE QUANTIES OF FRESH H2O FROM SALT AT LOW COST
ENGAGED IN WORK SINCE 1952-PRESENTLY RETIRED
VARIOUS TECHNICAL PUBLICATIONS ON PRODUCTIVITY OF SOLAR DISTILLERS SEE
SEA WAER CONV LAB,UNIV OF CALIF

ESP INCORPORATED — MEDIUM TEMPERATURE SOLAR COLLECTOR SYSTEMS
19 WEST COLLEGE AVE
PA 19067
EDOUARD JORDI

-PHONE-
215 4933659
DEV OF OUR VERTICAL SOLAR DISTILLATION CONCEPT-INCREASE EFF OF 3 TYPES
BASIC SYS: FIXED BASE STILL, GROUND SUCTION STILL,&FLOATING DISTILATON
MODELS APPEAR 3TIMES AS EFF IN OUTPUT THAN CURRENT MODELS

INSTITUTE D'ENERGIE SOLAIRE — SOLAR AGRICULTURAL DRYING; SOLAR SPACE HEATING & COOLING
OBSERVATOIRE
BOUZAREAH ALGERIA
D M DAMERDJI

-PHONE-
784229

U N E S C O — SOLAR AGRICULTURAL DRYING; SOLAR SPACE HEATING & COOLING
ELHARRACH B P 26
ALGER ALGERIA
DR Y ELMAHGARY

DEVELOP NEW METHODS ESP IN SOLAR DESALINATION ALSO: IMPROVING PRESENT
TOOLS & INCREASING KNOWLEDGE ON USE OF SOLAR ENGY IN COOLING HEATING &
DRYING OF FOOD

SUBCLASSES--SOLAR SUB SYSTEMS

BARI, UNIV OF
ENERGY SOURCES LAB
LARGO FRACCACRETA 1
1 70122 BARY ITALY
GIORGIO NEBBIA

-PHONE-
080 214495
WATER DESALINATION FOR SUPPLY OF ARID DEVELOPING AREAS WORK IN PROGRES
SINCE 1953

NATIONAL BLDG RESCH INST CSIR
ENVIRONMENT ENGR DIV
PO BOX 395
PRETORIA SOUTH AFRICA
C S GROBBELAAS
DIRECTOR
-PHONE-
PRT 746011

MEDIUM TEMPERATURE SOLAR COLLECTOR SYSTEMS
PUBLISH ORIGINAL RESEARCHED INFORMATION

UTILIZATION OF SOLAR BY FLAT PLATE COLL...CSIR RESEARCH REPORT NO 248
'SOLAR WATER HEATING IN SOUTH AFRICA' BY DNW CHINNERY COVERS ALL TYPES
FLAT PLATE COLL BUILT&TESTED INCLUDES BUILD SPECS AND OPERATION DATA

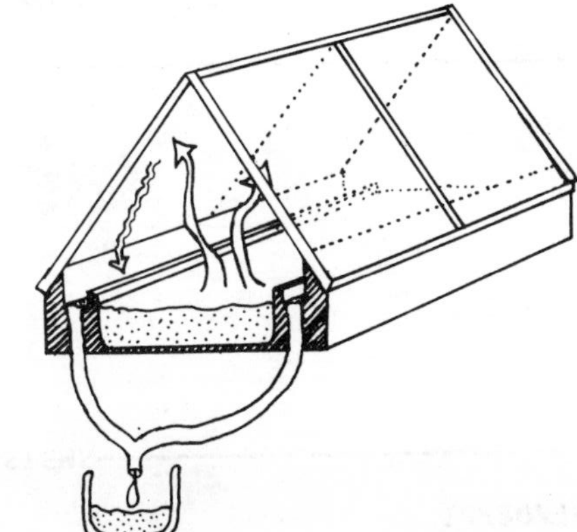

--------SUBCLASSES--------SOLAR SUB SYSTEMS--------

WILLIAM C ROOS
318 RENA DRIVE
LAFAYETTE LA 70501

-PHONE-
318 9843430
DEVELOPMENT OF PROTOTYPE HOT AIR SOLAR ENERGY MOTOR

CROWER CAMS & EQUIPMENT CO
3333 MAIN STREET
CHULA VISTA CA 92011
BRUCE CROWER
PRESIDENT

Solar sub systems:
COLLECTOR,MECHANICAL SYSTEMS
COLLECTOR,REFLECTIVE SELECTIVE SURFACES
COLLECTOR,ABSORPTIVE SELECTIVE SURFACES

SOLAR ENGINE OPERATING ON EXPANSION & CONTRACTION OF METAL: PROTOTYPE MODEL 18 IN DIA DRUM WITH 4 IN LONGITUDINAL ALUMINUM STRIPS PAINTED BLACK TO ABSORB OR RADIATE HEAT OF SUN SOME 50 TIMES---CURRENTLY RES-EARCHING TO IMPROVE EFFICIENCY

FLORIDA, UNIV OF
SOLAR ENGY/ENGY CONV
GAINESVILLE FL 92611
ERICH H FARBER *
DIRECTOR

Subclasses:
HIGH TEMPERATURE SOLAR COLLECTOR SYSTEMS
ALL SOLAR TECHNOLOGIES
SOLAR WATER DISTILLATION
SOLAR WATER HEATING
SOLAR SPACE HEATING & COOLING
SOLAR POWERED TRANSPORTATION

-PHONE-
904 3920820
ALSO RESEARCHING: SWIMMING POOL HEATING,SOLAR COOKERS, SOLAR POWER PLANTS,CLOSED&OPEN CYCLE HOT AIR ENGINES,SOLAR PUMPS,SOLAR TURBINES, SOLAR GRAVITY MOTORS,ELECTRIC CONVERSION,SOLAR HEAT SEWAGE DIGESTORS SOLAR HOUSE...ALL APPLICATIONS MENTIONED HAVE WORKING PROTOTYPES...PUB LICATIONS TO NUMEROUS TO LIST...FORFURTHER DETAILS WRITE,BUT SEND SOME CHANGE TO HELP THEM DEFRAY COSTS.THEY HAVE SOME NICE LITERATURE ON WHAT THEY ARE DOING...PLEASE SPECIFY YOUR INTEREST

SURCLASSES — SOLAR SUB SYSTEMS

HEAT ENGINE, RANKINE

INT'L RESEARCH&TECHNOLOGY CORP
1225 CONN AVENUE NW
WASHINGTON DC 20036
MARTIN O STERN

PUBLISHED REPORTS: 'A SURVEY ON STATE OF THE ART-RANKINE CYCLE WORKING FLUIDS' IRT-286-R, MAY 1972
'GOALS & GUIDELINES: RANKINE CYCLE PROPULSION SYSTEMS FOR APPLICATION TO URBAN BUSES & OTHER HEAVY-DUTY VEHICLES' IRT-301-R, DEC 1972
'THE CALIF STEAM BUS PROJECT: TECHNICAL EVALUATION' IRT-301-R, JAN 73

James

------SUBCLASSES------SOLAR SUB SYSTEMS------

SYSTEMS RESEARCH LABS INC
2800 INDIAN RIPPLE
DAYTON OH 45440
KEITH KLOPFENSTEIN

-PHONE-
513 2528758
INVESTIGATING SOLAR CONV TECH...SOLAR LASER HIGH POWER LEVELS...
RESEARCHING NEW POSSIBILITIES & HIGHER EFF OF OLD METHODS...BUILT SUN
PUMPED LASER TO USE IN CONJUNCTION WITH 1GHZ COMMUNICATION SYS

---

and behold
the droplets gleaming
send visions streaming
a multitude of suns
reflected
from
each
leaf.

Carolyn

--------------------SUBCLASSES--------------------SOLAR SUB SYSTEMS--------------------

ROBERT S LAWRENCE			MEDIUM TEMPERATURE SOLAR COLLECTOR SYSTEMS
PHYSICIST			HIGH TEMPERATURE SOLAR COLLECTOR SYSTEMS
SALINA STAR ROUTE			SOLAR SPACE HEATING
BOULDER	CO	80302	

-PHONE-
303 4991000 X6353
303 4427534
HAVE BUILT DOMESTIC HOT H2O HEATER USING CYLINDRICAL CONCENTRATORS...
PLANNING HOT H2O SYS TO HEAT MY HOUSE

JOSEPH H POPE			MEDIUM TEMPERATURE SOLAR COLLECTOR SYSTEMS
7328 MT SHERMAN RD			HIGH TEMPERATURE SOLAR COLLECTOR SYSTEMS
LONGMONT	CO	80501	SOLAR SPACE HEATING & COOLING

-PHONE-
303 4428452
DEVELOP PRACTICAL LOW COST COLLECTOR & SYSTEMS HAVE DEV FLAT PLATE
COLL FOR SWIMMING POOL HEAT MAINTAINS TEMP ABOUT 90F MAX 100F
DIRECTOR OF BOULDER COLORADO SOLAR ENERGY SOCIETY

FRANCIS DE WINTER			MEDIUM TEMPERATURE SOLAR COLLECTOR SYSTEMS
2940 THORNDIKE ROAD			HIGH TEMPERATURE SOLAR COLLECTOR SYSTEMS
PASADENA	CA	91107	PUBLISH ORIGINAL RESEARCHED INFORMATION

-PHONE-
213 7966414
DEVELOPMENT OF A HOME BUILT SOLAR SWIMMING POOL HEATER...MANUAL PRE-
PARED FOR DO-IT-YOURSELF MARKET...PROTOTYPE POOL HEATED BUILT&DEVELOP
FOR COPPER DEVELOPMENT ASSOC...PUBLISHED:'HOME BUILT SOLAR HEATER FOR
SWIMMING POOLS'PAPER PRESENTED JULY 1973 ISES CONF PARIS
'COMPUTER STUDIES OF PARABOLIC SOLAR CONCENTRATOR PERFORMANCE FOR
SOLAR THERMIONIC STUDIES,'JPL SPACE PROGRAM SUMMARY,37-49 PP99-102,
FEB 1968

SUBCLASSES — SOLAR SUB SYSTEMS

CALIF INST OF THE ARTS
SCHOOL OF DESIGN
24700 MCBEAN PKWY
VALENCIA CA 91355
STEPHEN SELKOWITZ
INSTRUCTOR
-PHONE-
805 2551050 X328

MEDIUM TEMPERATURE SOLAR COLLECTOR SYSTEMS
SOLAR SPACE HEATING & COOLING

DESIGNING & TESTING NEW CONFIGURATIONS FOR FLAT PLATE COLL/INSTALATION OF SAME COLL FOR SWIMMING POOL HEAT SYS- ULTIMATELY USE FOR DOMESTIC WATER HEAT & SPACE CONDITIONING VERY INTERESTED IN ENERGY CONSV IN RESIDENTIAL CONSTRUCTION

FEDERAL POWER COMMISSION
OFFICE CHIEF ENGR
825 N CAPITOL ST NE
WASHINGTON DC 20426
WALTER E CUSHEN

MEDIUM TEMPERATURE SOLAR COLLECTOR SYSTEMS
PUBLISH ORIGINAL RESEARCHED INFORMATION

-PHONE-
202 3765270
WORK WITH DEPT OF AG TO UPDATE 1938 SOLAR HOT H20 HEAT MANUAL.WANT TO STUDY PROBS OF LICENSING SOLAR POWER GEN...NEED FUNDS

U S NAVY
CODE L 80 CEL, NCBC
PORT HUENEME CA 93043
EARL BECK

INORGANIC SEMI-CONDUCTORS
SOLAR SPACE HEATING & COOLING
PUBLISH LAY + TECHNICAL BOOKS

-PHONE-
805 9824623
MISSION IS TO PRODUCE A NAVY SOLAR DESIGN CHAPTER FOR HANDBOOK TO BE READY BY 6/75...WILL DISCUSS SOLAR H2O HEATING IN DEPTH...WILL PREPARE HEAT & COOL CHAPTER FOR '76 AND ONE ON PHOTOVOLTAICS BY 1977

US DEPT OF AGRICULTURE
INDEPENDENCE AVE SW
WASHINGTON DC 20250
CHARLES BEER
DIR EXTENSION SER

MEDIUM TEMPERATURE SOLAR COLLECTOR SYSTEMS
SOLAR AGRICULTURAL DRYING
PUBLISH ORIGINAL RESEARCHED INFORMATION

WORK WITH FED POWER COMM TO UPDATE 1938 MANUAL ON SOLAR HOT H20 HEAT

------SUBCLASSES------SOLAR SUB SYSTEMS------

PHYSICS & ENGR LAB D S I R
PRIVATE BAG
LOWER HUTT NEW ZEALAND
D P MCAULIFFE

MEDIUM TEMPERATURE SOLAR COLLECTOR SYSTEMS

-PHONE-
WEL 699199 X818

WORKING TO PRODUCE A CHEAPER DOMESTIC THERMOSYPNON HOT WATER UNIT

--- SUBCLASSES --- SOLAR SUB SYSTEMS ---

ANDREW BLELLOCH PE
272 CLARKSVILLE RD
PRINCETON JCT NJ 08550

MEDIUM TEMPERATURE SOLAR COLLECTOR SYSTEMS
SOLAR WATER HEATING
RETROFITTING, TECHNOLOGIES

-PHONE-
609 7991475
INSTALLATION OF INDIRECTLY HEATED SOLAR DOMESTIC HOT WATER SYSTEMS &
HOT AIR SPACE HEATING EXISTING HOMES: FIRST INSTALLATION SUMMER 1974

DANIEL W DIXON *
ARCHITECT
PO BOX 797
GRANBY CO 80446

HIGH TEMPERATURE SOLAR COLLECTOR SYSTEMS

-PHONE-
303 8872200
FEASIBILITY OF SOLAR ENERGY FOR SINGLE FAMILY DWELLING

DR JOHN F ELTER *
130 LABURN CRESCENT
ROCHESTER NY 14620

MEDIUM TEMPERATURE SOLAR COLLECTOR SYSTEMS
BIO-CONVERSION

COLLECTOR,ALSO FUNCTION AS STRUCTURES SKIN
COLLECTOR,PASSIVE SYSTEMS

-PHONE-
716 4735314
716 8722000
2ND PHONE NO-EXTENSION 26128...COLLECTOR DESIGN FOR NATURAL CONVECTION
HEAT SYS, EMPHASIS ON TROMBE-MICHEL SOLAR WALL CONCEPT...ALSO SOLAR
POND STUDIES-SOLAR ENGY SUPPORT AQUACULTURE SYS FOR NORTHERN CLIMATES/
PONDS FUNCTIONING PRESENT RESEARCH AIMS AT EXTENDING OPERATION BY HEAT
STORAGE EFFECT OF POND EARTH SYS WITH AUX SOLAR HEATERS

BRYAN LEE
840 37TH
BOULDER CO 80303

MEDIUM TEMPERATURE SOLAR COLLECTOR SYSTEMS
HOMES

COLLECTOR,ABSORPTIVE SELECTIVE SURFACES

-PHONE-
303 4496278
TO PRODUCE INEXPENSIVE SYSTEM TO HEAT PRIVATE HOUSE USING FLAT PLATE
BASIC SYS OPERATING/IMPROVEMENTS BEING MADE

--------SUBCLASSES--------SOLAR SUB SYSTEMS--------

WILLIAM RASMUSSEN
BOX D
SUGAR HILL NH 03583

MEDIUM TEMPERATURE SOLAR COLLECTOR SYSTEMS

-PHONE-
603 8237713
FEASIBILITY OF SOLAR HEATING IN NORTH USING CRUSHED ROCK STORAGE & FLAT PLATE COLL USING AIR OR H2O...EFFICIENCY & COST ANALYSIS TO BE MADE

LINDA RHODES
JEFFERS POND
HOP BOTTOM PA 18824

MEDIUM TEMPERATURE SOLAR COLLECTOR SYSTEMS
WIND CONVERSION
HOMES

-PHONE-
717 2894743
RESEARCHING ALTERNATIVES TO BUILD OWN SOLAR HEATED DWELLING IN NORTH-EAST.....DEVELOPING WIND POWERED PUMPS & GENERATORS

BENJAMIN T ROGERS
CONSULTING ENGINEER
PO BOX 2
EMBUDO NM 87531

MEDIUM TEMPERATURE SOLAR COLLECTOR SYSTEMS
RETROFITTING, TECHNOLOGIES
NEW CONSTRUCTION, PASSIVE SYSTEMS
PROVIDE INFORMATION SEARCHES TO CLIENTS

-PHONE-
505 5794355
505 6675010
DESIGN & ANALYSIS OF SIMPLE INTEGRATED SOLAR-ENERGY CONSV SYS-SMALL PROTOTYPE BUILT...RESEARCHING NIGHT SKY RADIATION...PROTOTYPE CONVECTIVE SOLAR AIR HEATER IN OPERATION...CONSULT HEAT PIPE RECOVERY SYS 'PASSIVE HEAT RECOVERY AS AN ENERGYCONSV MEASURE,'BUILDING SYS DESIGN MAG, FEB MARCH 72, PP 27-30, 11-14.
'HEAT PIPES,INGENUITY & PASSIVE RECOVERY,'BUILDING SYS DESIGN MAG, MAY 73, PP 23-30.

	SUBCLASSES	SOLAR SUB SYSTEMS
WILLIAM A SHURCLIFF 19 APPLETON ST CAMBRIDGE MA 02138		
-PHONE- 617 4952816 617 8760764 DEVELOP SCHEMES FOR SOLAR HOUSE HEATING		
BRINKLEY THORNE 140 MAIN STREET NORTHAMPTON MA 01060	MEDIUM TEMPERATURE SOLAR COLLECTOR SYSTEMS WIND CONVERSION	
-PHONE- 413 5863212 ARCHITECTURAL INTEGRATION OF ENERGYCOLL SYS...DESIGNING & BUILD THE RENOVATION OF 1830 COMMERCIAL BLDG ON MAIN ST USA USING FLAT PLATE GREENHOUSE & WIND ENGY SYS		
ADRIAN MANUF & RESEARCH 619 19TH STREET GOLDEN CO 80401 FREDERICK T VARANI	MEDIUM TEMPERATURE SOLAR COLLECTOR SYSTEMS	STORAGE, METHANE
-PHONE- 303 2794536 ANAEROBIC DIGESTION SYSTEMS UTILIZING ANIMAL MANURES & PLANT WASTES SOLAR HEATING OF DIGESTOR, PILOT PLANT IN OPERATION.WE ARE NOW MEAS-URING PRODUCTION RATES & OTHER VARIABLES		
AJAX MAGNETHERMIC 1745 OVERLAND AVE NE WARREN OH 44482 THEODORE BURKE JR	MEDIUM TEMPERATURE SOLAR COLLECTOR SYSTEMS HOMES	
-PHONE- 216 3722529 WITHIN THE NEXT YEAR BUILD A SOLAR HEATED HOME,PRESENTLY GATHERING INFORMATION & TESTING DESIGN OF FLAT PLATE COLLECTOR USING ROUND FINN TUBING		

		SUBCLASSES	SOLAR SUB SYSTEMS
AMHERST COLLEGE PHYSICS DEPT AMHERST ROBERT H ROMER	MA 01002	MEDIUM TEMPERATURE SOLAR COLLECTOR SYSTEMS PROVIDE ECONOMIC ANALYSIS OF SOLAR TECH	

-PHONE-
413 5422258
413 5422251
FLAT PLATE COLLECTORS FOR SPACE HEATING/TECHNICAL & ECONOMIC EVALUATE OF SOLAR HEATED CONDOMINIUM PROJECT GRASSY BROOK VILLAGE,BROOKLINE VT

BENBOW RESEARCH 650 CALIFORNIA ST SAN FRANCISCO J BENBOW BULLOCK PRESIDENT	CA 99108	WIND CONVERSION SOLAR WATER HEATING RESIDENTIAL OTHER	

-PHONE-
415 9812516
PLANNING RESIDENCE USING SOLAR ENERGY FOR WATER & SPACE HEATING WILL USE STATE OF THE ART DEVICES. DEVELOPMENT OF HEAT PIPES FOR SOLAR HEATING APPLICATIONS ALSO INVESTIGATING USE OF WIND POWERED ELECTRIC SYS

CAUDILL ROWLETT SCOTT ARCH/ENG 111 W LOOP SOUTH HOUSTON MICHEL BEZMAN	TX 77027		COLLECTOR,PASSIVE SYSTEMS COLLECTOR,MECHANICAL SYSTEMS

-PHONE-
713 6219600
RESCH HEAT & COOL LARGE SCALE BLDGS...COMPLETE ENVIR SYSTEM USING SOLR WALL & MECH SYS

CLARK UNIVERSITY PHYSICS DEPT WORCESTER ROGER P KOHIN	MA 01610	WIND CONVERSION	

-PHONE-
617 7937711

Organization	State	Zip	Subclasses	Solar Sub Systems
COLORADO COLLEGE PHYSICS COLORADO SPRINGS DAVID F KERN STUDENT -PHONE- 303 6328168	CO	80903	MEDIUM TEMPERATURE SOLAR COLLECTOR SYSTEMS SOLAR WATER HEATING SOLAR POWERED TRANSPORTATION SOLAR TECHNOLOGY APPLICATION OTHER RETROFITTING, RESEARCHING	COLLECTOR,MECHANICAL SYSTEMS HEAT TRANSFER SYSTEMS, COLLECTOR TO STORAGE STORAGE, SOLID MATERIALS STORAGE, PHASE CHANGE MATERIALS STORAGE, OTHER ELECTROLYSIS
CRISS-CROSS BOX 173 BLACKHAWK -PHONE- 303 5829000 303 4437035	CO	80422	MEDIUM TEMPERATURE SOLAR COLLECTOR SYSTEMS BIO-CONVERSION WIND CONVERSION	STORAGE, METHANE

ALTERNATIVE RESOURCE DESIGN WITH EMPHASIS ON WIND-SUN-METHANE/BUILDING SOLAR HEATED GREENHOUSE ALSO EXPERIMENTING WITH SOLAR HEAT COLLECTORS

Organization	State	Zip	Subclasses	Solar Sub Systems
GEORGIA INST OF TECHNOLOGY ENGINEERING EXP STAT ATLANTA ATIP DEES -PHONE- 404 8943661	GA	30332	MOBILE HOMES	

FEASIBILITY STUDY: DEMONSTRATE COMBO OF ROOF MOUNT FLAT PLATE COLL & UNDERFLOOR COLLAPSIBLE THERMAL STORAGE TANKS PROVIDE 50 PERCENT MOBILE HOME HEAT REQUIREMENTS

Organization	State	Zip	Subclasses	Solar Sub Systems
LEAR SIEGLER INC MAMMOTH DIVISION 13120 B COUNTY RD 6 MINNEAPOLIS FLOYD H SCHENEEBERG VICE PRES -PHONE- 612 5442711	MN	55441		

---SUBCLASSES---SOLAR SUB SYSTEMS---

LISTON & ASSOC,MECH ENGRS
890 SARATOGA AVE
SAN JOSE CA 95129
THOMAS L LISTON

HIGH TEMPERATURE SOLAR COLLECTOR SYSTEMS

-PHONE-
408 2466281
OBJ: TO ACHIEVE ECONOMY OF TOTAL SYS THROUGH LOW COST PARABOLIC CONCEN
TRATORS FOR COLLECTING & STORING SOLAR HEAT IN H2O AT 350F..UNDER TEST
LOW COST OFF THE SHELF SYS COMPONENTS & CONTROLS ARE DESIGNED FOR LOW
COST TOTAL HEATING SYS

MARYLAND, UNIV OF
CAMPUS PLANNING DEPT
SERVICE BLDG 2ND FL
COLLEGE PARK MD 20740
WILLIAM F JONES
CAMPUS ARCHITECT

WIND CONVERSION
SOLAR ELECTRIC POWER GENERATION

-PHONE-
301 4545770
TO MAKE CAMPUS MORE EFFICIENT & LESS DEPENDENT ON GAS OIL & ELECTRIC
UTILITY CO

MASSACHUSETTS, UNIV OF
PHYSICS & ASTRONOMY
AMHERST MA 01002
ALLAN R HOFFMAN

MEDIUM TEMPERATURE SOLAR COLLECTOR SYSTEMS
PROVIDE ECONOMIC ANALYSIS OF SOLAR TECH

-PHONE-
413 5450933
TECHNICAL & ECONOMIC EVALUATION OF OPERATING SOLAR HEATING SYS...
WORKING RELATIONSHIP WITH DEVELOPER RICHARD BLAZEJ GRASSY BROOK VILLGE
NEWFANE VT

--------SUBCLASSES--------SOLAR SUB SYSTEMS--------

MINNESOTA, UNIVERSITY OF
ARCH&LANDSCAPE ARCH
110 ARCHITECTURE BLG
MINNEAPOLIS MN 55455
DENNIS R HOLLOWAY
ASSISTANT PROFESSOR
-PHONE-
612 3732198

MEDIUM TEMPERATURE SOLAR COLLECTOR SYSTEMS
BIO-CONVERSION
WIND CONVERSION
TOTAL SYSTEMS INTEGRATION
SOLAR WATER HEATING
HOMES

PROJECT OUROBOROS-ALTERNATIVE LOW-IMPACT TECHNOLOGIES IN HUMAN HABITAT PROJECT BY ENVIRO DESIGN CLASS SCH OF ARCH&LANDSCAPE ARCH...HOUSE USES FLAT PLATE H2O COLL STORAGE HOT ROCKS&H2O ALSO GREENHOUSE WIND MILL & CLIVUS TOILET

MINNESOTA, UNV OF
MECHANICAL ENGR
MINNEAPOLIS MN 55455
JOHN ILSE
RESEARCH ASSISTANT
-PHONE-
612 3764916
612 4233701

MEDIUM TEMPERATURE SOLAR COLLECTOR SYSTEMS
SOLAR WATER HEATING
NEW CONSTRUCTION, PASSIVE SYSTEMS
NEW CONSTRUCTION, MECHANICAL SYSTEMS
HOMES
PROVIDE AUDIO & VISUAL SERVICES

THESIS TITLE 'PERFORMANCE OF A RESIDENTIAL SOLAR HEATING SYSTEM', 'OUROBOROS'SOLAR HOUSE PROJECT AT UNIV OF MINN-SINGLE FAMILY EXP DWEL-LING COMPLETE

NATL CTR FOR ATMOS RESEARCH
BOX 3000
BOULDER CO 80303
PHILIP C BENEDICT
-PHONE-
303 4945151

MEDIUM TEMPERATURE SOLAR COLLECTOR SYSTEMS
WIND CONVERSION

SUPPLEMENTARY HEATING IN NEW HOUSE BEING BUILT

NETHERS COMMUNITY SCHOOL
BOX 41
WOODVILLE VA 22749
-PHONE-
703 9879011
703 9879041

MEDIUM TEMPERATURE SOLAR COLLECTOR SYSTEMS
HIGH TEMPERATURE SOLAR COLLECTOR SYSTEMS
SOLAR WATER HEATING

HAVE HUMAN WASTE COMPOSTING SYS & ORGANIC GARDEN BUILDING SOLAR HEATED HOUSE & SOLAR WATER HEATER

	SUBCLASSES	SOLAR SUB SYSTEMS
ROCHESTER INST OF TECH 50 WEST MAIN ROCHESTER NY 14614 PAUL IVOJCIECHOIVSKI -PHONE- 716 4642411		
SANGAMON STATE UNIV PHYSICAL SCI PROG SPRINGFIELD IL 62708 AL CASELLA ASST PROF -PHONE- 217 7866600 STUDENT PROG RESCH AND COURSE ON SOLAR ENERGY UNDERGRAD LEVEL		
SOLAR HYDRONICS P O BOX 4484 BOULDER CO 80302 TEAGUE VAN BUREN -PHONE- 303 4479828 LOW TEMP RADIANT FLOOR HEAT SYS FOR INTEGRATION WITH SOLAR SYS..DESIGN WORK ON BLDG THERMALLY TIGHTER STRUCTURES..ALSO WORKING ON COLLECTOR I NSTALLATION PROBLEMS I.E. EXPANSION & LEAKAGE HEATING CONTRACTING BUSINESS IN 3RD YR BUSINESS INSTALL HOT H2O BASE-BOARD HEAT SYS	MEDIUM TEMPERATURE SOLAR COLLECTOR SYSTEMS	
SONNEWALD SERVICE RD 1 BOX 457 SPRING GROVE PA 17362 H R LEFEVER -PHONE- 717 2253456 CONSULTATION & DESIGN...SOLAR WIND & ENERGY CONSERVATION	MEDIUM TEMPERATURE SOLAR COLLECTOR SYSTEMS WIND CONVERSION SOLAR ELECTRIC POWER GENERATION NEW CONSTRUCTION, PASSIVE SYSTEMS NEW CONSTRUCTION, MECHANICAL SYSTEMS HOMES	COLLECTOR,ALSO FUNCTION AS STRUCTURES SKIN COLLECTOR,PASSIVE SYSTEMS COLLECTOR,MECHANICAL SYSTEMS

	SUBCLASSES	SOLAR SUB SYSTEMS
U S NAVAL AMMUNITION DEPOT HAWTHORNE NV 89415 ERNERST J KIRSCHKE CAPT -PHONE- 702 9452451 X590	SOLAR WATER HEATING	COLLECTOR,MECHANICAL SYSTEMS
VANDERBILT UNIV ELECT ENGR 21ST AV S NASHVILLE TN 37203 ROBERT NASH ASSOC PROF -PHONE- 615 3227311 RESCH ON EFFECT OF HEAT LOSS IN SOLAR COLLECTORS		
VEXOCAVE COMPANY STAR ROUTE BOX 400 N TURKEY CREEK MORRISON CO 80465 RICHARD KITHIL -PHONE- 303 6978246 SOLAR HEAT COLLECTION & STORAGE PLUGGED INTO EXISTING LPG FIRED CIR- CULATED HOT H2O HEAT SYS...SEVERAL DESIGNS OF HEMICYLINDRICAL GLASS COVERED COLL WITH VARIOUS ABSORPTIVE COATINGS UNDER AGE TESTING	MEDIUM TEMPERATURE SOLAR COLLECTOR SYSTEMS INORGANIC SEMI-CONDUCTORS SOLAR ELECTRIC POWER GENERATION	
WISCONSIN, STATE OF FACLTY MGMT BUREAU 1 WEST WILSON ST MADISON WI 53702 PAUL L BROWN * DIRECTOR RM 180 -PHONE- 608 2661031	ALL SOLAR TECHNOLOGIES	

--------SUBCLASSES--------SOLAR SUB SYSTEMS--------

WISCONSIN, STATE OF — ALL SOLAR TECHNOLOGIES
FACLTY MGMT BUREAU
1 WEST WILSON ST
MADISON WI 53702
GORDON E HARMAN PE
ENVIROMT AFF OFFICER
-PHONE-
608 2662885
REVIEWING AGENCY PLANS & CONSULTANTS HEAT SYS FOR IMPLEMENTATION OF SOLAR TECHNIQUES IN BLDG DESIGN...WISCONSIN STATE BLDG PROG

XONICS INC — SOLAR WATER HEATING; SOLAR ELECTRIC POWER GENERATION
6837 HAYVENHURST AVE
VAN NUYS CA 91406
PAUL B SCOTT
PRINCIPAL SCIENTIST

-PHONE-
213 7877380

TOKYO UNIVERSITY — SOLAR RADIATION MEASUREMENTS
FAC OF ENGINEERING
BUNKYO-KU
TOKYO JAPAN
TAKASHI HIRAYAMA

-PHONE-
03 8122111
HOW MUCH HEAT WILL BE UTILIZED IN A BLDG A DAY MONTH&YEAR THROUGH THE SOLAR HEAT ON BLDG SURFACES FACTORED BY WINDOW DIRECTION GLASS THICK-NESS ETC

	SUBCLASSES	SOLAR SUB SYSTEMS
BARBER-NICHOLS ENGR CO 6325 W 55TH STREET ARVADA CO 80002 ROBERT BARBER	MEDIUM TEMPERATURE SOLAR COLLECTOR SYSTEMS SOLAR ELECTRIC POWER GENERATION	HEAT ENGINE, RANKINE

-PHONE-
303 4218111
DESIGN & FABRICATE SOLAR HEATED LOW TEMP RANKINE CYCLE AIR COND SYS WITH ELECTRIC GEN CAPABILITIES...SYS TO BE FUNCTIONAL MAY 1974

	SUBCLASSES	SOLAR SUB SYSTEMS
FLORIDA,UNIV OF/MECH ENGR 1222 N W 34 TERRACE GAINESVILLE FL 32605 JOHN ATKINS	HIGH TEMPERATURE SOLAR COLLECTOR SYSTEMS	COLLECTOR,MECHANICAL SYSTEMS

-PHONE-
904 3723117
AIDE TO SOLAR ERGY SECT DEPT MECH ENGR U OF FL/MAINTAINING SOLAR HOUSE PLOTTING DATA FOR SOLAR AIR-COND CALCULATING HEAT IMPUT REQ, HEAT REJECTION REQ, NH3 CONCENTRATION&EFFICIENCY FOR ABSORPTION A/C USING NH3 & H2O. HIGH TEMP=180F,REJECTIONTEMP=90F,EVAP TEMP=32F ALSO DESIGNING PARABOLIC MIRROR MOUNT TO TRACK SUN

	SUBCLASSES	SOLAR SUB SYSTEMS
HITTMAN ASSOCIATES INC ENGY & ENVIR SYS 9190 REED BRANCH RD COLUMBIA MD 21043 H M CURRAN DIRECTOR OF ENGR		HEAT ENGINE, RANKINE

-PHONE-
301 7307800
ECONOMIC CONSIDERATIONS OF SOLAR ENERGY

MARYLAND, UNIVERSITY OF
MECHANICAL ENGR
COLLEGE PARK MD 20740
REDFIELD W ALLEN

OBJECTIVE-EVALUATION OF EFFECTS OF SYSTEM OPTIONS & ACTUAL PROCESS FACTORS ON PERFORMANCE & OPTIMIZATION OF SOLAR ABSORPTION AIR CONDITN SYS...CURRENTLY PREPARING REPORTS ON THIS RESEARCH

	SUBCLASSES	SOLAR SUB SYSTEMS
MICHIGAN, UNIV OF DEPT OF ARCH&DESIGN TAPPAN & MONROE STS ANN ARBOR MI 48104 EDWARD J KELLY JR INSTRUCTOR ENV TECH -PHONE- 313 7641340	TOTAL SYSTEMS INTEGRATION SOLAR SPACE HEATING & COOLING NEW CONSTRUCTION, TECHNOLOGIES SINGLE FAMILY, EXPERIMENT PUBLISH ORIGINAL RESEARCHED INFORMATION INCENTIVES FOR SOLAR & ENERGY CONSERVATION	COLLECTOR,MULTI-FUNCTION IN DWELLING SYS

NSF GRANT GY10788 ENERGY EFFICIENT AIR CONDITIONING OF ARCH SPACES-'73
A LOW COST AUTOMATED LOW-TECH NATURAL COOLING SYSTEM-DOUBLE ROOF CONST

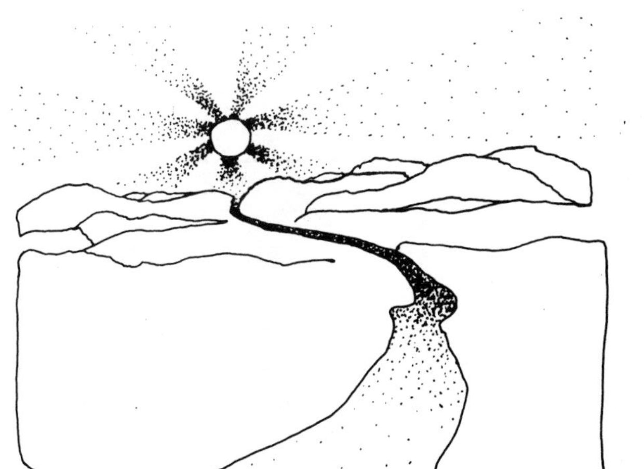

--------SUBCLASSES--------SOLAR SUB SYSTEMS--------

VICTOR J JOHNSON
1380 55TH STREET
BOULDER CO 80303

MEDIUM TEMPERATURE SOLAR COLLECTOR SYSTEMS

-PHONE-
303 4421581
303 4991000 X3523

SOLAR HEAT COLLECTION FOR ABSORPTION REFRIGERATION

WESTERN ONTARIO, UNIV OF
FAC OF ENGINEERING
LONDON N6A3K7 CANADA
ROBERT K SWARTMAN

MEDIUM TEMPERATURE SOLAR COLLECTOR SYSTEMS

-PHONE-
519 6793332

DEVELOPING SOLAR REFRIG DEVICES EXPERIMENTAL UNIT BUILT & MODIFIED TO INCORP IMPROVEMENTS SEE: 'SURVEY OFSOLAR-POWERED REFRIGERATION' ASME 73.WA/SOL-6...PAST WORK: TESTING PACKED BED COLLECTORS

---SUBCLASSES---SOLAR SUB SYSTEMS---

JEFFREY E ARONIN,AIA,FRIA
P O BOX 570
NEW YORK NY 10017

MEDIUM TEMPERATURE SOLAR COLLECTOR SYSTEMS
HIGH TEMPERATURE SOLAR COLLECTOR SYSTEMS
WIND CONVERSION

-PHONE-
516 3742394
SEVERAL PROJECTS IN DESIGN STAGES USEING SOLAR ENERGY, WIND ENERGY, & CONCENTRATORS

VERNON BURKE
BOX 2557
ASPEN CO 81611

MEDIUM TEMPERATURE SOLAR COLLECTOR SYSTEMS
HIGH TEMPERATURE SOLAR COLLECTOR SYSTEMS
SOLAR ELECTRIC POWER GENERATION
HOMES

PROPOSED SOLAR POWERED HOME WITH PASSIVE ENERGY UTILIZATION

WILLIAM A CLARK,IV *
SUGARLOAF STAR RTE
BOULDER CO 80302

MEDIUM TEMPERATURE SOLAR COLLECTOR SYSTEMS
HIGH TEMPERATURE SOLAR COLLECTOR SYSTEMS
WIND CONVERSION

-PHONE-
303 4477533
SOLAR & WIND SYS CAPABLE OF SUPPLYING ALMOST ALL ENERGY REQ/DEV COMBO COLLECTORS/DEV SUPERIOR SINKING CONTROLS

RICHARD E DORRINGTON *
11231 N 34TH DRIVE
PHOENIX AZ 85029

PRESENTLY WORKING ON H&C SYS...WOULD LIKE TO CORRESPOND WITH OTHER ORGANIZATIONS OF SIMILAR RESEARCH

------SUBCLASSES------SOLAR SUB SYSTEMS------

CARL T GRIMM
ARCHITECT
5 COOK LANE
CROTON-ON-HUDSON NY 10520

WIND CONVERSION
SOLAR WATER HEATING
RETROFITTING, RESEARCHING
NEW CONSTRUCTION, RESEARCHING

-PHONE-
914 2715471

EXPERIMENT WITH & INTEGRATE SUPPLEMENTARY SOLAR COMFORT CONDITIONING SYS TO NEW&EXISTING BLDG EMPHASIS ON RESIDENTIAL.....PROPOSAL TO NSF FOR RESEARCH: SOLAR DOMESTIC H2O HEATING & SWIMMING POOL H2O HEATERS CONSTRUCTED & TESTED.....ALSO RESEARCHING WIND CONVERSION

DWIGHT J HARRIS
ROUTE 1, BOX 249A
CHESNEE SC 29323

SOLAR WATER HEATING
GREENHOUSES

-PHONE-
803 5780768

PLANNING PRIVATE RESIDENCE HEAT&COOL BY SUN-PLANNING STAGE...SOLAR HEATING WATER FOR MOBILE HOME-TESTING PUMPS FOR H2O CIRCULATION OVER COLLECTORS

SOLAR HEATED GREENHOUSE WORKING WITH VOCATIONAL STUDENTS-EXP ON HEAT STORAGE, COLLECTORS BEING CONSTRUCTED

R GERALD IRVINE P E
1 TURNER LANE
LOUDONVILLE NY 12211

MEDIUM TEMPERATURE SOLAR COLLECTOR SYSTEMS
WIND CONVERSION

-PHONE-
518 4631334

TO PROMOTE USE OF NATURAL ENERGY RESOURCES IN MOST EFFICIENT ECONOMIC MANNER...PREPARATION OF ANALYSES FOR SOLAR ENGY UTILIZATION

--------SUBCLASSES--------SOLAR SUB SYSTEMS--------

			SUBCLASSES	SOLAR SUB SYSTEMS
RICHARD LA ROSA 317 OAK STREET SOUTH HEMPSTEAD	NY	11550	RETROFITTING, TECHNOLOGIES HOMES SINGLE FAMILY, UNDER CONSTRUCTION SINGLE FAMILY, DEMONSTRATION	

-PHONE-
516 2617000 X321
CONVERTING OWN HOME TO SOLAR HEATING-PRESENTLY HEAT 2 ROOMS EXPANDING SYSTEM
LA ROSA,'EXPERIMENTAL HEAT PUMP SYSTEM,'ALTERNATIVE SOURCES OF ENERGY, NO 11, JULY 1973, PP 6-7.

			SUBCLASSES	SOLAR SUB SYSTEMS
ALWIN B NEWTON CONSULTANT 136 SHELBOURNE DR YORK	PA	17403	MEDIUM TEMPERATURE SOLAR COLLECTOR SYSTEMS HIGH TEMPERATURE SOLAR COLLECTOR SYSTEMS WIND CONVERSION SOLAR ENERGY STORAGE SOLAR HEAT,COOL,&ELECTRIC POWER GENERATION NEW CONSTRUCTION, MECHANICAL SYSTEMS	HEAT TRANSFER SYSTEMS, COLLECTOR TO STORAGE COLLECTOR,ABSORPTIVE SELECTIVE SURFACES HEAT PUMPS HEAT ENGINE, STERLING STORAGE, SOLID MATERIALS STORAGE, MECHANICAL/PUMPED LIQUIDS STORAGE, GAS

-PHONE-
717 8435405
717 8430731 X223
WIND,HEATING COOLING STORAGE COLLRS BASIC & MODEL WORK TOTAL WIND ENGY COLLECTORS SEE PATENTS:2 342 211, 2 396 338, 2 969 788 ETC...EARLY WORK ON NIGHT COOLING RADIATION HYBRID FLAT PLATE-CONCENTRATING FOR MEDIUM TEMP USING DIRECT & DIFFUSE RADIATION GENERAL ANALYSIS TOTAL INTEGRATED WIND SYS WITH THERMAL & ELECTRIC COLLECTION & THERMAL & HYDROGEN CONVERSION & STORAGE

			SUBCLASSES	SOLAR SUB SYSTEMS
JOHN W RUCH 447 E DRIVE OAK RIDGE	TN	37830	MEDIUM TEMPERATURE SOLAR COLLECTOR SYSTEMS HIGH TEMPERATURE SOLAR COLLECTOR SYSTEMS WIND CONVERSION	

-PHONE-
615 4836436
HOME COMFORT HEATING

--------SUBCLASSES--------SOLAR SUB SYSTEMS--------

CHARLES SKELLY
BOX 971
USASAFSB APO NY 09742

MEDIUM TEMPERATURE SOLAR COLLECTOR SYSTEMS
BIO-CONVERSION

DESCRIBE & PRESCRIBE WHAT PARAMETERS WILL MAKE SOLAR SYS ECONOMICAL FOR VARIOUS LOCATIONS IN US...THEORETICAL CONSIDERATIONS AT INDIVIDUAL LEVEL...CURRENTLY CONDUCTING SHAPE FACTOR ANALYSIS FOR ANNUAL HEAT & COOLING LOAD

JOHN WATSON
1502 1/2 WEST 9TH
AUSTIN TX 78703

WIND CONVERSION

-PHONE-
512 4772042

DESIGN & IMPLEMENTATION OF SOLAR & CLIMATIC HOUSING ALSO WIND & METHNE LOW INCOME OWNER BUILT TECHNIQUES...MANUAL BEING PRODUCED TO ILLUSTRAT ENERGY & ARCH HOPEFULLY LEAD TO OWNER BUILDER MANUAL EXPLAINING PRINC-IPALS & TECHNIQUES...PLEASE NO INQUIRIES NOW

L L WORTES
550 E 12TH AVE #1801
DENVER CO 80203

MEDIUM TEMPERATURE SOLAR COLLECTOR SYSTEMS
HIGH TEMPERATURE SOLAR COLLECTOR SYSTEMS
SOLAR ELECTRIC POWER GENERATION
SOLAR TECHNOLOGY APPLICATION OTHER

-PHONE-
303 2556062

5 YR PROG-2 YRS DESIGN&BUILD PILOT PLANT FOR HEAT COLL&STORAGE 3RD YR FOR INDEPENDENT VERIFICATION ALSO INITIATE&SET UP 1 MILE SQ COMMUNITY FOR 1000 TO 5000 FAMILIES LAST 2YRS CONFIRM PROJECTIONS..PATENTS PENDG &DESIGNS INITIATED:PILOT SOLAR REFLECTOR-COLL-HEAT TRANSF SYS TO PROD ELECT & STEAM FOR INDUST KILNS/HIGH TEMP SOLAR REFLECTOR FOR SCRAP METAL MELTING&PROD OF PIG IRON & OR RE-BAR/LOW TEMP SOLAR REFLECT& ABSORBTION SYS FOR GREENHOUSES/MED TEMP REFLECT & COLLECT WITH HEAT EXCHANGE & SYS FOR HEAT COMMERCIAL GYPSUM WALL-BOARD DRYING KILN/INTE-GRATED CENTRAL HEATING & COOL SYS FOR COMMUNITIES 200-1000 DWELLINGS WITH SHOPPING CTR SML APT HIGH RISE & SCHOOL

---SUBCLASSES---SOLAR SUB SYSTEMS---

ALTERNATIVE SOURCES OF ENGY
928 2ND ST SW
ROANOKE VA 24016
EUGENE E CCLI
CO-EDITOR

-PHONE-
703 3448412
LOW ENERGY SOLAR HOUSING...COMPOSITE SYS ALLOWING FOR GREATEST ENERGY REDUCTION IN HOUSING

AM SOC HEAT REFRIG AIRCON ENGR
345 EAST 47TH ST
NEW YORK NY 10017
JOSEPH F CUBA
DIR OF RESEARCH

PUPLISH TECHNICAL PERIODICALS
PUPLISH ORIGINAL RESEARCHED INFORMATION

-PHONE-
212 7526800 X367
RESEARCH ON SOLAR ENERGY UTILIZATION FOR HEATING & COOLING BUILDINGS CURRENTLY INVOLVED IN 4 RESEARCH PROPOSALS SUBMITTED TO NSF WITH THE FOLLOWING PRINCIPAL CONTRACTORS: INSTALLATION & ENGR EVALUATION OF SOLAR HEAT&COOL MOBILE HOMES PURDUEUNIV/DEVELOP MARKETABLE FORCED FLOW SOLAR H2O HEAT SYS WITH AUTOMATIC CONTROLS SILVES LTD/DESIGN DEVELOP & ANALYSIS SOLAR AIRCOND UNIV OF FLA/DEVELOP PROCEDURES & OPTIMIZING COMM OFF BLDG SOLAR ENGY SYS BY UTILIZING RESULTS OF EVAL STUD ON A COMPLETED BLDG PENN STATE UNIV/...SOCIETY RECENTLY COSPONSORED NSF SOLAR COOLING WORKSHOP LA CALIF...SEE ALSO ASHRAE PUBLICATION SALES INFO DISSEM GROUPS

AMERICAN GAS ASSOCIATION
ENGY&ENVIROMENT SYS
1515 WILSON BLVD
ARLINGTON VA 22209
PETER E SUSEY
MANAGER

MEDIUM TEMPERATURE SOLAR COLLECTOR SYSTEMS
SOLAR WATER HEATING

-PHONE-
703 5242000
POTENTIAL APPLICATIONS FOR AUGMENTING GAS SYSTEMS-ECONOMIC FEASIBILITY CENTRAL HEATING COOLING & HOT WATER SYSTEMS

	SUBCLASSES	SOLAR SUB SYSTEMS
AMERICAN SCIENCE&ENGINEER INC SOLAR CLIMATE CONTRL 995 MASS AVE CAMBRIDGE MA 02139 DAVID A BOYD MANAGER -PHONE- 617 8681600	MEDIUM TEMPERATURE SOLAR COLLECTOR SYSTEMS HIGH TEMPERATURE SOLAR COLLECTOR SYSTEMS	

DEVELOP OF HIGH EFF FLAT PLATE & CONCENTRATING COLL...DEVELOP SIMPLIF-
IED TECHNIQUES FOR COMPUTER SIMULATION & PERFORMANCE PREDICTION OF
SOLAR ENERGY SYS...DESIGN & EXPERIMENTAL TESTS NOW BEING CONDUCTED
DR RYSZARD GAJEWSKI--RESEARCH DIRECTOR
RALPH E ABBOTT--MARKETING MGR, SYSTEMS DIVISION--PLAN TO DEV PRODUCT

	SUBCLASSES	SOLAR SUB SYSTEMS
ARGONNE NATIONAL LABORATORY ARF 9700 SOUTH CASS AVE ARGONNE IL 60439 VACLAV J SEVCIK MECH ENGR -PHONE- 312 7397711 X4182	MEDIUM TEMPERATURE SOLAR COLLECTOR SYSTEMS HIGH TEMPERATURE SOLAR COLLECTOR SYSTEMS WIND CONVERSION TOTAL SYSTEMS INTEGRATION	COLLECTOR,OTHER COLLECTOR,REFLECTIVE SELECTIVE SURFACES COLLECTOR,ABSORPTIVE SELECTIVE SURFACES COLLECTOR,PASSIVE SYSTEMS COLLECTOR,MECHANICAL SYSTEMS

WORKING ON PROTOTYPE OF CONCENTRATING NON-FOCUSING COLLECTOR

	SUBCLASSES	SOLAR SUB SYSTEMS
ARIES CONSULTING ENGRS INC SUITE D 1810 N W 6TH STREET GAINESVILLE FL 32601 ROBLEY S PULLER -PHONE- 904 3726687	SOLAR ENERGY STORAGE	

ENERGY CONVERSION & STORAGE SYSTEMS-DESIGNS, PLANS, SPECIFICATIONS FOR
PRACTICAL APPLICATIONS OF SOLAR ENERGY TO COMMERCIAL,MUNICIPAL,RESIDET
INDUSTRIAL NEEDS SUCH AS POWER FOR HVAC SYSTEMS

	SUBCLASSES	SOLAR SUB SYSTEMS
ARIZONA STATE UNIVERSITY COLL OF ARCHITECTURE TEMPE AZ 85281 JOHN I YELLOTT -PHONE- 602 9655562 602 9653216	MEDIUM TEMPERATURE SOLAR COLLECTOR SYSTEMS	

PARTICIPATING IN NSF STUDY OF SOLAR HEATING & COOLING OF BUILDINGS WITH TRW SYSTEMS,REDANDO BEACH CA MANY PUBLICATIONS ON SOLAR SEE: ASHRAE LISTING & SOLAR BIBLIOGRAPHY PLEASE NO INQUIRIES I AM ALREADY DELUGED WITH REQUESTS

	SUBCLASSES	SOLAR SUB SYSTEMS
ARMSTRONG CORK CO. R&D CENTER BUILDING SYSTEMS 2500 COLUMBIA AVE LANCASTER PA 17603 ALAN S GLASSMAN RESEARCH ARCHITECT -PHONE- 717 3970611 X7128		

PRESENTLY ONE OF SPONSORS IN PHASE I OF SOLAR CLIMATE CONTROL PROJECT BEING CONDUCTED BY AURTHUR D LITTLE INC CAMBRIDGE MASS PLEASE CONTACT ADL FOR FURTHER INFO

	SUBCLASSES	SOLAR SUB SYSTEMS
ARTHUR D LITTLE,INC ENGINEERING SCIENCES 20 ACORN PARK CAMBRIDGE MA 01240 DR.PETER F GLASER VICE PRESIDENT -PHONE- 617 8645770	SOLAR ELECTRIC POWER GENERATION HOMES ACADEMIC BUILDINGS RELIGIOUS/PUBLIC/AMUSEMENT BUILDINGS PUBLISH ORIGINAL RESEARCHED INFORMATION PROVIDE ECONOMIC ANALYSIS OF SOLAR TECH	COLLECTOR,MULTI-FUNCTION IN DWELLING SYS COLLECTOR,ALSO FUNCTION AS STRUCTURES SKIN COLLECTOR,ABSORPTIVE SELECTIVE SURFACES COLLECTOR,MECHANICAL SYSTEMS HEAT ENGINE, RANKINE STORAGE, LIQUIDS STEAM ENGINES

TERRESTRIAL & EXTRA-TERRESTRIAL SYSTEMS; RESEARCH PROGRAM SPONSORED BY 74 MAJOR FIRMS TO DEVELOP A SOLAR CLIMATE CONTROL INDUSTRY; SOLAR TECH NOLOGY ASSESSMENT FOR NSF; LARGE SCALE SCC DEMONSTRATION PROJECT FOR THE DESERT RESEARCH INSTITUTE; SOLAR CLIMATE CONTROL DESIGN FOR THE MASS AUDUBON SOCIETY; FEASIBILITY STUDY OF A SATELLITE SOLAR POWER STATION FOR NASA

	SUBCLASSES	SOLAR SUB SYSTEMS
ARTHUR D LITTLE,INC PUBLIC REALTIONS 20 ACORN PARK CAMBRIDGE MA 01240 ELLEN MCCAULEY * -PHONE- 617 8645770	ALL SOLAR TECHNOLOGIES SOLAR ELECTRIC POWER GENERATION HOMES ACADEMIC BUILDINGS RELIGIOUS/PUBLIC/AMUSEMENT BUILDINGS PUBLISH ORIGINAL RESEARCHED INFORMATION	COLLECTOR,MULTI-FUNCTION IN DWELLING SYS COLLECTOR,ALSO FUNCTION AS STRUCTURES SKIN COLLECTOR,ABSORPTIVE SELECTIVE SURFACES COLLECTOR,MECHANICAL SYSTEMS HEAT ENGINE, RANKINE STORAGE, LIQUIDS STEAM ENGINES
ASHLAND OIL INC RESEARCH & DEVELOP P O BOX 391 ASHLAND KY 41101 NORMAN W HALL -PHONE- 606 3293333 X8374	ALL SOLAR TECHNOLOGIES	
ASHLAND OIL INC P O 391 ASHLAND KY 41101 HARRY T WILEY * -PHONE- 606 3293333	ALL SOLAR TECHNOLOGIES	
AYRES&HAYAKAWA ENERGY MGT CONSULTING ENGINEERS 1190 SOUTH BEVERLY LOS ANGELES CA 90035 JEROME WEINGART SOLAR CONSULTANT -PHONE- 213 8794477		

ARTHUR D LITTLE,INC — (no description)

ASHLAND OIL INC (NORMAN W HALL): POTENTIAL NEW BUSINESS OPPORTUNITIES R+D-PARTICIPATING IN A D LITTLE PROGRAM TO DEVELOP SOLAR CLIMATE CONTROL INDUSTRY TO DETERMINE OPPORTS FOR ASHLAND OIL

ASHLAND OIL INC (HARRY T WILEY *): POTENTIAL NEW BUSINESS OPPORTUNITIES R+D-PARTICIPATING IN A D LITTLE PROGRAM TO DEVELOP SOLAR CLIMATE CONTROL INDUSTRY TO DETERMINE OPPORTS FOR ASHLAND OIL

AYRES&HAYAKAWA ENERGY MGT: CONSULTING TO SOLAR HEAT & COOL CONFERENCE CTR & LIBRARY AT LOS ALAMOS INVOLVED IN SOLAR RESCH SEVERAL YRS...CONDUCTED STUDIES SOCIAL&ENVIRO ASPECTS OF SOLAR ENERGY

	SUBCLASSES	SOLAR SUB SYSTEMS
BATTELLE NORTHWEST LABS BATTELLE BLVD RICHLAND WA 99352 KIRK DRUMHELLER -PHONE- 509 9462349 R & D PRELIMINARY STAGES	INORGANIC SEMI-CONDUCTORS	
BOEING AEROSPACE CO MAIL STOP 8C-64 P O BOX 3999 SEATTLE WA 98124 RONALD B CAIRNS RESEARCH & ENGR DEPT -PHONE- 206 7730464 IMPROVE FLAT PLATE HEATING & COOLING SYSTEMS FOR RESIDENTIAL PURPOSES RESEARCH PHASE COMPLETED SURVEY OF POTENT'L SOLAR-THERMAL ELECTRIC GEN FINAL REPORT: TO SEATTLE CITY LIGHT JUNE 1973	MEDIUM TEMPERATURE SOLAR COLLECTOR SYSTEMS INORGANIC SEMI-CONDUCTORS SOLAR ELECTRIC POWER GENERATION	HEAT TRANSFER SYSTEMS, COLLECTOR TO STORAGE COLLECTOR,MECHANICAL SYSTEMS HEAT RECOVERY, STORAGE TO UTILIZATION
BORG WARNER CORP. RESEARCH CENTER ALGONQUIN&WOLF RDS DES PLAINES IL 60018 RICHARD L KUEHNER STAFF SCIENTIST -PHONE- 312 8273131 COMPANY CONFIDENTIAL R&D IN ALL ASPECTS OF ENERGY CONSERVATION AND NON -CONVENTIONAL ENERGY SOURCE UTILIZATION WHICH CAN RESULT IN IMPROVED COMMERCIALLY USEFUL DEVICES FOR YEAR-ROUND COMFORT CONDITIONING AND BUILDING OPERATION	ALL SOLAR TECHNOLOGIES SOLAR WATER HEATING SOLAR HEAT,COOL,&ELECTRIC POWER GENERATION RETROFITTING, RESEARCHING NEW CONSTRUCTION, RESEARCHING HOMES	HEAT RECOVERY, STORAGE TO UTILIZATION HEAT TRANSFER SYSTEMS, COLLECTOR TO STORAGE HEAT PUMPS HEAT ENGINE, RANKINE

	SUBCLASSES	SOLAR SUB SYSTEMS
BRADLEY UNIV MECH ENGR DEPT PEORIA IL 61606 Y B SAFDARI	MEDIUM TEMPERATURE SOLAR COLLECTOR SYSTEMS HIGH TEMPERATURE SOLAR COLLECTOR SYSTEMS	

-PHONE-
309 6767611
DESIGN COOL SYS USEING SOLAR ENGY/RESEARCH SEMI-TRANSPARENT MEDIA/OPT-
IMIZATION SOLAR HEATING SYS/COLLECTORS BEING TESTED/ABSORPTION COOLING
SYS COMMERCIALLY AVAIL BEING MODIFIED

	SUBCLASSES	SOLAR SUB SYSTEMS
CALIF POLYTECHNIC STATE UNIV ARVH&ENVIRO DESGN&EN SAN LUIS OBISPO CA 93401 KEN HAGGARD	MEDIUM TEMPERATURE SOLAR COLLECTOR SYSTEMS	COLLECTOR,MULTI-FUNCTION IN DWELLING SYS

-PHONE-
805 5462573
RESEARCH & EVALUATION OF SKYTHERM SOLAR HOUSE IN ATASCADERO,CA
PRESENTLY COMPLETING 2ND QUARTER REPORT TO HUD
SEE: SKY THERM, HAROLD HAY

	SUBCLASSES	SOLAR SUB SYSTEMS
CALIF, UNIV OF--LOS ALAMOS LOS ALAMOS SCI LAB P O BOX 1663 LOS ALAMOS NM 87544 WILLIAM A RANKEN GROUP LEADER Q-25	MEDIUM TEMPERATURE SOLAR COLLECTOR SYSTEMS HIGH TEMPERATURE SOLAR COLLECTOR SYSTEMS OCEAN THERMAL GRADIENT SYSTEMS RETROFITTING, RESEARCHING MOBILE HOMES, PROPOSED	COLLECTOR,ALSO FUNCTION AS STRUCTURES SKIN

-PHONE-
505 6676578
WORK IN PROGRESS INCLUDES LOW COST INTEGRATED SOLAR COLLECTOR DEVELOP-
MENT/APPLICATION OF SOLAR HEATING TO NEW & EXISTING STRUCTURES/DEVELOP
OF SOLAR HEATED & COOLED MOBILE HOMES & PRELIMINARY INVESTIGATIONS OF
SOLAR THERMOCHEMICAL HYDROGEN PRODUCTION & OCEAN THERMAL GRADIENT CONV

	SUBCLASSES	SOLAR SUB SYSTEMS
CALIF, UNIV OF-L A ENGR & APPLIED SCIEN ENGY&KINETICS 5531BH LOS ANGELES CA 90024 HARRY BUCHBERG	MEDIUM TEMPERATURE SOLAR COLLECTOR SYSTEMS	HEAT TRANSFER SYSTEMS, COLLECTOR TO STORAGE HEAT RECOVERY, STORAGE TO UTILIZATION

-PHONE-
213 8255313
HEAT TRANSFER RELEVANT TO SOLAR CONV PROC FOR HEATING & COOL BLDG
SEVERAL TECHNICAL PUBLICATIONS ON OPTIMIZATION OF SOLAR COLLECT&STORAG
FOR HOUSE HEATING
PUBLISHED:'SIMULATION & OPTIMIZATION OF SOLAR COLLECTION & STORAGE FOR
HOUSE HEATING,'SOLAR ENERGY, VOL 12, 1968, PP 31-50.

CALIF, UNIV OF-LOS ALAMOS LOS ALAMOS SCI LAB Q-DIVISON (Q-DOT) LOS ALAMOS NM 87544 DR J D BALCOMB ASST ENERGY DIV-LDR	MEDIUM TEMPERATURE SOLAR COLLECTOR SYSTEMS SOLAR ENERGY STORAGE NEW CONSTRUCTION, INTEREST MOBILE HOMES, PROPOSED ACADEMIC BUILDINGS PUBLISH ORIGINAL RESEARCH + REPRINTS	COLLECTOR,ALSO FUNCTION AS STRUCTURES SKIN HEAT TRANSFER SYSTEMS, COLLECTOR TO STORAGE STORAGE, MECHANICAL/PUMPED LIQUIDS STORAGE, LIQUIDS

-PHONE-
505 6676441
RESEARCH & DEVELOPMENT LEADING TO A STRUCTURALLY INTEGRATED STEEL
SOLAR COLLECTOR UNIT, WORK FUNDED BY AEC & NSF (JOINT PROJECT WITH U S
STEEL CORP) ALSO SUBMITTED PROPOSALS FOR MOBILE HOME SOLAR HEATING, A
CONTROLS SIMULATION STUDY OF SOLAR HEATED & COOLED BUILDINGS, & SOLAR
THERMOCHEMICAL HYDROGEN GENERATING PLANT STUDY

CALIF, UNIV OF-SEA H2O CON LAB SEA WATER CONV LAB 1301 SOUTH 46TH ST RICHMOND CA 94804 BADAWI W TLEIMAT	SOLAR WATER DISTILLATION	HEAT PUMPS

-PHONE-
415 2356000 X231
SYSTEM DESIGN & TEST OF SOLAR ASSISTED HEAT PUMPS FOR HEAT & COOLING
INITIAL PLANNING BEGUN
DESIGN & TEST SMALL FAMILY SIZE SOLAR STILLS/MANY PUBLISHED WORKS ON
OPERATING SOLAR DISTILLERS-SEE BIBLIOGRAPHY

--------SUBCLASSES--------SOLAR SUB SYSTEMS--------

CARRIER CORP
RESEARCH DIVISION
CARRIER PKWY
SYRACUSE NY 13201
DR WENDELL J BIERMAN

-PHONE-
315 4638411 X3481
ASCERTIAN BUSINESS OPPORTUNITIES SUITABLE TO CORPORATE EXPERIENCE & RESOURCES

COLORADO STATE UNIV — SOLAR ELECTRIC POWER GENERATION
SOLAR ENGY APPL LAB
FORT COLLINS CO 80521
DAN S WARD

-PHONE-
303 4918211
BUILDING SOLAR HEATED BLDG TO BE COMPLETE APRIL 1974-THEN 1YEAR DATA TAKING/ANALYSIS OF SOLAR-THERMAL ELECTRIC POWER SYS-COMPLETE AUG 1974-GOAL:DEVELOP ANALYTICAL RELATIONSHPS TO DESCRIBE FUNCTION OF SYSTEMS. PROVIDE INFO REQ TO SELECT BEST METHODS&SYS FOR PRACTICAL GEN OF ELECT BY SOLAR ENGY

COLORADO STATE UNIVERSITY — MEDIUM TEMPERATURE SOLAR COLLECTOR SYSTEMS; NEW CONSTRUCTION, MECHANICAL SYSTEMS
SOLAR ENGY APPL LAB
ENGR RESEARCH CENTER
FT COLLINS CO 80521
SUSUMU KARAKI
ASSOC DIRECTOR
-PHONE-
303 4911101
PROGRAM CHAIRMAN 1974 US SEC ISES ANNUAL MEETING HELD AT COLO ST UNIV

SUBCLASSES—SOLAR SUB SYSTEMS

COLORADO STATE UNIVERSITY
SOLAR ENGY APPL LAB
ENGR RESEARCH CENTER
FORT COLLINS CO 80521
DR GEORGE LOF
DIRECTOR
-PHONE-
303 4918632
303 4918325
CENTER FOR RESEARCH & STUDIES ON SOLAR ENERGY UTILIZATION PROJECTS INCLUDE 18 MONTH STUDY OF METHODS FOR CONVERTING SOLAR ENERGY TO ELECT POWER & TO DESIGN CONSTRUCT & TEST A SOLAR HEATED & COOLED RESIDENCE ON CSU CAMPUS THIS BLDG WILL FUNCTION AS US TEST SITE FOR SOLAR EQUIP

CONSOLIDATED EDISON — OFFICE/BANK BUILDINGS
4 IRVING PLACE
NEW YORK NY 10003
R A BELL

-PHONE-
212 4604600
CONDUCTING FEASIBILITY STUDY OF RETROFITTING 910 FT CITICORP BLDG IN MANHATTAN...SOLAR COLLECTORS WOULD BE MOUNTED ON THE UPPER 40 STORIES OF THE BUILDING...SOLAR POWERED DEHUMIDIFICATION WOULD PROVIDE 30 PER-CENT OF THE TOTAL COOLING LOAD...STUDY DONE IN CONJUNCTION WITH ENERGY LAB MIT FIRST NATIONAL CITY BANK AND CUSHMAN AND WAKEFIELD

DARWIN TEAGUE INC — MEDIUM TEMPERATURE SOLAR COLLECTOR SYSTEMS; STORAGE, FUEL CELLS; SOLAR HEAT,COOL,&ELECTRIC POWER GENERATION
375 SYLVAN AVE
ENGLEWOOD CLIFFS NJ 07632
W D TEAGUE
PRES

-PHONE-
201 5685124
DESIGN LOW COST,HIGH EFF COLLECTOR/INCORPORATE INTO PROTYPE HOME/IN PLANNING STAGES

	SUBCLASSES	SOLAR SUB SYSTEMS
DEPT OF HOUSING & URBAN DEVEL BUILDING TECH DIV 451 7TH STREET S W WASHINGTON DC 20410 ORVILLE LEE SOLAR ENGY PROG MGR -PHONE- 202 7575356	RETROFITTING, TECHNOLOGIES NEW CONSTRUCTION, TECHNOLOGIES GOVERNMENT AGENCIES BUILDING CODES FOR SOLAR UTILIZATION	
ENVIRONMENTAL DESIGN PETTIGREW AR 72752 JOEL DAVIDSON RESEARCH, DESIGN, & TESTING LOW-COST SOLAR HOME HEATING SYS/PROTOTYPE DESIGN & TESTING STAGE	HOMES	
FRANKLIN INST RESEARCH LABS MECH&NUCLEAR ENGR DP 20TH & RACE STREETS PHILADELPHIA PA 19103 W H STEIGLEMANN MANAGER ENERGY SYS LABS -PHONE- 215 4481138 ASSISTING IN DESIGN OF ENERGY-EFFICIENT HOUSE USING SOLAR COLLECT,HEAT STORAGE,&HEAT PUMPS/IMPROVED PRIME MOVER FOR SOLAR-POWERD RANKINE CYCLE ENGINE	MEDIUM TEMPERATURE SOLAR COLLECTOR SYSTEMS WIND CONVERSION HOMES	HEAT ENGINE, RANKINE HEAT PUMPS
GEORGE O G LOF CONS ENGR 158 FILLMORE ST R204 DENVER CO 80206 GEORGE O LOF CONSULTING ENGINEER -PHONE- 303 3220446 303 4918632 APPROXIMATELY 100 TECHNICAL PAPERS BOOKS ARTICLES PATENTS IN SOLAR ENERGY RESEARCH DEVELOPMENT DESIGN TESTING ANALYSIS SOLAR COLLECTOR DEVELOPMENT AND PRACTICAL OPERATION SOLAR HEATING COOLING DEVELOPMENT SOLAR HEATED HOUSE DESIGN CONSTRUCTION & TESTING	MEDIUM TEMPERATURE SOLAR COLLECTOR SYSTEMS ALL SOLAR TECHNOLOGIES SINGLE FAMILY, PROPOSED SINGLE FAMILY, COMPLETED SINGLE FAMILY, DEVELOPMENT	HEAT TRANSFER SYSTEMS, COLLECTOR TO STORAGE COLLECTOR,REFLECTIVE SELECTIVE SURFACES COLLECTOR,ABSORPTIVE SELECTIVE SURFACES COLLECTOR,MECHANICAL SYSTEMS HEAT RECOVERY, STORAGE TO UTILIZATION STORAGE, SOLID MATERIALS STORAGE, LIQUIDS

-----SUBCLASSES-----SOLAR SUB SYSTEMS-----

HONEYWELL INC			MEDIUM TEMPERATURE SOLAR COLLECTOR SYSTEMS
SYSTEMS&RESEARCH CTR			HIGH TEMPERATURE SOLAR COLLECTOR SYSTEMS
2600 RIDGWAY PARKWAY			SOLAR ENERGY STORAGE
MINNEAPOLIS	MN	55413	SOLAR WATER HEATING
DR M K SELCUK			SOLAR ELECTRIC POWER GENERATION

-PHONE-
612 3314141
R & D RESIDENTIAL HEAT & COOL SYS...SOLAR HOUSE AUTOMATIC CONTROL SYS MOBILE SOLAR HEAT & COOL UNIT SPONSORED BY NSF WILL BE OPERATED IN NORTH & SOUTH CLIMATES IN VARIOUS PARTS OF THE COUNTRY FULL INSTRUMENT FOR DATA COLLECTION & COMPARATIVE SYS TESTS SEE ALSO RODGER SCHMIDT HONEYWELL INC

INSTITUTE OF GAS TECHNOLOGY			MEDIUM TEMPERATURE SOLAR COLLECTOR SYSTEMS
3424 S STATE			
CHICAGO	IL	60616	
WILLIAM F RUSH			

-PHONE-
312 2259600 X685
DESIGN CONSTRUCT & TEST EFF HEAT & COOL SYS...HAVE DEV SIMPLE GAS FUEL DEVICE MEC CAPABLE OF HEAT COOL VENTILATING & HUMIDITY CONTROL IN HOME AT HIGH EFF...CAN BE EASILY ADAPTED TO USE AVAIL SOLAR COLLECT TO REDUCE FUEL CONSUMP BY AT LEAST 50 PERCENT

INTERTECHNOLOGY CORPORATION			MEDIUM TEMPERATURE SOLAR COLLECTOR SYSTEMS
MARKETING OPERATIONS			BIO-CONVERSION
PO BOX 340			INORGANIC SEMI-CONDUCTORS
WARRENTON	VA	22186	SOLAR ELECTRIC POWER GENERATION
NORRIS L BEARD *			
DIRECTOR			

-PHONE-
703 3477900
SOLAR ENERGY USE FOR HVAC SYSTEMS...PHOTOSYNTHETIC CONVERSION TO PROD ELECTRICITY & SYNTHETIC NATURAL GAS PRODUCTION

----------SUBCLASSES----------SOLAR SUB SYSTEMS----------

INTERTECHNOLOGY CORPORATION	MEDIUM TEMPERATURE SOLAR COLLECTOR SYSTEMS
PO BOX 340	BIO-CONVERSION
WARRENTON VA 22186	WIND CONVERSION
GEORGE C SZEGO	INORGANIC SEMI-CONDUCTORS
PRESIDENT	SOLAR ELECTRIC POWER GENERATION

-PHONE-
703 3477900
703 3471113 X4121
FIRST SOLAR HEATED HIGH SCHOOL IN US...ENERGY PLANTATION, SOLAR FUEL GENERATION

IOWA STATE UNIV	MEDIUM TEMPERATURE SOLAR COLLECTOR SYSTEMS
DEPT OF ARCHITECTURE	
AMES IA 50010	
RAY D CRITES FAIA	

-PHONE-
515 2948460
DEVELOPING ENERGY CONSERVING HABITATION UTILIZING COMBO SOLAR ENERGY FLAT PLATE COLL & ELECTRIC HEAT PUMP FOR HEATING & COOLING

LOCKHEED PALO ALTO RESRCH LAB	MEDIUM TEMPERATURE SOLAR COLLECTOR SYSTEMS
DEPT 52-01 BLDG 201	
3251 HANOVER ST	
PALO ALTO CA 94304	
WAYNE SHANNON *	

-PHONE-
415 4934411
EXTENSION 45827

	SUBCLASSES	SOLAR SUB SYSTEMS
LOCKHEED PALO ALTO RESRCH LAB DEPT 52/21 BLDG 205 3251 HANOVER ST PALO ALTO CA 94304 ELMER R STREED -PHONE- 415 4934411	MEDIUM TEMPERATURE SOLAR COLLECTOR SYSTEMS	

EXTENSION 45867.DESIGN & ADMINISTER CONSTRUCTION & OPERATION OF SOLAR HEAT & COOL SYS FOR BLDGS...PERFORM R&D ON COMPONENT DEVELOPEMENT... ANALYTICAL CAPABILITY HAS BEEN DEVELOP & FEASIBILITY&PRELIM DESIGN OF SEVERAL PROJ PERFORMED...PUBLISHED:'EXPERIMENTAL PERFORMANCE OF A HONEYCOMB COVERED FLAT PLATE SOLAR COLLECTOR,'G CUNNINGTON&E STREED, PRESENTED AT 1973 INTERNTL SOLAR ENGY CONF, CLEVELAND, OHIO, OCT 1973. 'THE DEVELOPMENT OF A RESIDENTIAL HEATING & COOLING SYS USING NASA-DERIVED TECHNOLOGY,' M J O'NEIL, A J MCDONALD, W H SIMS, NOV 1972, CONTRACT NAS8-25986 (HREC-5986-3).

	SUBCLASSES	SOLAR SUB SYSTEMS
M I T ARCHITECTURE 3-407 MIT 77MASSAVE CAMBRIDGE MA 02139 DAY CHAHROUDI RESEARCH AFILLIATE -PHONE- 617 2537647	BIO-CONVERSION TOTAL SYSTEMS INTEGRATION ALL SOLAR TECHNOLOGIES GREENHOUSES PUBLISH LAY + TECHNICAL PERIODICALS PUBLISH ORIGINAL RESEARCHED INFORMATION	COLLECTOR,ALSO FUNCTION AS STRUCTURES SKIN COLLECTOR,MULTI-FUNCTION IN DWELLING SYS COLLECTOR,PASSIVE SYSTEMS HEAT TRANSFER SYSTEMS, COLLECTOR TO STORAGE STORAGE, SOLID MATERIALS STORAGE, PHASE CHANGE MATERIALS

TRANSPARENT INSULATION HEAT STORAGE MATERIALS CLIMATIC ENVELOPE GREEN-HOUSES SOLAR HVAC SOLAR DEHUMIDIFY PAST PROJ: NON ELECT SOLAR TRACKER H2O HEATER SOLAR PUMPS GREENHOUSES

MCFALL AND KONKEL CONSULT ENGR
2160 S CLERMONT ST
DENVER CO 80222
TEMPLE LOONEY

-PHONE-
303 7574976

	SUBCLASSES	SOLAR SUB SYSTEMS
MECHANICAL ENGINEER 3881 S GRAPE ST DENVER CO 80237 FRED L TRAUTMAN PE OWNER -PHONE- 303 7570416	NEW CONSTRUCTION, MECHANICAL SYSTEMS CONDOMINIUMS HOMES 15 FAMILY&UP LOW RISE, DEVELOPMENT LOW RISE MULTI-STRUCTURE,DEVELOPEMENT	COLLECTOR,MECHANICAL SYSTEMS COLLECTOR,ABSORPTIVE SELECTIVE SURFACES HEAT TRANSFER SYSTEMS, COLLECTOR TO STORAGE HEAT RECOVERY, STORAGE TO UTILIZATION HEAT PUMPS STORAGE, LIQUIDS
MIAMI, UNIV OF ARCHITECT& PLANNING CORAL GABLES FL 33124 RALPH WARBURTON ASSOCIATE DEAN -PHONE- 305 2843438	MEDIUM TEMPERATURE SOLAR COLLECTOR SYSTEMS	
APPLICATIONS OF THERMAL & NONTHERMAL COLLECTORS TO RESIDENTIAL LIVING ENVIRONMENTS		
NASA JOHNSON SPACE CRAFT CTR URBAN SYS PROJ OFF HOUSTON TX 77053 MARTIN B KEOUGH MAIL CODE EZ -PHONE- 713 4835557	ALL SOLAR TECHNOLOGIES SOLAR HEAT,COOL,&ELECTRIC POWER GENERATION PROVIDE ECONOMIC ANALYSIS OF SOLAR TECH	
ANALYZE & PERFORM TRADE STUDIES, HEAT & MASS TRANSFER ANALYSIS,& COST STUDIES UTILIZING SOLAR ENERGY IN INTEGRATED URBAN UTILITY SYS... PRELIM ANALYSIS COMPLETE ON SPECIFIC BLDG NETWORKS (JOHNSON SPACECRAFT CENTER HEAT&COOL FACILITY&PROTOTYPESHOPPING CTRS& OFFICE BLDGS)... FUTURE WORK:DESIGN FABRICATION&TESTING OF SML SOLAR COLL SYS FOR IN-HOUSE TESTING		
NASA LEWIS RESEARCH CTR SOLAR SYS SECTION 21000 BROOKPARK ROAD CLEVELAND OH 44135 FREDERICK SIMON M S 500-201 -PHONE- 216 4334000 X726	MEDIUM TEMPERATURE SOLAR COLLECTOR SYSTEMS	
PRACTICAL COLL TO HEAT&COOL BLDGS...INDOOR TEST FACILITY COMPLETE		

SUBCLASSES — SOLAR SUB SYSTEMS

NASA LEWIS RESEARCH CTR
SOLAR SYS SECTION
2100 BROOKPARK RD
CLEVELAND OH 44135
RONALD THOMAS
500-201
-PHONE-
216 4334000

MEDIUM TEMPERATURE SOLAR COLLECTOR SYSTEMS
SOLAR ELECTRIC POWER GENERATION

NASA LEWIS RESEARCH CTR
SOLAR SYS SECTION
2100 BROOKPARK ROAD
CLEVELAND OH 44135
RICHARD W VERNON

MEDIUM TEMPERATURE SOLAR COLLECTOR SYSTEMS

-PHONE-
216 4334000 X6857
DEVELOP&DEMO BUILDING IN LANGLEY VA HEAT & COOL BY SUN...HAVE SOLAR
COLLECTOR TEST FACILITY OPERATING TO EVALUATE COLLECTOR DESIGNS,
FACILITY INDOORS INCLUDES SOLAR SIMULATOR & CAN TAKE 4 X 4 FT COLLECT

NATIONAL ACADEMY OF SCIENCES
BLDG RES ADVISORY BD
2101 CONSTITUT AV NW
WASHINGTON DC 20418
BEN H EVANS AIA

MEDIUM TEMPERATURE SOLAR COLLECTOR SYSTEMS
WIND CONVERSION

-PHONE-
202 3896515
INVENTORY OF SOLAR ENERGY NON-GOVT RESEARCH
NATIONAL TECH CONF ON SOLAR EFFECTS ON BLDG DESIGN-STATE-OF-ART

	SUBCLASSES	SOLAR SUB SYSTEMS
NATIONAL BUREAU OF STANDARDS CTR FOR BLDG TECH BLDG 226 RM B104 WASHINGTON DC 20234 JAMES E HILL THERMAL ENGR SYS SEC -PHONE- 301 9213503	SOLAR RADIATION MEASUREMENTS MEDIUM TEMPERATURE SOLAR COLLECTOR SYSTEMS SOLAR ENERGY STORAGE HOMES	

DEVELOP STANDARD METHOD OF TEST FOR RATING SOLAR COLLECTORS & THERMAL STORAGE UNITS FIRST VERSION OF DOCUMENTS COMPLETE BY JUNE 74 DESIGNING SOLAR HEATING & COOLING SYS FOR 4 BEDROOM HOUSE COMPLETELY INSTRUMENTED FOR TESTING...RETROFITTING DONE BY SUMMER 74 THEN COMPRE-HENSIVE TEST PROG FOR ONE YEAR...PAST RES: DEVELOPED COMPUTER PROG CALLED NBSLD TO DETERMINE HEAT & COOL LOADS OF BLDGS...COMPLEMENTARY PROG TO DECODE WEATHER TAPES FROM US WEATHER BUREAU TO OBTAIN SOLAR DATA

	SUBCLASSES	SOLAR SUB SYSTEMS
NATIONAL BUREAU OF STANDRDS OFFICE OF HOUS TECH BLDG 226 ROOM B146 WASHINGTON DC 20234 ROBERT D DIKKERS MGR SOL ENG DEM PROG -PHONE- 301 9213285	GOVERNMENT AGENCIES LEGISLATIVE, FEDERAL	

RESPONSIBLE FOR ESTAB CRITERIA FOR HUD & NASA TO SELECT SOLAR PROJS FOR FUNDING THROUGH 1974 SOLAR HEAT & COOL DEMO ACT

	SUBCLASSES	SOLAR SUB SYSTEMS
NATIONAL BUREAU OF STNDRDS CTR FOR BLDG TECH ROOM 100B BLDG 226 WASHINGTON DC 20234 T KUSUDA -PHONE- 301 9213522 301 6566556	SOLAR RADIATION MEASUREMENTS MEDIUM TEMPERATURE SOLAR COLLECTOR SYSTEMS SOLAR ENERGY STORAGE HOMES PUBLISH TECHNICAL PERIODICALS PUBLISH TECHNICAL BOOKS	COLLECTOR,MULTI-FUNCTION IN DWELLING SYS COLLECTOR,PASSIVE SYSTEMS COLLECTOR,MECHANICAL SYSTEMS STORAGE, SOLID MATERIALS STORAGE, PHASE CHANGE MATERIALS STORAGE, LIQUIDS STORAGE, FUEL CELLS

MEASUREMENT OF BUILDING ENERGY CONSUMPTION...COMPUTERIZED CALCULATION OF BUILDING ENERGY REQUIREMENTS...SEVERAL PUBLICATION ON ASHRAE MEDIA

SUPCLASSES — SOLAR SUB SYSTEMS

NATIONAL SCIENCE FOUNDATION — GOVERNMENT AGENCIES PROVIDING GRANTS
ADVNCED ENGY RES&TEC
WASHINGTON DC 20550
HAROLD HOROWITZ
PROG MGR HEAT&COOL B

-PHONE-
202 6327364

NATIONAL SCIENCE FOUNDATN — GOVERNMENT AGENCIES PROVIDING GRANTS
PUBLIC TECH PROJECTS
WASHINGTON DC 20550
CHARLES CHEN
PROG MGR HEAT&COOL B

-PHONE-
202 6324175

NATIONAL SCIENCE FOUNDATN — GOVERNMENT AGENCIES PROVIDING GRANTS
PUBLIC TECH PROJECTS
WASHINGTON DC 20550
JOHN DEL GOBBO
PROG MGR HEAT&COOL B

-PHONE-
202 6324175

NATIONAL SCIENCE FOUNDATN — GOVERNMENT AGENCIES PROVIDING GRANTS
PUBLIC TECH PROJECTS
WASHINGTON DC 20550
RAYMOND FIELDS
DEPUTY DIRECTOR

-PHONE-
202 6324175

NATIONAL SCIENCE FOUNDATN — GOVERNMENT AGENCIES PROVIDING GRANTS
EXPLOR RESCH&PROB AS
WASHINGTON DC 20550
CHARLES R HAUER *
DIRECTOR

-PHONE-
202 6324175

	SUBCLASSES	SOLAR SUB SYSTEMS
NATL CNTR FOR ATMOS RESEARCH PLANT FACILITIES BOX 3000 BOULDER CO 80303 L PAUL MOORE PLANT ENGINEER -PHONE- 303 4945151 X547	SOLAR SPACE HEATING RETROFITTING, MECHANICAL SYSTEMS NEW CONSTRUCTION, PASSIVE SYSTEMS NEW CONSTRUCTION, MECHANICAL SYSTEMS ACADEMIC BUILDINGS UP TO 50 THOUSAND SQ FT,PROPOSED	COLLECTOR,MECHANICAL SYSTEMS COLLECTOR,ALSO FUNCTION AS STRUCTURES SKIN COLLECTOR,REFLECTIVE SELECTIVE SURFACES COLLECTOR,PASSIVE SYSTEMS HEAT RECOVERY, STORAGE TO UTILIZATION STORAGE, SOLID MATERIALS STORAGE, MECHANICAL/PUMPED LIQUIDS
NATL CONCRETE MASONRY ASSOC PO BOX 9185 ROSLYN S ARLINGTON VA 22209 THOMAS B REDMOND MGR RESEARCH & DEVEL -PHONE- 703 5240815	SOLAR RADIATION MEASUREMENTS	
NATL CTR FOR ATMOS RESEARCH BUDGET PLAN & INFO BOX 3000 BOULDER CO 80303 HENRY H LANSFORD * INFORMATION OFFICER -PHONE- 303 4945151	SOLAR SPACE HEATING RETROFITTING, MECHANICAL SYSTEMS NEW CONSTRUCTION, PASSIVE SYSTEMS NEW CONSTRUCTION, MECHANICAL SYSTEMS ACADEMIC BUILDINGS UP TO 50 THOUSAND SQ FT,PROPOSED	COLLECTOR,MECHANICAL SYSTEMS COLLECTOR,ALSO FUNCTION AS STRUCTURES SKIN COLLECTOR,REFLECTIVE SELECTIVE SURFACES COLLECTOR,PASSIVE SYSTEMS HEAT RECOVERY, STORAGE TO UTILIZATION STORAGE, SOLID MATERIALS STORAGE, MECHANICAL/PUMPED LIQUIDS
NEW MEXICO STATE UNIV MECHANICAL ENGR DEPT LAS CRUCES NM 88003 DR ROBERT L SAN MARTIN BOX 3450 -PHONE- 505 6463501	MEDIUM TEMPERATURE SOLAR COLLECTOR SYSTEMS HIGH TEMPERATURE SOLAR COLLECTOR SYSTEMS SOLAR REFRIGERATION SOLAR FURNACE NEW CONSTRUCTION, RESOURCES NEW CONSTRUCTION, TECHNOLOGIES	

NATL CONCRETE MASONRY ASSOC: HAVE BEGUN STUDY OF EFFECT OF SOLAR RADIENT HEAT ON HEAT/COOL REQUIRE FOR BUILDINGS WITH WALL MATERIALS WITH VARYING UNIT WEIGHT & SPECIFIC HEAT PROPERTIES

NEW MEXICO STATE UNIV: DEVELOP MED TEMP COLL,DESIGN INTEGRATED SOLAR HEAT & COOL SYS FOR RESIDENCES,DESIGN OF COMPETIVE PRICED RESIDENCES FOR UTILIZATION OF SOLAR,TESTING SOLAR COLLECTORS,DESIGN OF ENERGY EFF SOLAR RESIDENCE USING ENERGY CONSV MEASURES IN NEW CONSTRUCTION

	SUBCLASSES	SOLAR SUB SYSTEMS
OHIO STATE UNIVERSITY NORTH HIGH ST COLUMBUS OH 43210 CHARLES F SEPSY -PHONE- 614 4226898	NEW CONSTRUCTION, MECHANICAL SYSTEMS SINGLE FAMILY, DEMONSTRATION	

DEMO HOME SPONSORED BY HOMEWOOD CORP...ROOF MOUNTED H2O-GLYCOL COLL... TWO 2000 GAL STOR TANKS BELOW GRADE...SUPP ELECT HEAT...ARKLA ABSORP COOL UNIT

	SUBCLASSES	SOLAR SUB SYSTEMS
PENN, UNIV OF N CTR ENGY MGT&POWER 260 TOWNE BLDG D3 PHILADELPHIA PA 19174 NOAM LIOR ASSISTANT PROFESSOR -PHONE- 215 5944803	ALL SOLAR TECHNOLOGIES NEW CONSTRUCTION, RESEARCHING	COLLECTOR,MULTI-FUNCTION IN DWELLING SYS COLLECTOR,ALSO FUNCTION AS STRUCTURES SKIN HEAT TRANSFER SYSTEMS, COLLECTOR TO STORAGE HEAT PUMPS STORAGE, PHASE CHANGE MATERIALS STORAGE, LIQUIDS

SOLAR COLLECTORS...SOLAR HEATING & COOLING OF BUILDINGS

	SUBCLASSES	SOLAR SUB SYSTEMS
PENN, UNIV OF N CTR ENGY MGT&POWER 260 TOWNE BLDG D3 PHILADELPHIA PA 19174 HAROLD G LORSCH -PHONE- 215 5947181	MEDIUM TEMPERATURE SOLAR COLLECTOR SYSTEMS	

DEVELOPED SOLAR HEAT SYS.THERMAL ENGY STORAGE DEVICES.OFF PEAK AIR-CONDITION SYS
PUBLISHED:'SOLAR HEATING SYSTEMS ANALYSIS,'REPORT NO: NSF/RANN/GI27976/TR72/19./ 'THERMAL ENERGY STORAGE IN SODIUM SULFATE DECAHYDRATE MIX' REPORT NO NSF/RANN/SE/GI/27976/TR72/11./'RECONDENSER OFF PEAK AIR COND SYSTEM DESIGN & OPERATION' REPORT NO NSF/RANN/SE/GI27976/TR73/3. ALL REPORTS AVAILABLE THROUGH NATIONAL TECHNICAL INFO SERVICE

------SUBCLASSES------SOLAR SUB SYSTEMS------

RICE MAREK HARRAL HOLTZ INC — PROVIDE INFORMATION SEARCHES TO CLIENTS
PO BOX 12037
DENVER CO 80212
CLIFT EPPS

-PHONE-
303 4204455
DEVELOPED DESIGN CONCEPTS & GATHERED PERTINENT LITERATURE TO UTILIZE
SOLAR TO HEAT&COOL COMMERCIAL BLDGS-CONSULTING ENGINEERS

RICHARD P MUELLER & ASSOC INC — PROVIDE INFORMATION SEARCHES TO CLIENTS
ENVTL ENGRG DEPT
1900 SULPHUR SPRG RD
BALTIMORE MD 21227
ANDREW J PARKER
MANAGER
-PHONE-
301 2745666
202 7839159
PERFORM RESEARCH RELATED SYS STUDIES DEVELOP BASIC ANALYTICAL METHODS
PERFORM ENGINEERING EVALUATIONS OF IN-USE & TEST FUTURE SOLAR HEATING
& COOLING SYS FOR BUILDINGS

ROBERT G WERDEN&ASSOCIATES INC — SOLAR ELECTRIC POWER GENERATION; PROVIDE INFORMATION SEARCHES TO CLIENTS — HEAT PUMPS
PO BOX 414
JENKINTOWN PA 19046
ROBERT G WERDEN

-PHONE-
215 8852500
RESEARCH & APPLIED ENGINEERING-PATIENTS,DESIGNS,PILOT HARDWARE & SYS

SANDHU AND ASSOCIATES — PROVIDE INFORMATION SEARCHES TO CLIENTS
309 S RANDOLPH
CHAMPAIGN IL 61820
MARTIN D IGNAZITO

-PHONE-
217 3522999
DEVELOP PRACTICAL HARDWARE & SYS FOR HEAT & COOL BLDGS
CALCULATIONS,COMPUTER MODELING,RESEARCH & CONSTRUCT OF ECUIP,SYSTEMS
DESIGN-CONSULTING MECH ENGR

	SUBCLASSES	SOLAR SUB SYSTEMS
SEA PINES COMPANY ENVIRONMENTAL SERV PO BOX 5608 HILTON HEAD ISLAND SC 29928 RICHARD F WILDERMAN -PHONE- 803 7853333 X3144	MEDIUM TEMPERATURE SOLAR COLLECTOR SYSTEMS	
FOLLOW PROGRESS OF SOLAR ENERGY TO IDENTIFY TECHNIQUES APPLICABLE TO OUR ACTIVITIES-HEAT & COOL		
SUN MOUNTAIN DESIGN 960 CAMINO SANTANDER SANTA FE NM 87501 HERMAN G BARKMANN -PHONE- 505 9831680		COLLECTOR,PASSIVE SYSTEMS COLLECTOR,MECHANICAL SYSTEMS
TENNESSEE, UNIV OF MECH&AERO ENGR DEPT KNOXVILLE TN 37916 DR EDWARD LUMSDAINE ASSOCIATE PROFESSOR -PHONE- 615 9743035	MEDIUM TEMPERATURE SOLAR COLLECTOR SYSTEMS	
OPTIMIZE FLAT PLATE COLL EFF (THEORETICAL & EXP)...THEORETICAL STUDY DIFF TYPES CONCENTRATORS & TARGETS...HEAT & COOL LOAD CALCULATIONS FOR LARGE BUILD TO BE CONST ON UT CAMPUS..HAVE DEVEL COMPUTER PROG FOR THESE CALCULATIONS		
UNITED AIRCRAFT RESEARCH LABS ENERGY RESEARCH DEPT SILVER LANE EAST HARTFORD CT 06108 DR SIMION C KUO PRINCIPAL SYS ENGR -PHONE- 203 5658758	WIND CONVERSION ALL SOLAR TECHNOLOGIES SOLAR ELECTRIC POWER GENERATION SINGLE FAMILY, DEMONSTRATION	TURBINES HEAT TRANSFER SYSTEMS, COLLECTOR TO STORAGE HEAT ENGINE, RANKINE STORAGE, PHASE CHANGE MATERIALS STORAGE, MECHANICAL/COMPRESSED GAS STORAGE, LIQUIDS STORAGE, FUEL CELLS
SOLAR ENERGY UTILIZATION PROJECT...ESTIMATE CRITICAL TECHNOLOGIES IN-HERENT WITH EFFECTIVE INTERCEPTION,TRANSFER,STORAGE, & UTILIZATION OF SOLAR ENERGY FOR HEATING COOLING & POWER GENERATION SOLAR POWER STATIONS, UAR-K134...SOLAR CLIMATE CONTROL, UAR-N171107-3		

SUBCLASSES — SOLAR SUB SYSTEMS

VIRGINIA, UNIV OF — SOLAR REFRIGERATION
DEPT OF MECH ENGR
CHARLOTTESVILLE VA 22901
DR JAMES J KAUZLARICH

-PHONE-
804 9243661
SOLAR HEAT & REFRIG FOR PUBLIC BLDGS

WESTINGHOUSE ELECTRIC CORP. — TOTAL SYSTEMS INTEGRATION
SPECIAL ENERGY SYST — SOLAR ENGINES
PO BOX 1693 MS-999 — RETROFITTING, RESEARCHING
BALTIMORE MD 21203 — NEW CONSTRUCTION, RESEARCHING
ALBERT WEINSTEIN — ACADEMIC BUILDINGS
MANAGER
-PHONE-
301 7653454
NSF STUDY: PHASE 0 - SOLAR HEATING AND COOLING OF BUILDINGS
ARTICLE: FEASIBILITY OF SOLAR HEATING AND COOLING OF BUILDINGS, PROFES
SIONAL ENGINEER, FEB 74, WITH DR C.CHEN, NSF

WISCONSIN, UNIV OF
ENGINEER EXP STATION
1500 JOHNSON DRIVE
MADISON WI 53706
J A DUFFIE

COMPUTER MODELING & SIMULATION OF SOLAR HEAT&COOL SYS...OBJECTIVE TO
IDENTIFY & OPTIMIZE PRACTICAL SYS..DEVELOP DESIGNS OF VARIOUS COMBO OF
HEAT&COOL FACILITIES..SELECT MORE PROMISING SYS OPTIMIZE THEIR DESIGN
PARAMETERS & DETERMINE PERFORMANCE & COSTS IN DWELLINGS OF VARIOUS
SIZE IN CHOSEN LOCALITIES...DET DEVEL NECESSARY TO MOVE RESIDENTIAL
SOLAR ENERGY USE INTO PUBLIC SECTOR

NEWCASTLE UNIV OF
DEPT OF ELECT ENGR
NEW SOUTH WALES AUSTRALIA
JOHN P MOORE

	SUBCLASSES	SOLAR SUB SYSTEMS
INDIAN INSTITUTE OF TECHNOLOGY MADRAS 600036 INDIA DR MAKAM C GUPTA PROF MECH ENGR -PHONE- 802742 X235 TEACHING & RESEARCH DESIGN & DEVELOP SOLAR HEATERS & COOLERS/SOLAR ROOM AIR DEHUMIDIFIER DESIGNED DEVELOPED & TESTED		
ROORKEE, UNIV OF DEPT OF MECH ENGR ROORKEE 247667 INDIA R K MEHROTRA DESIGN & DEVELOPMENT OF ROCKPILE THERMAL STORAGE SYS FOR AIR COND OF BUILDING	SOLAR ENERGY STORAGE	
TECHNION-ISRAEL INST OF TECH BLDG RESEARCH STATN TECHNION CITY HAIFA ISRAEL BARUCH GIVONI HD BLDG CLIMATOLOGY -PHONE- 225111 X285 UTILIZATION OF NATURAL ENERGIES/SOLAR RADIATION NIGHT OUTGOING RAD EVAPORATION ETC/ FOR HEATING & COOLING BLDGS...STUDYING INTERACTION BETWEEN DIFF TECH SYS & TYPE OF BLDG & CLIMATIC COND		
TECHNION-ISRAEL INST OF TECH MECHANICAL ENG DEPT TECHNION CITY HAIFA ISRAEL GERSHON GROSSMAN SENIOR LECTURER -PHONE- 04 235105 04 235106 NATURAL AIR CONDITIONING OF BUILDINGS DESIGN OF SOLAR COLLECTORS	MEDIUM TEMPERATURE SOLAR COLLECTOR SYSTEMS HIGH TEMPERATURE SOLAR COLLECTOR SYSTEMS SOLAR WATER HEATING APARTMENTS HOMES	COLLECTOR,MULTI-FUNCTION IN DWELLING SYS COLLECTOR,ALSO FUNCTION AS STRUCTURES SKIN COLLECTOR,MECHANICAL SYSTEMS

--------SUBCLASSES--------SOLAR SUB SYSTEMS--------

TECHNION-ISRAEL INST OF TECH	MEDIUM TEMPERATURE SOLAR COLLECTOR SYSTEMS
DEPT MECH ENGR	HIGH TEMPERATURE SOLAR COLLECTOR SYSTEMS
TECHNION CITY	
HAIFA ISRAEL	
AVRAHAM SHITZER	

-PHONE-
04 235105

WASEDA UNIVERSITY	HIGH TEMPERATURE SOLAR COLLECTOR SYSTEMS
DEPT OF ARCHITECTURE	SOLAR WATER HEATING
4-170 NISHIOKUBO	
SHINJUKU-KU TOKYO JAPAN	
KEN-ICHI KIMURA	

-PHONE-
03 2093211 X420

EXPERIMENTS WITH EAST WEST HORIZONTAL PARABOLIC CYLINDER TYPE COLLECT IN PREFABRICATED UNITS ARE BEING MADE IN COMBINATION WITH LIBR ABSORPTION REFRIGERATION MACHINE. PAST WORK WITH SOLAR WATER HEATERS OF GRAVITY CIRCULATION

---SUBCLASSES--- ---SOLAR SUB SYSTEMS---

MICHAEL A AIMONE
GRADUATE STUDENT-U F
3461 SW 2ND AVE #425
GAINESVILLE FL 32607

HYBRID SOLAR ENERGY ELECTRICAL HOME SERVICE CONCEPT:I AM INVESTIGATING THE ELECTRIC UTILITIES RESPONSE TO INTEGRATED ELECTRIC ENERGY SUPPLY-SOLAR PLUS CONVENTIONAL

T L SWANSEN PE
RT 2 BOX 345
EAST TROY WI 53120

HIGH TEMPERATURE SOLAR COLLECTOR SYSTEMS
SOLAR ENGINES

HEAT ENGINE, RANKINE
HEAT ENGINE, STERLING

-PHONE-
414 6427487

HEAT ENGINES WITH HIGH TEMP CONCEN TO USE SOLAR FOR ELECTRIC PRODUCTON FOR SOME OPERATIONS SML FARM OR FACTORY...HAVE BUILT RANKIN CYCLE NOW REBUILDING TO STIRLING CYCLE...NOT READY FOR PUBLIC INQUIRY PLEASE

JOHN A WARING
RES WRITER&CONSULTNT
8502 FLOWER AVENUE
TAKOMA PARK MD 20012

-PHONE-
202 6938675
301 5887738

TO ASSEMBLE AVAIL DATA ON SOLAR FORELECT POWER & INCORPORATE INTO N AMERICAN CONTINENTAL HYDROLOGY & ENERGY DESIGNS FOR PLANNING CONTIN-ENTALS FUNCTIONAL ENERGY COORDINATION NOW UNDER PREP BY TECHNOLOGY INC

ATOMICS INTERNATIONAL
P O BOX 309
CANOGA PARK CA 91304
W V BOTTS

-PHONE-
213 3411000 X2062

DEVELOP ECONOMICAL SOLAR ELECTRIC POWER PLANT/INITIAL STUDIES COMPLETE OPTIMIZATION STUDIES UNDERWAY

SUBCLASSES — SOLAR SUB SYSTEMS

BABCOCK & WILCOX CO
RESEARCH CENTER
1562 BEESON
ALLIANCE OH 44601
LARRY LANDIS *
PERSONNEL&PUBLIC REL
-PHONE-
216 8219110 X387
SOLAR THERMAL POWER SUBCONTRACTOR TO UNIV OF MINN RESPONSIBLE FOR:
SURVELLANCE & EVALUATION OF DIRECT ENERGY CONVERSION

BABCOCK & WILCOX CO
RESEARCH CENTER
1562 BEESON
ALLIANCE OH 44601
ROBERT A LEE
STEAM GEN TECH SECT
-PHONE-
216 8219110 X376
SOLAR THERMAL POWER SUBCONTRACTOR TO UNIV OF MINN
COAUTHOR NSF/RANN REPORT 'RESEARCH APPLIED TO SOLAR THERMAL POWER SYS'
PREPARED BY U OF MINN & HONEYWELL NSF/RANN/SE/GI-34871/PR/73/2

BABCOCK & WILCOX CO
RESEARCH CENTER
1562 BEESON STREET
ALLIANCE OH 44601
KARL H SCHULZE
RESEARCH PHYSICIST
-PHONE-
216 8219110 X227
PROJECT RESPONSIBILITY: SURVEILLANCE AND EVALUATION OF DIRECT ENERGY
CONVERSION -TECHNOLOGICAL FORECASTING FOR ADVANCED ENERGY CONCEPTS
PAPERS:'DIRECT ENERGY CONVERSION STATUS FOR LARGE SCALE POWER GENERA-
TION', IECEC, 1968.
'POTENTIAL OF ELECTRGASDYNAMICS FOR BULK POWER GENERATION', ENERGY
CONVERSION, VOL 9, PP 47-53.

Subclasses: ALL SOLAR TECHNOLOGIES; HIGH TEMPERATURE SOLAR COLLECTOR SYSTEMS

Solar sub systems: HEAT TRANSFER SYSTEMS, COLLECTOR TO STORAGE; MAGNETOHYDRONAMICS

---SUBCLASSES---SOLAR SUB SYSTEMS---

BLACK & VEATCH
1500 MEADOW LAKE PKY
KANSAS CITY MO 64114
J C GROSSKREUTZ

HIGH TEMPERATURE SOLAR COLLECTOR SYSTEMS

-PHONE-
816 3617000
PROJECT ENGR SOLAR STUDY PROGRAMS...DESIGN OPTIMAL SYSTEM SOLR THERMAL CENTRAL POWER STATION NASA-LEWIS...TECH ASSESS SOLR THERMAL ELECT GENERATION FOR OTA

DENVER, UNIV OF
DENVER RESEARCH INST
UNIVERSITY OF DENVER
DENVER CO 80210
FRED P VENDITTI
SR RES ENGR
-PHONE-
303 7532241

SOLAR WATER HEATING

COLLECTOR,ABSORPTIVE SELECTIVE SURFACES
HEAT TRANSFER SYSTEMS, COLLECTOR TO STORAGE
HEAT RECOVERY, STORAGE TO UTILIZATION
STORAGE, LIQUIDS

PHOTOVOLTRIC CONVERSION
TECHNOLOGY ASSESSMENT FOR SOLAR WATER HEATING

FOSTER WHEELER CORP
APP THERMODYNAMICS
110 S ORANGE AVE
LIVINGSTON NJ 07039
R J ZOSCHAK
MANAGER
-PHONE-
201 5331100

HIGH TEMPERATURE SOLAR COLLECTOR SYSTEMS

STUDY DIRECT INPUT SOLR ENGY TO FOSSIL FUELED CENTRAL POWER STATION

GENERAL ATOMIC COMPANY
P O BOX 81608
SAN DIEGO CA 92138
DR JOHN RUSSELL
HEAD,SOLAR RESEARCH

HIGH TEMPERATURE SOLAR COLLECTOR SYSTEMS

-PHONE-
714 4531000
FIXED MIRROR EMBEDDED IN GROUND-MOVABLE HEAT COLLECTION PIPE SUPPORTED ABOVE SYS,ROTATES WITH SUN, HEATED GAS OR LIQUID INSIDE CIRCULATES TO STEAM GENERATOR TO BE USED AS IN CONVENTIONAL POWER PLANTS TO GENERATE ELECTRICITY...DIRECT INQUIRY TO: DAVE WALCK, GENERAL ATOMIC CO

SUBCLASSES—SOLAR SUB SYSTEMS

GENERAL ATOMIC COMPANY
P O BOX 81608
SAN DIEGO CA 92138
DAVE WALCK *

HIGH TEMPERATURE SOLAR COLLECTOR SYSTEMS

-PHONE-
714 4531000

FIXED MIRROR SOLAR ENERGY CONCENTRATOR WITH MOVEABLE HEAT COLLECTION PIPE TO PRODUCE ELECTRICITY...SPONSORED BY:ARIZONA PUBLIC SERVICE & SOUTHERN CALIF EDISON CO...FABRICATION BY:GENERAL ATOMIC...TESTING BY: ARIZONA STATE UNIV...SEE ALSO:DR JOHN RUSSEL, GENERAL ATOMIC

GEORGIA INST OF TECHNOLOGY
ENGINEERING EXP STAT
ATLANTA GA 30332
JESSE D WALTON JR

HIGH TEMPERATURE SOLAR COLLECTOR SYSTEMS
SOLAR FURNACE

-PHONE-
404 8943661

HIGH TEMP CONCENTRATORS-DESIGN OF SOLAR ELECT GENERATING PLANT USING SOLAR HEATED STEAM BOILER & CONVENTIONAL STEAM TURBINE DRIVEN GENERATR PROG BEGUN JAN 1974,WORKING WITH MARTIN MARIETTA CORP-DENVER DIV UNDER NSF GRANT.....FOR FURTHER DETAIL ON SOLAR FURNACE & RESEARCH WORK AT GIT SEE: DR STEVE BOMAR, GEORGIA INST OF TECH, ENGR EXP STATION

HONEYWELL INC
SYSTEMS&RESEARCH CTR
2600 RIDGWAY PARKWAY
MINNEAPOLIS MN 55413
ROGER N SCHMIDT

MEDIUM TEMPERATURE SOLAR COLLECTOR SYSTEMS
HIGH TEMPERATURE SOLAR COLLECTOR SYSTEMS
INORGANIC SEMI-CONDUCTORS

-PHONE-
612 3314141 X4078

CENTRAL ELECT POWER RESEARCH WITH UNIV OF MINN-CONCENTRATING HIGH TEMP SYS PAST RESEARCH: SELECTIVE SOLAR ABSORBER COATINGS DEVELOP HIGH TEMP SPACE STABLE COATINGS & DEMO DURABILITY & CAPABLE OF BEING UNIFORMLY DEP ON SPHERICAL SURFACES

----------SUBCLASSES----------SOLAR SUB SYSTEMS----------

HOUSTON, UNIV OF HIGH TEMPERATURE SOLAR COLLECTOR SYSTEMS STORAGE, HYDROGEN
PHYSICS DEPT
HOUSTON TX 77004
ALVIN HILDEBRANDT
PROFESSOR

-PHONE-
713 7492834
CONCENTRATOR COMPOSED OF LARGE NUMBER MIRRORS THAT REFLECT SOLAR ENGY
TO SINGLE COLLECTOR ATOP LARGE TOWER. ENERGY THEN CONVERTED TO ELECT
BY STEAM CYCLE OR CLOSED CYCLE MHD GENERATOR...STORAGE HYDROGEN COMP-
RESSED OR CRYOGENIC LIQUIDS...COVERS 1 SQ MILE-NSF GRANT
SEE:L VANTHULL,HOUSTON UNIV OF FOR FURTHER INFO

HOUSTON, UNIV OF HIGH TEMPERATURE SOLAR COLLECTOR SYSTEMS
PHYSICS DEPT
HOUSTON TX 77004
DR LORIN L VANT-HULL

-PHONE-
713 7493809
FEASIBILITY STUDY OF SOLAR THERMAL POWER SYS BASED ON OPTICAL TRANS
SEE:A HILDEBRANDT,HOUSTON UNIV OF FOR COLLECTOR DETAILS
PROJECT COMBINED EFFORT WITH MCDONNELL DOUGLAS ASTRONAUTICS CO
PUBLISHED:HILDEBRANDT & VANT-HULL, 'A TOWER TOP FOCUS SOLAR ENERGY
COLLECTOR,'ASME PUBLICATION 73-WA/SOL-7.

LAWRENCE LIVERMORE LAB SOLAR WATER HEATING STORAGE, LIQUIDS
SOLAR ENERGY GROUP 1MILLION SQ FT + UP,PROPOSED COLLECTOR,OTHER
P O BOX 808
LIVERMORE CA 94550
ARNOLD F CLARK
PHYSICIST
-PHONE-
415 4471100 X8693

----------SUBCLASSES----------SOLAR SUB SYSTEMS----------

MARTIN MARIETTA AEROSPACE — HIGH TEMPERATURE SOLAR COLLECTOR SYSTEMS — TURBINES
MAIL # E0455
PO BOX 179
DENVER CO 80201
FLOYD A BLAKE

-PHONE-
303 7945211 X3677
SYSTEMS ANALYSIS & COMPONENT DESIGNFOR A SOLAR POWER PLANT NSF SPONSOR WITH GEORGIA TECH...SUN FOCUSED ON BOILER WITH HELIOSTAT MIRRORS REACH 1000F CONVERT TO ELECT BY CONVENTIONAL STEAM TURBINES...TECH & ECON ANALYSIS TO BE PREPARED

MARYLAND DEPT OF NAT RESOURCES — ALL SOLAR TECHNOLOGIES; GOVERNMENT AGENCIES
PWR PLT SITING PROG
TAWES ST OFFICE BLDG
ANNAPOLIS MD 21401
DR PAUL MASSICOT

-PHONE-
301 2675384
INTEREST IN ALL TERRESTRIAL SYS RELATED TO ELECT GEN...WE EVALUATE PROPOSED ELECTRIC POWER PLANTS, ESPECIALLY THE ENVIRONMENTAL IMPACT & THE NEED FOR POWER

MCDONNELL DOUGLAS ASTRONAUTICS — MEDIUM TEMPERATURE SOLAR COLLECTOR SYSTEMS; HIGH TEMPERATURE SOLAR COLLECTOR SYSTEMS
ADV SOL&NVC ENGY SYS
5301 BOLSA AVE
HUNTINGTON BEACH CA 92646
RAYMON W HALLET JR
DIRECTOR
-PHONE-
714 8963664
DEFINE ECONOMIC ATTRACTIVNESS OF SOLAR THERMAL ELECT PWR SYS IN TERMS OF:COLLECTOR ENGY TRANSPT STORAGE POWER CONVERSION SYS HEAT REJECTION SYS & USER REQUIREMENTS

---SUBCLASSES---SOLAR SUB SYSTEMS---

MINN, UNIV OF
DEPT MECH ENGR
MINNEAPOLIS MN 55455
F R ECKERT

HIGH TEMPERATURE SOLAR COLLECTOR SYSTEMS — HEAT TRANSFER SYSTEMS, COLLECTOR TO STORAGE

-PHONE-
612 3733315
DEVELOP TECHNOLOGY OF PHOTOTHERMAL CONVERSION FOR EFFICIENT UTILIZATON SOLAR ENGY IN LARGE SCALE PROD OF ELECTRICITY...STUDYING SEVERAL HEAT TRANSFER FLUID SYS

MINN, UNIV OF
DEPT OF MECH ENGR
MINNEAPOLIS MN 55455
R C JORDAN *

HIGH TEMPERATURE SOLAR COLLECTOR SYSTEMS

-PHONE-
612 3733302
RESEARCH APPLIED TO SOLAR-THERMAL POWER SYS RESEARCH WITH HONEYWELL-NSF-RANN SPONSORED...BUILDING SCALE MODEL 4FT DIAMETER 15FT LONG TROUGH SURFACE REFLECTS SOLAR ENGY TO HEAT PIPE TO CONVERT ENERGY TO STEAM

MINN, UNIV OF
DEPT OF MECH ENGR
MINNEAPOLIS MN 55455
E M SPARROW

HIGH TEMPFRATURE SOLAR COLLECTOR SYSTEMS

-PHONE-
612 3733034
DEVELOPMENT OF COLLECTION TRANSFER & STORAGE SYS FOR TERRESTRIAL SOLAR THERMAL ELECTRIC POWER PLANTS

MINN, UNIV OF
ELECT ENGINEERING
MINNEAPOLIS MN 55455
G K WEHNER

HIGH TEMPERATURE SOLAR COLLECTOR SYSTEMS — COLLECTOR,REFLECTIVE SELECTIVE SURFACES
COLLECTOR,ABSORPTIVE SELECTIVE SURFACES

-PHONE-
612 3737831
SOLAR THERMAL POWER SYS-COATINGS RESEARCH

	SUBCLASSES	SOLAR SUB SYSTEMS
MOTOROLA INC SOLAR ENERGY GROUP 8201 E MCDOWELL ROAD SCOTSDALE AZ 85252 DR ARNOLD LESK HEAD -PHONE- 602 9493355	INORGANIC SEMI-CONDUCTORS SOLAR WATER HEATING SOLAR SPACE HEATING SOLAR REFRIGERATION	
MOTOROLA INC NEW VENTURES LABS 8201 E MCDOWELL ROAD SCOTSDALE AZ 85252 STEVE LEVY MANAGER -PHONE- 602 9493355	INORGANIC SEMI-CONDUCTORS SOLAR WATER HEATING SOLAR SPACE HEATING SOLAR REFRIGERATION	
NASA LEWIS RESEARCH CTR POWER SYSTEMS DIV 21000 BROOKPARK ROAD CLEVELAND OH 44135 GEORGE M KAPLAN -PHONE- 216 4334000 X6845	MEDIUM TEMPERATURE SOLAR COLLECTOR SYSTEMS HIGH TEMPERATURE SOLAR COLLECTOR SYSTEMS WIND CONVERSION	
PHILADELPHIA ELECTRIC CO ENGY CONVERSON RESCH 2301 MARKET ST S10-1 PHILADELPHIA PA 19101 OLIVER D GILDERSLEEVE JR -PHONE- 215 8414856	INORGANIC SEMI-CONDUCTORS	HEAT PUMPS

Notes:

- MOTOROLA INC, SOLAR ENERGY GROUP: RESEARCHING PHOTOVOLTAICS SHORT TERM-FOR REMOTE POWER SUPPLYS LOW ELECT NEEDS LONG TERM-DOMESTIC POWER SUPPLY & LARGE SCALE ELECTRIC GENERATION
- NASA LEWIS RESEARCH CTR: DETERMINE RELATIVE POTENTIAL & ECONOMICS OF UTILIZING THERMAL COLL SYS IN GENERATION OF ELECT...CONTRACT WITH HONEYWELL TO STUDY VARIOUS COLLECTOR CONVERSION SYS

	SUBCLASSES	SOLAR SUB SYSTEMS
REDE RESEARCH&DESIGN INSTITUTE P O BOX 307 PROVIDENCE RI 02901 RONALD BECKMAN DIRECTOR	WIND CONVERSION INORGANIC SEMI-CONDUCTORS TOTAL SYSTEMS INTEGRATION RETROFITTING, PRACTICING OFFICE/BANK BUILDINGS	HEAT PUMPS STORAGE, CHEMICAL BATTERIES

-PHONE-
401 8615390
SEE RALPH BECKMAN REDE FOR DETAILS

TEXAS TECH UNIVERSITY DEPT OF CIVIL ENGRG PO BOX 4089 LUBBOCK TX 79409 GEORGE A WHETSTONE	WIND CONVERSION	

-PHONE-
806 7421234
HYDRO-SOLAR POWER INVESTIGATED AS MEANS OF UTILIZING HYDROELECT POWER NO WORK AT PRESENT...PUBLISHED: WHETSTONE,'WATER POWER FOR THE DESERT' WATER POWER, OCT 1961, PP388-390.

WALTER V STERLING INC TECH&MANAGMT CONSULT 16262 E WHITTIER B WHITTIER CA 90603 CHESTER W YOUNG	HIGH TEMPERATURE SOLAR COLLECTOR SYSTEMS INORGANIC SEMI-CONDUCTORS SOLAR FURNACE	

-PHONE-
213 9438703
FEASIBILITY STUDIES IN SOLAR FURNACE & HIGH TEMP PHOTOVOLTAIC GEN COMPUTER CORRELATION OF SOLAR & TERRESTRIAL EVENTS...STUDIES JUST BEGINING...PAST WORK SOLAR RADIO ASTRONOMY PROG SPECTRAL ANALYSIS OF SUN BETWEEN 500 & 950 MHZ...3 NEW SOLAR RAD FEATURES OBSERVED PUB: 'PRELIMINARY STUDY OF DYNAMIC SPECTRA OF SOLAR RADIO BURSTS IN FREQ RANGE 500-950 MC/S,' ASTROPHYSICAL JOURNAL, JAN 1961.

SUBCLASSES	SOLAR SUB SYSTEMS
WESTINGHOUSE ELECT CORP 8401 BASELINE RD BOULDER CO 80303 DR CHARLES D BEACH -PHONE- 303 4946363	MEDIUM TEMPERATURE SOLAR COLLECTOR SYSTEMS HIGH TEMPERATURE SOLAR COLLECTOR SYSTEMS

SOLAR THERMAL ELECTRIC POWER SYS ANALYSIS TO DETERMINE LOWEST COST SYS CONFIGURATION FOR PROD OF ELECT POWER FOR SOLAR USING TODAYS TECH

MINISTRY OF INT'L TRADE ENERGY TRNSPT SECT ELECTROTECHNICAL LAB TOKYO JAPAN TAKASHI HORIGOME ENERGY DIVISION -PHONE- 042 6612141	HIGH TEMPERATURE SOLAR COLLECTOR SYSTEMS WIND CONVERSION

SOLAR THERMAL POWER STATION DEVELOPING SYSTEM MODEL BY USING HIGH EFF SELECTIVE FILMS&TROUGH CONCENTRATORS

"... it is interesting to note that the energy in sunshine falling on the surface of Lake Mead is five times the output of the generators at Hoover Dam..."

Halacy, The Coming Age of Solar Energy

			SUBCLASSES	SOLAR SUB SYSTEMS
MORRIS LIECHTY R R 1 GRAPILL	IN	46741	MEDIUM TEMPERATURE SOLAR COLLECTOR SYSTEMS HIGH TEMPERATURE SOLAR COLLECTOR SYSTEMS WIND CONVERSION	

-PHONE-
219 6272780
219 4858095
TO PERFECT & BUILD SOLAR HEATED HOMES WITH WIND & WATER ENERGY FOR ELECTRICITY...GETTING READY TO BUILD SOLAR HOME IN 1974-SEMI UNDER-GROUND

JAMES R MIHALOEW 13463 ATLANTIC RD STRONGSVILLE	OH	44136	MEDIUM TEMPERATURE SOLAR COLLECTOR SYSTEMS HIGH TEMPERATURE SOLAR COLLECTOR SYSTEMS WIND CONVERSION NEW CONSTRUCTION, RESOURCES HOMES SINGLE FAMILY, PROPOSED	COLLECTOR,MULTI-FUNCTION IN DWELLING SYS COLLECTOR,ALSO FUNCTION AS STRUCTURES SKIN COLLECTOR,ABSORPTIVE SELECTIVE SURFACES HEAT TRANSFER SYSTEMS, COLLECTOR TO STORAGE HEAT PUMPS STORAGE, LIQUIDS

-PHONE-
216 2384184
216 4334000 X6819
ARCHITECT AND ENGINEER (OWNER).....DESIGNED HOME FEATURING ADVANCED TECHNOLOGY SYSTEMS INCLUDING SOLAR AND WIND POWER.NASA TECHNOLOGY UTIL IZATION...IN DESIGN PHASE...CONSTRUCTION PERIOD 1975...SYSTEM & COMPONENT MANUFACTURER INQUIRIES INVIDED

EDWARD N WHITNEY PE 1978 E 2ND AVE DURANGO	CO	81301	WIND CONVERSION SOLAR WATER HEATING SOLAR SPACE HEATING & COOLING SOLAR ELECTRIC POWER GENERATION HOMES RELIGIOUS/PUBLIC/AMUSEMENT BUILDINGS	COLLECTOR,ALSO FUNCTION AS STRUCTURES SKIN COLLECTOR,MULTI-FUNCTION IN DWELLING SYS COLLECTOR,REFLECTIVE SELECTIVE SURFACES COLLECTOR,ABSORPTIVE SELECTIVE SURFACES COLLECTOR,PASSIVE SYSTEMS COLLECTOR,MECHANICAL SYSTEMS HEAT TRANSFER SYSTEMS, COLLECTOR TO STORAGE

-PHONE-
303 2473584
VERTICAL SOLAR PANEL ON HOUSE

	SUBCLASSES	SOLAR SUB SYSTEMS
ARGONNE NATIONAL LAB ENERGY & ENVIRONMENT 9500 SOUTH CASS AVE ARGONNE IL 60439 RALPH C NIEMANN MECH ENGR -PHONE- 312 7397711 X5007 SOLAR COLLECTOR DEVELOPMENT DESIGN OF SOLAR HEATED AND COOLED STRUCTURES	TOTAL SYSTEMS INTEGRATION RETROFITTING, RESEARCHING NEW CONSTRUCTION, RESEARCHING APARTMENTS 2-5 FAMILY, PROPOSED ACADEMIC BUILDINGS	COLLECTOR,MULTI-FUNCTION IN DWELLING SYS COLLECTOR,MECHANICAL SYSTEMS HEAT TRANSFER SYSTEMS, COLLECTOR TO STORAGE HEAT RECOVERY, STORAGE TO UTILIZATION HEAT PUMPS STORAGE, MECHANICAL/COMPRESSED GAS STORAGE, LIQUIDS
CALIF, UNIV OF-LIVERMORE LAWRENCE LIVER LAB P O BOX 808 LIVERMORE CA 94550 S SZYBALSKI ENERGY CONSV SECT -PHONE- 415 4471100 X7721 WISH TO USE FLATPLATES & WIND GENERATORS TO REDUCE FUEL BILLS-LOOKING FOR SUITABLE PRODUCTS	MEDIUM TEMPERATURE SOLAR COLLECTOR SYSTEMS WIND CONVERSION	
DOW CHEMICAL USA ROCKY FLATS DIV P O BOX 888 GOLDEN CO 80401 DEVELOPMENT PROJECTS -PHONE- 303 4942951 GOAL:TO ESTAB ALTERNATIVE ENERGY SOURCE FOR POPULOUS FRONT RANGE AREA PROPOSAL SUBMIT TO AEC: USE WIND & SOLAR FLAT PLAT GENERATORS IN 5YR STUDY TO DET PLACEMENT, ENVIRO IMPACT, EFFICIENT DESIGN,&OPERATIONAL DATA/COMP DATE 1979	WIND CONVERSION SOLAR SPACE HEATING & COOLING	

	SUBCLASSES	SOLAR SUB SYSTEMS
ENVIRONMENTAL CONSULT SERV INC 1485 SIERRA DRIVE BOULDER CO 80302 DR JAN F KREIDER CHIEF ENGR -PHONE- 303 4472218 303 4431235	HIGH TEMPERATURE SOLAR COLLECTOR SYSTEMS SOLAR ENERGY STORAGE RETROFITTING, MECHANICAL SYSTEMS NEW CONSTRUCTION, MECHANICAL SYSTEMS HOMES SINGLE FAMILY, PROPOSED	

DEVELOPMENT OF A NEW TYPE OF SOLAR CONCENTRATOR FOR HOMES & OFFICE BLDG HEATING/ CONSULTING ADVICE ON INSTALLING CONVENTIONAL SOLAR HEAT SYS IN RES & COMM BLDGS/ PERIODICALLY OFFER SOLAR ENERGY COURSES THRU UNIV OF COLORADO

ENVIRONMENTAL ENERGY FOUNDATN ROOM 1001 12000 EDGEWATER DR LAKEWOOD OH 44107 G GILBERT OUTTERSON DIRECTOR -PHONE- 216 2218130	MEDIUM TEMPERATURE SOLAR COLLECTOR SYSTEMS HIGH TEMPERATURE SOLAR COLLECTOR SYSTEMS SOLAR ELECTRIC POWER GENERATION	COLLECTOR,MULTI-FUNCTION IN DWELLING SYS COLLECTOR,ALSO FUNCTION AS STRUCTURES SKIN HEAT ENGINE, STERLING

HELIO ASSOCIATES INC 8230 E BROADWAY TUCSON AZ 85710 ROBERT E GERLACH VICE PRESIDENT -PHONE- 602 8865376	SOLAR RADIATION MEASUREMENTS MEDIUM TEMPERATURE SOLAR COLLECTOR SYSTEMS HIGH TEMPERATURE SOLAR COLLECTOR SYSTEMS SOLAR AGRICULTURAL DRYING SOLAR WATER HEATING SOLAR SPACE HEATING & COOLING	COLLECTOR,ABSORPTIVE SELECTIVE SURFACES COLLECTOR,PASSIVE SYSTEMS COLLECTOR,OTHER HEAT TRANSFER SYSTEMS, COLLECTOR TO STORAGE

DESIGN,DEVELOP,MANUFACTURE,&TEST SOLAR ENERGY COLLECTORS,CONTROLS,& STORAGE SYSTEMS FOR BUILDING HEATING & COOLING & ELECTRICAL POWER PRODUCTION BY PHOTOTHERMAL CONVERSION. SOLAR FLUX MEASUREMENT INSTRUMENTATION & DATA REDUCTION
CONSULTING & DESIGN SERVICES

			SUBCLASSES	SOLAR SUB SYSTEMS
HONEYWELL INCORP URBAN & ENVIRO SYS 2700 RIDGWAY PARKWAY MINNEAPOLIS NEIL C SHER	MN	55413	MEDIUM TEMPERATURE SOLAR COLLECTOR SYSTEMS SOLAR SPACE HEATING & COOLING SOLAR ELECTRIC POWER GENERATION	

-PHONE-
612 3314141 X4033
R & D & DEMO PROJECTS ON TERRESTRIAL APP OF SOLAR ENERGY FOR CENTRAL ELECT POWER & BLDG CLIMATE CONTRACT SEE ALSO RODGER SCHMIDT HONEYWELL

			SUBCLASSES	SOLAR SUB SYSTEMS
OAK RIDGE NATIONAL LAB REACTOR DIVISION PO BOX Y OAK RIDGE SAM E BEALL JR	TN	37830	MEDIUM TEMPERATURE SOLAR COLLECTOR SYSTEMS	

-PHONE-
615 4838611
STUDYING RESIDENTIAL SOLAR HEAT-ELECTRIC SYS & LOW TEMP POWER CONVER-SION

			SUBCLASSES	SOLAR SUB SYSTEMS
REDE RESEARCH&DESIGN INSTITUTE P O BOX 307 PROVIDENCE RALPH BECKMAN	RI	02901	WIND CONVERSION INORGANIC SEMI-CONDUCTORS TOTAL SYSTEMS INTEGRATION OFFICE/BANK BUILDINGS	HEAT PUMPS STORAGE, CHEMICAL BATTERIES

-PHONE-
401 8615390
INVESTIGATING ALTERNATIVE SOURCES OF ENERGY INCLUDING SOLAR TO BE USED IN ELECTRIC ENERGY CONSERVATION STATION THE STILLMAN WHITE FOUNDARY IN PROVIDENCE PLAN TO USE WIND GENERATOR SOLAR HEATING WATER TURBINE GEN HEAT PUMPS & PHOTOVOLTAICS...MAINTAIN EXTENSIVE ENERGY LIBRARY LISTING

	SUBCLASSES	SOLAR SUB SYSTEMS
WRIGHT-INGRAHAM INSTITUTE 1228 TERRACE ROAD COLORADO SPRINGS CO 80904 ELIZABETH W INGRAHAM DIRECTOR	TOTAL SYSTEMS INTEGRATION ALL SOLAR TECHNOLOGIES 50-500 THOUSAND SQ FT,PROPOSED 50-500 THOUSAND SQ FT,EXPERIMENT 50-500 THOUSAND SQ FT,DEMONSTRATION	COLLECTOR,ALSO FUNCTION AS STRUCTURES SKIN COLLECTOR,REFLECTIVE SELECTIVE SURFACES COLLECTOR,ABSORPTIVE SELECTIVE SURFACES HEAT TRANSFER SYSTEMS, COLLECTOR TO STORAGE STORAGE, SOLID MATERIALS STORAGE, LIQUIDS STORAGE, GAS

-PHONE-
303 6337011

INST IS A NON-PROFIT CORP ESTABLISHED TO PROMOTE DIRECT ENCOURAGE & DEVELOP EDUCATIONAL OPPORTUNITIES CONTRIBUTING TO THE CONSERVATION PRESERVATION & USE OF HUMAN&NATURAL RESOURCES...FIRST MAJOR PROJ IS THE BUILDING OF RUNNING CREEK FIELD STATION AN ENVIRONMENTAL UNIV ON 640 ACRES OF HIGH PLAINS GRASSLANDS IN COLORADO

--------SUBCLASSES--------SOLAR SUB SYSTEMS--------

ALABAMA, UNIVERSITY OF
CTR ENVIRONMENT STUD
P O BOX 1247
HUNTSVILLE AL 35807

RESEARCH PROPOSAL TO DEPT OF TRANSPORTATION-INVESTIGATE USE OF SOLAR TO SUPPLY INTERSTATE MASS TRANSIT & ELECTRIC RAILWAY POWER REQUIREMENT AVAILABLE RIGHT OF WAYS USED FOR COMBINED SOLAR CONVERSION & STRUCTURE SYSTEMS

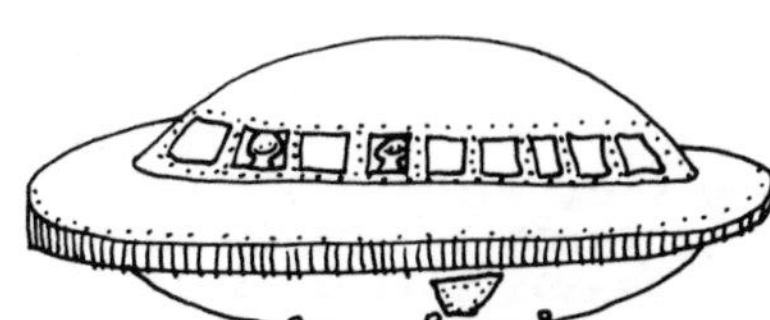

SUBCLASSES — SOLAR SUB SYSTEMS

ILLINOIS, UNIV OF-CHICAGO
ARCHITECTURE&ART CLG
BOX 4348
CHICAGO IL 60680
ROGER G WHITMER AIA
ASSOC PROF ARCH DEPT
-PHONE-
312 9963335

MEDIUM TEMPERATURE SOLAR COLLECTOR SYSTEMS
BIO-CONVERSION
WIND CONVERSION
INORGANIC SEMI-CONDUCTORS
PROVIDE INFORMATION SEARCHES TO CLIENTS

RESEARCH & SUPERVISION OF STUDENT THESES RELATING TO ENGY CONVERSION

SUBCLASSES		SOLAR SUB SYSTEMS
GEOCENTRIC SYSTEMS INC ARCHITECTURAL PO BOX 49 TAOS NM 87571 WILLIAM A MINGENBACH PRESIDENT -PHONE- 505 7584111 505 7582864	TOTAL SYSTEMS INTEGRATION SINGLE FAMILY, UNDER CONSTRUCTION SINGLE FAMILY, DEMONSTRATION GREENHOUSES UP TO 50 THOUSAND SQ FT,EXPERIMENT	COLLECTOR,MULTI-FUNCTION IN DWELLING SYS COLLECTOR,PASSIVE SYSTEMS STORAGE, PHASE CHANGE MATERIALS STORAGE, MECHANICAL/PUMPED LIQUIDS STORAGE, LIQUIDS ELECTROLYSIS

SOLAR HEATED RESIDENCES ALBUQUERQUE & ANGEL FIRE N M RESEARCH & DESIGN FACILITIES IN TAOS N M EMPHASIS ON PLAN ORGANIZATIONS FOR PASSIVE SUPPLEMENTAL PROTOTYPES

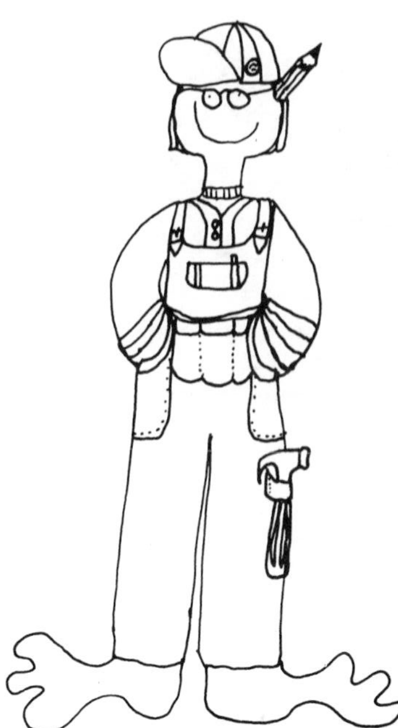

---SUBCLASSES---SOLAR SUB SYSTEMS---

DOWNING-LEACH ASSOC ARCH&PLANN	MEDIUM TEMPERATURE SOLAR COLLECTOR SYSTEMS
2305 BROADWAY	15 FAMILY&UP LOW RISE, PROPOSED
BOULDER CO 80302	
R B DOWNING	

-PHONE-
303 4437533
PROJECT DESCRIPTION: 80 UNIT CONDOMINIUM PROJECT UTILIZING SOLAR ENGY &ENGY CONSERVATION CONCEPTS.....PROJECT CONCEPTUALLY DESIGNED..BEGIN CONSTRUCTION BY 1975.

----------SUBCLASSES----------SOLAR SUB SYSTEMS----------

WILLIAM L BENGTSON
406 OAK BROOK ROAD
OAK BROOK IL 60521

SOLAR SPACE HEATING & COOLING
MEDIUM TEMPERATURE SOLAR COLLECTOR SYSTEMS
WIND CONVERSION
SINGLE FAMILY, PROPOSED

-PHONE-
312 6542492
PLAN TO BUILD COUNTRY HOUSE INCORPORATING SEVERAL ENERGY CONSERVING SYSTEMS SUCH AS SOLAR HEAT & COOL IN PLANNING STAGE

NED D CHERRY JR *
533 EAST 87TH STREET
NEW YORK NY 10028

MEDIUM TEMPERATURE SOLAR COLLECTOR SYSTEMS
WIND CONVERSION
INORGANIC SEMI-CONDUCTORS
SOLAR SPACE HEATING & COOLING
APARTMENTS

-PHONE-
212 8792503
212 4896666
DETERMINE FEASIBILITY TO HEAT&COOL NEW CHILD DAY CARE CTR NYC/RETROFIT 130 YR OLD FARMHOUSE GUILFORD NY-FLAT PLATE COLL HOT AIR-THIS HAS BEEN IN PLANNING FOR 3YEARS/3 4STORY BROWN STONE HOUSES NYC HEAT & COOL/ DESIGN HEAT&COOL FOR DOME UNDER CONST GEORGETOWN ISLAND MAINE

MICHAEL CURD *
BOX 865
FRISCO CO 80443

MEDIUM TEMPERATURE SOLAR COLLECTOR SYSTEMS
HIGH TEMPERATURE SOLAR COLLECTOR SYSTEMS
SOLAR WATER HEATING
SOLAR SPACE HEATING & COOLING
SINGLE FAMILY, PROPOSED

-PHONE-
303 4686004
BUILD LOW HEAT LOSS-FREE FORM HOUSE IN DESIGN STAGE RESEARCHING COMPONENTS & DATA ON SOLAR.CONVERSION SYSTEMS

-----SUBCLASSES-----SOLAR SUP SYSTEMS-----

ARIEL R DAVIS *
3476 FLEETWOOD DR
SALT LAKE CITY UT 84109

MEDIUM TEMPERATURE SOLAR COLLECTOR SYSTEMS
SOLAR ENERGY STORAGE
SOLAR SPACE HEATING & COOLING
RETROFITTING, TECHNOLOGIES

-PHONE-
801 2774140
801 2724225

HAVE SMALL SUPER INSULATED HOME WITH GAS HEAT-ELECT HEAT IN ALL OUTSDE WALLS TO MEASURE COMFORT ZONES&RADIATION FACTORS/WILL ADD SOLAR COLL-ECTION&HEAT STORAGE IN 1974
INVESTIGATING USE OF LOCAL COPPER SLAG AS BLACK BODY COLLECTOR

PETER VAN DRESSER
EL RITO NM 87530

SOLAR AGRICULTURAL DRYING
SOLAR WATER HEATING
SOLAR SPACE HEATING
SINGLE FAMILY, COMPLETED

SOLAR HOUSE IN OPERATION SINCE 1958,SIMPLE HEAT&COOL SYS, EXPERIMNTING WITH NEW INGENOUS MATERIALS...PAPER ON SOLAR HOUSE PRESENTED AT :1972 ECOLOGIC LIFE SUPP TECH CONF GHOST RANCH, ALBUQUERQUE NM,SEE BIO TECH-NIC PRESS TO ORDER

CHAFFEY COLLEGE
ALTA LOMA CA 91701
HOMER W DAVIS

WIND CONVERSION
TOTAL SYSTEMS INTEGRATION
SOLAR WATER HEATING
SOLAR SPACE HEATING & COOLING
SINGLE FAMILY, UNDER CONSTRUCTION

-PHONE-
714 9871737 X240

DESIGN&BUILDING SELF-SUF HOME NEAR WRIGHTWOOD CA/700 SQFT WALLS 1 FT THICK CEMENT BLOCK/ HOT AIR COLLECTOR WITH HOT ROCK STORAGE&FLAT PLATE WATER COLL WITH STORAGE INSIDE ROCKS/HOT WATER FOR DOMESTIC USE STORGE IN H2O BED IN CEILING/WIND GENERATOR FOR ELECT REQ

----------SUBCLASSES----------SOLAR SUB SYSTEMS----------

			SUBCLASSES	SOLAR SUB SYSTEMS
CHARLES W MOORE ASSOCIATES			MEDIUM TEMPERATURE SOLAR COLLECTOR SYSTEMS	
ESSEX	CT	06426	HIGH TEMPERATURE SOLAR COLLECTOR SYSTEMS	
WILLIAM H GROVER			SINGLE FAMILY, PROPOSED	
MANAGER				

-PHONE-
203 7670101
3 PROTOTYPES OF RESIDENTIAL HOUSING IN DESIGN-1 RESIDENCE DESIGNED &
READY FOR CONSTRUCTION
BOOKLET ON SOLAR HEAT BEING PREPARED FOR CONN SAVINGS BANKS

			SUBCLASSES	SOLAR SUB SYSTEMS
COLORADO, UNIV OF			SOLAR SPACE HEATING	
MEDICAL SCHOOL			SINGLE FAMILY, COMPLETED	
DENVER	CO	80220		
JOHN C COBB MD				

-PHONE-
303 3947150
HAVE A MOUNTAIN HOME SOLAR HEATER-HOME MADE/MAKING OBSERVATIONS ON
EFFICIENCY ETC/NEEDS MORE WORK

			SUBCLASSES	SOLAR SUB SYSTEMS
DELAWARE, UNIV OF			MEDIUM TEMPERATURE SOLAR COLLECTOR SYSTEMS	COLLECTOR,MULTI-FUNCTION IN DWELLING SYS
INST OF ENGY CONVSN			INORGANIC SEMI-CONDUCTORS	COLLECTOR,ALSO FUNCTION AS STRUCTURES SKIN
NEWARK	DE	19711	SOLAR HEAT,COOL,&ELECTRIC POWER GENERATION	COLLECTOR,ABSORPTIVE SELECTIVE SURFACES
DR KARL W BOER *			SINGLE FAMILY, COMPLETED	HEAT PUMPS
			SINGLE FAMILY, DEMONSTRATION	STORAGE, PHASE CHANGE MATERIALS
				STORAGE, CHEMICAL BATTERIES

-PHONE-
302 7388481
PROJECT DESC:SOLAR DEMO BLDG, NSF-RANN, DELMARVA POWER&LIGHT OF WILMIN
GTON DEL&DUANE C TAYLOR ASST VP OF ENGR DELMARVA, 1400 SQFT HOUSE
ENGY CONSV MEASURES:PAINT & FABRIC WITH DESIRED ABSORPTIVITY/REFLECT-
IVITY,INSULATION,SITE ORIENTATION,ROOF OVERHANG CONTROL WINTER/SUMMER
SUN.....SOLAR SYS:CADIUM SULFIDE SOLAR CELLS SUPP 20 KW HR INCORP INTO
FLAT PLATE COLL HOT AIR TYPE,STORAGE 75F & 120F EUTECTIC SALTS & LED
ACID BATT/EXCESS POWER GEN SENT BACK TO UTILITY, WORKS WITH UTILITY
FOR PEAK SHAVING...LOCATION: UNIV OF DEL, NEWARK DEL...COMPLETION DATE
JUNE 73...ALSO SEE:L R CRITTENDON,DEL, DELAWARE UNIV OF & MARIA TELKS,
DELAWARE UNIV OF.

			SUBCLASSES	SOLAR SUB SYSTEMS
DELAWARE, UNIV OF INST OF ENGY CONVSN NEWARK L R CRITTENDON	DE	19711	MEDIUM TEMPERATURE SOLAR COLLECTOR SYSTEMS INORGANIC SEMI-CONDUCTORS SOLAR HEAT,COOL,&ELECTRIC POWER GENERATION SINGLE FAMILY, COMPLETED SINGLE FAMILY, DEMONSTRATION	COLLECTOR,MULTI-FUNCTION IN DWELLING SYS COLLECTOR,ALSO FUNCTION AS STRUCTURES SKIN COLLECTOR,ABSORPTIVE SELECTIVE SURFACES HEAT PUMPS STORAGE, PHASE CHANGE MATERIALS STORAGE, CHEMICAL BATTERIES

-PHONE-
302 7388481
DEMONSTRATE:TECHNICAL&ECONOMIC FEASIBILITY TO APPLY SOLAR ENERGY AS SUPPLEMENTAL RESOURCE FOR THERMAL &ELECTRIC REQ OF BUILINGS/HOME IN OP ERATION USING SOLAR ENGY TO SUPPLY 80 PERCENT ELECTRIC & THERMAL ENGY
SEE:KARL BOER, DELAWARE,UNIV OF FOR SOLAR PROJECT DETAILS

DO LITTLE INDUSTRIES 23 TERRACE PLACE DANBURY SAM SAHLY	CT	06810	SOLAR SPACE HEATING & COOLING SINGLE FAMILY, PROPOSED

GATHERING SOLAR DATA TO BUILD SOLAR HOUSE, CONSTRUCTION TO BEGIN SPRNG 1974...WILL NOT BE MANUF THIS YEAR

ECOTEC FOUNDATION INC 2923 WOR LD AVENUE CINCINNATI PETER SEICLEL	OH	45206	SOLAR WATER HEATING SOLAR SPACE HEATING & COOLING SINGLE FAMILY, PROPOSED SINGLE FAMILY, DEMONSTRATION

-PHONE-
513 8618324
PROPOSALS TO BUILD EXPERIMENTAL HOUSES UTILIZING SOLAR & ENGY CONSV TECHNIQUES

	SUBCLASSES	SOLAR SUB SYSTEMS
DESERT RESEARCH INSTITUTE OFFICE OF PRESIDENT RENO NV 89507 JOHN R DOHERTY	SOLAR WATER HEATING SOLAR SPACE HEATING & COOLING UP TO 50 THOUSAND SQ FT,PROPOSED	COLLECTOR,MECHANICAL SYSTEMS HEAT TRANSFER SYSTEMS, COLLECTOR TO STORAGE HEAT RECOVERY, STORAGE TO UTILIZATION STORAGE, LIQUIDS

-PHONE-
702 7846131
PROJ DES: 8800 SQFT FACILITY TO HOUSE LABS&OFFICE HEADQTS FOR INSTITUT
DESERT BIOLOGY LAB/SOLAR SYS: 4000 SQFT FLAT PLATE COLL,WATER TRANSFER
SYS, SUPPLY ALL HEATING, HOT H2O AND 1/2 COOL NEEDS
FINANCED BY: MAX C FLEISHMAN FOUNDATION-$450,000 & SULO O&AILEEN MAKI
OF LAS VEGAS-$48,000 SEE:NEVADA SYSTEM, UNIV OF

	SUBCLASSES	SOLAR SUB SYSTEMS
NEVADA SYSTEM, UNIV OF DESERT RESEARCH INST RENO NV 89507 JOHN R DOHERTY *	SOLAR WATER HEATING SOLAR SPACE HEATING & COOLING UP TO 50 THOUSAND SQ FT,PROPOSED	COLLECTOR,MECHANICAL SYSTEMS HEAT TRANSFER SYSTEMS, COLLECTOR TO STORAGE HEAT RECOVERY, STORAGE TO UTILIZATION STORAGE, LIQUIDS

SEE:DESERT RESEARCH INSTITUTE FOR FURTHER DETAILS

	SUBCLASSES	SOLAR SUB SYSTEMS
ENERGY POLICY PROJECT 1776 MASS AV NW WASHINGTON DC 20036 DAVID S FREEMAN DIRECTOR -PHONE- 202 8721064 STUDY INSTITUTIONAL CONSTRAINTS OF IMPLEMENT SOLAR HEAT & COOL IN BLDG INDUSTRY...PUBLICATION FALL 1974...FUNDED THROUGH FORD FOUNDATION		
STANFORD RESEARCH INSTITUTE ENERGY TECHNOLOGY 333 RAVENSWOOD AVE MENLO PARK CA 94025 JOHN A ALICH JR ENGINEER ECONOMIST -PHONE- 415 3266200 X2343 415 3266200 X2704 EFFECTIVE UTILIZATION OF SOLAR ENERGY TO PRODUCE CLEAN FUELS..A TECHNICAL & ECONOMIC ASSESSMENT OF BIOMASS (PLANT MATERIAL) AS AN ENERGY FEEDSTOCK...ANALYSIS INCLUDES DEVELOPMENT OF FARMING & BIOLOGICAL FACT AS WELL AS CONVERSION TECH FOR ELECTRIC POWER GEN & SYTHETIC NATURAL GAS PRODUCTION	BIO-CONVERSION TOTAL SYSTEMS INTEGRATION SOLAR WATER HEATING SOLAR SPACE HEATING & COOLING SOLAR HEAT,COOL,&ELECTRIC POWER GENERATION SOLAR TECHNOLOGY APPLICATION OTHER	COLLECTOR,OTHER STORAGE, OTHER
STANFORD RESEARCH INSTITUTE 333 RAVENSWOOD AVE MENLO PARK CA 94025 RONALD I DEUTSCH * GEN MGR OFF PUB REL -PHONE- 415 3266200 X3754	BIO-CONVERSION TOTAL SYSTEMS INTEGRATION SOLAR WATER HEATING SOLAR SPACE HEATING & COOLING SOLAR HEAT,COOL,&ELECTRIC POWER GENERATION SOLAR TECHNOLOGY APPLICATION OTHER	COLLECTOR,OTHER STORAGF, OTHER

	SUBCLASSES	SOLAR SUB SYSTEMS
STANFORD RESEARCH INSTITUTE	BIO-CONVERSIQN	COLLECTOR,OTHER
ECONOMICS DIVISION	TOTAL SYSTEMS INTEGRATION	STORAGE, OTHER
333 RAVENSWOOD AVE	SOLAR WATER HEATING	
MENLO PARK CA 94025	SOLAR SPACE HEATING & COOLING	
JOHN P HENRY JR	SOLAR HEAT,COOL,&ELECTRIC POWER GENERATION	
DIR ENERGY TECH DEPT	SOLAR TECHNOLOGY APPLICATION OTHER	
-PHONE-		
415 3266200 X2358		
415 3266200 X2704		

EFFECTIVE UTILIZATION OF SOLAR ENERGY TO PRODUCE CLEAN FUELS..A TECH-NICAL & ECONOMIC ASSESSMENT OF BIOMASS (PLANT MATERIAL) AS AN ENERGY FEEDSTOCK...ANALYSIS INCLUDES DEVELOPMENT OF FARMING & BIOLOGICAL FACT AS WELL AS CONVERSION TECH FOR ELECTRIC POWER GEN & SYTHETIC NATURAL GAS PRODUCTION

---SUBCLASSES---SOLAR SUB SYSTEMS---

WILLIAM H JEWELL			WIND CONVERSION
R 4			RETROFITTING, RESOURCES
WEST BRATTLEBORO	VT	05301	RETROFITTING, PASSIVE SYSTEMS
			NEW CONSTRUCTION, RESOURCES
			NEW CONSTRUCTION, PASSIVE SYSTEMS

-PHONE-
802 2542403
LANDSCAPE ARCHITECT-ENERGY CONSERVATION & SOLAR GAIN THRU SITE PLANING TRYING TO USE TOPO & VEGETATION AS GUIDES FOR WINDMILL LOCATIONS... RESEARCH APPLIED TO MY OWN HOUSE & FARM & GRASSY BROOK VILLAGE...HAVE PERFORMED SOLAR GAIN & WIND CONTROL ON MANY HOUSING PROJECTS IN & AROUND PHILADELPHIA & SOUTHERN NJ

SUBCLASSES — SOLAR SUB SYSTEMS

COLORADO ENERGY RESEARCH INST GOVERNMENT AGENCIES
COLO SCHOOL OF MINES
GOLDEN CO 80401
THOMAS VOGENTHALER
DIRECTOR

-PHONE-
303 2790300
DIRECTOR OF RECENTLY ESTABLISHED STATE INSTITUTE WHICH WILL PROVIDE
FUNDS FOR ENERGY RESEARCH INCLUDING SOLAR

--SUBCLASSES--SOLAR SUB SYSTEMS--

SYSTEM PLANNING CORP
1500 WILSON BLVD
ARLINGTON VA 22209
BRUCE W MACDONALD

-PHONE-
703 5254904
COST BENEFIT ANALYSIS & SYSTEMS ANALYSIS OF SOLAR ENERGY SYS...HAVE
DEVELOPED ECONOMIC EVAL METHODOLOGY FOR SOLAR ENERGY

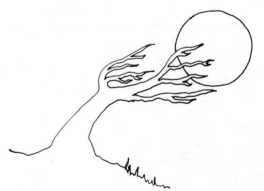

SUBCLASSES	SOLAR SUB SYSTEMS
LEGISLATIVE ENERGY COMMISSION	ALL SOLAR TECHNOLOGIES

NEW YORK STATE
ROOM 718 L.O.B.
ALBANY NY 12207
JAMES I MONROE
SCIENTIFIC ADVISOR
-PHONE-
518 4722700
THIS COMMISSION IS CHARGED WITH EXAMINING OVERALL ENERGY NEEDS OF NEW YORK STATE WITH OBJECTIVE OF PROPOSING LEGISLATION TO MEET THESE NEEDS MAJOR OBJECTIVE OF MY PROJECT IS TO EXAMINE ALTERNATIVE ENERGY SOURCES SUCH AS SOLAR FOR INCLUSION IN THE FUTURE ENERGY MIX

SEE APPENDICES FOR FURTHER STATE LEGISLATION

--SUBCLASSES-- --SOLAR SUB SYSTEMS--

SPENSLEY HORN LUPITZ AND JUBAS	LEGISLATIVE, LOCAL
PATENT LAW	INCENTIVES FOR SOLAR & ENERGY CONSERVATION
1880 CENTURY PK EAST	BUILDING CODES FOR SOLAR UTILIZATION
LOS ANGELES CA 90066	
DANIEL L DAWES	

-PHONE-
213 5535050
MANUSCRIPT SUBMITTED TO UCLA ON 'LEGAL BARRIERS TO URBAN SOLAR RIGHTS'

my hand glows
my mind grows
my heart flows
out to you all
for my soul knows
all that
you
are

James

INTRODUCTION TO SOLAR PROJECTS

The following section is a compilation of brief descriptions of solar heated and/or cooled buildings presently in existence, under construction, or planned for the near future. These buildings range from those employing mechanical systems to those which incorporate passive solar collection. The purpose of this inventory is to familiarize the reader with the various techniques and equipment being utilized as well as what individuals or firms are working in the field. A wide variety of projects are included, covering skyscrapers to small do-it-yourself efforts.

For information on a particular architect, engineer, contractor, etc., a project by project search is necessary. Most of these individuals or firms are not listed in section 1 in deference to the regulations of certain professional societies.

Projects are listed alphabetically by state, type of building (residential and then commercial types), owner's last name, and when the owner's name has been withheld, by the project director's last name. Foreign listings follow the United States entries and are also listed alphabetically.

Format for entries:

State or country

Type of structure

Location of project

Description: When available, information on square footage, type of construction, details of design, and funding source are given.

solar system: Type of collector, heat transfer system, and storage.

Project director

Owner

Architect

Engineer

Contractor

Completion date

Special consultants or notes are often added.

Whenever possible, photographs or drawings of projects have also been included. Captions identify to which project an illustration belongs.

We encourage readers to send any corrections, updates, or additional information needed for the particular projects listed. Any photos, plans, and written material on projects that we have omitted are most welcome so that we may continue to update our information.

Also see section 1, solar research: (heating), (cooling), (heating and cooling), (condominiums), (homes), (academic buildings), for listings of persons involved in the formative stages of other projects.

Special thanks is given to William Shurcliff, Margot Villecco and Architecture Plus, and all the individuals who provided information and helped us compile this section.

For a listing of solar buildings that were built in the past but are presently inoperative or demolished see Shurcliff's Survey of Solar Heated Buildings, which is distributed by Solar Energy Digest, P.O. Box 17776, San Diego, Ca. 92117. The price of the report is $6.00.

NASA SOLAR TEST HOUSE

COPPER DEVELOPMENT ASSOCIATION'S DECADE 80 SOLAR HOUSE

DR. MEINEL'S HOME

ALABAMA
house

Marshall Space Flight Center, Solar Test House
Huntsville, Alabama

DESCRIPTION: House consists of three office-type trailers joined together side-by-side, with long axis E-W. Trailer lengths and widths are: 58 ft x 10 ft, 58 ft x 10 ft, 30 ft x 12 ft. The roof has 45° slopes to N and S.

Solar system: 1300 ft^2 flat-plate, water-type collector with 45° slope. It includes 31 segments, each of which contains seven 2 ft x 3 ft panels hydraulically in series. The small (2 ft x 3 ft) size was used in order to conform to dimensions of available electroplating facility. Each panel is an Olin Brass Co. Roll-Bond panel of Type 1100 aluminum. The integral, expanded, parallel-connected channels of the panel are 1.5 inches apart on centers; each channel is 0.375 inch wide ID and 0.065 inch high ID, with wall thickness 0.030 inch at the channel and twice this in regions between channels. Cost of bare panel: about $0.70/$ft^2$. Highly selective, NASA-type, black coating applied to panel by electroplating. One layer of 0.004 inch glazing. Tedlar supported by a 2-inch x 4-inch mesh steel screen. Storage system: 4700 gallon horizontal cylindrical aluminum tank containing 3600 gallons of deionized water. Tank is insulated by 2 to 3 ft of 3 lb/ft^3 fiberglass. Cooling in summer: LiBr water absorption system like was made by Arkla Engineering Co. before 1968 but with 210^oF water supplied from storage tank instead of from gas-fired heater. Electric heaters for auxiliary heat for storage system in winter and for operating cooling system in summer.

Director: J. W. Wiggins
NASA Headquarters
Washington, D.C.
(202) 755-8617

Owner: J. W. Wiggins
address as above

Architect: Honeywell, Inc.
2600 Ridgeway Parkway
Minneapolis, MN 55413

Engineer: Honeywell, Inc.
address as above

Solar Systems Engineer: Barber-Nichols Engineering Company
6325 W. 55th Avenue
Denver, CO 80002
(303) 421-8111

Arkla Industrial Development
509 Marshall
Shreveport, LA 71151
(318) 425-1271

Completion date: February, 1974

ARIZONA
house

Decade 80 Solar House
Tucson, Arizona

DESCRIPTION: Single-story, four-bedroom, 3 bath, 3200 ft^2 residence. Sponsored by Copper Development Association.

Solar system: 2000 ft^2, water cooled, flat plate solar collector integrated with roof. 2500 gallon water storage tank. Cooling to be accomplished utilizing 2 three-ton Arkla absorption units with cooling tower.

Director: Paul Anderson
Manager of Building Construction
Copper Development Association
1011 High Ridge Road
Stamford, CT 06905
(203) 322-7639

Owner: Copper Development Association
1011 High Ridge Road
Stamford, CT 06905
mailing address:
P.O. Box 3346
Ridgeway Station
Stamford, CT 06905
(203) 322-7639
(New York information: 212-953-7323)

Architect: M. Arthur Kotch, AIA
6440 Hillcroft Road
Houston, TX 70036

Engineers: John E. Boretz
TRW Systems
One Space Park
Redondo Beach, CA 90278
(213) 536-3039

Contractor: McLaughlin Contractors & Engineers
3150 E. Lincoln
Tucson, AZ 85714
(602) 889-3371

Completion date: December, 1974

ARIZONA
house

10121 Catallena Highway
Tucson, Arizona 85715

DESCRIPTION: One-story, 2000 ft^2 house with 8-inch burnt adobe walls and a horizontal roof.

Solar system: Solar radiation passing through 300 ft^2 vertical, double-glazed south-southeast wall. Strikes black absorbing surface, and energy is carried by fan-driven air to adjacent, insulated, outdoor storage bin that contains 1000 spaced, water-filled, 1-gallon, polyethylene milk bottles. Roof of bin serves as reflector to increase amount of radiation incident on SSE wall. System supplies 30% of winter's heat-need. In summer, evaporative cooling is used for bin; cool dry air from bin then cools rooms.

Director: Drs. Aden & Marjorie Meinel
c/o Helio Assoc.
10121 Catallena Highway
Tucson, AZ 85715
(602) 886-5376

Owner: Drs. Aden & Marjorie Meinel
address as above

Architect: Drs. Aden & Marjorie Meinel
address as above

Engineers: Drs. Aden & Marjorie Meinel
address as above

Contractor: Drs. Aden & Marjorie Meinel
address as above

Completion date: November, 1973

ARIZONA
environmental research laboratory

Tucson International Airport
Tucson, AZ 85706

DESCRIPTION: Funded by Arizona Foundation. 576 ft^2 test building.

Solar system: 400 ft^2 flat plate air collector. 2000 gallon hot water storage.

Director: John F. Peck
Environmental Research Lab.
Tucson International Airport
Tucson, AZ 85706
(602) 884-2931

Owner: Environmental Research Lab.
address as above

Architect: John F. Peck
address as above

Engineers: John F. Peck
address as above

Contractor: John F. Peck
address as above

Completion date: October, 1974

ARKANSAS
apartment competition winner

DESCRIPTION: Winning project in Misawa Homes International Competition. Shapes of apartment units determined by their relationship to the sun.

Solar system: Units would use reflective lens roofs to focus solar radiation onto a moving collector rod, producing about 1100° F. Heat would be used for space conditioning and electricity generation. Storage underneath dwelling units.

Architect: James Lambeth
1591 Clark Street
Fayettesville, AR 72701
(501) 521-2693

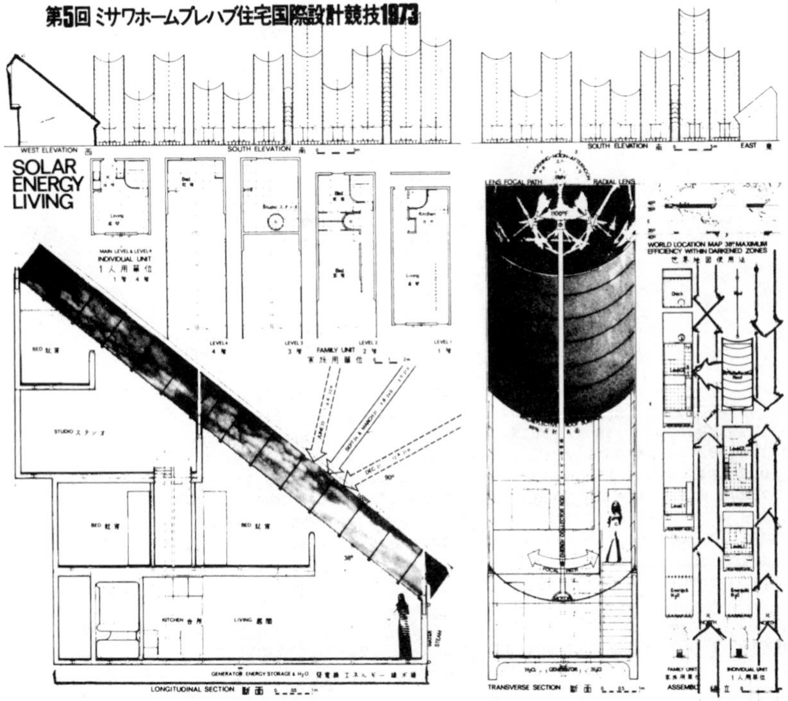

第5回 ミサワホームプレハブ住宅国際設計競技1973

SOLAR
ENERGY
LIVING

JAMES LAMBETH'S APARTMENT DESIGN

CALIFORNIA
apartments

Project SAGE (Solar Assisted Gas Equipment)
El Toro, California

DESCRIPTION: 3-story, 32 unit garden apartments.

Solar system: 1728 ft^2 of water/glycol cooled flat plate solar collectors will supply 70% of yearly hot water requirements. Solar hot water heating system will be owned and serviced by Southern California Gas Company and leased to apartment owner.

Director: Vince V. Fiore
Southern California Gas Company
810 South Flower Street
Los Angeles, CA 90017
(213) 689-2968

Owner: Southern California Gas Company
address as above

Architect: Retrofit

Engineers: Edgar S. Davis
Environmental Quality Lab
Jet Propulsion Lab
Building 79-239
4800 Oak Grove Drive
Pasadena, CA 91103
(213) 354-7004

Contractor: Fredricks Development Corp.
2035 Ball Road, East
Anaheim, CA 92806
(714) 956-1010

Completion date: Late March, 1975

CALIFORNIA
house

13792 Pine Needles Drive
Del Mar, California 92014

DESCRIPTION: Two-story, one-family custom residence, 2200 ft^2, approximately $80,000.

Solar system: 500 ft^2 collector, stainless steel tubing on aluminum sheet, full line pressure. Storage 2000 gallon steel hot water tank at low pressure. Storage served by fan coil unit using conventional forced air ducting. Domestic hot water extracted by conventional heat exchanger. Back-up auxiliary heater is 75 gallon gas water heater. Differential thermostat and mixing valve for hot water supply. 70% solar space heating, 90% water heating. Also solar still. All components off-the-shelf items, materials approved by ICBO. No deviations from standard building codes required.

Director: Jack Schultz Enterprises
6986 La Jolla Boulevard
Los Angeles, CA 92037
(714) 454-1312

Owner: John and Barbara Beckstrand
13792 Pine Needles Drive
Del Mar, CA 92014

Architect: Jack Schultz Enterprises
address as above

Engineer: Jack Schultz and Shayne McDaniels
4464 Alvarado Canyon Road
San Diego, CA 92120
(714) 283-3181

Completion date: November 10, 1974

CALIFORNIA
house

DESCRIPTION: Designed in 1966, still in schematic form. The Solterra home...enough solar heating capacity so that no back-up system is necessary.

Solar system: Collectors on roof. Solar air heater that circulates heated air through a closed system of ducts to an earthen bin in the basement. Foam concreted bin can retain sufficient heat for two months. Max. temperature 130-140°F. Absorption type air conditioner.

Director: William B. Edmondson
Editor
Solar Energy Digest
P.O. Box 17776
San Deigo, CA 92117
(714) 277-2980

Architect: William B. Edmondson
address as above

Engineer: William B. Edmondson
address as above

Completion date: Seeking funds.

ENERGY SYSTEMS INC.- SOLAR HOME

CALIFORNIA
homes - housing development

'Vista del Colinas'
2400 Euclid Avenue
San Diego, California 92021

DESCRIPTION: A total of 23 homes, single and multi-story, ranging from 3000 to 5000 ft^2. Price range $105,000 to $300,000 on 1 to 3-acre plots.

Solar system: Integrated roof-mounted flat plate collectors, 400 ft^2 mean. Some water/ethyl-glycol, others air co led for heating. Thermal storage accomplished by rock bed or 1000 to 1500 gallon systems. Designed to provide 50% to 80% of space heating needs in south California climate zone.

Director: Terrence R. Caster
Energy Systems, Inc.
634 Crest Drive
El Cajon, CA 92021
(714) 447-1000
(714) 440-4646

Owner: Energy Systems, Inc.
address as above
(Gary Beals, Public Relations)

Architect: J. Flotz
8810 La Mesa Boulevard
La Mesa, CA 92041
(714) 469-0157

Solar Engineer: Robert Gallagher
Energy Systems, Inc.
address as above

Contractor: Energy Systems, Inc.
address as above

Completion date: Now under construction, 6 units complete.

Note: Public inquiry should be made to Gary Beals.

CALIFORNIA
house

Skytherm Demonstration Home
7985 Santa Rosa Road
Atascadero, California

DESCRIPTION: One-story, split-level house with seven rooms, carport, and patio. No basement. 1140 ft^2 living area.

Solar system: Four 8-ft-wide, 38-ft-long water-filled bags of transparent 20 mil polyvinyl chloride (with overlying air bags for insulation) on black plastic sheet rest on horizontal metallic roof. During night the bags are covered by horizontal insulating panels. Heat flows downward through steel ceiling into rooms. The room walls, of sand-filled concrete blocks, have large thermal capacity. In summer the system is operated in reverse, so insulating panels are open at night, and cooling is achieved by radiation to the night sky. 100% solar heated. Performance evaluated by Professor K. Haggard with $42,000 HUD funding.

Director: Harold R. Hay
Sky Therm Processes and Engineering
2424 Wilshire Boulevard
Los Angeles, CA 90057
(213) 389-2300

Owner: Harold R. Hay
address as above

Architect: Kenneth L. Haggart and
John W. Edmisten
California Polytechnic State University
San Luis Obispo, CA

Thermal Engineer: Professor Philip Niles
Department of Environmental Engineering
California Polytechnic State University
San Luis Obispo, CA

Solar System Engineer: Harold R. Hay
address as above

Contractor: Anret Corp.
1524 Country Club Drive
Paso Robles, CA 93446
(805) 238-1224

Completion date: September, 1973

HAY'S SOLAR PONDS & MOVEABLE INSULATION PANELS

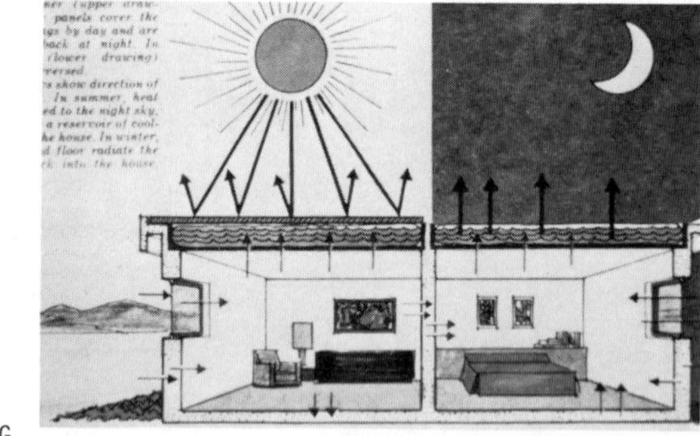

HAY SYSTEM - SUMMER COOLING

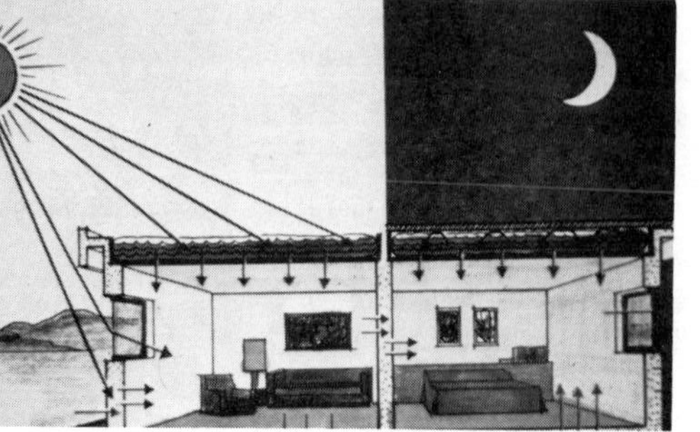

HAY SYSTEM - WINTER HEATING

CALIFORNIA
house

San Gabriel Academy
San Gabriel, California

DESCRIPTION:	Frame house, one-room, plus attic.
Solar system:	Hot air flat plate collector system 200 ft^2 of fiberglass panels covered the slanted roof to catch the sun rays. Thermostats and blowers keep the 10 by 20 foot living space with its 8 ft. high ceiling, at a constant 70°.
Director:	Theodore M. Bauer, Principal * East Valley Elementary School 3554 North Maine Baldwin Park, CA 91706
Owner:	San Gabriel Academy 3554 North Maine San Gabriel, CA 91706 (213) 444-7502
Designer:	Theodore M. Bauer, and students of San Gabriel Academy address as above
Engineers:	Theodore M. Bauer address as above
Contractor:	Theodore M. Bauer address as above
Completion date:	April, 1974

* Note: Mr. Bauer previously held a position at San Gabriel Academy.

SAN GABRIEL ACADEMY STUDENT BUILT HOME

CALIFORNIA
mobile home

48780 Eisenhower Drive
La Quinta, California 92553

DESCRIPTION:	Solar/Sonic Mobile Home - 720 ft^2 (60 x 12) experimental solar home. There are transducers planted in walls and electrostatic speakers in curtains for total sound. Private funding.
Solar system:	360 ft^2 of water-type collectors, 2500 gallon storage. 2 1/2 kw of wind generator. 12V florescent light system. Battery storage Globe Union, Inc. 100% solar heated.
Director:	Fred Rice La Quinta Palm Springs, CA 92553 (714) 564-4823
Owner:	Fred Rice Productions 6313 Peace Avenue Van Nuys, CA 91401 (213) 786-3860
Architect:	John Marzicola College of the Desert Palm Desert, CA 92260 (714) 346-8041
Engineer:	Larry Fredricks College of the Desert Palm Desert, CA 92260 (714) 346-8041
Contractor:	Mission Village Mobile Homes 6014 Riverside Corona, CA 91720 (714) 686-4241
Completion date:	March, 1975

CALIFORNIA
condominium office development

Messineo Plaza
Lake and Cordova Streets
Pasadena, California 91001

DESCRIPTION: 28-story office condominium structure, approximately 300,000 ft^2. This will be the first unit of a 3-unit complex having a total area of 770,000 ft^2.

Solar system: "Heat pipes" will be utilized to transfer heat energy from warmer areas of the building to either the building heating system for redistribution to cooler areas or to the building sprinkler system storage tank. Hot water from the storage tank will then be cooled by night sky radiation via flat plate solar collectors located on building curtain walls.

Director: Peter Messineo
924 E. Green Street
Pasadena, CA 91001
(213) 449-2232

Owner: Peter Messineo
address as above

Architect: Harrison, Beckhart & Mill
844 W. Colorado Boulevard
Los Angeles, CA 90041
(213) 254-7141

Engineers: Helms and Associates
Mr. d'Autremont
6311 North Figueroa Street
Los Angeles, CA 90042
(213) 255-7121

Contractor: Not selected yet.

Completion date: Late 1976.

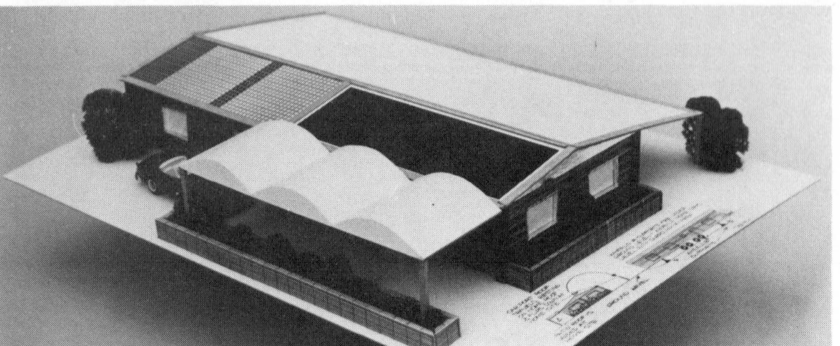

FRED RICE PRODUCTIONS' SOLAR SONIC MOBILE HOME

MESSINEO PLAZA CONDOMINIUM OFFICE DEVELOPMENT
LAKE & CORDOVA STREETS PASADENA, CALIFORNIA

CALIFORNIA
office building

Edwards, California

DESCRIPTION: Preliminary engineering study and report completed for retrofitting existing office building, 2 stories, 45,000 ft^2.

Solar system: Solar collectors and storage systems for heating and cooling using hot water coils in air handling units for heating; and absorption refrigeration for cooling. Collectors are approximately 20,000 ft^2 in area, and storage is approximately 100,000 gallons of water. An unusual feature is storage of chilled water as well as hot water to take advantage of off-peak cooling at low condensing temperatures. 8/12/74 has been selected for final design.

Owner: National Aeronautics and Space Administration
Flight Research Center
Edwards, CA

Engineer: Dubin-Mindell-Bloome Associates
Consulting Engineers and Planners
42 West 39th Street
New York, NY 10018
(203) 236-1654

COLORADO
condominium

Sun-Ra Condominium Building
Copper Mountain, Colorado

DESCRIPTION: 90,000 ft^2 gross building area, funded by Copper Development Association (CDA).

Solar system: 10,800 ft^2 copper collector system. Reverse acting heat pump delivery. 2000 gallons water/ehylene glycol.

Director: William Coffey
P.O. Box 2614
Lictleton, CO 80120
(303) 770-1830

Owner: Copper Development Association (CDA)
1011 High Ridge Road
Stamford, CT 06905
(203) 322-7639

Architect: Danny Cervants
Taniguchi Assoc.
1126 Spruce Street
Boulder, CO 80501
(303) 449-2450

Engineers: Beckett-Harmon Association, Inc.
2049 Broadway
Boulder, CO 80501
(303) 449-3302

Contractor: Not selected yet.

Completion date: December, 1975

COPPER DEVELOPMENT ASSOCIATION'S CONDOMINIUM

COLORADO
house

Mesa College Physics Department
P.O. Box 75A, Route #1
Palisade, Colorado 81526

DESCRIPTION: Experimental solar home, private funding, 53 ft x 36 ft. 1 1/2 stories, frame construction.

Solar system: 800 ft^2 combined air and water collector, single glazing, non-selective, corrugated aluminum, trickle water type. Storage: 1600 gallon plus 50-ton granite rock and surrounding earth. Heating 57%.

Director: Hermann C. Allmaras
Mesa College Physics Department
P.O. Box 75A, Route #1
Palisade, CO 81526
(303) 248-1354

Owner: Hermann C. Allmaras
address as above

Architect: Hermann C. Allmaras
address as above

Engineer: Hermann C. Allmaras
address as above

Contractor: Hermann C. Allmaras
address as above

Completion date: December, 1974

COLORADO
homes

Beulah, Colorado

DESCRIPTION: Single-story house.

Solar system: Water collector system is presently operating in this house, but plans are to build several other houses with air systems.

Director: Joseph A. Bullen, Jr.
3504 North Ridge Drive
Pueblo, CO 81008
(303) 544-8281

Owner: Joseph A. Bullen, Jr.
address as above

Designer: Joseph A. Bullen, Jr.
address as above

Engineer: Joseph A. Bullen, Jr.
address as above

Contractor: Joseph A. Bullen, Jr.
address as above

Completion date: Late 1974

COLORADO
house

Above the town of Alice, Colorado,
off of Fall River Road

DESCRIPTION: Three-story, 20' x 40', weekend and vacation cabin. Two rooms solar heated to keep pipes from freezing (pipes run through the two rooms).

Solar system: 1000 ft^2 hot air type flat plate collector with rock storage. Wood stove used as auxiliary heat.

Director: Dr. John C. Cobb, M.D.
Director of Preventive Medicine
University of Colorado Medical Center
4200 E. 9th Street
Denver, CO 80220
(303) 394-8744

Owner: Dr. John C. Cobb, M.D.
address as above

Designer: Steve Baer
Box 712
Albuquerque, NM
(505) 242-5354

Engineer: Steve Baer
Zomeworks
address as above

Contractor: Student labor

Completion date: Fall, 1972

DR. JOHN COBB'S MOUNTAIN RETREAT

RICHARD CROWTHER'S PASSIVE SOLAR·ENERGY CONSERVING HOME

CROWTHER'S RETROFIT & TOWNHOUSE

COLORADO
homes--Duplex

500 Cook Street
Denver, Colorado

DESCRIPTION: 2600 ft^2 residence with adjoining 1200 ft^2 apartment. Structure utilizes many energy conserving features.

Solar system: House designed to act as a passive solar collector with solar heated roof plenums and vent stacks to induce air convection for ventilation and passive temperature moderation. Still air solar radiation through south glass builds up heat in slab on grade during winter months, and large plants provide additional humidity.

Director: Richard L. Crowther, A.I.A.
Crowther-Kruse-McWilliams
2830 E. Third Avenue
Denver, CO 80206
(303) 355-2301

Owner: Richard L. Crowther, A.I.A.
address as above

Architect: Richard L. Crowther, A.I.A.
address as above

Engineer: Richard L. Crowther, A.I.A.
address as above

Contractor: Shaw Construction Company
123 S. Kalamath
Denver, CO 80223
(303) 744-1454

Completion date: June 1, 1972

COLORADO
house

419 St. Paul Street
Denver, Colorado

DESCRIPTION: One-story, 1000 ft^2, brick house with full basement(retrofit). Extensively remodeled for energy conservation and solar heating.

Solar system: 500 ft^2 flat plate, air-type collector, 53° sloping from horizontal. Double glazed.

Director: Richard L. Crowther, AIA
Crowther-Kruse-McWilliams
2830 E. Third Ave.
Denver, CO 80206
(303) 355-2301

Owner: Richard L. Crowther, AIA
address as above

Architect: Richard L. Crowther, AIA
address as above

Engineer: Richard L. Crowther, AIA
address as above

Contractor: Columbia Construction Co.
2480 W. 26th, #146B
Denver, CO
(303) 458-7221

Completion date: April, 1974

COLORADO
townhouse

421 St. Paul Street
Denver, Colorado

DESCRIPTION: Existing 2000 ft^2 townhouse, remodeled for minimum heat loss.

Solar system: Demonstration of passive use of solar energy for environmental comfort control, utilizing a combination of techniques such as control of solar absorptive surfaces by shading, surface coating absorptivity, window area, natural ventilation, etc. Flat plate roof collector with thermal water storage, solar heating of greenhouse, and domestic hot water.

Director: Richard L. Crowther, AIA
Crowther-Kruse-McWilliams
2830 E. Third Ave.
Denver, CO 80206
(303) 355-2301

Owner: Richard L. Crowther, AIA
address as above

Architect: Richard L. Crowther, AIA
address as above

Engineer: Richard L. Crowther, AIA
address as above

Contractor: Columbia Construction Co.
2480 W. 26th, #146B
Denver, CO
(303) 458-7221

Completion date: April, 1974

COLORADO
house

503 South El Paso Street
Colorado Springs, Colorado

DESCRIPTION: Retrofit of 1800 ft^2, two-story home built in 1890. Utilized as field office for Colorado Springs Urban Renewal.

Solar system: 180 ft^2 collector area, 250 ft^3 gravel storage. Provides domestic hot water and 25-30% of heating load. Will have instrumentation.

Director: Dr. Val Viers
c/o Physics Department
Colorado College
Wood Avenue
Colorado Springs, CO 80226
(303) 591-9040 or (303) 473-2233, Ext 300

Owner: Colorado Springs Urban Renewal
Colorado Springs, CO 80901

Engineers: Dr. Va. Viers
address as above

Contractor: Eric Swab
Rehabilitation Specialists
Colorado Springs Urban Renewal
503 South El Paso Street
Colorado Springs, CO

Completion date: September, 1974

Note: Roger W. Taylor, a physics student at Colorado College (permanent address - 9000 W 2nd Avenue, Lakewood, CO, 80226, phone - (303) 591-9040) was also involved in this project.

COLORADO
house

West of Boulder, Colorado

DESCRIPTION: Plan to build solar heated house west of Boulder; will include a number of structures utilizing alternative technologies.

Director: James E. Davis
Alternative Development Association
Boulder, CO 80302
(303) 443-5157

Architects: Peter Slack
2804 Marine Street
Boulder, CO 80302
(303) 449-1684

Vern Stone
Earth Dynamics, Inc.
Box 1175
Boulder, CO 80302
(303) 443-6548

Teague Van Buren
Solar Hydronics, Inc.
P.O. Box 4484
Boulder, CO 80302
(303) 447-9828

Completion date: Project on hold until financing can be obtained.

COLORADO
house

11 Karlann Road, Rt. 4
Golden, Colorado

DESCRIPTION: Within a 50-ft diameter hemispherical geodesic dome space is a dynamic, down-right far out living home experience. On one half of the dome is a living, responsive house sculpture, on the other half is an open natural landscape and garden. The dome will be solar heated. 4000 ft^2, privately funded.

Solar system: The dome incorporates passive solar heating through the use of a large south opening under which is an indoor/outdoor pond(s) which reflects and absorbs heat. To actively heat the dome are two 12 ft diameter concentrator collectors. (The active solar system is in design stage.)

Director: Phillip J. Tabb and Tom Clark
1140 Detroit Street
Denver, CO 80206
(303) 333-8997

Owner: Lee John Droege
11 Karlann Road, Rt. 4
Golden, CO
(303) 572-0770

Architect: Joint Venture: architects, environmentalists, and visionaries
1140 Detroit Street
Denver, CO 80206
(303) 333-8997

Solar Engineer: Dr. Jan Kreider, Mechanical Engineer
Environmental Consulting
Walnut Street
Boulder, Colorado 80302
(303) 447-2218

Structural Systems Engineer: John Giltner
Boulder, Colorado 80301
(303) 494-0380

Contractor: Lee John Droege
address as above

Completion date: Spring, 1975

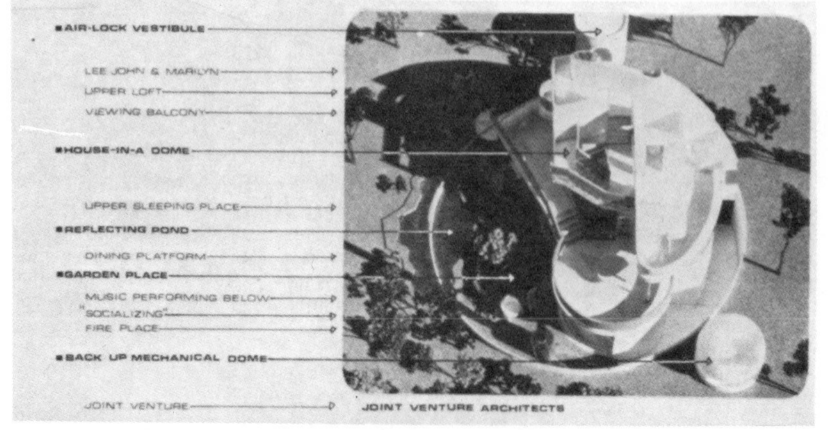

SECTION through Dome Environment
JOINT VENTURE, ARCHITECTS

DROEGE DOME HOME

AIR-LOCK VESTIBULE
LEE JOHN & MARILYN
UPPER LOFT
VIEWING BALCONY
HOUSE-IN-A DOME
UPPER SLEEPING PLACE
REFLECTING POND
DINING PLATFORM
GARDEN PLACE
MUSIC PERFORMING BELOW
"SOCIALIZING"
FIRE PLACE
BACK UP MECHANICAL DOME
JOINT VENTURE

JOINT VENTURE ARCHITECTS

COLORADO
house

Lookout Mountain
Colorado

DESCRIPTION: Personal financing, 1700 ft^2, three story house.

Solar system: Flat plate water system 500 ft^2, pyramid shape.

Director: Charles Hartling, AIA
1300 Canyon
Boulder, CO
(303) 442-1342

Owner: M. M. Frederick
1722 14th Street
Boulder, CO 80203
(303) 442-1342

Architect: Charles Hartling, AIA
address as above

Engineers: Jan Kreider
Environmental Consulting Service
Boulder, CO 80203
(303) 443-1235

Completion date: December, 1974

COLORADO
house

Friendship Lane East
Colorado Springs, Colorado

DESCRIPTION: 2400 ft^2 duplex, 2-story, 4-bedroom, 1 1/2 baths in each. $55,000

Solar system: 6 ft x 9 ft, 5-1/4-ft-deep spherical collector, single axis, vertical attitude. 5000 gallon fiberglass storage tank, 30% ethylene glycol.

Director: Bill Baker
1910 Hercules Drive
Colorado Springs, CO 80906
(303) 635-4354

Owner: Dan Howells
559 Castle Road
Colorado Springs, CO 80904

Architect: Bill Baker
address as above

Engineer: Bill Baker
address as above

Contractor: Dan Howells
address as above

Completion date: Drawings are complete, working model.

COLORADO
house

1066 E. Woodman Road
Colorado Springs, Colorado 80918

DESCRIPTION: Privately funded 4400 ft^2 residence; cathedral ceilings, 3-bedroom, 4 baths. Utilizes skylights for natural lighting. Heavily insulated with rock wool and urethane. Cost of home $95,000; solar system cost $2,500.

Solar system: Thomason system using 384 ft^2 collector area, double glazed. Main storage 1000 gallon tank for water/ethylene glycol mixture. Also uses 300 gallon tank which has 50 gallon hot water tank within for domestic hot water. Water tanks are surrounded by 29 cubic yards of 3-5 inch aggregate iron-pyrite rocks. Forced air (3000 c.f.m.) over rocks to heat home. Three fireplaces and 500,000 BTU natural gas furnace as auxiliary heat. System will be used for both heating and cooling. Predicted 80% of heating load solar.

Director: Roy H. Jackson (USAF retired)
1066 E. Woodman Road
Colorado Springs, CO 80918
(303) 598-6853

Owner: Roy H. Jackson
address as above

Designer: Roy H. Jackson
address as above

Engineer: Roy H. Jackson
address as above

Solar System Engineer: Dr. Harry Thomason
Thomason Solar Homes, Inc.
6802 Walher Mill Road, S.E.
Washington, D.C. 22027
(301) 336-4042

Contractor: Roy Jackson
address as above

Completion date: October 25, 1974

COLORADO
house

Pinebrook Hills
Boulder, Colorado

DESCRIPTION: $60,000, 2000 ft^2, two-story, five bedroom, 2 1/2 bath residence. Private funding.

Solar system: 500 ft^2 double glazed ABS plastic absorber plate. (2) 850 gal steel storage tanks. Dual distribution/thin wall-PVC. Coil in 5-inch concrete floor slab, and radiant heat up-stairs.

Director: Jay Lynch
Route 3, Box 330
Longmont, CO 80501
(303) 442-0372

Owner: Mr. R. B. Kraft
Chicago, IL

Architect: Jay Lynch
address as above

Engineer: Jay Lynch
address as above

Solar System Engineer: Peter Slack & John Brown
Earth Dynamics, Inc.
Box 1175
Boulder, CO 80302
(303) 443-6548

Teague Van Buren
Solar Hydronics
P.O. Box 4484
Boulder, CO 80302
(303) 447-9828

Contractor: Cooperative Construction Company
Route 3, Box 330
Longmont, CO 80501
(303) 442-0372

Completion date: December, 1974.

COLORADO
house

6 Parkway Drive, Cherry Hills
Denver, Colorado 80110

DESCRIPTION: One-story, 40 ft x 50 ft house with horizontal roof.

Solar system: 600 ft^2 air-type in-depth collector consisting of two 50-ft-long, 6-ft-high, 45°-sloping arrays on roof. Radiation is absorbed in black glass plates arranged at six depths within the airstream. Two-speed, 1-HP motor drives centrifugal blower which produces 770 scfm (at 1.8 inch water pressure) flow of air to the bases of the two energy-storage towers. Storage system consists of two vertical cylindrical, 3-ft-diameter, 18-ft-high towers, each of which contains 6 tons of granite rock with 1 to 1 1/2-inch diameter. Bulk density of rock is 96 lb/ft^{-3}. Specific heat is 0.18 $Btu\ lb^{-1}\ {}^oF^{-1}$.

Director: Dr. George O. G. Löf
Executive Vice President
Solaron Corporation
4850 Olive
Denver, Colorado 80216
(303) 289-1221

Owner: Dr. George O. G. Löf
address as above

Architect: James M. Hunter
(retired)
1505 Mariposa Street
Boulder, CO 80302
(303) 442-5183

Engineer: Dr. George O. G. Löf
address as above

Completion date: 1958

DR. GEORGE LÖF'S HOME

COLORADO
house

CSU Solar House
Colorado State University
Fort Collins, Colorado 80521

DESCRIPTION: Funded through National Science Foundation, $238,000. One story, 3 bedrooms, 1500 ft^2 residence with 1500 ft^2 basement. Insulation conventional. Built-in heated two-car garage. South windows are shaded in summer by roof overhang.

Solar system: Flat plate collectors using aluminum Roll-Bond; water/ethylene glycol cooled. 650 ft^2 collector area. 1100 gallon steel tank in basement for storage. System employs heat exchanger. Rooms are heated by air coils supplied with hot water from storage tank or auxiliary gas boiler. A 3-ton LiBr absorption cooling system powered by 200°+F. water from collector utilized for summer cooling. 75% solar heated (predicted). Domestic hot water also provided via heat exchanger.

Director: George G. O. Löf
Solar Energy Applications Lab
Colorado State University
Ft. Collins, CO 80521
(303) 289-1221

Owner: Solar Energy Applications Lab
National Science Foundation

Architect: Richard Crowther, A.I.A.
Crowther, Kruse, McWilliams
2830 East Third Avenue
Denver, CO 80206
(303) 355-2301

Engineers: George O. G. Löf and Dan Ward
Solar Energy Applications Lab
address as above

Contractor: Howard DeLozier
Construction Company
3034 E. Mulberry
Ft. Collins, CO 80521
(303) 482-5344

Completion date: May, 1974

COLORADO
house

The Phoenix Home
5925 Del Paz Drive
Colorado Springs, Colorado 80918

DESCRIPTION: Two-story demonstration dwelling. Floor area, 2200 ft^2, no attic, 1/2 basement, 6 inch mineral wool in ceiling, 3 1/2 inches bats in wall. Heat loss at -10°F with 7 mph wind is 55,000 Btu/hr. Heating season degree days (rel. to 65° F): 6305. Overall cost of building $81,000; included in this is $9000 cost of solar heating system. Joint venture of Colorado Springs, the Phoenix Corp, and National Science Foundation; financing through Pikes Peak Regional Clearing House Association.

Solar system: Collector: 810 ft^2 net, flat-plate, liquid-type, consisting of 30 panels arranged in two roof-top arrays each at 55° from horizontal and each containing 15 side-by-side panels. Each panel is 36 inches x 104 inches, is double glazed (with 1/8 inch PPG tempered plate glass in Oldach Products frame), and employs Olin Brass Co. Roll-Bond #1100 aluminum plate with integral expanded channels in which Dowtherm J flows. The plate has a PPG non-selective black coating. Back of panel is insulated with Rigid Rock-Wool. White quartz gravel on horizontal roof reflects additional radiation to the north collector array. 25 GPM pump circulates liquid to collector. Liquid temperature may sometimes reach 200° F. Storage system: 8000 gallons of water in underground, 9 1/2 ft diameter, 15 ft long steel tank. Dowtherm J supplies energy to it via interchanger. Rooms are heated usually by circulation of water from tank; circulation is by 50 GPM pump; fan-coil sends warm air to rooms. If temperature of tank is below 90° F, rooms are warmed by energy supplied by heat pump. Auxiliary heat from heat pump. 80% solar heated (predicted). Carrythrough is 5 days (predicted). Cooling in summer is also provided by heat pump.

Director: Douglas M. Jardine
P.O. Box 7246
Colorado Springs, CO 80933
(303) 633-2633

Solar Systems Engineer: Douglas M. Jardine
address as above

Owner: Phoenix Corporation
P.O. Box 7246
Colorado Springs, CO 80933
(303) 633-2633

Contractor: Dan Howells and Sons
Castle Road
Colorado Springs, CO 80904
(303) 475-2611

Architect: Michael Lane, AIA, Charles England, and John Wheeler
The Design Group
1624 S. 21st Street
Colorado Springs, CO 80904
(303) 475-1112

Completion date: July, 1974

CROWTHER · KRUSE · McWILLIAMS ARCHITECTS
ARCHITECTURE
INTERIORS
PLANNING
DENVER, COLORADO

NATIONAL SCIENCE FOUNDATION
SOLAR HOUSE · COLORADO STATE UNIVERSITY

COLORADO
house

Happy Canyon
north of Castle Rock

DESCRIPTION: Single-family residence. Private funding.

Solar system: Water-cooled flat plate collectors. 1000 gallon storage in insulated water tanks in basement. Carrythrough (predicted) two to three days. Predicted 70-90% solar heated. Approximate cost installed $4000.

Owner: Bob Roberts
Happy Canyon, CO
(north of Castle Rock)
(303) 688-2403

Engineer: Haiko Eichler
Solar Engineering Company
8395 Elati Street
Denver, CO 80221
(303) 428-8524

Completion date: Fall, 1974

SOLAR SYSTEM DRAWINGS OF TYPE USED IN ROBERTS' HOME

COLORADO
house

Route #1
Livermore, Colorado 80536

DESCRIPTION: 350 ft^2 aspen log structure.

Solar system: 84 ft^2 air-type black aluminum collectors at 55°, double glazed with (outer) clear Filon tedlar, (inner) 2 mil. mylar. Uses 105 ft^2 hinged reflector doors and 5-inch urethane insulation on back and sides of wood frame. Air transferred through 20 ft 2-way conduit box to be stored in 280 gallons water in 3700 beer cans stacked 1 inch apart enclosed in stone and earth.

Director: Paul W. Shippee
President
Colorado Mountain Builders
Route 1
Livermore, CO 80536
(303) 881-2372

Owner: Paul W. Shippee
address as above

Designer: Paul W. Shippee
address as above

Builder: Paul W. Shippee
address as above

Completion date: Early 1975

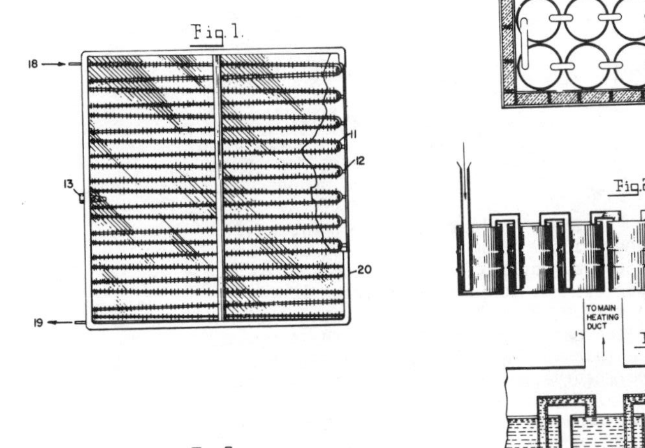

PAUL SHIPPEE'S SOLAR HEATED LOG HOME & SYSTEM DIAGRAM

COLORADO
house

Old Snowmass, Colorado
(14 miles west of Aspen)

DESCRIPTION: 1400 ft^2 private residence. One bedroom, kitchen-dining area, studio, living room, and one bath. Also 400 ft^2 loft which acts as another bedroom. Has 12 inches of fiberglass in walls and 2 inches of Dow SM insulation underneath a 7-inch floor slab. Walls 5 ft below grade on north side, 3 ft below grade on south. 7200 ft elevation.

Solar system: 556 ft^2 water-cooled, corrugated aluminum collectors at 50^0 angle on south roof. System drained at night to avoid freezing. Storage in a 5300 gallon cement tank with a butyl rubber membrane. Tank is under building and its top functions as a floor for one bedroom. Heat is distributed to house by radiant heating coils; 80^0 water can meet heating requirements of the house. Meets 100% (predicted) of heating requirements.

Director: Ron Shore
P.O. Box 238
Snowmass, CO 81654
(303) 927-4122

Owner: Ron Shore
address as above

Designer: Ron Shore
address as above

Engineer: Ron Shore
address as above

Contractor: Ron Shore
address as above

Completion date: November 9, 1974

RON SHORE'S SOLAR HOME

COLORADO
house

Swiss Peaks, Colorado
(near Boulder)

DESCRIPTION: 3400 ft^2 structure, private funding. Erected in 1969.

Solar system: Now installing system utilizing a spherical, stationary, reflective concentrator. The 10 ft x 10 ft mirror consists of a reflecting surface of polished aluminum mounted on a rigid substrate (foam or concrete, for example). Will cost \$12.00/ft^2. The absorber is a blackened copper cylinder within a glass envelope which acts as a convection shield. Absorber costs are estimated at 70¢/ft^2 of reflector. The tracking mechanism consists of either a gear track or a cable spool, either of which is controlled by a solid state sun sensor made from off-the-shelf components for less than \$100. Five-day storage utilizing Al_2O_3 briquettes.

Director: Dr. Gene Steward
Sugarloaf Road
Swiss Peaks, CO
(303) 444-0875

Owner: Dr. Gene Steward
Sugarloaf Road
Swiss Peaks, CO
(303) 444-0875

Architect: Charles Haertling, A.I.A.
1300 Canyon
Boulder, CO 80302
(303) 442-1342

Engineers, Contractors: Dr. Gene Steward and Dr. Jan Kreider

Completion date: Late 1974

Note: SRTA = Stationary Reflector Tracking Absorber

DR. GENE STEWARD'S HOME WITH SRTA SYSTEM

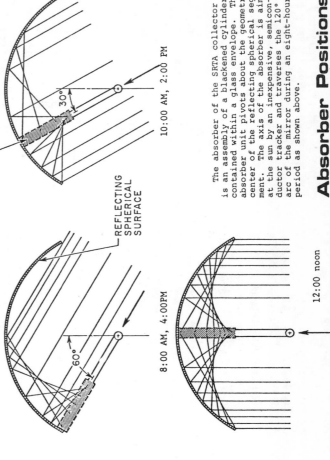

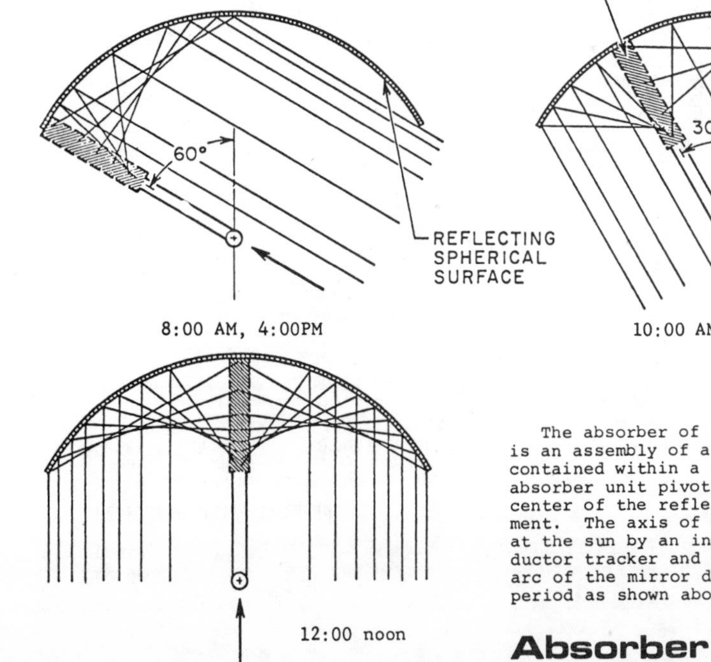

The absorber of the SRTA collector is an assembly of a blackened cylinder contained within a glass envelope. The absorber unit pivots about the geometric center of the reflecting spherical segment. The axis of the absorber is aimed at the sun by an inexpensive, semiconductor tracker and traverses the 120° arc of the mirror during an eight-hour period as shown above.

Absorber Positions
.......during a day's operation

COLORADO
house

DESCRIPTION: Retrofit of existing 2000 ft^2 home.

Solar system: 240 ft^2 of vertically mounted "fence style" water/glycol cooled flat plate collectors in (1) 4 ft x 6 ft modules, with adjacent aluminum reflector panels of equal size. System design is "open ended", i.e. additional collectors can be added at a later period if required. Thermal storage accomplished by (2) 1000 gallon storage tanks installed below grade.

Director: Byron Bloomfield
President
Energy Systems Corporation
1767 Woodmore Drive
Monument, CO 80132
(303) 481-3737

Owner: Name withheld on request

Architect: Byron Bloomfield
address as above

Engineer: Byron Bloomfield
address as above

Contractor: Byron Bloomfield
address as above

Completion date: March, 1974

SOLAR FENCE BEFORE REFLECTORS WERE INSTALLED

COLORADO
house

near Gold Hill, Colorado

DESCRIPTION: Extensive energy conservation and passive solar design. 2440 ft^2 house, orientation due south, and wall slope 35° off the vertical. All windows are insulated . Roof and south wall have 2 inches of insulation on top of 1 1/2 inches of wood decking. All walls have 3 inches of sprayed urethane foam. North wall has only one window and is half in the ground. House will heat itself on clear dav in 10° weather. Propane auxiliary

Director: John M. Storyk
Sugarloaf View, Inc.
Salina Star Route
Boulder, CO 80302
(303) 442-1253

Owner: John M. Storyk
address as above

Architect: John M. Storyk
address as above

Engineer: John M. Storyk
address as above

Contractor: John M. Storyk
address as above

Completion date: October, 1974

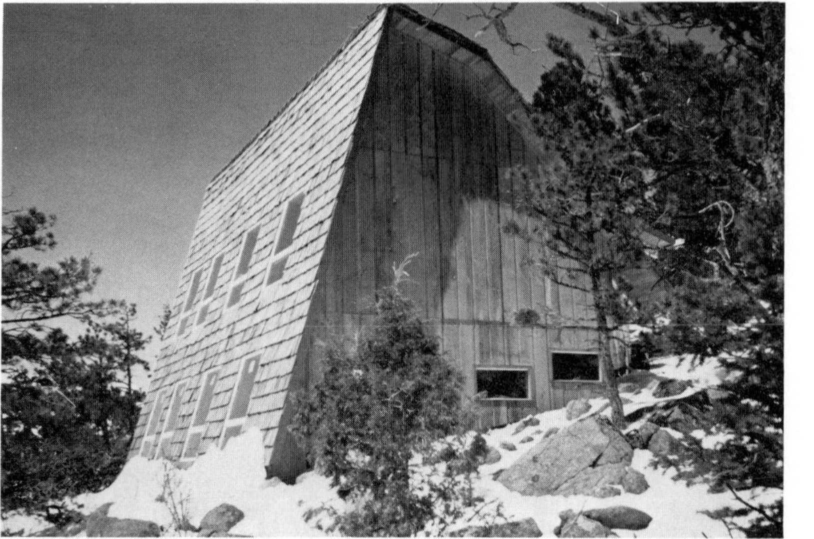

STORYK'S PASSIVE SOLAR HOME

COLORADO
house

5932 Del Paz Drive
Colorado Springs, Colorado

DESCRIPTION: Two-story, 3-bedroom, 2 1/2 bath, 2400 ft^2, full basement. Private funding through Security Savings and Loan; approximately $57,000 including solar system.

Solar system: Flat plate collector based on Thomason system. 30 ft x 25 ft, 2000 gallon storage tank, 25 ton of river bed rock.

Director: Peter Wood
408 Canyon Avenue
Manitou Springs, CO 80829
(303) 685-1155

Owner: Peter Wood
address as above

Architect: Peter Wood
address as above

Engineer (solar system): Harry Thomason
Thomason Solar Homes, Inc.
6802 Walker Mill Road, S.E.
Washington, D.C. 22027
(301) 336-4042

Contractor: Majestic Developers
2411 Providence Circle
Colorado Springs, CO 80909
(303) 591-0383

Completion date: July, 1974

COLORADO
academic building

Community College of Denver
North Campus
Westminster, Colorado 80203

DESCRIPTION: 278,815 ft^2 building, designed to minimize environmental impact and maximize humane utilization of space. The building will have 1300 ft-long double windows, extra wall and roof insulation, and waste heat recovery. The cost is estimated at $8,393,465; solar heating increment over conventional system is $667,000. Funded by the Colorado Legislature.

Solar system: 50,000 ft^2, water-cooled flat plate collectors facing winter noonday sun at angle at 60^0 to the horizontal. Collectors are integrated into folded plate roof on south, combining the structure and insulation of the collector and the roof. Thermal storage in underground tanks with capacity of approximately 400,000 gallons. When temperature of stored water exceeds 100^oF, it can be pumped directly to the heating coils' air-handling units, where heat is transferred to air and distributed. When water temperature is 55^oF to 100^oF, it is pumped through the chiller-heat pump units where its heat is extracted, consolidated, and sent out in 100^oF water to air handling units.

Director: Robert Kula
Planning Director
Community College of Denver
1009 Grant Street
Denver, CO 80203
(303) 892-3496

Owner: Colorado Department of Public Works
1525 Sherman, Rm 712
Denver, CO 80203
(303) 892-2626

Architect: The A.B.R. Partnership
1200 Walnut Street
Denver, CO 80204
(303) 825-8123

Design team: Phillip Tabb (Director)
Alan Brown
Joint Venture
1140 Detroit Street
Denver, CO 80206
(303) 322-6587

COLORADO
academic building
(continued)
Community College of Denver
North Campus
Westminster, Colorado 80203

Design team (continued):
Ronald Mason
Salvatore DiDomenico
A.B.R. Partnership
address as above
(303) 322-6587

Contact:
John Anderson, Partner
(A.B.R. Partnership)

Energy Consultant:
Alan Gass, A.I.A.,
602 South Harrison Lane
Denver, CO 80209
(303) 744-0583

Engineer:
Bridgers and Paxton
Consulting Engineers
213 Truman Street, N.E.
Albuquerque, NM 87108
(505) 265-8577

Contractor:
Pinkard Construction
P.O. Box 26227
Lakewood, CO 80226
(303) 986-4555

Completion date: Summer of 1976

COMMUNITY COLLEGE OF DENVER-NORTH CAMPUS

COLORADO
greenhouse

10075 E. County Line Rd.
Longmont Colorado 80501

DESCRIPTION: 1320 ft.^{2} solar heated greenhouse

Solar system: 1850 ft.^{2} of collector area at a 55 degree angle from the horizontal. Collector and storage are integrated in an A-frame structure 96 ft. long and 20 ft. on the side. Collector is comprised of three layers of high density U.V. retardent polyethylene plastic. A black absorber sheet (4 mills) is sprayed with water to absorb heat (Thermo-Spray process patent pending). Two clear 6 mill sheet that are inflated by an 80 watt blower to prevent wind buffating act as a glazing. The storage system is a concrete tank with a maximum capacity of 20,000 gals. Styrofoam sheets float on top of water surface to reduce heat loss. Hot water is pumped from storage to the greenhouse during night time and extended cloudy periods. Truck radiators are employed as water to air heat exchangers. Auxiliary heat supplied by natural gas furnace. Projected 80-85% of load can be carried by solar in coldest months.

DIRECTOR:
James B Wiegand, President
Solar Energy Research Corp.
10075 E. County Line Rd.
Longmont, CO. 80501
(303) 442-5105

OWNER:
Solar Energy Research Corp.
address as above

DESIGNER:
Solar Energy Research Corp.
address as above

ENGINEER:
John M. Freeman Jr.
Solar Energy Research Corp.
address as above

CONTRACTOR:
Solar Energy Research Corp.
address as above

COMPLETION DATE: April, 1974

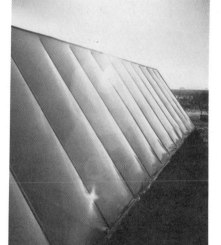

SOLAR ENERGY RESEARCH CORP. COLLECTOR

COLORADO
greenhouse

Denver, Colorado

DESCRIPTION: Solar heated-solar heater greenhouse

Solar system: A 75 ft.^{2} south facing window of tedlar coated, U.V. stabilized fiberglass, transmits the energy to the interior of the greenhouse where it is converted to heat. All other walls in the A-frame structure are 4 inches of expanded polystyrene with a coating of concrete mixed with fiberglass. A movable door consisting of 4 inches of expanded polystyrene is employed to reduce energy losses through the south window at night and during cloudy and cold days. Automatic controls operate the movable insulation and the fan which is used to transfer the hot air from the interior to storage area below the growing surface. Water, dirt and cinderblocks are utilized as storage mediums. Supplementary lighting is used as the auxiliary heat source and to extend photo-period of plants: When heat storage is full, feed water preheating may be used for other energy loads (i.e. domestic hot water preheating, space conditioning, etc.)

DIRECTOR: Richard S. Speed, President, SOLTEC
70 Adams St.
Denver, Colorado 90206
(303) 333-8869

ENGINEER: Kent M. Johnson
SOLTEC

OWNER: SOLTEC
address as above

CONTRACTOR: SOLTEC

COMPLETION DATE: Feburary, 1974

DESIGNER: Richard S. Speed

Note: Models to be constructed soon will employ cooling provided by night sky radiation from north wall, external dew ponds, and night time rejection of stored heat. For further information on SOLTEC and solar heated greenhouses, write or call Richard S. Speed.

SOLTEC GREENHOUSE

COLORADO
barn

Nelson Road
Longmont, Colorado 80501

DESCRIPTION: Approximately 1000 ft^2 barn, privately funded.

Solar system: 500 ft^2 collector (water-type); storage and distribution integrated into 5 inch concrete slab floor (PVC thin wall pipe coils imbedded in floor).

Director: Jay Lynch
Route 3, Box 330
Longmont, CO 80501
(303) 442-0372

Owner: Community on Nelson Road, Inc.
3743 Nelson Road
Longmont, CO 80501

Architect: Jay Lynch
address as above

Engineer: Joy Lynch
address as above

Contractor: Cooperative Construction Co.
Route 3, Box 330
Longmont, CO 80501
(303) 442-0372

Completion date: January, 1975

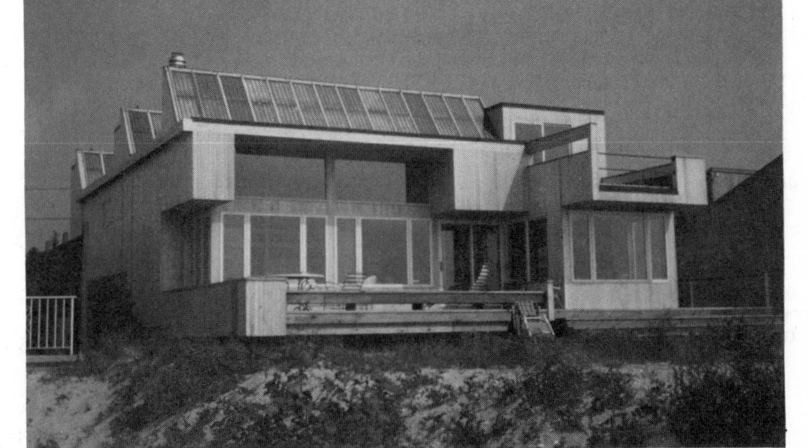

SUNWORKS SOLAR HOME

COLORADO
office building

119 Madison
Denver, Colorado 80206

DESCRIPTION: Three-level modern office building, with one level below grade opening to below-grade courtyard. 7000 ft^2 total floor area. Very well insulated with many other energy conserving features. 180 ft^2 total, double glazed, regressed, view window area on first level.

Solar system: 600 ft^2 flat plate, air-type collector sloping 53° from horizontal. Double glazed with tempered glass. Storage is 4000 gallons of water in two insulated concrete tanks in basement. 80% (predicted) solar heated.

Director: Richard L. Crowther, A.I.A.
Crowther-Kruse-McWilliams
2830 E. Third Avenue
Denver, CO 80206
(303) 355-2301

Owner: Richard L. Crowther, A.I.A.
address as above

Architect: Richard L. Crowther, A.I.A.
address as above

Engineer: Richard L. Crowther, A.I.A.
address as above

Contractor: Undecided at present

Completion date: Will commence construction as soon as a tenant has been located and lease is signed.

RICHARD CROWTHER'S OFFICE BUILDING

CONNECTICUT
house

Unlisted address
Contact project director

DESCRIPTION: 1 1/2-story, 3-bedroom, $60,000 house with built-in 2-car garage, basement, and horizontal roof. Living area: 1900 ft^2. House aims 10° W of S. Excellent insulation. Many energy conserving features.

Solar system: South sides of first and second stories are 60% and 70% window, double-glazed. Ground floor windows are shaded in summer by roof overhang. Collector: 450 ft^2 of flat-plate, water-type, 57°, sloping panels on roof arranged in 3 parallel rows each of 24 to 28 ft length. Each panel is a standard Sunworks, Inc. assembly, 2 ft x 6 ft, single-glazed. It employs a 0.017-in. copper sheet with selective black coating having a/E ratio of about 5. Copper tubes (1/4-inch in diameter and 5 inches apart on centers) are soldered to copper sheet. Panel backing: 3-inch fiberglass. Water and propylene glycol (non-toxic) flows in (parallel-connected) tubes to heat exchanger incorporated in storage tank. Storage system: 2000-gallon steel tank (insulated 5 ft 4 in. in diameter, 12 ft long, axis horizontal; underground). Carrythrough: 2 days. Coil and fan system sends hot air to rooms. When tank is cold and sun is shining, hot liquid from collector can be made to bypass tank and (via coil and fan system) contribute immediately to room heating. 55% solar heated (predicted). Auxiliary heat: from oil furnace. Cooling in summer to be arranged later.

Director: Everett Barber, Jr.
Sunworks
669 Boston Post Road
Guilford, CT 06437
203-453-6191

Owner: Name and address of owner withheld on request

Architect: Donald Watson, A.I.A.
Box 401
Guilford, CT 06437
(203) 453-6388

Engineer: Everett Barber, Jr.
address as above

Completion date: Late 1974

Note: Various articles are available from Sunworks, including "Energy Conservation in Building Design" and "Preliminary Design of Solar Heated Buildings." Write Sunworks for further information.

CONNECTICUT
house

Guilford, Connecticut

DESCRIPTION: Floor area 28 ft x 28 ft = 1300 ft^2. Two stories, 3 bedrooms, no basement. Below floor slab is 2-ft-thick layer of 4-inch-diameter stones, below which is a 1-inch insulating layer. Excellent insulation, double-glazed windows, north windows small, insulating shutters (3-inch fiberglass) for some windows. Concrete-block walls have 3-inch sprayed polyurethane foam insulation on outside; thus the concrete blocks help stabilize house temperature. Some fireplace chimney heat is intercepted and sent to storage system. Two 10-ft-diameter windmills produce 80% of electrical power needed. Estimated cost, near $50,000.

Solar system: Collector: 400 ft^2, flat-plate, water-type, on roof sloping 57° from horizontal. Sunworks panels fit between rafters. Typical panel: 38 3/4 in. x 86 in. with fiberglass-backed 0.017-in. copper sheet to which 3/8-in. diameter copper tubes (5-in. apart on centers) are soldered on underside. Selective black coating has a/E ratio of 10. Glazing: one 1/8-in. sheet of glass. Liquid: water plus 40% ethylene glycol. Collector is not drained at night. Circulation by 1/12 HP pump.
Storage system--3 kinds: (1) 1500 gallon, vertical cylindrical steel tank, 5 ft diameter, 15 ft high, behind (to east of) fireplace; heat distributed to second story by fan-coil system, or by gravity convection of water if electricity fails; preheated domestic hot water via pipe through the 1500 gallon tank; (2) the below-floor, 2-ft layer of stones; heat extracted by airflow; heat input from hot air pumped from near ceiling on sunny day; and (3) mass of concrete blocks, etc. of house itself, the blocks having an outside insulating layer.
60 to 70% solar heating (predicted). Auxiliary heat from oil-fired, 110,000 Btu/hr. domestic water heater.
Cooling in summer: Eaves exclude much sunlight. Sliding shutters block sunlight approaching south glass doors. South side clerestory and roof belvedere encourage natural venting of hot air. 2-ft layer of stones is cooled at night by forced air circulation, and helps to cool building during day.

Director: Everett Barber, Jr.
Sunworks
669 Boston Post Road
Guilford, CT 06437
(203) 453-6191

CONNECTICUT
house
continued

Guilford, Connecticut

Owner: Everett Barber, Jr.
address as above

Architect: Charles W. Moore and Associates
Essex, CT 06426
(203) 767-0101

Engineers: Everett Barber, Jr.
address as above

Contractor: Richard Riggio, Builders
Ivory Town, CT 06442
(203) 767-8494

Completion date: April, 1975

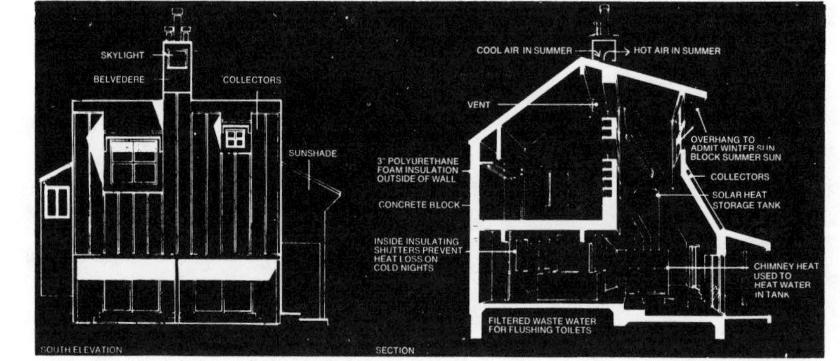

BARBER HOUSE

CONNECTICUT
house

Guilford, Connecticut 06437

DESCRIPTION: One-story, 2-bedroom, 24 ft x 32 ft (750 ft^2), well insulated house with basement and unheated attic. House aims 5^0 E of S. House is contiguous to 3-car garage.

Solar system: Water-type collector on 45^0 sloping roof. Comprises 18 panels, each 38 3/4 inches x 86 inches, with fiberglass-backed copper sheet (with moderately selective black coating) to which 3/8 inch diameter copper tubes are soldered on underside. Uses water without antifreeze; liquid is drained from collector before freezing danger arises. Collector and pipes contain 25 gallons of water, which is circulated by 15 gpm pump when T-collector exceeds T-tank. Storage system: 1500 gallon, rectangular water-filled concrete tank with 6-inch fiberglass insulation in basement, for heating house and preheating domestic hot water. Also 30-gallon tank with Carlin oil-fired heating, for domestic hot water, and when temperature of main tank is below 95^0 F, for heating house. Rooms are heated by fan and two-coil system, one coil being fed from 1500-gallon tank and one from 30-gallon tank; each coil has its own centrifugal pump and the appropriate pump is invoked automatically. Carrythrough is 2 days (predicted). 50-60% solar heated (predicted).

Director: Carlton Granbery
123 York Street
New Haven, CT 06511
(203) 865-3132

Owner: Dr. Gifford B. Pinchot
Guilford, CT 06437

Architect: Carlton Granbery
address as above

Engineers: Dubin-Mindell-Bloome Associates
312 Park Road
West Hartford, CT
(203) 236-1654

Everett Barber
Sunworks
669 Boston Post Road
Guilford, CT 06437
(203) 453-6191

Contractor: Erwin C. Griffiths, Inc.
Dromara Road
Guilford, CT 06437 (203) 453-2164

Completion date: December, 1974

CONNECTICUT
house

Westover Road
Stamford, Connecticut

DESCRIPTION: Split level, 2000 ft^2 Retrofit.

Solar system: Pyramidal optical solar energy collection system. Collector area 2 ft x 8 ft, optical aperture 8 ft x 12 ft. Solar hot water heating only at present stage. Plan on doubling collection area in future.

Director: Gerald Falbel
Wormser Scientific Company
88 Foxwood Road
Stamford, CN 06983
(203) 322-1981

Owner: Wormser Scientific Company
address as above

Engineers: Gerald Falbel
address as above

Contractor: Wormser Scientific Company
address as above

Completion date: Early 1974

Note: Wormser plans on doing complete solar heating at two other locations in future.

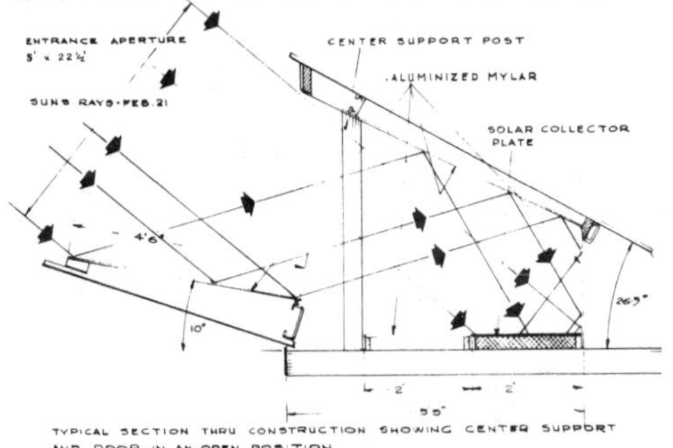

WORMSER SCIENTIFIC CORPORATION

DELAWARE
house

Solar One
190 S. Chapped St.
Neward, Delaware

DESCRIPTION: Experimental house. First story - 1400 ft^2 - contains living and dining room, two bedrooms, 1 1/2 baths, kitchen, and two protruding bays on south side. Second story is unfinished; may become two bedrooms. Full basement and 2-car garage.

Solar system: Utilizes cadmium sulfide (CdS) photovoltaic cells as well as flat plate collectors. At pressent, CdS cells occupy 3 out of 24 roof panels, each 4 ft x 8 ft and total net area 75 ft^2. Conversion efficiency is 5%. Cells are used to charge a set of 180 amp/hour lead acid batteries from which current is drawn to operate electric lights, etc. An inverter is used to convert DC to AC. It should be noted that the system is used on an experimental basis and that full loads are computed.

Space heating is provided by six air-cooled panels, each 4 ft x 6 ft, on southmost vertical faces of the two bays. Main storage is 8000 lbs of eutectic salt (sodium thiosulfate pentahydrate), the operating temperature (phase change temperature) of which is 120°F. The salt is contained in plastic trays, each 22 inches x 22 inches x 1 inch, arranged in stacks, enclosed in thermally insulated housing. Supplementary storage in 1400 lbs of eutectic salt consisting mainly of sodium sulfate decahydrate, with an operating temperature of 75°F. 1973-74 percent solar heated: 67%. Predicted for 1974-75: 80%. 9 kw electric heater auxiliary heating supply.

Director: Dr. Karl Boer
Institute of Energy Conversion
University of Delaware
Newark, DE 19711

Owner: University of Delaware
address as above

Contractor: Fred G. Krapf
Walnut Street
Wilmington, DE
(302) 656-6686

Architect: Harry M. Weese & Assoc.
10 W. Hubbard
Chicago, IL 60610
(312) 467-7030

Completion date: July 20, 1973

Engineer: Dr. F. C. Schwarz
Tech U.
Delft, Holland

Solar System Engineers: Dr. Maria Talkes
Institute of Energy Conversion
University of Delaware
Newark, DE

Dr. M. K. Selcuk
Honeywell Corp.

DELAWARE
amusement building

Wilmington, Delaware

DESCRIPTION: 30,000 ft^2 building being designed for Boy's Club in an energy conservative manner.

Solar system: Solar heated swimming pool, hot water for the showers and locker rooms and tempering the outdoor air for ventilation for the gymnasium and swimming pool.

Owner: Boy's Club
Wilmington, DL

Architect: George Whiteside

Engineer: Dubin-Mindell-Bloome Associates
Consulting Engineers and Planners
42 West 39th Street
New York, NY 10018
(203) 236-1654

UNIVERSITY OF DELAWARE'S SOLAR ONE

DISTRICT OF COLUMBIA
houses

Washington, D.C.

DESCRIPTION: Several homes now built, all based on original design of Thomason's private residence. First home is 3 bedroom, 1064 ft^2. (Harry Thomason now lives in a solar home he built a few years after this one completed.)

Solar system: (first home) 850 ft^2 collector area. Collector is on two sloping positions of roof, the slopes being 45^0 and 60^0 from the horizontal. Water trickles down 1-inch-wide valleys of blackened corrugated aluminum sheet. Corrugations are 2 1/2 inches apart on centers. Originally glazed with 5-mil mylar film, replaced with glass three years later. Water from the valley is collected in a gutter and flows to a 1600 gallon steel tank surrounded by 50 tons of fist sized stones. Air flows through stones and thence to rooms. At night a 1/3 HP water pump used to send water from tank to distribution pipe is turned off, and all water drains to tank. Cooling accomplished by sending water to north slope at night to cool by evaporation.

Director: Dr. Harry E. Thomason
6802 Walker Mill Road, S.E.
Bethesda, MD
(301) 336-0009
(301) 336-4042
(301) 336-5329

Owner: Dr. Harry E. Thomason
address as above

Designer: Dr. Harry E. Thomason
address as above

Engineer: Dr. Harry E. Thomason
address as above

Completion date: 1959 (first). Several more built in 1960's. Currently building two homes, one off Route 4, just across the Calvert County line from Prince Georges, and the other near Indian Head Highway, six miles south of the Capital Bettway.

Note: Edmund Scientific Company distributes plans for Thomason's solar homes. Address:

Edmund Scientific Company. 150 Edscorp Bldg.
Barrington, N.J. 08007 (609) 767-8483

Illustration see page 2.033.

FLORIDA
houses

Lake Podgett
Florida

DESCRIPTION: Solar housing development.

Solar system: Water system with collector on south half of sloping roof; storage tank in basement.

Developer: Covington Properties, Inc.
Land O' Lakes
Florida
(813) 949-4258

Solar Consultants: Dr. Erich Farber
Solar Energy Laboratory
238 Mechanical Engineering Building
University of Florida
Gainesville, FL 32611
(904) 392-0820

H. Ingley and C. A. Morrison

UNIVERSITY OF FLORIDA'S SOLAR HOUSE PROJECT DESCRIPTION PAGE 2.032

FLORIDA
house

The Solar House of the
University of Florida
Gainsville, Florida 32611

DESCRIPTION: One-story, built in 1955. Partly converted to solar heating in 1963, fully converted in 1974.

Solar system: Collector on roof, water type. Consists of 10 rectangular panels, each 5 ft x 8 ft, sloping 30° from horizontal. Three types of absorber plate are used. Single glazing is of ordinary, double-strength, low-iron-content window glass. Coating on absorber plates is common, non-selective, flat-black paint. For heating domestic hot water, collector on ground with steeper slope (40°) to provide much energy capture the year around. Storage system: 3000 gallon lined steel tank insulated with 4 inches of styrofoam. Situated above ground so visitors can inspect it. Rooms are heated by hot-water flow to radiators. 100% solar heated. Additional to this project are solar cooking facility, solar generation of electricity, solar powered car, and more.

Director: Dr. E. A. Farber
338 Mechanical Engineering Building
University of Florida
Gainsville, FL 32611 (904) 392-0820

Owner: University of Florida
Gainsville, FL 32611

Designer: Dr. E. A. Farber
address as above

Engineer: Dr. E. A. Farber
address as above

Contractor: Dr. E. A. Farber
address as above

Completion date: 1974

GEORGIA
academic building

George A. Towns Elementary School
760 Bolton Road
Atlanta, Georgia 30331

DESCRIPTION: National Science Foundation funding with cost sharing by contractor participants. School is one-story, 32,000 ft^2. Plan to heat and cool entire building.

Solar system: 13,000 ft^2 water-cooled flat plate collector area using aluminum Roll-Bond and two laminated sheets aluminized mylar as reflector in front of collector. Use drainage to avoid freezing. 45,000 gallon hot and cold storage. Will be monitored and analyzed for approximately one year after operation begins.

Director: Richard T. Duncan
Westinghouse Electric Corp.
P.O. Box 1693 MS-999
Baltimore, MD 21203
(301) 765-3545

Owner: After 2 yr. NSF study is complete system will be owned by:
George A. Towns Elementary School
760 Boulton Road, Atlanta, Georgia 30331

Architect: P. Richard Rittelmann, A.I.A.
Burt, Hill, and Associates
610 Mellon Bank Building
Butler, PA 16001
(412) 285-4761

Engineer: Dubin-Mindell-Bloome Associates
42 West 39th Street
New York, NY 10018
(212) 868-9700

Solar Engineers: Burt, Hill, and Associates
address as above

Contractor: (Cost Manager) Burt, Hill, and Associates
Construction Services Division
address as above

Completion date: April, 1975

THOMASON'S FIRST SOLAR HOME

GEORGE A. TOWNS ELEMENTARY SCHOOL

GEORGIA
bank

Fulton Federal Savings and Loan
Atlanta, Georgia

DESCRIPTION: 3000 ft^2 single-story bank building. Goal of the project is to demonstrate the feasibility and aesthetically pleasing qualities of solar heating, cooling and energy conservation methods to both the local community and the building developer/owners. Modular construction designed to adapt to different types of lots with different sun angles.

Solar system: 2000 ft^2 solar collector. 80% solar heated and cooled.

Director: Bob Frederic
Fulton Federal Savings and Loan
P.O. Box 1077
Atlanta, GA 30301
(404) 522-2300

Owner: Fulton Federal Savings and Loan
address as above

Architect: Nichols and Young
979 Tennyson Drive
Atlanta, GA 30318
(404) 355-4577

Engineer: Dr. Erich Farber
Solar Energy Laboratory
University of Florida
288 Mechanical Engineers Building
Gainesville, FL 32611

Completion date: Design stage.

ILLINOIS
house

Eureka, Illinois

DESCRIPTION: Mass-produced, 3-bedroom, with basement, 1200 ft^2.

Solar system: 800 ft^2, 3 ft x 8 ft, water/glycol collectors using Roll-Bond aluminum baseline. Storage 1250 gallon steel water tank. Provides space heating and domestic hot water.

Director: Dr. Y. B. Safdari, President
Sun Systems, Inc.
P.O. Box 155
Eureka, IL 61530
(309) 467-3632

Owner: Sun Systems, Inc.
address as above

Architect: Tom Landers and Associates
Pine Street
Chillicothe, IL
(309) 274-2898

Engineer: Dr. Y. B. Safdari
address as above

Contractor: Lakeview Realty Company
RR 1
Eureka, IL
(309) 467-2321

Completion date: December 1, 1974

SUN SYSTEMS HOUSE

ILLINOIS
academic building

Sangamon State University
Springfield, Illinois 62708

DESCRIPTION: 24-ft-diameter dome, 5/8 sphere. Used as play area for day care center, solar experimental lab, and teaching and demonstration aid. National Science Foundation funded.

Solar system: 210-ft, Roll-Bond , double-glazed, non-selective 3M Nextel Black collector. Storage in above-ground, 2000 gallon, insulated steel water tank.

Director: John Drabanski, student director
Dr. Al Casella, faculty advisor
Physical Sciences Program
Sangamon State University
Springfield, IL 62708

Owner: Sangamon State University
address as above

Designer, Engineer, and Contractor: Students designed, engineered, and built the structure.

Completion date: September, 1974

SANGAMON STATE UNIVERSITY'S SOLAR HEATED DAY CARE CENTER

ILLINOIS
office building

Chicago, Illinois

DESCRIPTION: Office building in preliminary design stage. Designed for energy conservation.

Solar system: Solar collectors on roof and south wall. Collectors on south wall replace conventional building construction, providing a savings in cost. Energy used for this facility is expected to be less than 1/10 of the energy consumed in existing facilities which this building will replace.

Owner: Argonne National Laboratory
Chicago, IL

Engineer: Dubin-Mindell-Bloome Associates
Consulting Engineers and Planners
42 West 39th Street
New York, NY 10018
(203) 236-1654

Note: A concentrating flat plate solar collector of a novel design is also being developed at Argonne in conjunction with the Enrico Fermi Institute & Department of Physics, University of Chicago, Chicago, Illinois 60637. Dr. Roland Winston is principal investigator.

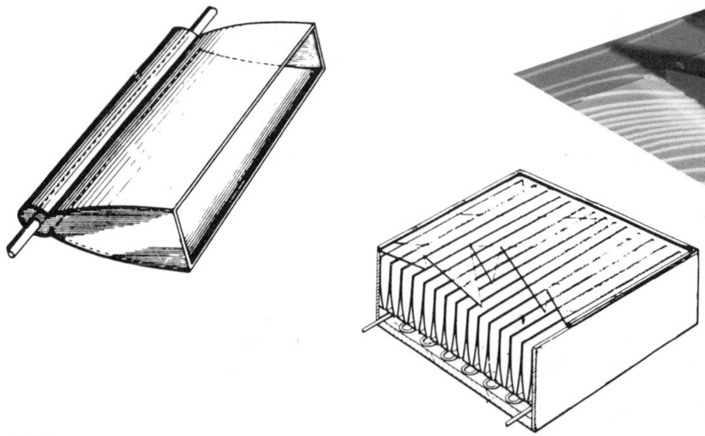

ROLAND WINSTON'S COMPOUND PARABOLIC CONCENTRATOR

MID-AMERICA INDUSTRIES' IOWA HOUSE

IOWA
house

Lot No. 24, Equestrian Hill Development
Highway 6, South of Ashland, Nebraska

DESCRIPTION: House funded by Iowa State University Research Institute. 2516 ft^2 usable floor area. Northeast wall shaded and insulated by earth berm. Minimum window area, movable insulation panels instead of curtains. No wall faces directly east or west to reduce summer cooling load. Northeast wall shaded and insulated by adjoining garage. Structural system consists of triangular modules joined with rectangular connectors, thus allowing large collector areas on the hypotenuse, while limiting the enclosed volume.

Solar system: Utilizes two types of solar collector. One is "solar wall system," 1100 ft^2 hot air collectors mounted on south vertical walls to intercept maximum solar radiation during winter, minimum during summer. Hot air collectors contain semi-transparent plates which permit visibility and allow light to enter structure. Storage material used is 200 cubic ft encapsulated sodium sulfate decahydrate, to be housed in new type of container which will hopefully solve problem of reduced heat storage capacity due to phase change cycling of the eutectic salts. Second type is hot water type to be added at later date for domestic hot water and space conditioning.

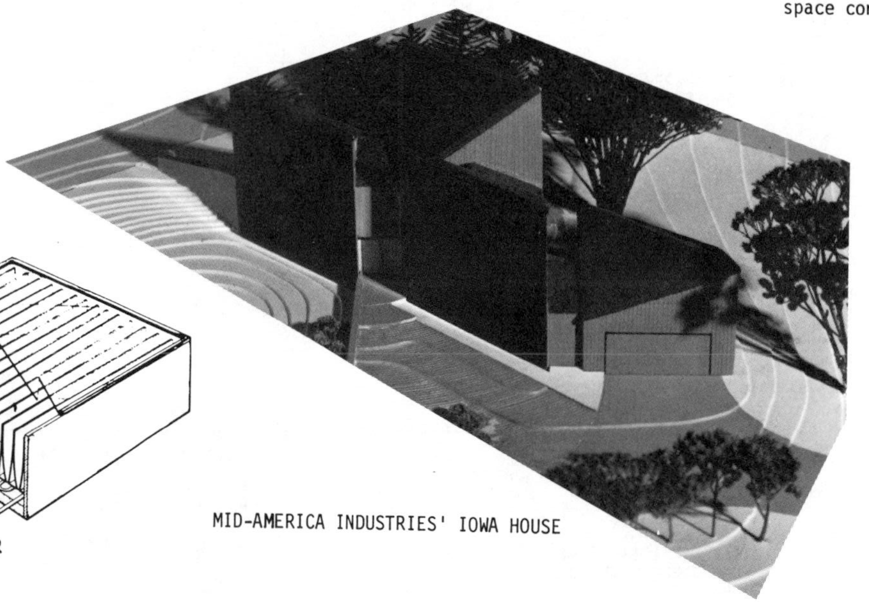

Director: Carl D. Meyer
Hansen-Lind-Meyer
116 South Linn Street
Iowa City, IA
(319) 338-7555

Owner: John McLaughlin, President
Mid-America Industries
Mead, Nebraska
(402) 624-6611

Architect: Richard Kruse
Hansen-Lind-Meyer
address as above

Engineer: James Schoenfelder
Hansen-Lind-Meyer
address as above

Contractor: Thornton Construction Co.
8998 "L" Street
Omaha, NB 68127

Completion date: December, 1974

IOWA
office building

Mead, Nebraska

DESCRIPTION: One-story, 1880 ft^2 office building employing energy conservation features, including: light weight concrete insulating roof tile (product of mid-America Industries), task lighting, vestibule as a "thermal shock absorber," earth excavated for construction as windbreak at northwest corner, and trees and climbing vines on west wall.

Solar system: 800 ft^2 perforated plate-type hot air collectors mounted on south-facing vertical walls. Storage: eutectic salts ($NaSO_4 \cdot 10\ H_2O$) placed in specially designed containers. Cooling done with standard residential air conditioning unit. An economizer cycle incorporated to provide 100% free cooling when outside conditions favorable. Electric furnace as backup heating.

Director: James Schoenfelder
Hansen-Lind-Meyer
116 Linn Street
Iowa City, IA 52240

Contractor: Thornton Construction
8998 "L" Street
Omaha, NB 68127
(402) 331-1700

Owner: Solar, Inc.
Box 246
Mead, NB 68041
(402) 624-6611

Completion date: November, 1974

Architect: James Schoenfelder
address as above

Engineers: James Schoenfelder
address as above

SOLAR, INC.'S OFFICE BUILDING

KANSAS
house

205 S. Pershing
Liberal, Kansas 67901

DESCRIPTION: Solar heating system designed for one room of house that needed supplementary heating. Experimental projects planned including redesigning solar collectors.

Solar system: 8 ft x 15 ft collector, triple plastic glazing, aluminum corrugated, non-selective flat black paint. Storage, 1200 gallon water.

Director: Elmo P. Atwell
205 S. Pershing
Liberal, KS 67901
(316) 624-6566

Owner: Elmo P. Atwell
address as above

Designer: Elmo P. Atwell
address as above

Engineer: Elmo P. Atwell
address as above

Contractor: Elmo P. Atwell
address as above

Completion date: September, 1974

MAINE
house

New Harbor, Maine

DESCRIPTION: $18,000 privately financed, five-story combination of windmill, house, workshop. Vertical axis windmill on top of tetrahedron shaped structure.

Solar system: 1300 ft^2 collector wall - beadwall insulation.

Director: David Seller
Dimetrodon Corporation
Prickly Mountain
Warren, VT 05674
(802) 496-2907

Owner: Nancy Barrett
Box 83
Durham, NH 03824

Architect: David Seller
address as above

Engineer: David Seller
address as above

Contractor: David Seller
address as above

Completion date: December, 1974

MAINE
house

Bar Harbor, Maine

DESCRIPTION: Private funding, $35,000 mortgage by Depositor's Trust Co. and Rockland Savings Bank. One-story, 1340 ft^2 dwelling situated on bare, rocky ledge. Faces 10° W of S. No attic. Small, lower-level, daylight-lit utility room. Many energy saving features. Heavy insulation: 7 inch fiberglass in walls, 9 inch in roof. Foundations insulated with styrofoam. Thermopane windows; Thermopane sliding doors. Small window areas on E and N. Small greenhouse on S side humidifies house, thus making lower temperature acceptable.

Solar system: Collector: 540 ft^2, trickling water type, 26 ft long, 20 ft high. The corrugated aluminum sheets, with valleys 2 1/2 inches apart, and with non-selective black coating, slopes 54° from horizontal. It is double-glazed from aluminum sheet and sealed with silicone. Large area of white stones adjacent to lower edge of collector reflects additional solar radiation to collector. Large south view window area also receives much radiation; is shuttered at night.
Storage system: 2000-gallon horizontal cylindrical steel tank (fiberglass lined) in utility room. Tank contains 1700 gallon plain water (glycol may be used later). Tank is surrounded by crushed rock in insulated wooden bin. Crawl space under 2/3 of house serves as plenum for distribution of hot air to rooms. Domestic hot water tank resides inside main tank and is heated by it. Auxiliary heat 8500 Btu Valley Comfort wood burning stove.

Director: Ernest McMullen
Canoe Point
Bar Harbor, ME 04690
(207) 288-4969

Owner: Richard D. Davis
Seal Harbor, ME 04675
(207) 276-5161

Designer: Ernest McMullen
address as above

Engineer: Ernest McMullen
address as above

Contractor: Ernest McMullen
address as above

Completion date: November, 1974

MARYLAND
house

Gaithersburg, Maryland

DESCRIPTION: Two-story, 4-bedroom house and/or laboratory. Prefabricated elsewhere, purchased, brought to site by truck in 1972, assembled, thermally tested in giant coldroom in 1973 by NBS and HUD cosponsorship.

Solar system: By spring of 1974 orders were placed for PPG Baseline flat plate, water-type collectors to be mounted at two slopes: 12° and 55° from horizontal; also for 1000 gallon and 500 gallon storage tanks; and an Arkla air conditioner.

Director: Dr. James Hill
National Bureau of Standards
Building 226, Room B104
Washington, D.C. 20234
(301) 921-3503

Owner: National Bureau of Standards
Building 226, Room B104
Washington, D.C. 20234
(301) 921-1000

Engineers: National Bureau of Standards
address as above

Contractor: National Bureau of Standards
address as above

Completion date: December, 1974

MARYLAND
academic building

Timonium Elementary School
201 East Ridge Road
Timonium, Maryland

DESCRIPTION: Existing one-story elementary school building retrofitted with solar system. Funded by AAI, Inc. ($73,000 of its own funds) and National Science Foundation ($427,000).

Solar system: 5000 ft^2 water-type collector (with honeycomb) on one of three wings of school building. The 180 4 ft x 7 ft panels are arranged in 10 banks (18 panels per bank), each of which slopes about 45°. Storage in 15,000 gallon water tank providing 4-day carrythrough.

Director: Irwin Barr
AAI, Inc.
Weapons and Aero Systems Division
P.O. Box 6767
Baltimore, MD 21204
(301) 666-1400, Ext. 232

Owner: After 2 yr. NSF study is complete the system will be owned by:
The Baltimore County School System
Timonium Elementary School
201 East Ridge Road
Timonium, MD
(301) 252-3932

Solar System Engineer: AAI, Inc.
address as above

Contractor: AAI, Inc.
address as above

Completion date: 1974

TIMONIUM ELEMENTARY SCHOOL

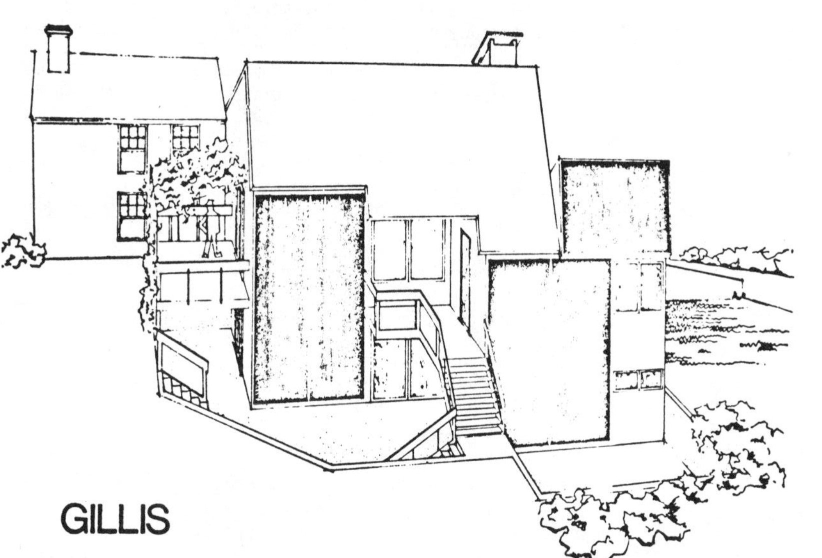

GILLIS

MASSACHUSETTS
house

Hudson, Massachusetts

DESCRIPTION: In design phase. Combination of new and old construction. Portion of 200 year old house to be relocated on same plot and 1700 ft^2 of new construction. Designed to use solar energy for space heating, domestic hot water, and a swimming pool. Massive concrete walls with the insulation located between the concrete and the ground. Earth brought up around building to reduce possible heat loss during the winter and to help keep the building cool during the summer. A Clivus Multrum, a dry composting toilet, proposed for use.

Solar system: Solar collectors on south walls and sloping roofs. Large areas of south facing glass transmit sunlight directly. A sunken patio, on the south side, acts as a "solarium" which would be heated by the sun during the winter and temper the outdoor air.

Owner: Nancy and John Gillis & Family
Hudson, MA

Architect: Bruce Anderson
Total Environmental Action
Box 47
Harrisville, NH
(603) 827-3087

Engineer: Total Environmental Action
address as above

Completion date: Designed spring 1974, not built yet. Redesigned with reduced square footage being considered.

MASSACHUSETTS
house

Weston, Massachusetts

DESCRIPTION: Two-story, 4-bedroom house. Floor area is 2600 ft^2.

Solar system: Collector, originally in two parts: (1) Vertical south wall, 460 ft^2, double glazed, with vertical baffles and roof overhang to reduce wind and block sun's rays in summer. (2) Roof collector, sloping 3° from horizontal toward north, had area of 1200 ft^2 and was double-glazed during some years of operation; the energy was carried to storage by air flow. No separate storage system used, but the massive concrete floor and 12-inch concrete exterior block walls and outlets provided thermal capacity of 75,000 BTU 40-60 KWH/C°. Complete system provided 60% of winter's heat need. When roof collector was abandoned (due to leakage), the vertical south wall collector provided 40% of heat need. Now there is a 25 m^3 (270 ft^2) auxiliary collector on the south face, and system is back to 60%.

Director: Norman B. Saunders
15 Ellis Street
Weston, MA 02193
(617) 894-4748

Owner: Norman B. Saunders
address as above

Architect: Architects Collabrative
46 Brattle Street
Cambridge, MA 02138
(617) 868-4200

Contractors: Norman B. Saunders
address as above

Dick Soule
16 Hancock
Lexington, MA 02173
(617) 862-1548

Completion date: 1960. Still in use.
Testing of new ideas has been continuous. Newest full-scale collector is number 11.

Note: "A Dozen Years in the Solar Heat System of Experimental Manor" is a 9 page paper describing the system and is available, as well as many other papers, from Norman B. Saunders. Write for further information; include self-addressed stamped envelope.

MASSACHUSETTS
academic building

11 Chase Street
Boston, Massachusetts 02125

DESCRIPTION: Retrofit of existing school by General Electric Co. with National Science Foundation funding of $335,000. Installed in March, 1974.

Solar system: Water-type collector on roof. Collectors, 4 ft x 8 ft. Olin Brass, Inc. aluminum Roll-Bond panels, double glazed with Lexan, comprised of three 45° -sloping arrays. Water and anti-freeze solution carries energy to 2000 gallon tank. Rooms heated by air-flow system served by heat exchangers. System is expected to supply 20% of heat needed.

Director: Mr. Hurlebaux
General Electric Company
King of Prussia, PA 19406
(215) 962-4294

Owner: After 2 yr. NSF study is complete the system will be owned by:
Grover Cleveland School
11 Chase Street
Boston, MA 02125

Engineer: General Electric Co.
address as above

Completion date: March, 1974.

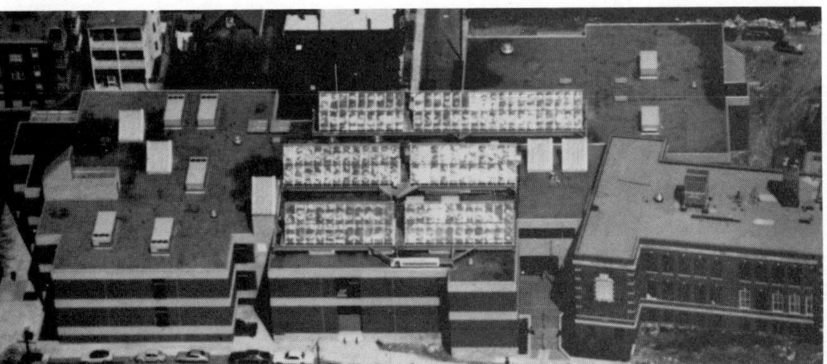

GROVER CLEVELAND SCHOOL

MICHIGAN
experimental dwelling

Annex, University of Michigan
321 Plymouth Road
Ann Arbor, Michigan 48104

DESCRIPTION: 10'4" x 18' experimental dwelling, funded by the National Science Foundation.

Solar system: Automated low-tech natural cooling system. Double roof construction, evaporative cooling done on the lower roof.

Director: Edward J. Kelly, Jr.
Arch. Res. Lab.
Tappan and Monroe Streets
University of Michigan
Ann Arbor, MI 48104
(313) 764-1340

Owner: Department of Architecture and Design
University of Michigan
Tappan and Monroe Streets
Ann Arbor, MI 48104
(313) 764-1340

Architect: Edward J. Kelly, Jr.
address as above

Engineer: W. Appleyard
424 Benjamin
Ann Arbor, MI 48104
(313) 665-0350

Solar Engineer: R. MacMath
424 Benjamin
Ann Arbor, MI 48104
(313) 665-0350

Completion date: July 31, 1973

MICHIGAN
office building

GSA Environmental Demonstration Project
Federal Building
Saginaw, Michigan

DESCRIPTION: One-story, split level structure using energy-conserving design and planning with the environment. Approximate gross area 59,500 ft^2. Estimated construction cost $7,256,000. Included features: park-like roof section for public use; use of recycled construction materials; toilet waste system using recycled mineral oil; greatly reduced water consumption, including collection and use of rain water for lawn irrigation; parking and postal trucking operations screened from view.

Solar system: 8000 ft^2 flat plate solar collector. Will provide all the domestic hot water and approximately 60% of space heating requirements.

Director: Ray Whitley
OCM Project Coordinator
Government Services Administration/Public Building Service
18th and F Streets
Washington, D.C. 20405
(202) 343-5256

Owner: General Services Administration
18th and F Streets
Washington, D.C. 20405

Architect: Smith, Hinchman & Grylls Associates, Inc.
455 W. Fort Street
Detroit, MI 48226
(313) 964-3000

Engineer: Smith, Hinchman & Grylls Associates, Inc.
address as above

Solar System Engnr. (Consultant): Dr. Erich Farber
Solar Energy Laboratory
238 Mechanical Engineering Building
University of Florida
Gainesville, FL 32611

Contractor: Not selected yet

Completion date: October, 1976

HIGH BIDS FORCE GSA TO REVIEW PLANS FOR SOLAR-HEATED BUILDING

GSA's Public Buildings Service is reviewing plans for its Saginaw, Mich., solar office building after high contract bids halted the contracting process.

The proposed Federal building utilizes a 40 ft. X 200 ft. collector thrust upward from the roof of the 51,600 sq. ft. structure. The solar system is expected to provide 70% of building heating and all domestic hot water. (Solar Energy Washington Letter, 3/4/74.)

Charles Law, director of GSA's Professional Services Division, says every effort will be made to retain the environmental concepts of the building, including the solar system. Changes in building design may be necessary to lower the bids, which were 20% higher than authorized.

According to Law, some decisions on the project are expected next month. GSA has submitted an NSF proposal for the solar systems, and failing that and other possibilities, may have to return to Congress for more money.

Bids for the solar system for the Manchester, N. H., office building will be solicited next spring.

SOLAR ENERGY INDUSTRY REPORT 12-9-74 Vol.1 No.22

MICHIGAN
office building

Ann Arbor, Michigan

DESCRIPTION: Consulting and design work on a solar heating system for a renovated office building in Ann Arbor, MI.

Designers: Sun Structures
Integrated Environmental Design
424 Benjamin Street
Ann Arbor, MI 48104

FEDERAL OFFICE BUILDING
ENVIRONMENTAL DEMONSTRATION PROJECT
SAGINAW, MICHIGAN
SMITH, HINCHMAN & GRYLLS ASSOCIATES
ARCHITECTS ENGINEERS PLANNERS

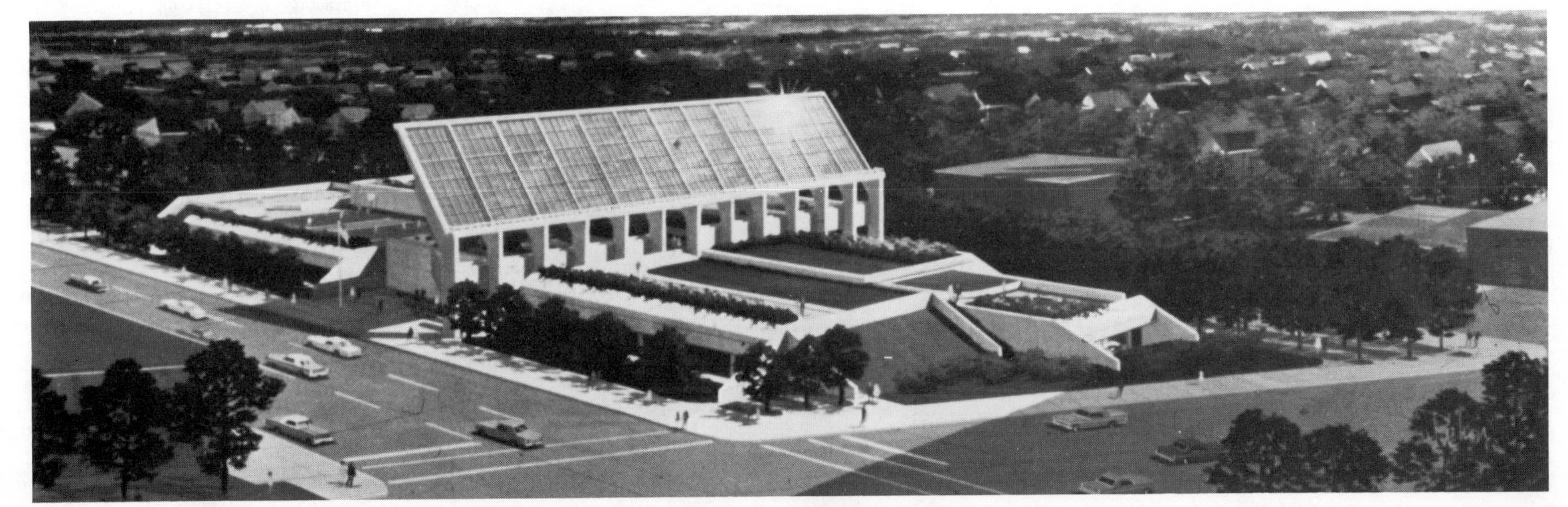

MINNESOTA
house

Rosemount, Minnesota

DESCRIPTION: Two-story, 2000 ft^2 house with sod roof and many energy saving features. Floor plan is a trapezoid with sides 48 ft (S), 16 ft (N), and 32 ft (E & W). Earth berms cover most of north wall and half of east and west walls. Small greenhouse at southeast corner. Roof includes 9 1/2 inches of fiberglass and 3 inches of sod.

Solar system: Collector: Thomason trickling-water type, on second story. It employs 728 ft^2 of corrugated galvanized steel behind two sheets of glass. Slope 60° from horizontal. Water runs down corrugations, becomes 20 to 50°F hotter (on sunny day), enters gutter, flows (via domestic hot water pre-heater) to 2000 gallon steel tank in basement. Tank is surrounded by 50 tons of fist-size stones. When rooms need heat, blower drives air through array of stones and to rooms. The pump that sends water to top-of-collector distribution pipe is powered by tower-mounted, 2-blade, 15-ft-diameter aerogenerator (or if no wind, by batteries).

Director: Dr. Dennis R. Holloway
School of Architecture
and Landscape Architecture
University of Minnesota
Minneapolis, MN
(612) 373-2198

Owner: University of Minnesota
Minneapolis, MN

Architect: John Ilse
Center for the Studies
of the Physical Environment
Institute of Technology
124 Space Science Center
Minneapolis, MN 55455
(612) 376-4916

Engineer: John Ilse
address as above

Contractor: John Ilse
address as above

Completion date: December, 1974

UNIVERSITY OF MINNESOTA ROSEMOUNT HOUSE

MINNESOTA
house

St. Paul, Minnesota

DESCRIPTION: Two-story, 2000 ft^2 house (with basement and attic) built in 1910 and now made available by HUD. Retrofitted in 1974-1975 with high-quality insulation and a special water-type solar heating system to provide 80% of winter's heat need.

Solar system: Collector is in two parts: 600 ft^2, 45° sloping S roof, and 500 ft^2 on vertical S wall. Total: 1100 ft^2. Water is pumped upward in ¼-in. space between two sheets of galvanized steel behind double glazing. Storage system: 2000 gallon steel water tank in basement. Rooms are heated by hot air, via heat exchanger and blower.

Director: Dr. Dennis R. Holloway
110 Architecture Building
School of Architecture and
Lanscape Architecture
University of Minnesota
Minneapolis, MN 55455
(612) 373-2851

Owner: University of Minnesota

Architect: Patrick Starr
School of Architecture
University of Minnesota
Minneapolis, MN 55455

Engineer: Patrick Starr
address as above

Contractor: Patrick Starr
address as above

Completion date: May, 1975

MINNESOTA
academic building

North View Junior High School
5869 69th Avenue, North
Brooklyn, Minnesota 55429

DESCRIPTION: Retrofit of a junior high school with 166,000 ft^2 of floor area by Honeywell. National Science Foundation granted $358,000.

Solar system: In March 1974 Honeywell, Inc., installed water-type collector: 5000 $ft.^2$ (246 panels, 3 ft x 8 ft x ½ ft; sloping 55° from horizontal); double-glazed with 3/16-in. tempered glass and 0.004-in. mylar; water & glycol. Collector is free standing in open area beside school. Storage, 3000 gallon water. Estimate system will supply 6% of winter heat load.

Director: Rodger Schmidt
Honeywell
2600 Ridgeway Parkway
Minneapolis, MN 55413
(612) 331-4141 x 4078

Owner: After 2 yr. NSF study is complete the system will be owned by:
North View Junior High School
5869 69th Ave. North
Brooklyn, MN 55429

Engineer: Honeywell
2600 Ridgeway Parkway
Minneapolis, MN 55413
(612) 331-4141

Contractor: Honeywell
address as above

Completion date: March, 1974

NORTH VIEW JUNIOR HIGH SCHOOL

MINNESOTA
laboratory

Honeywell Mobile Solar Lab
Minneapolis, Minnesota

DESCRIPTION: Mobile solar laboratory developed by Honeywell, Inc., and funded by Honeywell and National Science Foundation.

Solar system: Solar heated and cooled lab consists of an 8 ft x 9 ft x 45 ft trailer to which two large areas of collector are attached. Each of the two areas includes 32 panels, dimensions of which are 3 ft x 4 ft. The flat-plate water-type panels are of steel plates welded together to form channels. Selective black coating is used. Outer layer of double-glazed window is 3/16-inch iron-free tempered glass; inner layer is of 0.004-inch Tedlar. Total area of collector is 570 ft^2. The two large areas can be given any desired slope and are folded close to trailer when trailer is to travel along highway. Storage: two 450-gallon tanks. Total amount of fluid used almost 1000 gallons. Extensive instrumentation is provided, and more than ten modes of operation are available. A second trailer (office-type trailer with 12 ft x 50 ft floor) serves as thermal load. Summer cooling by Arkla LiBr absorption air-conditioner. Equipment calibrated by National Bureau of Standards in spring, 1974.

Director: Rodger N. Schmidt
Honeywell, Inc.
2600 Ridgeway Parkway
Minneapolis, MN 55413
(612) 331-4141, Ext. 4078

Owner: National Science Foundation/Honeywell, Inc.

Architect: Honeywell, Inc.
2600 Ridgeway Parkway
Minneapolis, MN 55413
(612) 331-4141

Engineer: Honeywell, Inc.
address as above

Contractor: Honeywell, Inc.
address as above

Completion date: May, 1974.

MINNESOTA
laboratory and office facilities

Minneapolis, Minnesota

DESCRIPTION: 570 acres, 10 million ft^2 of research and office facilities. All energy conservative measures for the site and buildings are being considered in integrated systems with solar energy and total energy systems for energy and power supply, along with conventional sources.

Owner: Minnesota Mining & Mfg. Co.
3M Center, St. Paul, MN 55101

Engineers: Dubin-Mindell-Bloome Associates
Consulting Engineers and Planners
42 West 39th Street
New York, NY 10018
(203) 236-1654

MISSOURI
schools

Three schools for the mentally retarded
Missouri

DESCRIPTION: In the design stage.

Solar system: The prototype model employs water-type, roof-mounted collectors. Water will be distributed to a heat coil for space heating or to storage. An absorption cooling unit which operates at over 200° F will also be employed; boiler will be used to raise water to this temperature if necessary. Electric heating as a back-up.

Architect: Hoffman/Saur & Associates
Bonhomme Avenue
St. Louis, MO
(314) 862-6363

Engineer: PGA Engineers
The Equitable Building
10 Broadway
St. Louis, MO 63102
(314) 231-7318

David Lord
Lindell & Skinker Boulevard
Washington University
St. Louis, MO 63110
(314) 863-0100

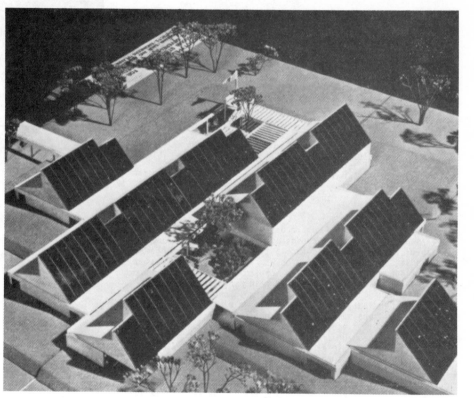

MISSOURI SCHOOLS DESIGNED BY HOFFMAN/SAUR & ASSOCIATES

NEVADA
houses

U.S. Naval Ammunitions Depot
Hawthorne, Nevada 89415

DESCRIPTION:	Retrofit of two one-story, two-bedroom, one-bath family residences. Standard insulation (3½ in/wall-6 in./ ceiling).
Solar system:	Ground-mounted flat plate air-cooled collectors, 675 ft^2 each, about 100 ft from houses. Back side insulated with 2 inches plastic foam. Double glazing 5 mil mylar. All interior surfaces painted dull black, additional absorber layers inside made of wire mesh. 2100 gallon water storage.
Director:	Robert Dempsey Head of Station Resources and Planning U.S. Naval Ammunition Depot Hawthorne, NV 89415 (702) 945-2451, Ext. 772 Capt. Ernest James Kirshke U.S. Naval Ammunition Depot Hawthorne, NV 89415 (702) 945-2451, Ext. 590
Owner:	U.S. Navy
Solar Systems Engineer:	Dr. Jerry Plunkett Materials Consultants, Inc. Denver, CO 80205 (no public inquiry please)
Completion date:	January 29, 1974

U.S. NAVY'S SOLAR COLLECTOR AT HAWTHORN, NEVADA

NEVADA
laboratory building

Boulder City, Nevada

DESCRIPTION:	Reinforced concrete and concrete block structure 6374 ft^2. Energy conservative design utilizing styrofoam insulation outside of concrete block walls which face the building interior. Budgeted cost $648,000.
Solar system:	6000 ft^2 of water/glycol cooled flat plate solar collectors roof mounted at 45° south. Thermal storage provided by 2 10,000 gallon storage tanks with electrical immersion heater for back up. Storage sufficient for 3 days. Cooling will be provided by solar power absorption chiller.
Director:	Dr. B. Sniley Desert Research Institute Reno, NV 89507 (702) 784-6131
Owner:	University of Nevada 405 Marsh Avenue Reno, NV 89502
Architect:	Robert Fieldon Jack Miller & Associates 522 E. 20th Street Las Vegas, NV 89101 (702) 735-5222
Engineers:	John Bartly and Ralph Joeckel Johnson, Joeckel & Bartly 953 E. Sahara Avenue Las Vegas, NV 89105 (702) 734-1312
Solar system Engineer:	Arthur D. Little, Inc. Alan Balfour 25 Acorn Park Cambridge, MA 02140 (617) 864-5770
Contractor:	Not yet decided
Completion date:	First need to raise funds.

NEW ENGLAND
house

Transportable, built for NE climate

DESCRIPTION: Prototype residence presently in design stage, to be built somewhere in NE. Geometric, 2 story, 3 post foundation, portable, inexpensive, self-contained, and self sustaining structure. Private funding. Each unit is designed for comfortable living, in any climate, for two occupants.

Solar system: Solar heating, hot water, air conditioning. Waste reprocessing. Rainwater to be collected and stored. Electricity unplanned.

Director: Bret N. Brierley
Artist, Designer
P.O. Box 92
Deep River, CT 06417
(203) 488-4638

Owner: Bret N. Brierley
address as above

Architect: Mr. David C. Grunwald
P.O. Box 1215
Annex Station
Providence, RI
(401) 861-3188

Completion date: Spring, 1975

NEW HAMPSHIRE
house

Nelson, New Hampshire

DESCRIPTION: 970 ft^2, 1-1/2-story, wood frame home. Will minimize energy usage. Salvaged materials will be used including flooring, plumbing fixtures, and doors. House is set into the south side of a hill to utilize relatively constant ground temperature; helps keep house warm in winter and cool in summer.

Solar system: System is passive in nature. Large expanses of south-facing glass are shaded by deciduous trees during summer but are exposed to winter sun. Storage is the thermal mass of walls and floor of structure. Foundation of dry laid concrete block plastered on both sides with a fiberglass reinforced cement binder and insulated from the earth with sheets of rigid foam, which are in turn protected by cement asbestos panels. This increases the thermal inertia of the house.

Director: Bruce Anderson
Total Environmental Action
Box 47
Harrisville, NH 03450
(603) 827-3087

Owner: Elizabeth Cohn
Box 99
Harrisville, NH 03450
(603) 876-4084

Architect: Total Environmental Action
address as above

Engineer: Total Environmental Action
address as above

Contractor: Total Environmental Action
address as above

Completion date: October 1, 1974

Cohn House
Nelson, N.H.

NEW HAMPSHIRE
house

Marlow Hill Road
Marlow, New Hampshire 03456

DESCRIPTION: 30-ft diameter dome with approximately 1200 ft^2 of floor area. Two story. Concrete block wall is dry, laid, and plastered with a fiberglass reinforced cement binder. Most openings face south to emit direct sunlight. Insulated glass for all glazing with insulating shutters for nighttime use. Dome insulated with 3 inches of urethane foam on outside, covered with a cement-like material which providés waterproof skin.

Solar system: 400 ft^2 air-cooled solar collectors placed adjacent to dome. Storage in tall, insulated, cylindrical container of rocks in center of dome. Some of heat normally lost from fireplace will help heat storage container. Solar is primary source of heat with fireplaces and wood stoves and very small electrical resistance heater auxiliary.

Director: Doug Coonley
Total Environmental Action
Box 47
Harrisville, NH 03450
(603) 827-3087

Owner: Barry Grossberg
422 W. 22nd Street
New York, NH 10011
(212) 989-2426

Architect: Total Environmental Action
address as above

Engineer: Total Environmental Action
address as above

Contractor: Total Environmental Action
address as above

Completion date: November 30, 1974

NEW HAMPSHIRE
house

Bedford, New Hampshire

DESCRIPTION: 1920 ft^2, single family residence. Earth cover on east, north and west walls. Roof is wood frame with 8 inches of fiberglass insulation. Poured concrete walls with two inches of rigid urethane insulation between walls and roof.

Solar system: Combination of 256 ft^2 of vertical water-cooled solar collectors and 200 ft^2 of double glazed (Kalwall panel) windows in front of a 12 inch concrete wall painted black. At night the space between the glazing is filled with tiny polystyrene pellets (Dave Harrison's "beadwall"). Percent solar heated: 70-90 (predicted). Auxiliary heat is two wood stoves, but the house is wired for electric heating if required.

Owner: Ralph Tyrrell and Holly Anderson & Family
Bedford, NH

Architect: Bruce Anderson
Total Environmental Action
Box 47
Harrisville, NH 03450
(603) 827-3087

Engineer: Total Environmental Action
address as above

Contractor: Armand & Ronald Ouellette
40 Moore Street
Manchester, NH
(603) 623-1672

Completion date: February 1975

TYRRELL HOUSE

NEW HAMPSHIRE
house

Hanover, New Hampshire

DESCRIPTION: Single family residence. Private funding. Utilizes many energy conservation measures.

Solar system: Air-cooled, vertical south facing flat-plate collectors. Storage fist sized stones beneath house.

Owner: Stuart White
Hanover, NH

Consultants: Total Environmental Action
Box 47
Harrisville, NH 03450
(603) 827-3087

NEW HAMPSHIRE
mercantile building

White River Junction, New Hampshire

DESCRIPTION: 200 ft^2 Custom Leather Boutique building at the Howard Johnson complex in White River Junction. Privately funded.

Solar system: 1000 ft^2 flat plate collectors on south roof. Data on system's performance will be gathered by Dartmouth College through an NSF grant being administered by the MITRE Corporation and the Vermont Public Service Board. System predicted to save 2/3 to 3/4 of conventional electric heating and cooling costs.

Director: Sol-R-Tech
(division of Quechee Construction Co.)
Hartford, VT 05047

Designer: Sol-R-Tech
address as above

Engineer: Sol-R-Tech
address as above

Completion date: Late 1974, early 1975

Note: Sol-R-Tech has a second solar heated building under construction in Hanover, NH and several more are on the drawing boards.

NEW HAMPSHIRE
office building

Energy Conservation Demonstration Project
Federal Office Building
Manchester, New Hampshire

DESCRIPTION: Seven-story GSA office building with two-level basement parking garage. 175,000 ft^2 gross area. Approximate cost $9,682,000. Purpose of project to dramatize commitment of PBS/GSA to conserve energy in design, construction, and operation of federal buildings and to inspire others to do likewise. Different heating, lighting, and air-conditioning systems will be utilized on different floors to determine relative efficiency. Some energy conserving features include windowless north wall; minimal window area on other walls using double glazing fins and overhangs; exterior shell designed to reduce heat transfer. Extensive instrumentation.

Solar system: 15,000 ft^2 water-cooled flat plate collector to be erected on roof. Will utilize same hot water tanks, etc., already included for waste heat. Weather station for simultaneous collection of local weather information.

Director: Ray Whitley
OCM Project Coordinator
Government Services Administration/Public Building Service
18th and F Streets
Washington, D.C. 20405
(202) 343-5256

Owner: Government Services Administration
18th and F Streets
Washington, D.C. 20405

Architect: Isaak-Isaak
616 Beach Street
Manchester, NH 03104

Engineers: Richard D. Kimball Company
201 Monsignor O'Brian Highway
Cambridge, MA 02141
(617) 492-7550

Dubin-Mindell-Bloom Assoc.
Energy Conservn Consultants
42 W. 39th Street
New York, NY

Contractor: Not yet selected

Completion date: July, 1976.

Illustration see page 2.050

FEDERAL OFFICE BUILDING
ENERGY CONSERVATION DEMONSTRATION PROJECT
MANCHESTER, N.H.
NICHOLAS ISAAK & ANDREW ISAAK
ARCHITECTS

NEW JERSEY
academic buildings

Blairstown Camp Project
Blairstown, New Jersey 07825

DESCRIPTION:	Education Center, total of 29,000 ft^2 including laboratories, classrooms, maintenance and recreational facilities, kitchen, and dining hall.
Solar system:	About 8,600 ft^2 of solar collectors with 30,000 gallons of hot water storage. Domestic hot water also provided by 1,600 ft^2 of solar collectors with 1,600 gallons of storage.
Director:	Harris Fraker 245 Nassau Street Princeton, NJ 08540 (609) 924-9238
Owner:	Princeton University Princeton, NJ 08540
Architect:	Harris Fraker address as above
Engineers:	ECS Associates Division of Flack and Kurtz 29 W. 38th Street New York, NY 10018 (212) 736-3733
Completion date:	Spring, 1975

NEW MEXICO
house

"Sun King" Solar Heated Home
Albuquerque, New Mexico

DESCRIPTION: Privately funded, 1200 ft^2 3-bedroom test residence. Developing for southern part of state.

Solar system: 96 ft^2 water-type collector (6 modules). Water is pumped at 5 gallon/minute through collectors and is returned to 2000 gallon storage tank. Thermostat activated blower forces air through heat extractor and heats house. Plan to install water/air heat pump for auxiliary heating and summer cooling.

Director: Mr. T. L. White
Albuquerque Western Industries, Inc.
612 Comanche, N.E.
Albuquerque, NM 87107
(505) 344-7224

Owner: Albuquerque Western Industries, Inc.
address as above

Architect: Albuquerque Western Industries, Inc.
address as above

Engineer: Dikewood Corporation
1009 Bradbury Drive, S.E.
Albuquerque, NM 87106
(505) 243-9781

Contractor: Albuquerque Western Industries, Inc.
address as above

Completion date: February, 1974

NEW MEXICO
house

1330 Cerro Gordo
Santa Fe, NM 87501

DESCRIPTION: Solar heated log cabin (all recycled materials), 1200 ft^2, 1 bedroom, 1 bath, brick floor, pane windows. (Original log cabin was constructed in late 1800's.)

Solar system: 800 ft^2 air collector with fist-size rock storage under floor. Also has solar H_2O heater for domestic hot water and a wind driven generator (115V).

Director: David Wright
Sun Mountain Design
960 Camino Santander
Santa Fe, NM 87501
(505) 893-1680

Owner: Carolyn Allers
1330 Cerro Gordo
Santa Fe, NM 87501

Architect: David Wright
address as above

Engineer: Herman Barkman
107 Cienica
Santa Fe, NM 87501
(505) 982-8907

Solar Systems Engineers: Sun Mountain Design
address as above

Zomeworks Corporation
P.O. Box 712
Albuquerque, NM 87103
(505) 242-5354

Contractor: Ben Zeller
Santa Fe, NM 87501
(505) 982-5791

Completion date: July, 1974

NEW MEXICO
house

Zomeworks
P. O. Box 712
Albuquerque, New Mexico

DESCRIPTION: One-story, rambling-style house, 1925 ft^2. A curved chain of about 10 single rooms (many of which are not rectangular parallelepipeds but have 10 or more faces and are called zomes) all connected by doorways so air can circulate from one to another (except when doorway-curtains are closed). Many of the walls are adobe. Slab floor of 5-in. thick concrete. No basement or attic.

Solar system: Collector: The four south walls of the four south-most rooms are vertical, 10ft x 10ft area, and are single glazed with single strength common window glass. Behind each of the four windows there is a stack of 56-gal. steel, 95%-water filled barrels, each of which is oriented with axis horizontal (N-S). Barrels are stacked 4 or 5 high, in metal support frame. South end of barrel (2 ft^2 area) is black (common, non-selective black paint). 90 barrels in all. Total amount of water: 20 ton. At night, window exterior is covered by large, flat, aluminum-faced, insulating cover operated by handcrank and 3/8-in. nylon rope. During day, cover lies close to ground just outside of window and acts as a crude mirror to direct additional radiation to window. Inside house, curtains partially control flow of radiation and warm air from barrels to rooms Cooling:Cool night air flows past barrels, cooling them. During day the barrels cool the rooms. Rooms can be kept at 80°F even when outdoor temp reaches 100°F. Covers stay closed on hot days. 90% solar heated (for the 4 solar-heated rooms) if a large temp range 55 to 80°F is accepted Durability of barrels: Two zinc-bunged barrels leaked (at bung). No steel-bunged barrels have leaked. Drums sometimes "bong" when changing temp. Cost of drums: $9 if new, $2 to $4 otherwise.

Director:	Steve Baer Zomeworks P.O. Box 712 Albuquerque, NM (505) 242-5354	Architect:	Steve Baer Zomeworks P.O. Box 712 Albuquerque, NM (505) 242-5354
Owner:	Steve Baer address as above	Engineer:	Steve Baer address as above
		Contractor:	Steve Baer address as above
		Completion date:	Fall 1972

Note: plans are available from Steve. Write for further infor mation .

STEVE BAER'S DRUM WALL

SKYLID IS USED WITH SKYLIGHTS OR LARGE WINDOWS. BALANCED, FOAM CORE, ALUMINUM SKINNED, INSULATING LOUVERS ROTATE TO OPEN IN SUN SHINE & CLOSE AT NIGHT OR WHEN CLOUDY. WEIGHT OF FREON, WHICH FLOWS FROM WARM TO COOL CANNISTER, TILTS OPEN OR CLOSED. WORKS WITH 1°F DIFFERENCE. MANUAL OVERRIDE ALLOWS SKYLID TO BE KEPT CLOSED DURING SUMMER.

ZOMEWORKS' SKYLID SIDE SECTION

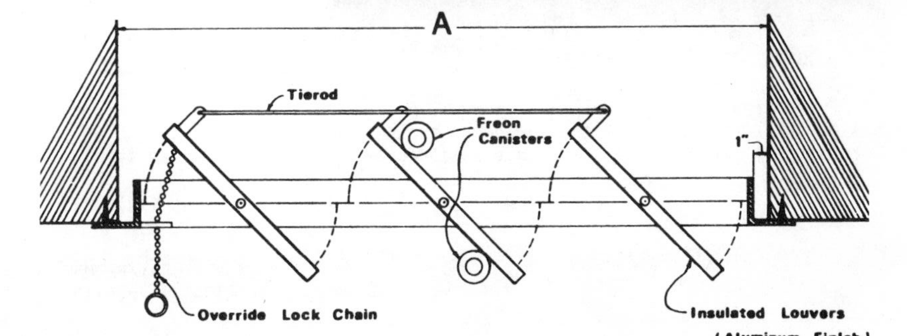

NEW MEXICO
house

Nom Be, New Mexico

DESCRIPTION: 350 ft^2 adobe residence/office addition. Walls are 16 inches thick (U-factor = 25). Embedded coil in mud floor for radiant heat.

Solar system: 100 ft^2 Roll Bond double-glazed (double-strength) collector with 120 gallons storage.

Director: Herman Barkman
107 Cienega
Santa Fe, NM 87501
(505) 982-8907

Owner: Herman Barkman
address as above

Completion date: December 15, 1974

NEW MEXICO
house

Santa Fe, New Mexico

DESCRIPTION: 1400 ft^2, one-bedroom adobe residence.

Solar system: Direct gain solar system. Manual styrofoam shutters inside.

Director: Michael Watson
Bill Lumpkins & Associates
535 Camino Del Monte Sol
Santa Fe, NM 87501
(505) 983-2618

Owner: Mr. and Mrs. Al Dasbury
825 El Caminito
Santa Fe, NM 87501
(505) 982-0479

Architect: Bill Lumpkins
Bill Lumpkins & Associates
address as above

Contractor: Greg McFarland
Aztec Construction
Santa Fe, NM 87501
(505) 982-3088

Completion date: Late 1975

Note: Bill Lumpkins & Associates are also conducting a number of other solar projects.

NEW MEXICO
house

Corrales (suburb of Albuquerque)
New Mexico

DESCRIPTION: 1000 ft^2, one-story house.

Solar system: Gravity-convection air-type collector, 430 ft^2, on sloping ground. Storage approximately 45 tons of fist-size stone in insulated bin under floor. 75% solar heated.

Director: P. Davis
Corrales, NM

Owner: P. Davis
Corrales, NM

Designer: P. Davis
Corrales, NM

Solar Engineer: Steve Baer
Zomeworks
P.O. Box 712
Albuquerque, NM 87103

Contractor: P. Davis
Corrales, NM

Completion date: 1972

P. DAVIS SOLAR HOUSE

NEW MEXICO
house

DESCRIPTION: Hot air thermo siphon collector system with rock storage.

Designer and Builder: Fred Hapman
P.O. Box 171
Arroyoseco, NM 87514

Completion date: 1969

NEW MEXICO
house

120 Edith St., S.E.
Albuquerque, New Mexico 87103

DESCRIPTION: House utilizing a modified Thomason/thermosiphon water system.

Director: Jay Davis
Sunset Ranch
Rural Route
Vallecitos, NM 87581

Owner: Ouida Ma
Graduate Student, Architecture
University of New Mexico

Designer: Jay Davis
address as above

Engineer: Jay Davis
address as above

NEW MEXICO
house

Albuquerque, New Mexico

DESCRIPTION: Adobe residence which is solar heated with air/rock system, heat pump assisted.

Owner: Danny Martinez
Albuquerque, NM

Note: There are several houses in New Mexico that are partially solar heated that are not included here due to lack of adequate information and space.

NEW MEXICO
house

15 miles SW of Zuni National Monument
New Mexico

DESCRIPTION: 552 ft^2 mountain cabin, hexagon base with dome cover.

Solar system: 32 inches x 46 inches, 6 inches deep collector using copper tube coils. Domestic hot water only. Plans to eventually have 60% heating supplied by solar.

Director: Charles Mattox
820 Hermosa Avenue, N.E.
Albuquerque, NM
(505) 268-9751

Owner: Charles Mattox
address as above

Designer: Charles Mattox
address as above

Engineer: Charles Mattox
address as above

Contractor: Charles Mattox
address as above

Completion date: Spring, 1975

CHARLES MATTOX-HIS MOUNTAIN CABIN & COLLECTOR

NEW MEXICO
house

Tesuque, New Mexico

DESCRIPTION: $2300 ft^2$, 3-bedroom, 1-bath, earth cast adobe (1/3 pummace) construction coated with emulsified asphalt (water resistant), 2-inch thick slab on 3 1/2 ft-deep bed of rock (100 yds^3). Walls are 18 inches thick.

Solar system: $560 ft^2$ greenhouse on south wall acts as collector with direct heat gain to adobe wall and rock storage under floor slab. Monsanto "602 Poly" membrane is used as glazing in winter only. Instrumented data aquisition is being funded by National Science Foundation.

Director: Hamilton Migel
Box 392
Tesuque, NM 87574
(505) 982-8434

Owner: Hamilton Migel
address as above

Designer: Hamilton Migel
address as above

Solar System Engineer: Herman Barkman
107 Cienega
Santa Fe, NM 87501
(505) 982-8907

Contractor: Bingham/Migel, Inc.
Box 392
Tesuque, NM 87574
(505) 982-8435

Completion date: January 1, 1975

NEW MEXICO
house

New Mexico State University
Interstate 25
Las Cruces, New Mexico

DESCRIPTION: State funded, $1900 ft^2$, one-story, three-bedroom, 1½ bath residence.

Solar system: $700 ft^2$ water cooled collector system with 2500 gal. of underground water thermal storage.

Director: Dr. R. L. San Martin
New Mexico State University
Box 3450
Las Cruces, NM 88003
(505) 646-3424

Owner: New Mexico State University
address as above

Architect: Dean & Hunt Associates
201 Truman Street, NE
Albuqqerque, NM 87108
(505) 255-4864

Engineers: In house (NMSU)

Contractor: Cona Ana Vocation Center
New Mexico State University
(505) 646-3211

Completion date: April 1975

NEW MEXICO STATE UNIVERSITY'S SOLAR HOUSE

NEW MEXICO
house

Seton Village
Santa Fe, New Mexico 87501

DESCRIPTION: New home, approximately 1500 ft^2.

Solar system: Approximately 550 ft^2 air collector. Storage: 20 tons 3-inch diameter rocks. Electric auxiliary heating.

Director: Wayne Nichols
Seton Village
Santa Fe, NM 87501
(505) 893-1600

Owner: Wayne Nichols
address as above

Architect: D. Wright, T. Price
Sun Mountain Design
960 Camino Santander
Santa Fe, NM 87501
(505) 983-1680

Engineer: Herman Barkman
107 Cieniga
Santa Fe, NM 87501
(505) 982-8907

Solar System Engineer: Wayne Nichols
address as above

Contractor: Wayne Nichols
address as above

Completion date: November, 1974

SANTA FE SOLAR HOUSE DESIGNED BY DAVID WRIGHT

NEW MEXICO
house addition

631½ Old Pecos Trail
Santa Fe, New Mexico 87501

DESCRIPTION: Solar heated 400 ft^2 addition built on to an existing structure.

Solar system: Air collector. Radiant heat from storage beneath floor.

Director: Wayne Nichols
631 1/2 Old Santa Fe Trail
Santa Fe, NM 87501
(505) 983-3500

Owner: Wayne Nichols
address as above

Architect: D. Wright and T. Price
Sun Mountain Design
960 Camino Santander
Santa Fe, NM 87501
(505) 983-1680

Engineer: Herman Barkman
107 Cienega
Santa Fe, NM 87501
(505) 982-8907

Solar Systems Engineer: Sun Mountain Design
960 Camino Santander
Santa Fe, NM 87501
(505) 983-1680

Contractor: Wayne Nichols
address as above.

Completion date: March, 1974.

NEW MEXICO
house/laboratory

Tijeras, New Mexico

DESCRIPTION: 31-inch diameter igloo-type structure, steel construction with 3 inches of insulation. Energy conservation design consumes only 10% of thermal energy required by the average house of equivalent size. A demonstration home, $12,000. Funded by the John Muir Foundation.

Solar system: Ground mounted water/glycol cooled flat plate solar arrays and water storage tank for thermal storage. Collectors supply 100% of heating requirements. Electricity is supplied by a windmill.

Director: Robert Reines
Integrated Life Support (ILS) Labs
Star Route 103
Tijeras, NM 87059

Owner: Robert Reines
address as above

Architect: Robert Reines
address as above

Engineer: Robert Reines
address as above

Contractor: Robert Reines
address as above

Completion date: 1973

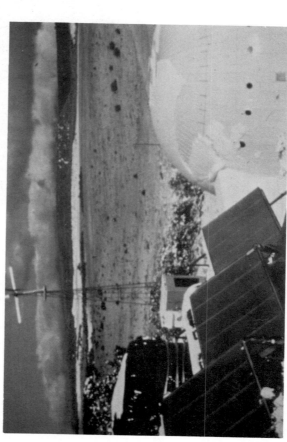

ROBERT REINES' ILS LABORATORY

NEW MEXICO
house

636 Camino Lejo
Santa Fe, New Mexico 87501

DESCRIPTION: 1000 ft^2, 1-bedroom adobe residence. Privately financed, $16,000.

Solar system: Direct-gain passive solar heating to adobe and 55 gallon water-filled drums from skylights. 416 ft^2 (32 ft x 13 ft) P.P.G. patio doors used in skylights. Manual insulating shutters made of polystyreen cover skylights at night and extended periods of cloudcover. Solar energy also provides domestic hot water via flat plate collector.

Director: David Wright
Sun Mountain Design
960 Camino Santander
Santa Fe, NM 87501
(505) 983-1680

Owner: Karen Terry
640 Camino Lejo
Santa Fe, NM 87501
(505) 982-8470

Architect: David Wright
address as above

Engineer: Sun Mountain Design
address as above

Solar Engineer: Sun Mountain Design
address as above

Contractor: Karen Terry
address as above

Completion date: August, 1975

KAREN TERRY'S HOME

NEW MEXICO
house

Canyon Road
Santa Fe, New Mexico

DESCRIPTION: Small adobe residence, rectangular with long axis running north and south, approximately 500 ft^2 floor area. Designed to use maximum indigenous low-cost materials and take advantage of architectural placticity of Pueblo adobe style.

Solar system: Air collector panels (one at front of house, one at back) utilizing ordinary galvanized sheet metal and window glass. Ducts to convey air to storage. Storage (front) fist-sized and larger rocks, storage (back) sand. Also provides domestic hot water using air-to-water heat exchanger, storing water in well insulated 30 gallon range boiler with electric immersion heater for auxiliary. Auxiliary heat for space heating from fireplace and small gas heater.

Director:	Peter van Dresser Santa Fe, NM	Engineer:	Peter van Dresser Santa Fe, NM
Owner:	Peter van Dresser Santa Fe, NM	Contractor:	Peter van Dresser Santa Fe, NM
Designer:	Peter van Dresser Santa Fe, NM	Completion date:	1958

Note: Detailed description - "A Pioneer Solar House in Santa Fe," by Peter van Dresser, proceedings of the conference at the Ghost Ranch, available from Biotechnic Press, Albuquerque, NM.

NEW MEXICO
academic building/greenhouse

Albuquerque, New Mexico

DESCRIPTION: Greenhouse.

Solar system: 30 water-filled barrels are against north side of greenhouse. South wall is a 276 ft^2 beadwall which allows sun to heat barrels during day. Beadwall can be filled with insulating styrofoam beads during overcast weather or nighttime. Vacuum cleaner motors move beads from storage tanks to beadwall windows. A thermostat inside a small collector on the west end of greenhouse switches on motors when temperature drops below 70° F. Automatic venting system regulates temperature, utilizing a pair of freon-filled and connected tanks, balanced, one inside and one outside greenhouse. Sun's heat causes freon to expand and move into cooler tank, thereby opening vent.

Director:	Steve Baer Zomeworks P.O. Box 712 Albuquerque, NM 87103	Designer:	David Harrison Zomeworks address as above
		Engineer:	David Harrison address as above
Owner:	Monte Vista Elementary School Monte Vista Avenue Albuquerque, NM 87106 (505) 255-4680	Contractor:	Zomeworks address as above

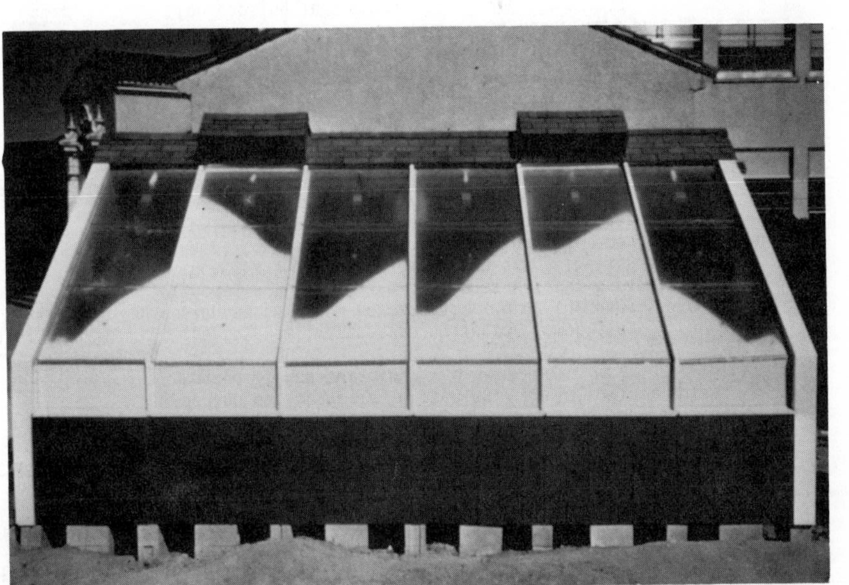

MONTE VISTA ELEMENTARY SCHOOL'S BEADWALL GREENHOUSE (BEADS ARE EMPTYING)

NEW MEXICO
mercantile building

DESCRIPTION: Demonstration project funded through grant from Four Corners Council of $20,000. 15,000-20,000 ft^2, 2-story commercial structure to be solar heated and energy conserving.

Director: Dr. Maurice Wildin
Professor of Mechanical Engineering
University of New Mexico

Architect: Robert Cohlmyer
4812 Madison Court, N.E.
Albuquerque, NM 87110
(505) 268-9212

BRIDGERS AND PAXTON'S SOLAR OFFICE BUILDING

NEW MEXICO
office building

Albuquerque, New Mexico

DESCRIPTION: One-story, well-insulated office building in two portions (N and S), with collector on south face of north portion. Useful floor area is 4300 ft^2. Drafting room ceiling behind collector is 14 ft high. World's first solar heated office building.

Solar system: 750 ft^2 net, water-type collector comprising 55 panels sloping 60° from horizontal. 26 of the panels are 22 inches x 9 ft; 26 are 22 inches x 7 ft; 3 panels are 22 inches x 5 ft. Each panel employs 3/16-inch aluminum sheet to the back of which 1/2-inch diameter copper tubes (6 inches apart on centers) have been soldered. Black coating is non-selective flat black paint. Glazing is one layer of 1/8-inch double-strength glass. There is a 4-inch-thick air space between black surface and glass. The fluid is water and ethylene glycol used in collector, but not in tank. Storage system: 6000-gallon insulated steel tank buried in ground beside building. Carrythrough: 3 days. 100% solar heated; 92% of energy from sun, 8% from energy driving water-to-water heat-pump motor. No auxiliary heat source. Water-to-water heat pump also used for cooling in summer.

Director: F. H. Bridgers (original builder)
Bridgers & Paxton
213 Truman Street, N.E.
Albuquerque, NM 87108
(505) 265-8577

Owner: Bridgers & Paxton
Consulting Engineers
address as above

Engineer: Bridgers & Paxton
address as above

Contractor: Paul Priest
8309 Dellwood Road, N.E.
Albuquerque, NM 87110
(505) 299-0612

Architect: Stanley Wright, A.I.A.
125 NW Carlito Road
Albuquerque, NM 87107
(505) 345-7652

Completion date: August, 1956; down 1962-1974; up again 1974

Note: S. F. Gillman, Pennsylvania State University, is conducting study under National Science Foundation grant and is responsible for reactivating system.

NEW MEXICO
office building

New Mexico State University
Las Cruces, New Mexico

DESCRIPTION: Combined offices and labs, New Mexico Dept. of Agriculture, New Mexico State Univ. One story, rectangular, well-insulated lab. building for study of crop diseases and pesticides. Floor area: about 25,000 ft^2. There is a large quantity (9500 cmf) of exhaust air from lab hoods; heat exchanger recovers energy from exhaust air.

Solar system: Collector: 2250 ft^2, flat-plate water-type, on horizontal roof. Collector arrays slope 30^0 from horizontal. Storage system: 35,000 gallon concrete tank, water-filled. Electric power driven chiller and heat pump used in cold weather to transfer energy from rooms to conventional cooling tower. Cooling in summer: via chiller and heat pump. 100% solar heated (predicted).

Director: Dr. R. L. San Martin
New Mexico State University
Box 3450
Las Cruces, NM 88003
(505) 646-3424

Engineer: Bridgers and Paxton
Frank Bridgers
213 Truman St, NE
Albuquerque, NM 87108
(505) 265-8577

Owner: New Mexico State University
Box 3450
Las Cruces, NM 99003
(505) 646-3424

Contractor: Wooten Construction Co.
100 South Archuleta Road
Las Cruces, NM 88001
(505) 526-5581

Architect: W. T. Harris and Associates
105 E. Lohman Ave.
Las Cruces, NM 88001
(505) 524-0912

Completion date: September, 1975

NEW MEXICO STATE UNIVERSITY'S OFFICE BUILDING

NEW MEXICO
office building

Los Alamos, New Mexico 87544

DESCRIPTION: 67,000 ft^2, 3-story National Security Research and Study Center.

Solar system: 10000 ft^2 continuous sheet collector at 35^0 from horizontal. Will be cooled by either a 200 ton absorption water chiller or Rankine Cycle vapor compression unit.

Director: J. D. Balcomb
Solar Coordinator
Los Alamos Scientific Laboratory
Los Alamos, NM 87544
(505) 667-6441

Owner: U.S. Atomic Energy Commission
Los Alamos, NM 87544

Architect: Sam Burnett
Project Manager
Charles Luckman Associates
9220 Sunset Boulevard
Los Angeles, CA 90069
(213) 274-7755

Solar Energy Consultant: Dr. Jerome Weingart
Ayers and Hayawaka
1180 S. Beverly Drive
Los Angeles, CA 90035
(213) 878-0880

Engineer: Ayers and Hayawaka
1180 S. Beverly Drive
Los Angeles, CA 90035
(213) 878-0880

Contractor: Not selected yet

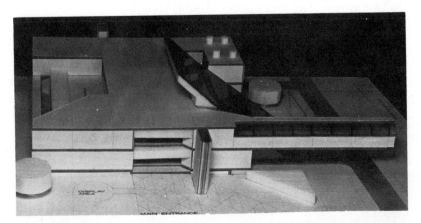

Completion date: Summer, 1976

U.S. A.E.C.-LOS ALAMOS OFFICE BUILDING

NEW MEXICO
office/laboratory building

Sandia Labs
Albuquerque, New Mexico

DESCRIPTION: AEC funded, $800,000 project. Existing office building that will have load equivalent of heat required for 12 typical houses; electric power for 13 houses. Test bed with low temperature and high temperature collectors.

Solar system: Linear parabolic focusing collector which has dual system, one for medium temperature (200°F) water heating and one for high temperature (600oF) collection using Dow Therminol as the heat transfer medium.

Director: R. P. Stromberg and Dr. R. H. Braasch
Solar Energy Systems, Division 5717
Sandia Laboratory
Albuquerque, NM 87115
(505) 264-8170, Ext. 3850

Owner: U.S. Atomic Energy Commission/Sandia Laboratory

Architect: In house (Sandia)

Engineers: In house (Sandia)

Contractor: In house (Sandia)

Completion date: Partially finished June, 1975, Totally finished June, 1976.

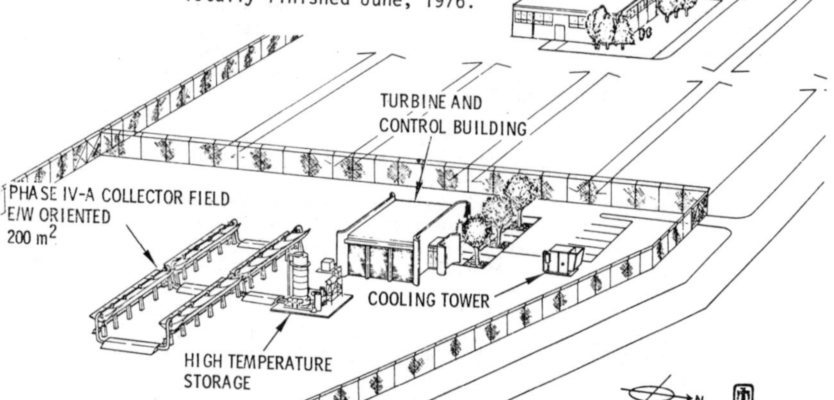

U.S. A.E.C.-SANDIA LABORATORY'S SOLAR TEST FACILITY

NEW YORK
academic building

Low Energy Utilization School
New York, New York

DESCRIPTION: This is a five-year, four-phase project to research, design, construct, and evaluate a low-energy utilization school The first phase (research) entailed preparing a report for the New York City Board of Education with the support of the National Science Foundation. Report finds that a properly designed new building could save 49.4% of energy that would normally be used; a retrofit could save 27.7%.

Solar system: Thermosiphoning collectors, glazed with fiberglass. Use of thermal-solar shutters on exterior walls which would be used to reflect light into the room while it's occupied. During summer, the shutters can be used to reflect heat off the building. System will contribute to heating requirements as well as provide domestic hot water.

Director: Richard G. Stein, FAIA
Richard G. Stein & Associates, Architects
588 Fifth Avenue
New York, NY 10036

Architect: Richard G. Stein
Carl Stein
address as above

Engineer: Peter Flack
Flack & Kurtz
29 W. 38th Street
New York, NY 10018
(212) 736-3733

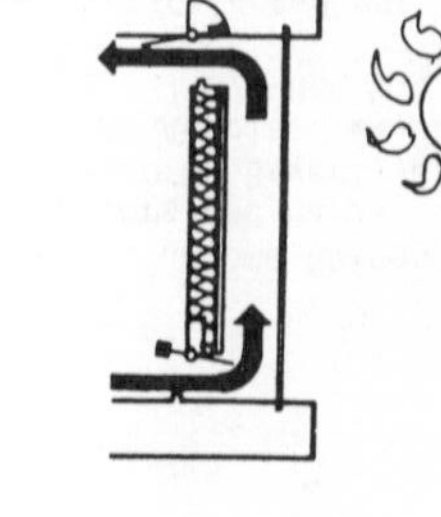

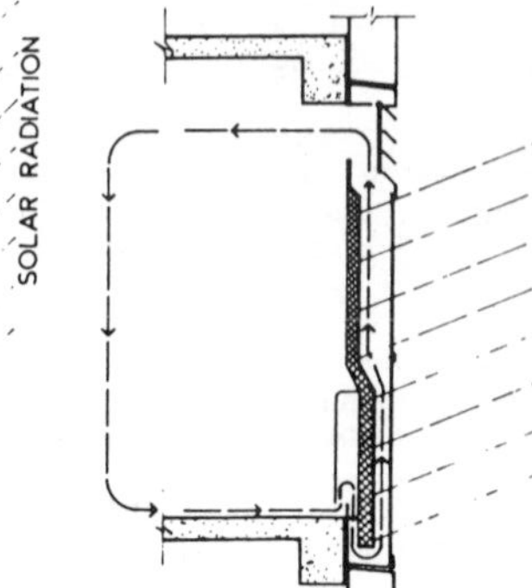

THERMO-SIPHONING PANEL VENTILATING POSITION

THERMO-SIPHONING PANEL HEATING POSITION

TO PREVENT REVERSE AIR FLOW, CHECK DAMPERS CAN BE PROVIDED WHICH PERMIT AIR CIRCULATION ONLY DURING SOLAR INPUT.

NEW YORK
arboretum

Cary Arboretum
New York Botanical Gardens, Box 609
Millbrook, New York 12545

DESCRIPTION: Environmentally compatible building consonant with basic ecological objectives of the institution. Will serve as administration, research, and education building. 25,000 ft^2, 2-story, using energy conservation features such as reflective earth banked north and west walls, masonry wall insulation, double glazed windows with shutters. Also utilizes sod roof, collection of rain water, "light tubes," low wattage lighting, and task lighting. Recycling of air, water, and wastes. Outside commercial energy requirements reduced by 75% and total energy requirements by at least 35%.

Solar system: Approximately 9000 ft^2 water/glycol type flat plate collectors, roof-mounted in 7 saw tooth arrays. Heat pumps employed. Thermal storage in compartmentalized hot water tanks in basement. Cool water from deep well will also be used for temperature control. Conventional furnace for auxiliary heat.

Director: Dr. Thomas Elias
Box AB
Millbrook, NY 12545
(914) 677-5070

Owner: New York Botanical Gardens
Dr. Howard S. Irwin, President
Bronx Park
Bronx, NY 10458
(212) 933-9400

Architect: Mr. Malcolm Wells
Box 183
Cherry Hills, NJ 08002

Engineer: Fred Dubin
Dubin-Mindell-Bloome
42 West 39th Street
New York, NY 10018
(212) 868-9700

Note: Public inquiry to:

Mr. Robert J. Goodland
The Cary Arboretum of the
New York Botanical Garden
Box AB
Millbrook, NY 12545
(914) 677-5071

Contractor: Not selected yet

Completion date: July, 1976

NEW YORK BOTANICAL GARDEN'S CARY ARBORETUM

NEW YORK
greenhouse

New Paltz, New York

DESCRIPTION: One-story greenhouse. Floor area 400 ft^2.

Solar system: Collector area 270 ft^2, of which about half is Thomason trickling-water type; the balance consists of a "greenhouse" with a massive floor. Insulating panels are employed at night. 70% solar heated (predicted).

Director: Jerome Kerner
Coordinator of Environmental Studies Program
New York State University
New Paltz, NY 12561

Owner: New York State University
New Paltz, NY 12561

Architect: Jerome Kerner
address as above

Engineer: Eugene Eccli
Design Alternatives with Nature
(DAWN Associates)
928 Second Street, S.W.
Roanoke, VA 24016
(703) 344-8412

Contractor: Eugene Eccli address as above

Completion date: March, 1974

NEW YORK STATE UNIVERSITY'S GREENHOUSE

NEW YORK
office complex center

Between Lexington and 3rd Avenue and 53rd and 54th Streets
New York, New York

DESCRIPTION: Office complex center, to be known as Citicorp Center, includes: 910 ft high, 46-story tower resting on a 112 ft high platform. Will contain over one million ft^2 of office space, a new free standing St. Peters Lutheran Church (a condominium partner in the development), a low rise office retail building, a sunken plaza at 53rd Street and Lexington, and a shopping area in the center of the block.

Solar system: Will utilize "state of the art" collectors, producing 130-150° water temperature. In summer cycle incoming air is dehumidified and pre-cooled, reverse in winter. Chemical desiccant triethelene glycol used in process. 80,000 gallon water storage already designed for heat reclamation system of complex; portion of this will be used for solar cycle. In Citicorp Center tower, an estimated 500 tons of equivalent energy could be provided by solar. Thus refrigeration tonnage reduced by 50%. Data indicates purchased energy consumed reduced by approximately 18%. Collector area to cover 1/2 acre of surface on tower's face which slopes to south. It is circulated via a stream of circulating TEG solution to 25,000 gallon storage tank. Air handling units, absorption machines on building's eighth floor, heat reclamation units, dehumidifier, regenerator, domestic hot water, and off-hour building heating can be fed on demand from this tank.

Director: Henry DeFord III
Senior Vice President, Area III, Operating Group
399 Park Avenue
New York, NY 10022
(212) 559-1000

Owner: First National City Bank
399 Park Avenue
New York, NY 10022
(212) 559-1000

Architect: Hugh Stubbins and Associates, Inc.
1033 Massachusetts Avenue
Cambridge, MA 02138
(617) 491-6450

Engineer: Joseph R. Loring & Associates
Two Pennsylvania Plaza
New York, NY 10001
(212) 563-7400

NEW YORK
office complex center
continued

Solar Systems Engineer:	Joseph R. Loring Associates address as above
	Meckler-Gerschon Associates 1150 17th Street, N.W. Washington, D.C. 20036 (202) 296-5131
Contractor:	HRH Construction, Inc. 515 Madison Avenue New York, NY 10022 (212) 751-7400
Completion date:	Fall, 1976.

Note: Dr. Robert Bell, Vice President of R&D for Consolidated Edison and President of the Council of St. Peter's Lutheran Church (a condominium partner in the development) was largely responsible for the idea to include solar collectors on the sloping face of the "crown" of the building.

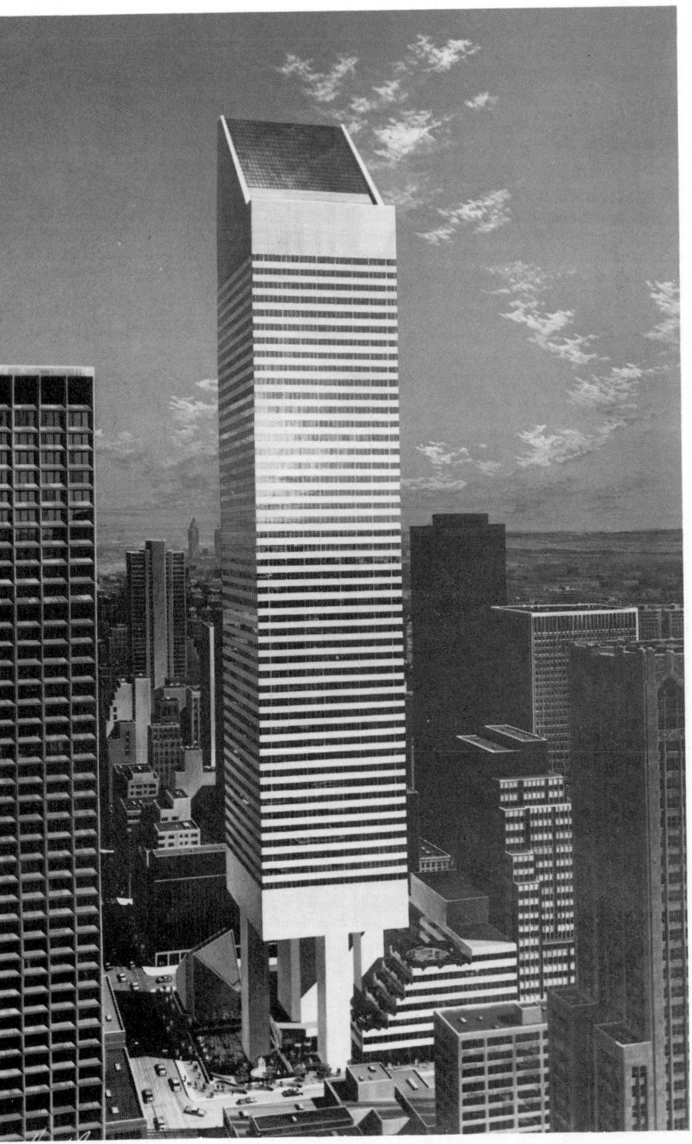

FIRST NATIONAL CITY BANK'S CITICORP CENTER

NEW YORK
office building

Management Conference Center
RCA Building
Rockefeller Center, New York City

DESCRIPTION: 1.2 million ft^2 center to be built on the 12th floor of the existing RCA building. Center will utilize solar energy and heat recovery systems. Cost is estimated to be $150,000 above conventional system cost.

Solar system: 3500 ft^2 of water cooled flat plate collectors to be mounted on the flat roof of the Center. System will have 7000 gallon thermal storage tank for the collectors and the heat recovery system. Cooling will be supplied by 15 ton absorption chiller.

Director: Howard C. Enders (RCA)
30 Rockefeller Plaza
New York, NY 10020
(212) 598-5900, Ext. 4522

Owner: RCA
address as above

Architect: Ford & Earl Design Assoc.
515 Madison Avenue
New York, NY 10022
(212) 751-4351

Engineer: Dr. Jagdish Prasad & Sital Daryanani
Syksa & Hennessy
110 West 50th Street
New York, NY 10020
(212) 489-9200

Completion date: Spring, 1975

For further information, contact James P. Maguire or Marlys Hann with Ford & Earl Associates.

RCA'S MANAGEMENT AND CONFERENCE CENTER

NEW YORK
office building

Encon Building
Third Avenue
New York, New York

DESCRIPTION: Approximately 1,100,000 ft^2, 43-story office building. Cost is calculated at \$44/ft^2, of which \$4 is the estimated cost of energy conservation features. Glazing limited to minimal amount needed to allow for natural lighting and contact with the outdoors. All glazing will be double panel. Task illumination will be used rather than uniform distribution of light fixtures throughout the ceiling area. Will also utilize individual fan rooms on each floor rather than the traditional mechanical floor. This allows shorter ducts and low velocity air distribution, which requires 35% less power.

Solar system: Large roof-mounted water-cooled collectors as well as horizontal overhangs on south and east facades and vertical fins on the west facade to alleviate cooling load. Flat plate collectors on overhangs on south and east at 45° angle. Storage in large below-grade tanks.

Director: Julien J. Studley
President
Julien J. Studley, Inc.
342 Madison Avenue
New York, NY 10017
(212) 697-7788

Owner: Julien J. Studley
address as above

Architect: Joseph H. Solomon
Emery Roth and Sons
850 Third Avenue
New York, NY 10022
(212) 753-1733

Interior Designer: Jack Freidin
Freidin, Kleiman, and Associates
342 Madison Avenue
New York, NY 10017
(212) 697-7797

Engineers: Peter Flack and Norman Kurtz
Flack and Kurtz
29 W. 38th Street
New York, NY 10018
(212) 736-3733

Stanley Goldstein
Lemessurier Associates
515 Madison Avenue
New York, NY 10022
(212) 371-8090

Contractor: Not yet selected

Completion date: 1976

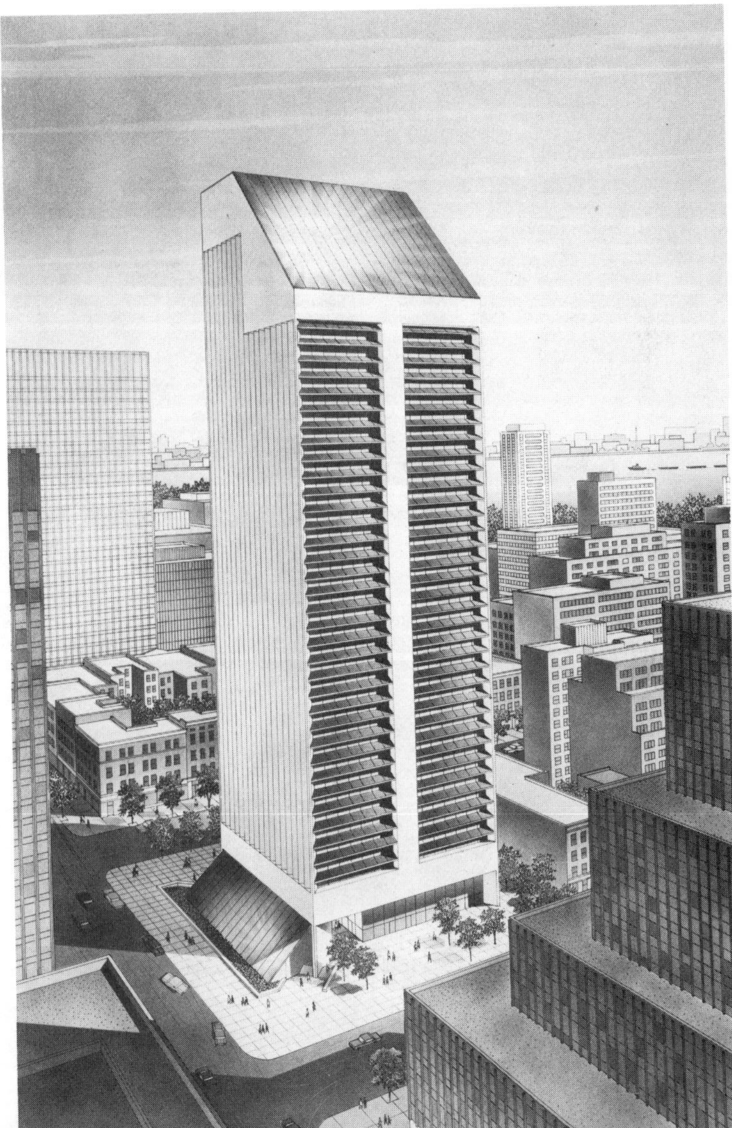

THE ENCON OFFICE BUILDING

SOUTH FACADE WITH SOLAR COLLECTORS
WEST FACADE (AT LEFT)

OHIO
house

The Sun House
Cleveland, Ohio

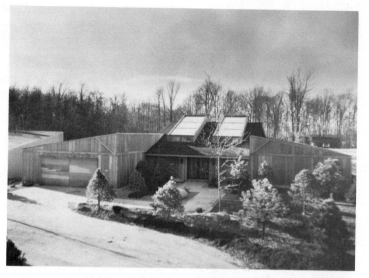

DESCRIPTION: 3600 ft^2 single story "ranch" style designed around 1100 ft^2 covered atrium. Energy conservative design with no glass area on exterior building envelope (all glass area faces atrium). Price range $80,000 to $100,000.

Solar system: Demonstration of passive use of solar energy through the use of louvered skylights in the atrium roof. Thermal storage provided by stone atrium patio. Supplementary heating and cooling will be provided by a heat pump. Solar heat supply 60 to 70%. Also a large freestanding fireplace in center of atrium.

Director: Edward A. Dirkse
Bob Schmitt Homes
13079 Falling Water
Strongville, OH 44136
(216) 238-6915

Owner: Edward A. Dirkse
address as above

Architect: Bob Schmitt Homes
13079 Falling Water
Strongville, OH 44136

Engineer: Bob Schmitt Homes
address as above

Contractor: Bob Schmitt Homes
address as above

Completion date: March, 1974. Have completed and sold four homes with more underway.

DIRKSE'S SUN HOUSE

OHIO
house

Columbus, Ohio

DESCRIPTION: One-story, 4-bedroom, 2 1/2 bath, demonstration residence with basement and adjacent 3-car garage. Heated living area 2200 ft^2. Well-insulated walls (U=0.07) and roof (U=0.05); small windows ("vision strips"); many energy saving features. Also there are three unheated interior enclosed tropical courtyards (600 ft^2 in all) with transparent roof and ventilation.

Solar system: Collector: 800 ft^2 flat-plate, water-type collector on 45^0 sloping roof areas; three such areas, employing a total of 40 panels. Each panel is 3 ft x 6 ft and employs an aluminum assembly made by arc-welding together two mating aluminum sheets in which 13 parallel tube halves have been formed by a stamping operation. Diameter of inlet and outlet tubes 3/8 inch. Black coating of Duracron enamel (non-selective). Double glazing of 1/8-inch PPG Herculite K glass. Fluid is water and ethylene glycol solution. Storage system: 4000 gallon in two 2000-gallon tanks; stratification used. Heat pump used to assist solar heating and as cooling back up. 50 to 70% solar heated (predicted) including domestic hot water. Summer cooling: LiBr and water absorption system powered by heat from tank. Back up electric heater and/or heat pump. Instrumentation: extensive; computerized. Test program to start in September, 1974.

Director: Dr. Charles F. Sepsy
Ohio State University
Columbus, OH 43210

Owner: Ohio State University
address as above

Architect: G. M. Clark
Ohio State University
address as above

Engineer: Dr. Charles F. Sepsy
address as above

Contractor: W. Goldman, Homewood Corporation
679 Northgate Road
Columbus, OH 43229
(614) 846-3400

Completion date: September, 1974

DR. CHARLES SEPSY & THE OHIO STATE UNIVERSITY'S SOLAR HOUSE

OHIO
house

"The Kansas City House at Raven Rocks"
Raven Rocks, Route 1
Beallsville, Ohio 43716

DESCRIPTION: Privately financed, two-story underground house, 6500 ft^2. House has 3 1/2 feet of earth cover. Minimal energy consumption achieved by combining buried location with complete wrap of rigid styrene insulation ranging from 2 inch thickness to 6 inches (with urethane used in a few spots for equivalent R factor with thinner appreciation), insulating panels for windows; further reduction for heating by placing insulating sheath outside the mass of concrete and stone which form structure, this mass then serving as giant heat storage "system." Other features: thermoelectric refrigeration systems which eliminate compressor and reduce energy demands, waterless Clivus toilets which will handle both toilet and garbage waste and convert them to fertilizer. Engineered and designed for long life.

Solar system: Up to 1000 ft^2 flat plate air collector, passive storage in walls. Home built savonious rotor windmill.

Director: Warren Stetzel
Raven Rocks, Route 1
Beallsville, OH 43716
(614) 926-1705

Owner: Warren Stetzel
address as above

Architect: Malcolm B. Wells
Box 183
Cherry Hills, NJ 08002
(609) 663-5423

Engineer: (Structural) I-t'an Yu
2107 Spruce
Philadelphia, PA 19103
(215) 546-5670

Solar & Wind System Engineers: (Consultants) Bruce Anderson & Douglas Coonley
Total Environmental Action
Box 47
Harrisville, NH 03450
(603) 827-3087

OHIO
house

"The Kansas City House at Raven Rocks"
continued

Contractor: Warren Stetzel, et al
Raven Rocks Builders
Raven Rocks, Route 1
Beallsville, OH 43716
(614) 926-1705

Note: Sinclair-Koppus of Monaca, PA 15061 (412-774-1000) is supplying rigid styrene board for project. Mr. Paul Nelson of that company is consulting on over-all matter of insulation, waterproofing, and vapor barrier considerations.

I-R-A Incorporated, Industrial Park, Hibbing, MN 55746 (218-263-7597) are assisting with Irathane 161 Elastromeric Polyurethane waterproofing membrane, with William Valeri and Donald Moore of that company consulting on waterproofing, vapor barrier, and insulation matters.

Kappy Wells, 200 Munn Lane, Cherry Hill, NJ 08034 (609-429-1390) is doing extensive sculpture that will be poured into the permanent structure of the house. The sculpture is a 74 ft long band depicting the evolution of life from a barren earth and sun scene to a final scene projecting a peaceable kingdom.

THE STETZEL'S KANSAS CITY HOUSE AT RAVEN ROCKS

OREGON
house

Route 3, Box 768
Coos Bay, Oregon 97420

DESCRIPTION: One-story, 3-bedroom house with heated area of 1650 ft^2. Roof slopes symmetrically 8° from horizontal. House aims 6° west of south.

Solar system: Solar collector: main portion is 5 ft high, 80 ft long, slopes 8° from vertical, runs along ridge of roof, single glazed. Radiation strikes non-selective black coating on corrugated aluminum; the corrugations run approximately east-west. Horizontal galvanized iron pipes, 1/2 inch in diameter, are tied by wire at locations 30 inches apart to the gutters of the corrugated aluminum sheet; pipes are 4 inches apart on centers. The 8° sloping south face of roof is covered with aluminum foil which reflects much radiation toward the collector. A supplementary collector, of similar construction and with 325 ft^2 area, is situated 65 ft north of house. Total collector area is 725 ft^2. Storage system: insulated, 8000 gallon steel water tank underground beneath heated portion of house. Rooms are heated by hot air flowing by gravity convection from 18-inch air space just outside tank and inside the tank insulation. Electric space heaters are sometimes used. 80% solar heated. Monitored by a group from the University of Oregon and Oregon State University headed by Professor John Reynolds. Plans are available from Henry Mathew for $10.

Director: Henry Mathew
Route 3, Box 768
Coos Bay, OR 97420
(503) 267-4762

Owner: Henry Mathew
address as above

Architect: Henry Mathew
address as above

Engineer: Henry Mathew
address as above

Contractor: Henry Mathew
address as above

Completion date: June, 1967

HENRY MATHEW'S COOS BAY HOUSE

PENNSYLVANIA
house

Old Woods Road
Green Lane, Pennsylvania 18054

DESCRIPTION: Privately funded 500 ft^2 single-story experimental dwelling.

Solar system: 60 ft^2 water-cooled collector area utilizing aluminum Roll-Bond, double glazing, organic base polymer as corrosion inhibitor, drainage to prevent freezing. Storage: 150 gallon water tank.

Director: James Wenton
Sun Earth Construction Co., Inc.
Box 99
Melford Square, PA 18935
(215) 536-8555

Owner: Howard S. Katz
Old Woods Road
Green Lane, PA 18054
(215) 536-5544

Architect: Steven Wierz
Pittsburgh, PA

Engineer (Storage and Heat Transfer): Burt, Hill and Associates
610 Mellon Bank Building
Butler, PA 16001
(412) 285-4761

Solar Systems Engineer: Sun Earth Construction Co., Inc.
address as above

Contractor: Sun Earth Construction Co., Inc.
address as above

Completion date: December 1, 1974

PENNSYLVANIA
house

Stoverstown, Pennsylvania

DESCRIPTION: Two-story, 3-bedroom dwelling. Second story is unheated, supports the collector and serves as attic. Heated area 1250 ft^2. House and solar heating system somewhat patterned after Peabody House in Dover, Massachusetts.

Solar system: Collector with gross area of 65 ft x 8 ft = 520 ft^2; net area is 450 ft^2. Covers entire vertical south face of second story. Double-glazed. Radiation strikes non-selective black surface, and fan-driven air carries heat from that surface, via ducts, to several 4-ft wide closets (bins) between rooms, and thence to rooms. First-story windows collect some solar radiation. No storage system other than general thermal capacity of house itself. 40 to 50% solar heated. Auxiliary heat from gas-fired furnace.

Director: Harold R. Lefever
Sonnewald Service
R.D. 1, Box 457
Spring Grove, PA 17362

Owner: Harold R. Lefever
address as above

Architect: Harold R. Lefever
address as above

Engineer: Harold R. Lefever
address as above

Contractor: Harold R. Lefever
address as above

Completion date: 1954

32" O.C.
1.75"
96" O.C.
ISOMETRIC
0 1 2 3 4
FEET

SUN EARTH COLLECTOR TO BE UTILIZED IN HOWARD KATZ'S HOUSE

HAROLD LEFEVER'S HOUSE

PENNSYLVANIA
house

Schnecksville, Pennsylvania

DESCRIPTION: Two-story, 3-bedroom, 1 1/2 baths, 1600 ft^2. Energy conservative design incorporating 1 inch thick styrofoam sheath insulation in walls and urethane around framing areas. In addition, waste heat will be recovered from septic tank, laundry, refrigerator and dryer; bath water. Design goal is to save from 1/3 to 1/2 of energy requirements of equivalent all-electric home. Target cost is $35,000. Sponsored by the Pennsylvania Power and Light Company.

Solar system: Vertically mounted fence-style water/glycol cooled flat plate solar collector, 200 ft^2 in 3 ft x 3 ft or 2 ft x 4 ft modules. Thermal storage will utilize 1000 gallon storage tank. Solar heating will be assisted by 2 heat pumps and 15 kw immersion heater in the storage tank. System design goal is to supply building heat requirements from solar energy reclaimed heat and heat pumps in equal amounts.

Director: Robert Romancheck
Pennsylvania Power & Light Co.
901 Hamilton Mall
Allentown, PA 18101
(215) 821-5534

Owner: Pennsylvania Power & Light Co.
address as above

Architect: Donald Dunck1ee
West Rock Road
Allentown, PA 18103
(215) 797-3200

Engineer: In house (Pennsylvania Power & Light Co.)

Contractor: Curtis Schneck
25 Oak Wood Circle
Schnecksville, PA 18078
(215) 767-5018

Completion date: September, 1974

P.P.L.'s FENCE COLLECTOR

PENNSYLVANIA
office building

Pittsburgh, Pennsylvania

DESCRIPTION: Commercial complex, sponsored by PPG, SOHIO, ALCOA, Oliver Tyrone Corp., and possibly others. Test building will be near PPG Research Center in Pittsburgh. Will ultimately be 6 to 10 stories, but will start with mock-up of 1 1/2 stories to test solar radiation in area. ($30,000)
Second phase: 30 ft^2, 2 1/2 floors costing $90,000.
Third phase: 6 to 10 stories (this phase not yet funded).

Solar system: Aluminum collectors with Alcoa selective surface. Water/ethylene glycol cooled. Windows on east angled to north and northeast; windows in west angled north and northwest. Windows on south will be shaded by collector panels.

Director: Donald W. Pulver
Executive Vice President
Oliver Tyrone Corp.
One Oliver Place
Pittsburgh, PA 15222
(412) 281-7400

Owner: Aluminum Company of America
425 6th Avenue
Pittsburgh, PA 15219
(412) 553-4545

PPG Industries, Inc.
One Gateway Center
Pittsburgh, PA 15222
(412) 434-3131

Standard Oil Company of Ohio Corporation
101 Prospect Avenue, N.W.
Cleveland, OH 44115
(216) 575-4141

Oliver Tyrone Corp.
address as above

possibly others

Architect: Oliver Tyrone Corp.
address as above

Engineer: Environmental Systems Design, Inc.
35 E. Wacker Drive
Chicago, IL 60601
(312) 263-0466

Solar System Engineer: Oliver Tyrone Corp.
address as above

Completion date: Second phase complete by spring, 1975.

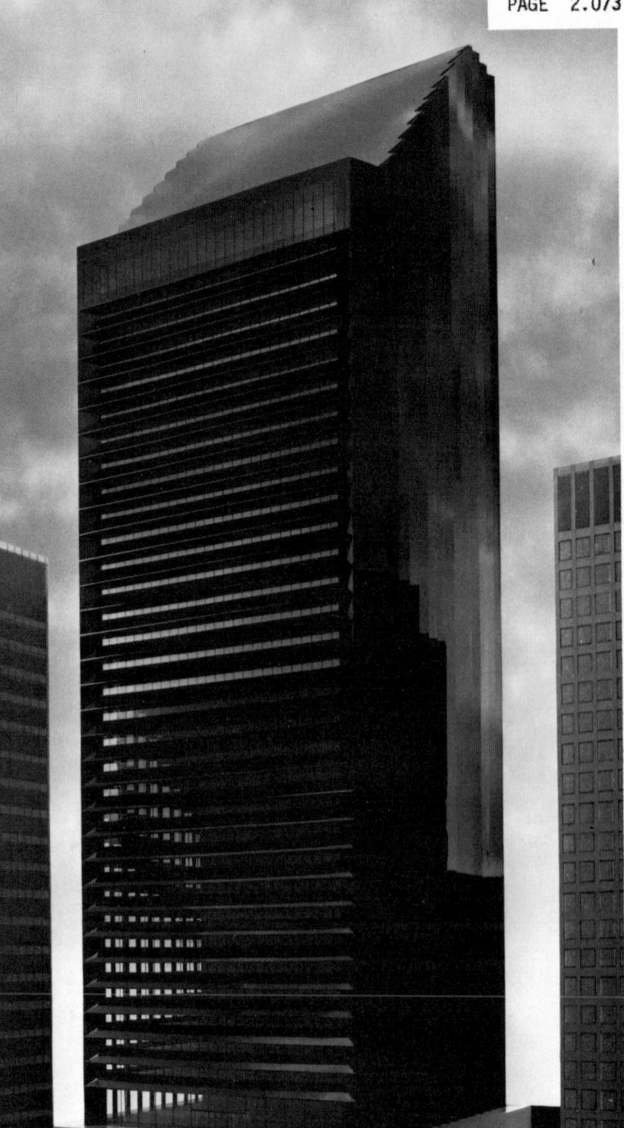

OLIVER TYRONE CORP. et al -- SOLAR OFFICE TOWER THAT MAY BE PHASE FOUR OF PROJECT

PENNSYLVANIA
post office

Branch Post Office
Ridley Park, Pennsylvania

DESCRIPTION: 5000 ft^2 branch post office. An experiment to determine whether cost effective for future post offices.

Solar system: 2500 ft^2 flat plate liquid type collector. Gas or oil auxiliary heat.

Director: Mr. Emerson Smith
Assistant Post Master General
U.S. Postal Service
Lefant West
Washington, D.C. 20260
(202) 245-4000, Ext. 4372

Owner: Postal Service
United States Government
Washington, D.C.

Architect: Environmental Design Collaborative
1832 Arch Street
Philadelphia, PA 19103
(215) 563-1802

Engineer: Honeywell
2600 Ridgeway Parkway
Minneapolis, MN 55413
(612) 331-4141

Completion date: 1976

RHODE ISLAND
office building

The Stillman White Foundary, #1 Bark Street
Electric Energy Conservation Station
Providence, Rhode Island

DESCRIPTION: Restoration of 125-year-old mill, which is 3 stories and has 5000 ft^2 of usable floor space. The building will be an experimental station and will demonstrate to the community alternative sources of electrical power generation and energy conservation. Some of the test projects include: development of experimental solar climate control hardware, systems instrumentation and evaluation, wind tunnel testing of wind generator designs and full-scale prototype development, comparative analysis of available high-efficiency lighting devices, energy storage systems, etc. The building will be insulated with urethane foam in newly constructed and exposed walls and in the roofing. Windows are highly insulated double pane, with adjustible venetian blinds incorporated in air space, manufactured by Amelco. Recycling "Cycle-Let" toilet, manufactured by Thefford Corp., Ann Arbor, MI, is also being used. Several companies contributed to the project, including Narragansett Electric Company. This "electric energy conservation station" will also provide a means of investigating how people react to the utilization of natural energy systems as opposed to conventional energy sources.

Solar system: A combination of solar heating, wind power, and water power (from Moshassuck River) to supply building's lighting and heating requirements. Roof-mounted solar collector array. Two types of wind generators being tested: one 15 ft diameter turbine coupled with 10 kilowatt alternator, to provide 30 kwh's per day; one vertical axis wind turbine of low complexity. Also two types of storage: thermal energy storage utilizing heavily insulated tank capable of maintaining water/anti-freeze solution at high temperatures (such as product developed by Megatherm Corp., E. Providence); and electrical storage using a battery system, as a lead-acid battery manufactured by the Mule Battery Co. Back-up electricity to be supplied by local utility when there are lapses in sun, wind, and water.

Director: Ronald Beckman
Research and Design Institute (REDE)
P.O. Box 307
Providence, RI 02901
(401) 861-5390

RHODE ISLAND
office building

The Stillman White Foundary #1 Bark St.
Electric Energy Conservation Station
continued

Owner: Stillman White Association
P.O. Box 307
Providence, RI 02901
(401) 861-5390

Engineer: Joseph J. Loferski
Division of Engineering
Brown University
Providence, RI 02901
(401) 863-2677

Contractor: Space Development Corporation
P.O. Box 246
Providence, RI 02901
(401) 861-7550

Completion date: Ready for occupancy in early 1975.

ROSTER OF SPONSORS DISPLAYED ON THE RHODE ISLAND ENERGY CONSERVATION STATION

REDE, the Research and Design Institute, announced grants which have been awarded by the State of Rhode Island and Federal agencies to support the establishment of the Rhode Island Energy Conservation Station at the newly restored Stillman White Foundry. This project is attracting a broad base of support from private and public sectors of the economy. Participating sponsors include the State of Rhode Island, the National Trust for Historic Preservation, the National Endowment for the Arts and the United States Department of the Interior. The Narragansett Electric Company provided the initial seed money for this project, to demonstrate energy conservation and low building maintenance in architecture and construction. Four Rhode Island banks, the Old Stone Bank, Industrial National Bank, People's Savings Bank and the Woonsocket Institute Trust Company have provided financing and direct grants to support this study. New England manufacturers of energy related products and systems are participating through contributions of goods and services in the "retrofitting" of this early American factory building. Their products will become weapons in the fight against inflation, demonstrating energy savings in an area which is most wasteful of resources -- building and construction.

THE STILLMAN WHITE ASSOCIATION'S ENERGY CONSERVATION STATION

TEXAS
house

Alpine, Texas 79830

DESCRIPTION: Privately funded 4700 ft^2 2-bedroom, 2-bathroom adobe structure. House has enclosed swimming pool, floor is 4 inches thick (brick on sand), walls are all adobe with 4 inches of fiberglass in between. Stucco outermost layer. Estimated $140,000.

Solar system: Built-in direct-gain solar collectors using bead walls and sky lids (Zomeworks).

Director: David Wright
Sun Mountain Design
960 Camino Santander
Santa Fe, NM 87501
(505) 983-1680

Owner: Dr. A. Tonton
Alpine, TX 79830
(915) 837-3484

Architect: David Wright
address as above

Engineer: Herman G. Barkman
107 Cienica
Santa Fe, NM 87501
(505) 982-8907

Solar Systems Engineer: David Wright
address as above

Zomeworks Corporation
P.O. Box 712
Albuquerque, NM 87103
(505) 242-5354

Contractor: Dr. A. Tonton, with some assistance from local builder.

Completion date: Undetermined due to high bids from professional contractors.

VERMONT
apartments

Shelburne, Vermont

DESCRIPTION: The west end of a barn totaling 5600 ft^2 will be converted into a three bedroom apartment (1500 ft^2.). The apartment is intended for the teachers of "The Schoolhouse", a private alternative elementary school across the road. The remaining barn space will be for school and community activities. Energy conservation measures include the use of six inches of fibergalss insulation in the walls and use of insulated glass windows with opaque insulating shutters for nightime use. Windows face predominantly south and west with almost none in other directions. Clivus toilet will be used and bath, dish, and wash water will be disposed of in a sand filter bed.

Solar system: Automatic thermosiphoning solar hot water heater which doubles as a canopy over the front entrance. A glassed in porch on the south side will collect heat which can be pumped into apartment during winter. The porch will be unglazed and will shade the southern exposure in the summer. Primary source of heat will be wood stoves and a small oil fired hot water heating system for back-up.

Architect: Bruce Anderson
Total Environmental Actions
Box 47
Harrisville, NH
(603) 827-3087

Completion date: Winter/Spring 1975

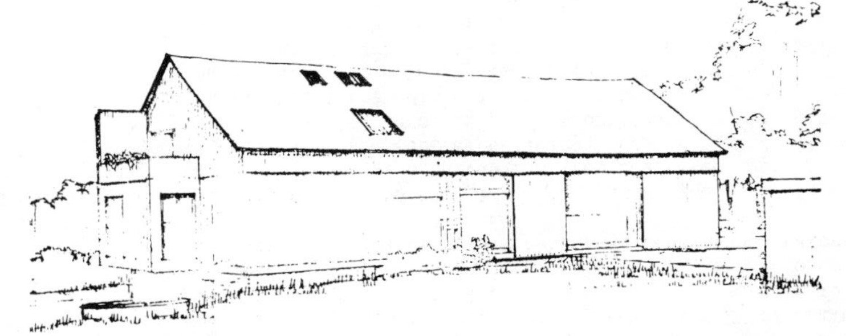

BROUGHTON RESIDENCE
rehabilitation of Vermont barn

DR. TONTON'S TEXAS HOME

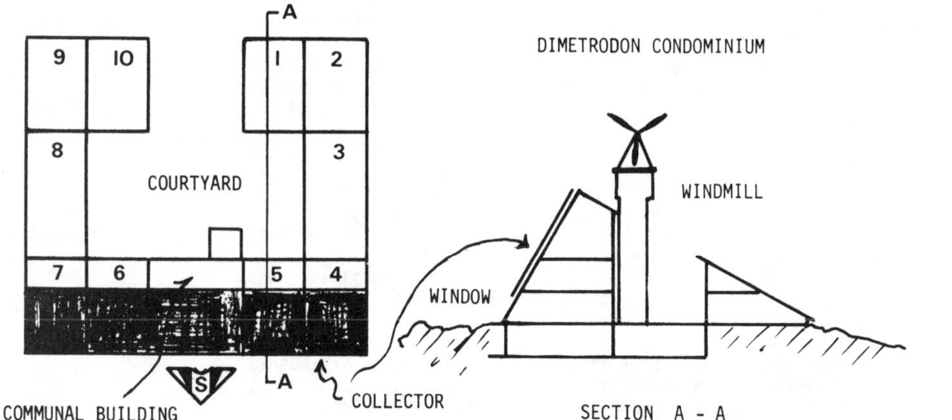

VERMONT
condominium

Dimetrodon Condominium
Prickly Mountain
East Warren, Vermont

DESCRIPTION: A ten-unit condominium on slope of Prickly Mountain. Each unit is 16 ft x 28 ft, and there is a 24 ft x 28 ft communal building around a central courtyard. Most units have full basements. Windmill for electricity generation on tower of courtyard. Buildings have many energy conserving features.

Solar system: 2500 ft^2 net of Thomason trickling water-type collector employing corrugated galvanized iron sheets on south sloping roof of large central building. Double glazed with sheet of Kalwall Corp. Sun-Lite fiberglass-reinforced polyester; sheets are held 1 1/2 inches apart by U-shaped spacer bar. The assembly is mounted with inner sheet 1 inch from corrugated metal; the spacing between established by blocks of rubber cut from automobile tires. Collector air space is vented in summer. Storage is 12,000 gallons of water in 3 steel storage tanks surrounded by approximately 100 tons of fist-size rocks in basement of units #6 and #7. Auxiliary heat from aerogenerator, wood stoves, and gas heaters.

Director: Bill Maclay
c/o Dimetrodon Corp.
Warren, VT 05674
(802) 496-2907

Owner: Dimetrodon Corp.
address as above

Designers: Bill Maclay, R. Travers, J. Sanford
Dimetrodon Corp.
address as above

Engineer: Dimetrodon Corp.
address as above

Contractor: Dimetrodon Corp.
address as above

Completion date: September, 1975 (now partially completed)

VERMONT
condominium

Grassy Brook Village
Grassy Brook Road
Brookline, Vermont

DESCRIPTION: A twenty-unit condominium complex, each with three bedrooms, clustered in groups of ten. Extra-thick insulation. Double glazing (skylights triple-glazed).

Solar system: Collector (4500 ft.^{2}) includes several rows of panels tilted 57° from horizontal. Storage system: 20,000 gal. water; compartmentation used to permit thermal segregation. Heat pump may be used. Auxiliary heating: central oil furnace, with small wood-burning stoves in individual units.

Director: Richard D. Blazej
RD #1, Box 39
Newfane, VT 05345
(802) 365-7901

Owner: Grassy Brook Village, Inc.
Richard D. Blazej
RD #1, Box 39
Newfane, VT 05345
(802) 365-7901

Architect: Robert F. Shannon
People/Space Co.
259 Marlborough Street
Boston, MA 02116
(617) 261-2064

Engineer: Dubin-Mindell-Bloome Associates
42 W. 39th Street
New York, NY 10018
(212) 868-9700

Contractor: Grassy Brook Village, Inc.
address as above.

Completion date: Construction to begin Spring, 1975; completion date late 1975.

Note: Alvin O. Converce, Professor of Engineering, Thayer School of Engineering, Dartmouth College, Hanover, NH is monitoring electrical energy demand patterns for Grassy Brook project. Purpose is to identify large users and reduce overall comsumption.

VERMONT
house

Mid-western Vermont

DESCRIPTION: Single family residence. Private funding. Designed for very low consumption of heating energy.

Solar system: 1000 ft^2 water-cooled flat-plate collectors tilted at 57° angle. Storage in 4000 gallon insulated concrete reservoir in the basement. Heat exchanger in reservoir to preheat domestic hot water. Considerable south facing glass to admit solar heat directly. Oil fired furnace and wood stoves for auxiliary heating system.

Owner: Steve and Alice Brown

Architect: Sue B. Jenner

Heating Consultant: Walter Voyes

For further information contact TEA

Solar System Consultant: Total Environmental Action
Box 47
Harrisville, NH 03450
(603) 827-3087

Completion date: Fall and Winter 1974

GRASSY BROOK VILLAGE • Brookline, Vermont

VERMONT
house

DESCRIPTION: Funding: 1/3 - Vermont Council of the Arts, 2/3 - private. Molded ferro cement. First they erected an air structure, then sprayed it with 3 inches of 6 lb urethane foam, covered that with steel mash and then added 1 inch of ferro cement. Most of structure underground.

Solar system: 16 ft x 12 ft collector on back of structure with two foam eye lid windows on front. Collectors are hot air type. Other natural energy features, windmill and water.

Director: David Seller
Dimetrodon Corporation
Prickly Mountain
Warren, VT 05674
(802) 496-2907

Owner: David Seller
address as above

Architect: David Seller
address as above

Engineer: David Seller
address as above

Contractor: David Seller
address as above

Completion date: April, 1975

VIRGINIA
mobile home

InterTechnology "Eccovan"
Warrenton, Virginia

DESCRIPTION: In development and design phase, this is a mobile home with reduced fuel requirements and pollutant impact. The Eccovan uses energy conservation measures, a fuel cell, a heat pump, solar energy with storage and catalytic combusters.

Solar system: Roof mounted flat plate collectors which can be retracted to be horizontal to the roof for transport. Storage is in tanks underneath the floor.

Director: Dr. George Szego
InterTechnology Corporation
P.O. Box 340
Warrenton, VA 22186
(703) 347-7900

Owner: InterTechnology Corporation
address as above

Architect: InterTechnology Corporation
address as above

Engineer: InterTechnology Corporation
address as above

VIRGINIA
academic building

Fauquier County Public High School
Waterloo Road and Van Roigen Street
Warrenton, Virginia 22186

DESCRIPTION: Existing high school retrofitted with solar heating system for five separate classrooms which comprise a 4100 ft^2 portion of the school. Funded by InterTechnology Corp. and a $173,000 grant by the National Science Foundation.

Solar system: Water-type solar collector 126 ft x 26 ft (2400 ft^2 net) on 530-from-horizontal scaffold adjacent to school. Double glazed with 1/8-inch double-strength glass. Radiation strikes a chemically etched and black coating on face of Olin Brass Company Roll-Bond aluminum sheet. Selectivity ratio: a/E=3, per measurement. Water with corrosion inhibitor flows in the integral expanded channels in aluminum sheet. No antifreeze is used, but liquid can be drained from collector into storage tank in 1 1/2 minutes. There are 105 collector panels, each 3 1/2 ft x 8 ft. Storage system consists of two separate water-filled, underground 5500 gallon tanks. Total capacity is 11,000 gallons. Each tank (an electric transformer bunker) is of reinforced concrete with L, W, H of 11-1/2 ft, 8-1/2 ft, 7-1/2 ft. Twelve-day carrythrough predicted.

Director: Dr. George Szego
InterTechnology Corporation
Box 340
Warrenton, VA 22180

Owner: After 2 yr. NSF study is complete the system will be owned by:
Fauquier County Public School
Waterloo Road & Van Roigen Street
Warrenton, VA. 22186

Engineer: InterTechnology Corporation
address as above

Contractor: InterTechnology Corporation
address as above

Completion date: March, 1974

Note: InterTechnology Corporation is to solar heat and cool their offices (a converted Safeway Food Store). Solar system is to include 4000 ft^2 of aluminum Roll-Bond collector with three or four 10,000 gallon steel or fiberglass tanks for storage. Project should be complete in September, 1975.

VIRGINIA
museum

Science Museum of Virginia
Richmond, Virginia

DESCRIPTION: 55,600 ft^2, multi-level building containing exhibit halls, laboratories, shops, planetarium, and auditorium. Building design will incorporate several energy conservation features such as extensive waste heat recovery, variable volume ventilation, extensive zoning through the use of reheating and air filtering to reduce the outside air needs. Recovered energy to supply 50% of the heating requirements and approximately 40% of cooling requirements.

Solar system: 28,000 ft^2 of water/glycol cooled flat plate solar collector in 4 x 8 ft modules in a single 112 x 240 ft array mounted over an aluminum sheet roof facing south at a 30^0 angle. Thermal storage will be provided by (2) 50,000 gallon below-grade tanks for storage at high and low temperatures respectively. Chilled water from the cooling units will be stored in a third 50,000 gallon below grade tank. Base cooling load will be handled by a 150 ton centrifugal unit. The balance of the cooling requirements will be handled by either stored chilled water or a 150 ton absorption unit powered by the solar collectors.

Director: Richard B. Prudhomme
209 W. Franklin
Richmond, VA 23220
(804) 649-9303

Owner: Science Museum of Virginia
217 Governor Street
Richmond, VA 23219
(809) 770-4133

Architect: Glave, Newman, Anderson & Assoc., Inc.
209 W. Franklin Street
Richmond, VA 23220
(804) 649-9303

Engineers: Hankins & Anderson
2117 N. Hamilton Street
Richmond, VA 23230
(804) 353-1221

Contractor: Undecided

Completion date: Early 1977

VIRGINIA
academic building

Madeira School
8328 Georgetown Pike
Green Way, Virginia

DESCRIPTION: Multi-level, 7000 ft^2 building to house physics, chemistry, biology classrooms and greenhouse.

Solar system: 4500 ft^2 of water/glycol cooled flat plate solar collector mounted on 26° S.E. built-up roof. Thermal storage provided by 10,000 gallon steel storage tank. Supplementary heating will be supplied by oil-fired boiler.

Director: George Rainer
ECS Associates
Division of Flack & Kurtz
29 West 38th Street
New York, NY 10019
(212) 736-3733

Owner: Madeira School
8328 Georgetown Pike
Green Way, VA 22067

Architect: Mr. Alexander
Arthur C. Moore Assoc.
1214 28th Street, N.W.
Washington, D.C.
(202) 337-9083

Engineer: ECS Associates
Division of Flack & Kurtz
29 West 38th Street
New York, NY 10019
(212) 736-3733

Contractor: Out for bid

Completion date: Summer, 1975

MEDEIRA SCHOOL

VIRGINIA
office building

NASA Engineering Building
Hampton, Virginia

DESCRIPTION: Solar heated and cooled engineering building, Langley Research Center, 53,000 ft^2.

Solar system: Collector area: 15,000 ft^2. Central steam auxiliary. Water storage, absorption refrigeration for cooling. May employ evacuated glass tubes. May be situated on field adjacent to building.

Director: Maurice Parker
Public Information Officer
Langley Research Center
Hampton, VA
(804) 827-3966

Owner: US-NASA Langley Research Center
Hampton, VA

Engineer: F. F. Simons and Colleagues
NASA-Lewis Research Center
Cleveland, OH 44135

Contractor: F. F. Simons and Colleagues
address as above

Completion date: Mid 1975

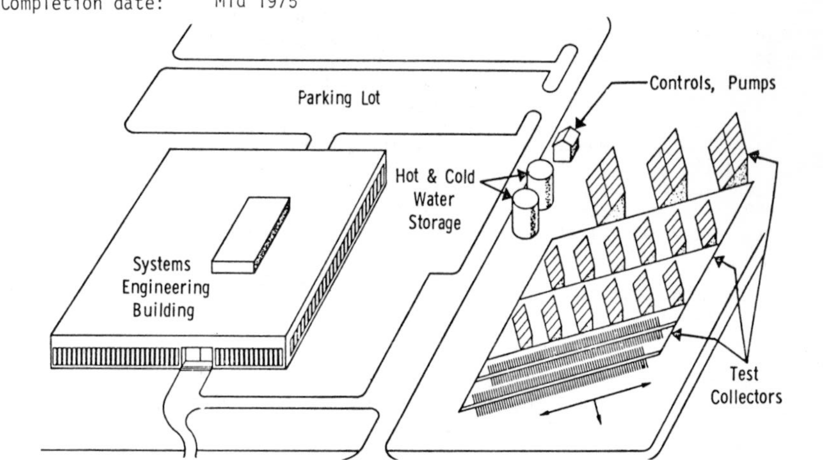

U.S. N.A.S.A.- LANGLY RESEARCH CENTER'S TEST BED

WEST VIRGINIA
house

Demonstration Home
Wilson Farm
Shanghai, West Virginia

DESCRIPTION: Two and a half story, 2-bedroom, modern-type building with 1400 ft^2 of living area plus adjacent greenhouse and garage. Energy conservative design with triple glazing, air lock entrance and centrally located fireplace.

Solar system: 588 ft^2 of water/glycol flat plate solar collectors integrally mounted with roof structure at a 45^0 angle. Thermal storage provided by (2) water tanks of 2400 gal. and 400 gal. respectively for thermal storage at two temperature levels. Supplementary heating to be provided by oil fired water heater. No provision for summer cooling has been made; however, the system has been designed to accommodate absorption air conditioning unit in the future.

Director: P. Richard Rittelmann
610 Mellon Bank Bldg.
Butler, PA 16001
(412) 285-4761

Owner: Mrs. Nicky Wilson
Martinsburg, WV 25401

Architect: P. Richard Rittelmann
address as above

Engineer: P. Richard Rittelmann
address as above

Completion date: Seeking funds

WISCONSIN
house

Comstock, Wisconsin

DESCRIPTION: Goal of project to construct a completely self-sustaining solar house on 70-acre site near Comstock. House will be built into side of a hill (with north side almost entirely below grade) and will contain a total of 4600 ft^2 on 3 levels. A swimming pool enclosure will contain 640 ft^2 and a combination storage building and garage will contain another 1560 ft^2. An entrance corridor and tunnel which connects the various elements will add an additional 690 ft^2; total area approximately 7500 ft^2. Financed through Northwestern Bank of Cumberland, Wisconsin.

Solar system: Will probably use conventional flat plate solar collector system, but investigating a unique method of storing the heat in a swimming pool. Also working on possibility of incorporating a reaction turbine wind energy system for creating all our electrical power. Will probably also incorporate a methane gas generation system.

Director: William A. Vievering, A.I.A.
1939 Munster Avenue
St. Paul, MN 55116
(612) 699-1294
also:
RFD 3, Cumberland, WI 54829
(715) 822-4895

Owner: Dr. and Mrs. John B. Draves
Box 149A
Comstock, WI 54826

Architect: William A. Vievering, A.I.A.
address as above

Engineer: (Mechanical, Electrical, and Solar System) Lundquist, Wilmar, Schultz, and Martin, Inc.
614 Endicott on Fourth Building
St. Paul, MN 55101
(612) 291-1293
also:
430 Oak Grove Street
Minneapolis, MN 55403
(612) 399-1542

Engineer: (Structural) Feyereisen and Boughton, Inc.
2361 West Highway 36
St. Paul, MN 55113

Contractor: B. A. Johnson and Sons
1170 Grove Street
Cumberland, WI 54829

Completion date: July, 1975 (tentative)

FOREIGN
Andorra
house

Padern College solar cabin
Padern, Andorra

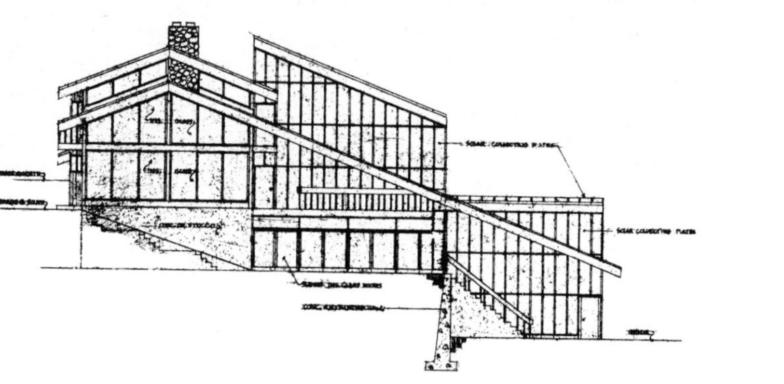

DESCRIPTION: First of five very-low-cost cabins, on south slope of a mountain in the Pyrenees (1500 meter altitude). Two stories, 19 m x 8 m, with long axes east-west. Roof slope is 30° from horizontal, to conform to local building code. Building is well insulated; typical U-value is 0.4 W m^{-2} $^{o}C^{-1}$. Many energy conserving features. Electrical supply from a 2 kw Electro-GmbH windmill and 12-v batteries. Building provides classroom, living room, 12 bedrooms, utility room, etc. for 24 students. Low-cost, on-site materials used. Construction mainly by students themselves.

Solar system: CNRS-type solar wall (see France). 80 m^2, air-cooled, covering 90% of vertical south wall. Heart of collector is a thick (35 cm), vertical, natural-dark-color, masonry wall made with soot-blackened mortar. Single glazing is used and there is a 45 cm air space between glass and wall. Other 10% of the south wall consists of 12 double-glazed windows. There is a 45 cm space between glass layers, occupied by flower-filled boxes. Another collector, consisting of (24) 2 1/2 m x 1/2 m panels is on a 30° sloping roof for domestic hot water. Auxiliary heat supplied by Catalan wood burning stove for cooking. Organic waste disposal system also provides some heat. 80% solar heated.

Owner: Padern College

Architects: I. Hogan
A. Gerrand
B. Ford
J. F. Robert (consultant on CNRS solar wall)

Completion date: 1975

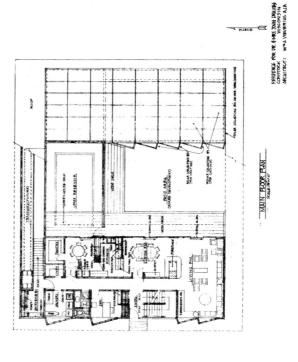

FOREIGN
Canada
house

Surrey, British Columbia
(near Vancouver)

DESCRIPTION: One-story, 1400 ft^2 plus basement (1300 ft^2 partially heated). House built in 1968; solar heating system in 1971.

Solar system: 413 ft^2 of water-cooled flat plate collectors on roof sloping 58° from horizontal. Double glazing, copper sheet with non-selective black coating, copper tubing soldered to sheet. Water circulated through collector by small centrifugal pump; drained automatically at end of day. Two vertical, cylindrical, uninsulated tanks (800 gallons total) in insulated basement room for storage. 40% solar heated. Also provides much heat for domestic hot water and swimming pool.

Director: Erich Hoffmann
5511 128 Street
Survey, British Columbia, Canada

Owner: Erich Hoffmann
address as above

Engineer: Erich Hoffmann
address as above

Completion date: 1971

C.N.R.S.- SOLAR HOUSE

FOREIGN
France
houses

Centre National de la Recherche
Scientific Solar Energy Laboratory
Odeillo, France

DESCRIPTION: Single-family dwellings. Volume of each house 300 m^3.

Solar system: Utilizes south facing vertical concrete wall, painted black and covered with double glazing. The wall acts as a heat sink. There are slots (20 to 30 cm high, 50 to 70 cm wide) through the concrete wall at top and bottom. Air circulates naturally upward between glass and concrete and downward through house. Fan or blower may be used. Cost of system $40/$m^2$ to install. Provides 66% of winter heat needs. Overheating in summer can be avoided by ventilating air using chimney-effect vent system with vent open at top of collector.

Director: Professor F. Trombe
Director
Centre National de la Recherche Scientific (CNRS)
Odeillo, France

Owner: CNRS
address as above

Engineer: CNRS
address as above

Completion date: 1962

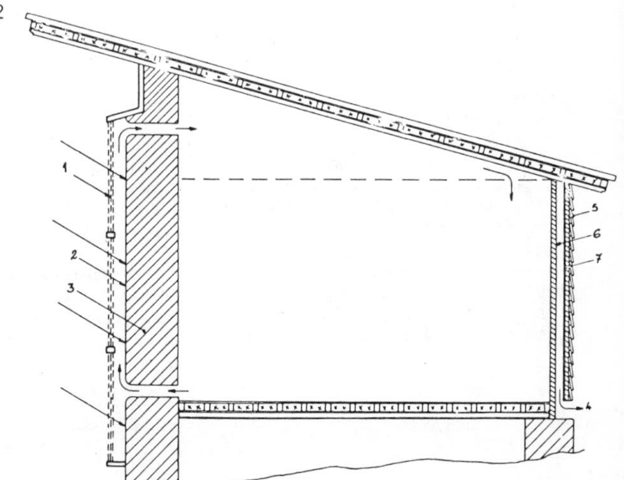

C.N.R.S.- SOLAR WALL CONCEPT

FOREIGN
France
house

Southern France

DESCRIPTION: A "traditional" Mediterranean house that will be totally autonomous. Platform and roof suspended from an external three dimensional steel spaceframe. Tubular elements and cast jointing units are supported on six piled foundation pads.

Solar system: Twelve flat plate collectors as well as a panel of solar cells that provide electric power for the pump in the heating system. Storage in underground water tanks. Designers expect to store enough heat in summer to heat house through winter. Electric appliances will be powered by a windmill.

Designer: Dominic Michaelis Assoc.
Bay 8
16 South Whars Road
W2 London, England
Phone: 01-402-5771

DOMINIC MICHAELIS ASSOCIATE'S
MEDITERRANEAN HOUSE

FORIEGN
France
laboratory and office building

Centre National de la Recherche
Scientific Solar Energy Laboratory
Odeillo, France

DESCRIPTION: Nine-story building which houses the Solar Energy Laboratory. North facing wall is a large reflector which concentrates solar radiation beamed into it by south facing, hill-mounted heliostats into a 1000 kw solar furnace in adjacent tower.

Solar system: Black corrugated metal panels are located behind glass panels which cover the east, south, and west walls of the building. As air is heated between the glass and metal, it flows upward out through a room vent. Receives 1/2 of heating requirements from solar energy.

Director: Professor F. Trombe
Director
Centre National de la Recherche (CNRS)
Scientific Solar Energy Laboratory
Odeillo, France

Owner: Centre National de la Recherche (CNRS)
address as above

Engineer: Centre National de la Recherche (CNRS)
address as above

Completion date: 1971

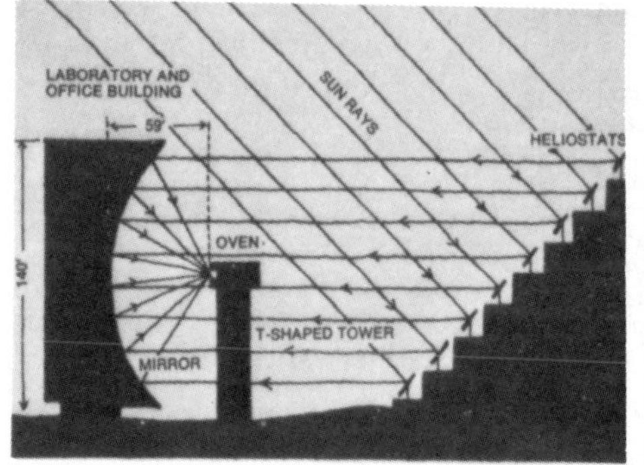

Solar furnace at Odeille, France, uses field of mirrors to direct sun radiation at reflective facade of structure, which then focuses it onto central tower top

FOREIGN
Great Britain
academic building

Waleasey
(near Liverpool), England

DESCRIPTION: Two-story, 230 ft long school, housing 320 pupils. North, east, and west walls are of brickwork; floors are 10 inches thick reinforced concrete. Roof is reinforced concrete with 5 inches of expanded polystyrene on outside. Insulation is protected by bituminous layers impervious to moisture.

Solar system: Major portion of south vertical wall, which is 27 ft high, consists of two layers spaced 24 inches apart to allow access for maintenance. Outer layer is in part (85%) a single layer of glass, and in part (15%) a concrete wall, the south face of which is painted black. (In summer, it is covered by white panels.) The minor portion of the south vertical wall (about 8%) consists of one layer only, this made up of single glazed, openable windows that have especially tight seals. Ventilation is via these windows and north wall windows. Storage is in the concrete wall, floor, and roof, and the brick wall. Daily temperature swing on typical day $3^{o}C$ and $10^{o}C$ under extreme conditions. No auxiliary heat is used except luminaries.

Director: R. E. Shaw
Director of Development
Municipal Offices
Brighton Street
Wallasey, Cheshire L44 8ED

Designer: A. E. Morgon (deceased)

Completion date: 1961

FOREIGN
Iran
city

DESCRIPTION: A proposed new city for 200,000 people in early master planning stage. Solar energy is being considered for heating, cooling with absorptive refrigeration, and in the future, for electric power generation.

Architect: Skidmore, Owings & Merill, Associates

Engineer: Dubin-Mindell-Bloome Associates
Consulting Engineers and Planners
42 West 39th Street
New York, NY 10018
(203) 236-1654

FOREIGN
Italy
laboratory

Swedish Solar-heated Laboratory
Capri, Italy

DESCRIPTION: Two-story laboratory with horizontal roof. Floor area is 1940 ft^2.

Solar system: 320 ft^2 of vertical, water-cooled flat plate collectors on southwest wall. Metallic radiator plates used with one layer of glass and one of plastic. 800 gallon storage; rate of flow from storage tank to collector 2.1 gallons/minute-1. 50 to 90% solar heated. Auxiliary heat from stove and electric heater.

Designers: G. Pleigel and B. Lindstrom
Sweden

Completion date: Approximately 1960

FOREIGN
Japan
house

Yanagimachi Solar House I
Tokyo, Japan

DESCRIPTION: Solar heated and cooled house. Heat pump employed.

Completion date: Before 1958

FOREIGN
Japan
house

Yanagimachi Solar House II
Tokyo, Japan

DESCRIPTION: 2460 ft^2 area heated.

Solar system: 1400 ft^2 water-cooled collectors. Employs black aluminum plate with integral channels. No glazing. Utilizes 9600 gallon water storage. Heat pump, operated by 3 HP motor, used for auxiliary heating and cooling.

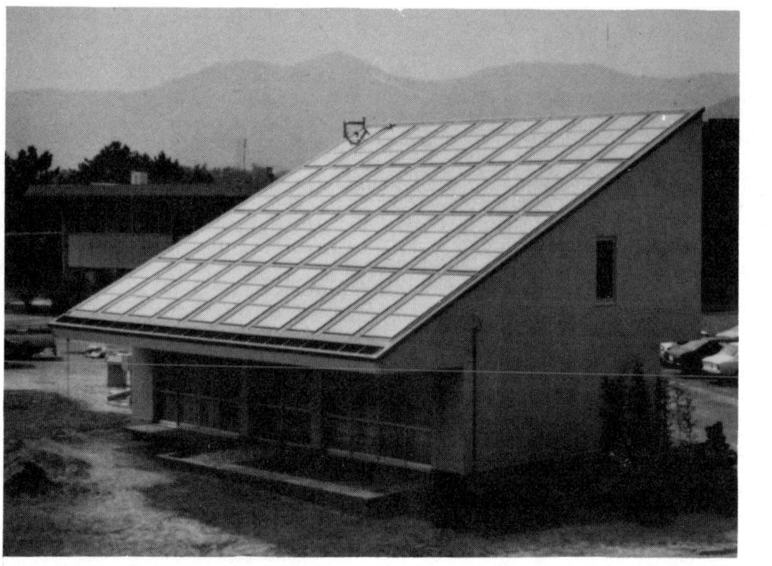

JAPANESE SOLAR LABORATORY

FOREIGN
Japan
laboratory

Nagoya, Japan

DESCRIPTION: Solar heated laboratory building with 880 ft^2 floor area.

Solar system: 300 ft^2 water-cooled flat plate with expanded channels in aluminum sheet. Storage is 1500 gallon water tank.

Completion date: Approximately 1960

FOREIGN
Virgin Islands
community

Anegada, Virgin Islands

DESCRIPTION: Feasibility study completed for community the size of small town with population of 10 to 20 thousand people. Would include 1500-room hotel, 2,000 houses, and supporting community facilities.

Solar system: Use of solar energy for water heating, water distillation. Electric power from wind and solar conversion.

Owner: Anegada Corporation
Anegada, Virgin Islands

Engineer: Dubin-Mindell-Bloome Associates
Consulting Engineers and Planners
42 West 39th Street
New York, NY 10018
(203) 236-1654

INTRODUCTION TO BIBLIOGRAPHY

The following bibliography constitutes an attempt to survey the type of published material available on the subject of solar energy. Publications herein are not geared to one specific audience; literature may be aimed at the novice or the professional. These entries should not be construed as the best material available on a particular subject, but rather, an indicator of the type of literature that has been published.

To access more comprehensive and specialized inventories of pertinent literature, the section entitled "Major Bibliographies and Inventories" should be utilized.

Citations are divided according to subject (see Table of Contents). Each entry includes author, title, date, publisher, and an annotation. Whenever possible, the price and the name of the source from which a particular publication is available is listed.

TYPICAL CITATION

AUTHOR——Daniels, F.

TITLE——Direct Use of the Sun's Energy

DATE——1964

PUBLISHER——Yale University Press
149 York St.
New Haven, CN 06511

PRICE——$10.00 (hardcover) $1.95(paperback-Ballentine Books)

AVAILABILITY*——Available: EARS (paperback)

ANNOTATION——An excellent basic book on solar energy utilization. Covers all aspects of solar energy applications and experiments, including: cooking, heating, agricultural drying, heat storage, furnaces and engines, cooling and refrigeration, distillation photochemical conversion, etc. Other sources are listed for those interested in greater technical details. 300 pages.
•

* AVAILABILITY--The following abbreviations are often used to indicate where publications can be obtained:

EARS ---------- Environmental Action Reprint Service
University of Colorado at Denver
1100 14th St.
Denver, CO 80202

GPO ---------- Superintendent of Documents
U.S. Government Printing Office
Washington,DC 20402

NTIS ---------- National Technical Information Service
U.S. Department of Commerce
5285 Port Royal Road
Springfield, VA 22151

OALS ---------- Office of Arid Land Studies
University of Arizona
Tuscon, AZ 85721

GENERAL INFORMATION

Baer, S.

Solar Energy

Mar.-Nov., 1973
Tribal Messenger
Sun Publishing Co.
P.O. Box 4383
Albuquerque, NM 87106
$3.50

Good introduction to solar applications, sun orientation and energy concepts. Describes in detail functioning low tech passive systems. Use of water and air as heat transfer fluids is discussed as well as storage utilizing water and/or rocks. Guide to junk yard heat transfer panels.
•

Bell, G. (ed.)

Energy Sources and Uses

Aug., 1974
Journal of Ekistics: Reviews on the Problems and Science of Human Settlements:
Vol. 38, No. 225

A special issue focussing on energy which includes articles on solar, geothermal, fission and fusion energy, as well as the energy picture in general. Includes a bibliography on energy, January 1971 to March 1974. Articles dealing specifically with solar energy include: "Energy for Millenium Three" by Earl Cook; "Solar Energy in Developing Countries: Perspectives and Prospects" by the National Academy of Sciences; "The Solar House and Its Portent" by K. W. Boer; "Plumbing the Ocean Depths" by Abraham Lavi and Clarence Zener.
•

Branley, F.

Solar Energy

1957
Thomas Y. Crowell Company
201 Park Ave. S.
New York, NY 10003

Easily understood by the young reader, this book explains the basic principles of solar energy applications. Includes discussion on water heaters, heat pumps, space heating, solar furnaces, cookers, distillers and wind power.
•

GENERAL INFORMATION

Brinkworth, B. J.

Solar Energy for Man

1972
Halsted Press
A Division of John Wiley and Sons, Inc.
605 Third Ave.
New York, NY 10016

Available: EARS
$8.95 (hardcover)

Suitable for the student or professional reader, this book provides a thorough, fundamental background on solar energy applications. Material ranges from the relationship of the sun to the earth to the different devices and methods employed when utilizing the sun's energy.

•

Clark, W.

Energy for Survival-The Alternative to Extinction

1974
Anchor Books
Anchor Press/Doubleday
Garden City, NY 11530
$12.50

Analysis of our socio-economic relationship with energy, current sources of energy, why energy is utilized the way it is, alternatives to present sources and affects of the changeover in energy usage. Large portion devoted to solar energy.

•

Daniels, F.

Direct Use of the Sun's Energy

1964
Yale University Press
149 York St.
New Haven, CN 06511
$10.00 (hardcover) $1.95 (paperback-Ballentine Books)
Available: EARS (paperback)

An excellent basic book on solar energy utilization. Covers all aspects of solar energy applications and experiments, including: cooking, heating, agricultural drying, heat storage, furnaces and engines, cooling and refrigeration, distillation, photochemical conversion, etc. Other sources are listed for those interested in greater technical details. 300 pages.

•

GENERAL INFORMATION

Daniels, F./ Duffie, J. (eds.)

Solar Energy Research

1955
The University of Wisconsin Press
Box 1379
Madison, WI 53701

Based loosely on a symposium on solar energy held in Wisconsin on Sept. 12-14, 1953, this is a compilation of papers which basically point the way toward needed research. Good reference.

•

Halacy, D. S., Jr.

Experiments with Solar Energy

1969
W.W. Norton and Co., Inc.
55 Fifth Ave.
New York. NY 10022

Excellent for the young scientist. Basic descriptions accompanied by photographs and diagrams provide a guide for the young reader to follow in building and experimenting with solar energy devices such as furnaces, stills, ovens, water heaters, and model airplanes.

•

Halacy, D. S., Jr.

The Coming Age of Solar Energy

1963
Harper & Row
10 East 53rd St.
New York, NY 10022
$7.95 (hardcover)
Available: EARS

Recounts the history of solar energy and discusses many concepts and hardware for utilizing the sun's energy, including: an orbiting solar collector; a desert solar-power complex; and a sea-thermal energy plant to tap the ocean's heat. 219 pages.

•

GENERAL INFORMATION

Hoke, J.

Solar Energy

1968
Franklin Watts, Inc.
555 Lexington Ave.
New York, NY 10022
$3.95

Easily understood primer suitable for all ages. Enumerates simple methods of putting the sun to work such as solar cells and solar house heating. Also discusses some school projects. Well illustrated.

•

Gamov, G.

A Star Called the Sun

1964
Bantom Books
The Viking Press, Inc.
625 Madison Ave.
New York, NY 10022
$.95 (paperback)

Explanation of fact and theory about the sun itself: the origin and nature of its chemical elements, sunspots, plasma, how the sun's mass is calculated, the thermonuclear reaction that takes place, and more.

•

Givoni, B.

Man, Climate and Architecture

1969
Applied Science Publishers Ltd.
Ripple Road
Barking, Essex
England
£5.50

Various chapters in the book deal with the description of the climatic factors affecting human comfort and building design, the biophysical relationship between man and thermal environment, physiological and sensory effects of heat and cold stress in relation to work and various thermal indices used to express the combined effect of various environmental factors.

•

GENERAL INFORMATION

Maddox, K.P.

Solar Energy Conversion

Sept., 1974
Mineral Industries Bulletin: Vol. 17, No.5
Available: Colorado School of Mines Research Institute
Golden, CO 80401
$1.00 (single issue)

Special issue which provides an overview of solar energy ranging from an historical view to the various conversion techniques.

•

McLaughlin, R.H.

Solar Energy Research- Can it Survive the Congressional Hot Air Treatment?

July-Aug., 1974
New Engineer: Vol. 3, No. 7, pp. 28-33

Provocative article which heavily criticizes congressional legislation on solar energy, particularly the McCormick bill.

•

Menzel, D.H.

Our Sun

1959
Harvard University Press
79 Garden St.
Cambridge, MA 02138
$8.50

A general text on the sun with an astronomy orientation. Covers physical characteristics and phenomena of the sun, including: light waves and spectra, solar chemistry, sunspots, geysers and volcanoes, the corona, eclipses, atomic energy and the solar interior, the sun and the universe, sun-earth relations, and solar power for man.

•

Pintauro, J./ Labiberte, N.

A Box of Sun

1970
Harper & Row
49 E. 33rd St.
New York, NY 10016
$3.95

Continued on the next page.

GENERAL INFORMATION

A Box of Sun (continued)

Recommended reading for anyone conducting solar research. A combination of artwork and verse on the sun and the season it is most pronounced- summer. Helpful in gaining your perspective and suitable for readers of all ages. The following quote illustrates:

God bless the sun
You can't vote for the sun
The sun is king
You can't plant your flag
On the sun
Or split it into townships

Amen.

•

Rau, H.

Solar Energy

1964
MacMillan Co.
866 Third Ave.
New York, NY 10022

Various aspects of solar energy are discussed, including: early myths about the sun, solar physics, solar energy pioneers, practical solar applications today, storage and conversion problems, and photosynthesis. Appendix of sun facts and figures.

•

Rudd, C.G. (ed.)

Solar Energy

Autumn, 1973
Naturalist: Vol. 24, Special Issue #3
Available: Natural History Society
315 Medical Arts Bldg.
Minneapolis, MN 55402

Special issue devoted to the subject of solar energy. Includes ample photographs and the following articles: "Two Centuries of Solar Energy" by John I. Yellott; "The National Energy Crisis" by George W. Rusler; "Solar Energy: It's Potential for Meeting World Energy Demands" by Erich A. Farber; "Water Supply and Sun Power" by Everett D. Howe.

•

GENERAL INFORMATION

Shafelva, K.A.

Sun Cults through the Ages and Early Attempts at the Practical Utilization of Solar Energy

1969
Applied Solar Energy(Geliotechnika): Vol. 5, No. 6, pp. 76-78
Available: NTIS

Review of how ancient peoples worshipped the sun and methods by which they utilized solar energy for practical purposes, such as Pliny the Elder's (79-23 B.C.) observation that sun rays passed through glass spheres could be used to cauterize a wound.

•

Sky, A./ Stone, M. (eds.)

On Site- On Energy

Fall, 1974
Available: Site, Inc.
60 Greene St.
New York, NY 10012
(Order # On Site 5/6)
$6.95

An examination of energy in the light of new conceptual ideas which may determine the shape of our environment. The energy question is treated as a psychological phenomena with potential impact on perception and habitat- solar energy included. The publication is divided into sections on resources, systems, mobility, habitat, synergy, and iconolgy. Partial list of contributors: Hugh Hardy, James S. Mellett, Rene Dubos, John Lobell, Alan Balfour, Jeffry Cook, Richard Stein, Ant Farm, Percival Goodman, Mimi Lobell, Fred Dubin, Denise Scott Brown, Lewis Mumford.

•

Weingart, J.

Everything You Always Wanted to Know About Solar Energy, But Were Never Charged Up Enough to Ask

1972
Environmental Quality: Vol. 3, No. 9, pp. 39-43
Available: EARS

It is possible for solar energy, although now under developed, to become an enormous resource of renewable energy. Discusses solar water heating, space heating, cooking, electric power generation, and bioconversion.

•

GENERAL INFORMATION

Williams, J.R.

Solar Energy- Technology and Applications

1974
Ann Arbor Science Publishers, Inc.
P.O. Box 1425
Ann Arbor, MI 48106
Available: EARS
$6.95 (paperback)

Includes various techniques for utilizing solar energy and covers the board spectrum of solar energy systems. Although technical, it is readily understood by anyone with very little background in the field.

•

Zarem, A.M./ Erway, D.D.

Introduction to the Utilization of Solar Energy

1963
McGraw-Hill Book Co.
330 W. 42nd St.
New York, NY 10036

Very good reference which provides fundamentals on applications of solar energy. Out of print.

•

ENERGY CONSERVATION/ PASSIVE SOLAR

Anonymous

Control the Climate Your House Lives In

Fall-Winter, 1974
House Beautiful's Building Manual: No. 68, pp. 168-193

Illustrated guidelines on how to utilize site orientation, vegetation, and other simple methods to minimize the amount of fuel your home requires.

•

Anonymous

Energy Conservation Hearings before the Committee on Interior and Insular Affairs- U.S. Senate- Pursuant to S. Res. 45, A National Fuels and Policy Study, 93rd Congress- Serial No. 93-7 (92-42)

March 22-3, 1973(Part I) August 1, 1973(Part II)

continued

ENERGY CONSERVATION/ PASSIVE SOLAR

Energy Conservation Hearings (continued)

Informative record of expert testimony on energy conservation with illustrative tables and references. The appendices include a list of booklets on energy conservation published by energy companies for their customers, and a list of companies which have undertaken energy conservation activities.

•

Anonymous

Utility Designs Energy-Saving Home

Sept., 1973
Electrical World: pp. 84-85

Description of project sponsored by Pennsylvania Power and Light which entails building an energy conservation demonstration home that uses heat pumps, solar collectors, and heat reclaim systems.

•

Aronin, J.E.

Climate and Architecture

1953
Van Nostrand Reinhold Co.
450 W. 33rd St.
New York, NY 10001

Reference book which provides information on the environmental considerations necessary in architectural design. Illustrated, includes bibliography. 304 pages

•

Citizen's Advisory Committee on Environmental Quality

Citizen Action Guide to Energy Conservation

1973
Available: GPO
$1.75

An account of our problem and some well-documented solutions. Of particular interest is Chapter 4, "Energy Conservation in Our Homes and Offices" and a home heating table on page 28 which provides estimated savings from energy conservation measures.

•

ENERGY CONSERVATION/ PASSIVE SOLAR

Conklin, G.

The Weather Conditioned House

1958
Van Nostrand Reinhold Co.
450 W. 33rd St.
New York, NY 10001

An illustrated manual for the "intelligent layman." Includes sections on moisture control, thermal control in the summer, special problems of hot climates, and heating and cooling systems.
•

Dubin, F.S.

GSA's Energy Conservation Test Building-A Report

Aug., 1973
Actual Specifying Engineer

Adapted from the author's testimony before the Subcommittee on Energy, this article reviews the GSA project, which includes analysis of the solar energy heating and cooling system as well as broader aspects of energy conservation in office buildings. The author then discusses energy conservation and solar energy utilization for commercial buildings throughout the country.
•

Eccli, E.

Conservation of Energy in Existing Housing

Feb., 1974
Alternative Sources of Energy: No. 13, pp. 5-13

Provides do-it-yourself information for implementing energy conservation techniques in existing structures. Includes directions on sealing, insulating windows, stopping air leaks, insulation, hot air circulation, obtaining solar heat through windows, house additions, and costs and benefits. Bibliography.
•

Educational Facilities Lab

The Economy of Energy Conservation in Educational Facilities

1973
Available: Educational Facilities Lab
477 Madison Ave.
New York, NY 10022
$2.00

continued

ENERGY CONSERVATION/ PASSIVE SOLAR

The Economy of Energy Conservation in Educational Facilities (continued)

A summary of energy conservation methods which can be used in educational facilities, both old buildings and new construction.
•

Kern, K.

The Owner-Built Home

1970
Specialty Printing Co.
Yellow Springs, OH
Available: Ken Kern
Ken Kern Drafting
Sierra Route
Oakhurst, CA 93644
$7.50

Complete manual on a variety of low-cost building methods gathered from the United States and around the world. The end of each chapter features a good bibliography for those wishing to do further research into the design of their own home, such as choosing a site and making best use of the prevailing climate. Complete and detailed information on natural heating and cooling systems.
•

Kinzey, B.Y., Jr./ Sharp, H.M.

Environmental Technologies in Architecture

1963
Prentice-Hall International, Inc.
Englewood Cliffs, NJ 07632

"A practical up-to-date treatment of architectural design and building equipment." Includes a detailed section on thermal atmosphere and environmental control, tables and instructions for calculating heat loss/gain from insulation, shading methods, cooling techniques, various heating and cooling systems, and a limited discussion of an adaptation for the future energy sources such as solar energy.
•

Malloy, J.F.

Thermal Insulation

1969
Van Nostrand Reinhold Co.
450 W. 33rd St.
New York, NY 10001

Continued on the next page.

ENERGY CONSERVATION/ PASSIVE SOLAR

Thermal Insulation (continued)

A large, complete "encyclopedic" volume on insulation. Fundamental heat transfer physics theory is discussed and the total package of thermal insulation applications, including: vapor weather barriers, economic thicknesses, inspection and contracting considerations, equipment-pipes, lines, and related factors.

•

Office of Emergency Preparedness

The Potential for Energy Conservation- A Staff Study

1972
Available: GPO
$2.00

A detailed report of methods for all facets of energy conservation. Includes excellent tables applicable to heating and cooling of buildings.

•

Olgyay, G.

Design with Climate: Bioclimatic Approach to Architectural Regionalism

1969
Princeton University Press
Princeton, NJ 08540

"New insights into the relationship of climatic environment to housing." Excellent illustrations and photographs depicting various architectural forms which emphasize climatic adaptations. Detailed instructions and tables on the applications of theories presented.

•

Peavy, B.A./ Burch, D.M./ Powell, F.J.

Thermal Performance of a Four Bedroom Wood-Frame Townhouse

1974
Available: National Bureau of Standards
Washington, DC 20234

Measurements of the dynamic heat transfer in a four bedroom townhouse were made under controlled conditions in a large environmental chamber to explore the validity of a computer program developed at NBS, labelled NBSLD, for predicting heating and cooling loads. This study is a part of a broader research program supported by both the Department of Housing and Urban Development and the National Bureau of Standards to improve performance test procedures and criteria for housing.

•

ENERGY CONSERVATION/ PASSIVE SOLAR

Roberts, R.

Your Engineered House

1964
M. Evans & Co., Inc.
216 E. 49th St.
New York, NY 10017
$4.95 (paperback)

One of the best all-around books on how to design a house tailored to your own desires. Excellent sections on how to orient your house with the sun and take advantage of its heating and cooling properties.

•

Stein, R.G.

Architecture and Energy

1973
Architectural Forum: Vol. 139, No. 1, pp. 38-58

Discusses how architects could use various design techniques which would greatly increase energy conservation. Some of the topics covered are energy and high-rise buildings, lighting systems, and the building envelope.

•

Theobald, R.

Habit and Habitat

1972
Prentice-Hall, Inc.
Englewood Cliffs, NJ 07632
$8.95

Analysis of methods of making conservation economically feasible, even profitable, for industry and government. The need for a systemic approach that looks beyond immediate objectives in a society in which everyone must think for himself is demonstrated. One of the rare "ecology" books that provides real understanding of our predicament.

•

Watson, D.

Energy Conservation in Architecture- Part I: Adapting Design to Climate, Part II: Alternative Energy Sources

1974
Connecticut Architect
Available: EARS

Continued on the next page.

ENERGY CONSERVATION/ PASSIVE SOLAR

Energy Conservation in Architecture (continued)

Both sections are well illustrated by reference to experimental house designs. Part I reviews the subject of climatology and the traditional and new designs that architects can use to conserve energy by relatively simple and natural building devices. Part II focusses on alternative energy sources, with particular attention to solar energy applications.
•

SOLAR R&D

Australian Academy of Science

Solar Energy Research in Australia- Australia Academy of Science Report No. 17

Sept., 1973
Australian Academy of Science
Canberra, Australia
$1.80

An ad hoc committee of the Australian Academy of Sciences comprised of individuals from a variety of disciplines was established in 1972 to determine whether or not a more extensive R&D program should be initiated in Australia. This report provides the conclusions and recommendations of the committee.
•

Donovan, P./ et al

An Assessment of Solar Energy as a National Energy Resource

1972
University of Maryland
Dept. of Mechanical Engineering
College Park, MD
Available: NTIS (PB-221 659/6)
$2.75

Areas assessed: all applications of direct solar energy, as well as wind power, ocean thermal difference, and energy from renewable organic materials. Panel concludes that if solar development programs are successful, building heating for public use is possible within 5 years, building cooling in 6 to 10 years, synthetic fuels from organic materials in 5 to 8 years, and electricity production in 10 to 15 years. Sponsored by NASA and NSF. 85 pages.
•

SOLAR R&D

Federal Energy Administration

Project Independence Blueprint: Final Task Force Report-Solar Energy

Nov., 1974
Prepared by the Interagency Task Force on Solar Energy under the direction of the National Science Foundation
Available: GPO

Large volume replete with charts, graphs, and diagrams, this report covers all aspects of solar energy conversion and recommends incentives for its use as well as a research and development thrust costing $1 billion over a five year period.
•

National Academy of Sciences

Solar Energy in Developing Countries: Perspectives and Prospects

1972
Available: NTIS (PB-208 550)
$4.50 (paper) $.95 (microfiche)

This is a report of an ad hoc advisory panel, made up of specialists from the U.S. and abroad to 1) assess the state of the art in utilizing solar energy for developing countries and review current practical applications; 2) identify promising areas for research and development; and 3) examine the desirability of establishing an international solar energy institute in North Africa to carry out solar R&D.
•

Yeager, P.B.

Solar Energy Research-A Multidisciplinary Approach

1972
U.S House of Representatives
Washington, D.C.
Committee on Science and Astronautics-Committee Print Serial Z-Staff Report

Report comprised of responses to requests by committee for information on research presently underway in areas of solar energy. Covers current status and future potential of solar energy research. Includes: 1) extent of need; 2) who is doing R&D; 3) level of effort being exerted; 4) technological, economic, political, and social obstacles; and 5) degree we should rely on solar energy. 119 pages.
•

SOLAR INSOLATION MEASUREMENT

Abbott, C.G.

Solar Variation and Weather: A Summary of Evidence, Completely Illustrated and Documented

Oct. 12, 1963
Smithsonian Miscellaneous Collections: Vol. 146, No. 13

Details solar-constant observations taken from 1924-1944 on mountain tops around the world.

•

Bennett, I.

Monthly Maps of Mean Daily Insolation for the U.S.

1965
Solar Energy: Vol. 9, No. 3, pp. 145-158

Maps showing monthly means of the daily totals of solar radiation. Based on data for the period 1950-1962 from 59 stations of the U.S. Weather Bureau plus estimated results based on cloudiness determinations on another 113.

•

Elterman, L.

Atmospheric Attenuation Model, 1964, in the Ultraviolet, Visible and Infrared Regions for Altitudes to 50 Km

1967
Air Force Camb. Research Labs
Environmental Research Paper No. 46

Refinement of past calculations done for a standard clear sky containing resonable amounts of dust, cloudiness, and water vapor. Data from sounding rockets and other modern methods yeild tables of scattering and attenuation coefficients as a function of altitude and wavelength.

•

Hotchkiss, C.H.B./ Strock, C.

The Degree Day Handbook

1937
The Industrial Press
200 Madison Ave.
New York, NY 10016

Temperature reference data for winter and summer air conditioning operation and design calculations. Includes the practical application of degree-day tables for checking heating plant efficiency and predicting or estimating fuel or steam consumption. Gives actual examples from various buildings and offers solutions to operating problems.

•

SOLAR INSOLATION MEASUREMENT

Liu, B.Y.H./ Jordon, R.C.

The Interrelationship and Characteristic Distribution of Direct, Diffuse, and Total Solar Radiation

1960
Solar Energy: Vol. 4, No. 3, pp. 1-19

Describes method for calculating hourly insolation values on inclined surfaces. A later work, "The Long Term Average Performance of Flat-Plate Solar Energy Collectors," Solar Energy: Vol. 7, No. 2, pp. 53-57, 1963, is also a good reference.

•

Löf, G.O.G./ Duffie, J.A./ Smith, O.

World Distribution of Solar Radiation

1966
Solar Energy: No. 10, pp. 27-37

Extensive report that gives a comprehensive record of solar radiation by months for the world.

•

Swartman, R.K./ et al

Effects of Pollution on Solar Energy

1971
American Society of Mechanical Engineers: ASME PAP 71-WA/SOL-12
Available: OALS (#293)

A review of solar radiation received by the earth, followed by discussion on attenuation of solar radiation by pollutants.

•

MEDIUM TEMPERATURE COLLECTORS

de Winter, Francis

Solar Energy and the Flat-Plate Collector: An Annotated Bibliography
(see major bibliographies and inventories for annotation)

•

MEDIUM TEMPERATURE COLLECTORS

Jordon, R.C. (ed.)

Low Temperature Engineering Application of Solar Energy

1967
Available: American Society of Heating, Refrigeration and Air Conditioning Engineers
345 East 47th St.
New York, NY 10017
$9.00

A booklet of six technical articles covering most aspects of flat-plate collector design and performance as well as engineering information on design and performance of solar water heaters. Excellent reference.

•

Lior, N./ Saunders, A.P.

Solar Collector Performance Studies

June, 1973- Revised August, 1974
The University of Pennsylvania
The National Center for Energy Management and Power
Philadelphia, PA 19174

Comprehensive evaluation of 24 types of flat plate collector configurations was performed. Twenty-one of these were actually built and tested in the Univ. of Penn. Solar Collector Test Facility. A dual-glass flat black collector served as the test reference. In addition to the insolation collection tests, separate experiments were conducted to determine heat losses. The experimental facilities used are described.

•

Vernon, R.W./ Simon, F.F.

Flat-Plate Collector Performance Determined Experimentally with a Solar Simulator

August, 1974
NASA Lewis Research Center
Cleveland, OH 44135
NASA-TMX-71602

Presented at the ISES U.S. Section annual meeting in August '74, this paper provides results of a collector technology program at NASA being conducted at Lewis to provide a basis for selecting collectors for use at the NASA Langley Research Center. Performance data on five collectors tested up to August 1974 are presented along with a description of the solar simulator at the experimental facility. NASA's facility is set up to be a national test center for flat-plate collectors. Anyone wishing to have a collector evaluated should contact the Lewis center.

•

HIGH TEMPERATURE COLLECTORS: SEE ALSO SOLAR POWER GENERATION

OCEAN THERMAL GRADIENT

Anderson, J.H., Jr.

Economic Power and Water from Solar Energy

1972
American Society of Engineers
ASME PAP 72-WA/SOL-2
Available: OALS (#202)

Electric power, fresh water, and food can be produced by solar energy without expensive collectors. The ocean is the collector and reservoir. Cost is competitive with today's power cost, and there is no environmental pollution.

•

Carnegie-Mellon University

Solar Sea Power Plant Conference and Workshop, June 27-28, 1973

1973
Carnegie-Mellon University
Schenley Park
Pittsburgh, PA 15213

Papers and conclusions of workshops on various aspects of solar sea power plants, ranging from "Some Views of NSF/RANN on Solar Sea Power Plants" to "Cold Water Pipe Design and Large-Scale Sea Water Piping." Sponsored by NSF/RANN, Division of Advanced Technology Applications.

•

Metz, W.D.

Ocean Temperature Gradients: Solar Power from the Sea

1973
Science: Vol. 180, No. 4092, pp. 1266-1267

Power plants utilizing ocean thermal gradients are expected to cost very little more than conventional power plants. Could also produce fresh water through vaporization and condensation of warm water used in the plant and produce hydrogen and oxygen through electrolysis. Solar power from the sea may well turn out to be a source of clean energy long overlooked.

•

OCEAN THERMAL GRADIENT

Othmer, D.F. / Rcels, O.A.

Power, Fresh Water, and Food from Cold, Deep Sea Water

1973
Science: Vol. 182, No. 4108, pp. 121-125

The stored heat of the ocean can be utilized to produce electric power as well as vast amounts of fresh water. Aquaculture using deep ocean waters is also feasible.
•

Zener, C.

Solar Sea Power

January, 1973
Physics Today: Vol. 26, No. 1, pp. 48-53

Describes heat engines operating in the tropical oceans, capitalizing on the temperature differential between upper and lower levels to provide a source of economical, pollution free electricity.
•

BIOCONVERSION

Biomass Energy Institute

Proceedings: International Biomass Energy Conference, May 13-15, 1973

1973
Available: Biomass Energy Institute
P.O. Box 129, Postal Station C
Winnipeg, Manitoba R3M 3S7
Canada

$10.00

Contains approximately 20 papers related to the subject of bioconversion. "Processing Organic Solids by Anaerobic Fermentation" and "Production of Methane Gas from Manure," two of the papers, include useful figures, instructions, and diagrams.
•

Grout, A.R.

Methane Gas Generation from Manure

1973
Available: Department of Agriculture, Commonwealth of Pennsylvania
2301 N. Cameron St., Harrisburg, PA 17120

Booklet describes methane generators built for the 1973 Pennsylvania Agricultural Progress Days at Hershey.
•

BIOCONVERSION

Lindsay, E.F.

Methane from Waste . . . How Much Power Can It Supply?

December, 1974
Popular Science: Vol 205, No. 6, pp. 52-55, 120-121

Discusses the feasibility of utilizing methane on a large scale for commercial purposes. Points out the problems connected with such technology and provides an overview of recent research and development.
•

New Alchemy West

Methane Digesters Background and Biology Using the Gas and Sludge Designs for Simple Working Models

for further information see:
section 1 - information dissemination (publish lay periodicals)
•

Singh, R.B.

Bio-Gas Plant Designs with Specifics
also Bio-Gas Plant Generating Methane from Organic Wastes

for further information see:
section 1 - information dissemination (bioconversion)
•

University of Massachusetts

Proceedings of Bioconversion Energy Conference, June 25th and 26th, 1973

1973
University of Massachusetts
Institute for Man and His Environment
Amherst, MA 01002

Provides summary statements provided by each principal speaker as well as the transcripts of the question and answer periods. Researchers in bioconversion and repesentatives of government and industries such as meat-packing, utilities, and waste handling were in attendance at the meeting.
•

WIND CONVERSION

Anonymous

Project No. 12: Windmills and Wind Power

Continued on next page.

WIND CONVERSION

Project No. 12: Windmills and Wind Power (continued)

April, 1973
Elements of Technology

Good article which provides brief background and historical information on windmills, a description of a Wind Works unit, and information regarding the power and efficiency that can be obtained from windmills.
•

Electrical Research Association

Publications List

Electrical Research Association
Cleeve Road
Leatherhead, Surrey, England

Offers list of approximately 40 entries on wind conversion. Good reference.
•

Golding, E.W.

The Generation of Electricity by Wind Power

1956
Philosophical Library, Inc.
15 East 40th Street
New York, NY

Out of print, but probably one of the best works on wind power written to date. Includes tables, illustrations, and descriptions of applications. Recommended.
•

Huetter, V.

Operating Experience Obtained With a 100 kw Wind Power Plant

1964
Available: NTIS (N73-29008/2)
$3.50

Report on experiments and experience associated with a wind power plant, covering design data and those aspects which were decisive in its layout and the type of regulating provisions used, as well as results of the detailed operational tests. The regulating system and the automatic cut-in provisions permit the system to be connected automatically to the public power supply network on the basis of a specific program. The various types of wind conditions considered in designing the power plant are described.
•

WIND CONVERSION

Kelinhenz, F.

Utilization of Wind Power by Means of Elevated Wind Power Plants

December, 1947
Available: NTIS (N73-23011)

Exploration of wind power by power plants at high altitudes is considered. A design of a wind power plant is proposed, and its efficiency and economy in the framework of present conditions in Germany is demonstrated. Although costs are higher than a coal fired steam plant, they compare favorably with hydroelectric power plants, and the saving of coal is a great advantage.
•

Kidd, S./ Garr, D.

Can We Harness Pollution-Free Electric Power from Windmills?

November, 1972
Popular Science: Vol. 201, No. 5, pp. 70-72

General overview of wind power for the layman.
•

NASA

Energy: A Special Bibliography with Indexes

(see major bibliographies and inventories for annotation)
•

NTIS

Solar Energy and Wind Power

(see major bibliographies and inventories for annotation)
•

Putnam, P.C.

Power from the Wind

1948
Von Nostrand Reinhold Company
450 W. 33rd Street
New York, NY 10001
$9.95 (hardcover)

Relates story of the installation of the Smith-Putnam wind turbine, a 175 ft, 1250 kilowatt experimental unit on Grandpa's Knob near Rutland, Vermont.
•

WIND CONVERSION

Savino, J.M. (ed.)

Wind Energy Conversion Systems, Workshop Proceedings

December, 1973
NASA, Lewis Research Center
Cleveland, OH
Available: NTIS (PB 231 341)
$6.50 (paper) $1.45 (microfiche)

Includes: Important Past Developments; Wind Characteristics and Siting Problems; Rotor Characteristics; Energy Conversion Systems; Energy Storage; Small Wind Power Systems; Wind Systems for Large-Scale Applications. Good discussion of working problems.

•

Solar Wind Company

Electric Power from the Wind

1973
Solar Wind Company
P.O. Box 7
East Holden, ME 04429

Detailed description of a wind system the company has had in operation for over a year. Illustrates basic operation of this home electrical system. Tables and charts are included by which you can determine you own wind system requirements. Bibliography.

•

UNESCO

Proceedings of the New Delhi Symposium on Wind and Solar Energy

1956
Available: UNESCO
c/o Department of State
Washington, D.C. 20520

Several papers ranging from research in Israel to utilization of wind power in arid lands.

•

United Nations

Proceedings of the 1961 Rome Conference on New Sources of Energy, Vol. 7, Wind Power

Continued

WIND CONVERSION

Proceedings of the 1961 Rome Conference on New Sources of Energy, Vol. 7, Wind Power

1964
Sales Section
United Nations
United Nations Plaza
New York, NY 10017
$3.50

Over 40 papers on various aspects of wind power authored by experts from all over the world.

•

Wind Works

Wind Bibliography

Wind Works
Box 329, Route 3
Mukwonago, WI 53149
$3.00

Excellent reference list with over 500 entries.

•

OTHER PUBLICATIONS OF INTEREST CONCERNING WIND CONVERSION (not annotated):

Coonley, D.R.
Design with Wind
1974
Total Environmental Action
Box 47
Harrisville, NH 03450

Hackleman, M.A./ House, D.W.
Wind and Windspinners
1974
Earthmind
Gosel · Saugus
California 91350

Merriam, M.F.
Is There a Place for the Windmill in the Less Developed Countries?
Technology and Development, East-West Center
1777 East-West Road
Honolulu, HI 96822

Reynolds, J.
Windmills and Watermills
1970
Praeger Publishers
P.O. Box 1323
Springfield, MA 01101

Thomas, P.H.
A Survey: Electric Power from the Wind
March, 1945
Federal Power Commission

WIND CONVERSION

Villecco, M.
Wind Power
May/June, 1974
Architecture Plus, pp. 66-74
•

PHOTOVOLTAICS

Berman, P.A.

Design of Solar Cells for Terrestrial Use

July-December, 1967
Solar Energy: Vol. 11, No. 3-4, pp. 180-185

Discusses considerations in design of solar cells for terrestrial systems. Mentions developments up to that date and factors affecting solar cell performance.
•

California Institute of Technology, Jet Propulsion Lab

Photovoltaic Conversion of Solar Energy for Terrestrial Applications, Workshop Proceedings, Vol. I, Working Group and Panel Reports; Vol. II, Invited Papers

1973
California Institute of Technology
Jet Propulsion Lab
Pasadena, CA

Vol. I covers introductory remarks by NSF, working group summaries and discussions, and panel discussions. Vol. II encompasses five sessions of technical presentations and discussions. Attended by 135 participants from manufacturing, marketing, user fields, and technical community, the conference purpose was to encourage exchange of information.
•

Chalmers, B./ et al

Continuous Silicon Solar Cells, Progress Report Covering Period October 1, 1973, to December 31, 1973

1974
Harvard University, Division of Engineering and Applied Physics
Tyco Laboratories, Inc
Harvard University, Cambridge, MA   Tyco Laboratories, Waltham, MA

Describes research on application of the EFG crystal growth technique to the continous growth of single silicon ribbons from the melt. The results of the ribbon growth experiments and of solar cell fabrications are presented.

PHOTOVOLTAICS

Defense Documentation Center

Solar Cells and Solar Panels, Vol. I, Report Bibliography, Jan., 1958-Oct., 1969

1970
Available: NTIS (AD-700 500)
$6.00 (paper) $.95 (microfiche)

An annotated bibliography is provided of documents in which performance characteristics of various solar cells, particularly types containing gallium arsenides, silicon, or cadmium sulfides, are evaluated. Other reports include solar-cell fabrication, development of solar-cell power systems generating higher electrical power levels, in-flight solar-cell degradation studies, and systems for orienting solar panels continuously toward the sun. 111 pages.
•

Free, J.R.

Solar Cells - When Will You Plug Into Electricity from Sunshine?

December, 1974
Popular Science, Vol. 205, No. 6, pp. 52-55, 120-121

Informative state-of-the-art overview. Discusses work currently being done in the field and insight into future prospects.
•

International Rectifier

Solar Cells and Photocells

1971
Available: International Rectifier
Semiconductor Division
233 Kansas Street
El Segundo, CA 90245
$2.00 (paperback)

A theoretical discussion coupled with practical considerations and illustrated data on the use of semi-conductor photovoltaic and photoconductive devices. 98 pages.
•

Lindmayer, J./ Allison, J.F.

The Violet Cell: An Improved Silicon Solar Cell

Spring, 1973
Constat Technical Review: Vol. 3, No. 1, pp. 1-22

Good technical discussion of the state-of-the-art of photovoltaics.
•

STORAGE

Close, D.J.

Rock Pile Thermal Storage for Comfort Air Conditioning

1964
Mechanical and Chemical Engineering Transactions of the Institution of Engineers
Melbourne, Australia
No. 1875

Heating and cooling air conditioning cycles using solar heaters and evaporative coolers respectively are described, together with examples to illustrate design methods based on present knowledge.
•

Löf, G.O.G.

Unsteady-State Heat Transfer Between Air and Loose Solids

June, 1948
Industrial and Engineering Chemistry: Vol. 40, No. 6, pp. 1061-1070

Investigation of design data in the form of unsteady-state transfer coefficients from air to a bed of granite gravel are presented.
•

FUEL CELLS

Crowe, B.J.

Fuel Cells-A Survey

1973
Computer Science Corporation *and* Technology Utilization Office
Falls Church, VA NASA
Washington, D.C.
Available: GPO

Attempt to discuss fuel cells factually and to describe their current status. Discusses the history and basic principles involved in the operation of the fuel cell and present and possible future applications.
•

ALL TECHNOLOGIES

Brand, Stewart (ed.)

Whole Earth Epilog

1974
Available: P.O. Box 99554
San Francisco, CA 94109
$4.00

Source book for tools, hardware, and information on a wide variety of subjects, including alternative energy sources. Has several listings on where to access solar equipment as well as a list of references. Includes index to *Last Whole Earth Catalog*.
•

ALL TECHNOLOGIES: SEE ALSO GENERAL INFORMATION

Davis, A.J./ Schubert, R.P.

Alternative Energy Sources in Building Design

October, 1974
Available: Alternative Energy Sources in Building Design
P.O. Box 499
Blacksburg, VA 24060
$5.00 postpaid

Provides information regarding basic design criteria for alternative natural energy sources. Subjects covered include energy conservation, natural ventilation, solar energy, wind energy, biological processes for waste treatment, and energy production and integration as primary determinants in building morphology.
•

Gage, T./ Merrill, R./ Missar, C./ Robert, J.

Energy Primer: Solar, Water, Wind, and Biofuels

December, 1974
Available: Energy Primer
558 Santa Cruz Ave.
Menlo Park, CA 94025
$4.50 (U.S.) $5.50 (elsewhere) $2.00 (microfiche-mailed anywhere)

A cooperative effort of the Whole Truck Store, New Alchemy West, Ecology Action/ Palo Alto, and Alternative Sources of Energy Newsletter, with the publishing help of Portola Institute. Aimed at architects, experimenters, students, and concerned citizens wanting to know more about alternatives to the depletion of fossil fuel reserves and the dangerous nuclear fuel options. Includes: reasonably technical articles; extensive book and hardware reviews (including where to order most ones); central source guide for access to additional information and resources; excursions into architecture and energy conservation; how solar, wind, and biofuels are being integrated to provide for energy needs.
•

Shuttleworth, J.

The Mother Earth News Handbook of Homemade Power

May, 1974
The Mother Earth News, Inc.
P.O. Box 70
Hendersonville, NC 28739
Available: EARS
$1.95 (paperback)

Compilation of articles from the Mother Earth News that deal with converting wind, sun, water, wood, and waste into energy. For the do-it-yourselfer.
•

ALL TECHNOLOGIES

Stoner, C. (ed.)

Producing Your Own Power- How to Make Nature's Energy Sources Work for You

September, 1974
Rodale Press, Inc.
Book Division
Emmaus, PA 18049

Excellent reference for the do-it-yourselfer. This book is a compilation of articles and plans for utilizing natural energy sources such as wind, solar, wood, and methane.

•

SOLAR PONDS

Dickinson, W.C./ Clark, A.F./ et al

The Shallow Solar Pond Energy Conversion System

1974
Lawrence Livermore Laboratory
Livermore, CA 94550

One of the papers at the August 1974, meeting of the U.S. Section of ISES. This paper presents the concept of a shallow solar pond energy conversion system as a cost effective way to produce large-scale electric power from solar energy. The use of these ponds to provide process hot water, up to the boiling point, for industrial and commercial purposes is also discussed.

•

Tabor, H.

Solar Ponds

June, 1968
Science Journal: pp. 66-71

Very good article on solar ponds, bodies of water in which convection currents are prevented by density zoning, using brine solution. Author relates problems associated with the technique and reports results of his experimental work in Israel.

•

SOLAR COOKING

Swet, C.J.

A Prototype Solar Kitchen

1973
American Society of Mechanical Engineers
ASME PAP 72-WA/SOL-4
Available: OALS (#294)

This prototype can provide high grade thermal energy for a variety of household uses. The heat is available indoors, in the evening, during periods of intermittent cloudiness and high winds without manual positioning to track the sun. Unit is designed for cooking and basic needs of a small family, but can be scaled up for larger requirements. Commonly available materials and components are used throughout; no unusual skills required.

•

Tabor, H.

A Solar Cooker for Developing Countries

1966
Solar Energy: Vol. 10, No. 4, pp. 153-157

A durable solar cooker designed specifically for fabrication in centralized workshops in developing countries is described. Over 500 watts can be delivered in bright sunlight. Cost including mirrors is under $8. (1966 prices)

•

Telkes, M.

Solar Cooking Ovens

January, 1959
Solar Energy: pp. 1-11

An excellent introduction to the history and basic design principles of solar cookers and ovens. The author favors the use of low-cost plane mirrors in the three designs presented. She also advocates utilization of such ovens in countries where waste products used as fuel could be used as fertilizer. Includes bibliography.

•

SOLAR FURNACES

Baum, W.A.

Basic Optimal Considerations in the Choice of a Design of a Solar Furnace

July-October, 1958
Solar Energy: Vol. 2, No. 3-4, pp. 37-45

Continued on next page.

SOLAR FURNACES

Basic Optimal Considerations in the Choice of a Design of a Solar Furnace (con't)

Examines simple relationships existing between the solar furnace aperature, target size, angle of convergence, maximum attainable concentration ratio, over-all efficiency, and amount of spill light surrounding target. The author's system and another two-mirror system are evaluated.

•

Davies, J.M./ Cotton, E.S.

Design of the Quartermaster Solar Furnace

April-July, 1957
Solar Energy: Vol. 1, No. 2-3, pp. 16-22

Discusses furnace built by the U.S. Army to produce radiation flux sufficiently high to destroy materials and burn protected skin. The concentrating element consists of spherical mirrors arranged so all images are superimposed at target.

•

Trombe, F./ et al

First Results Obtained from 1000 kw Solar Furnace

May, 1973
Solar Energy: pp. 63-66

Initial experimental data on solar furnace built by the National Center of Scientific Research in Odeillo, France.

•

DISTILLATION

Eibling, J.A./ Falbert, S.G./ Löf, G.O.G.

Solar Stills for Community Use-Digest of Technology

1971
Solar Energy: Vol. 13, No. 2, pp. 263-276

The many factors that influence the productivity of solar stills are discussed in three categories: atmosheric variables, design features, and operational techniques. Discusses productivity and desalted water costs.

•

Frick, G./ Hirschmann, J.

Theory and Experience with Solar Stills in Chile

Continued

DISTILLATION

Theory and Experience with Solar Stills in Chile (continued)

1973
Solar Energy: Vol. 14, No. 4, pp. 405-413

Discusses theoretical factors as well as operating results. Designs with evaporating trays of wood, metal, cement, and plastic and evaporating cloths of various designs were tested. Tests were made with glass and plastic covers.

•

Hay, H.R.

Plastic Solar Stills: Past, Present, and Future

1973
Solar Energy: Vol. 14, No. 4, pp. 393-404

A review of technical and patent literature on plastic solar still elements. Covers and explains successes as well as failures. Details properties of different plastics.

•

Talbert, S.C./ Eibling, J.A./ et al

Solar Distillation of Saline Water

1970
Battelle Memorial Institute
505 King Avenue
Columbus, OH 43201

A report prepared for the Office of Saline Water Conversion, U.S. Department of the Interior, which describes the construction and operation of most of the solar stills that have been built around the world. One section of the report was presented at the Solar Energy Society Conference in Melbourne (1970); 270 pages.

•

Tleimat, B.W./ Howe, E.

Nocturnal Production of Solar Distillers

April-June, 1966
Solar Energy: Vol. 10, No. 2, pp. 61-66

Production of deep-basin solar still with longitudinal baffles is compared with a tilted tray still. Results: in deep basin type evaporation continues the entire 24 hour period; tilted tray type ceases production shortly after sunset.

•

SOLAR ENGINES

Anonymous

The Engine That Runs on Sunshine

April, 1974
Popular Science: Vol. 204, No. 4, pp. 87, 146

Concise description of the Banks heat engine which can run and generate electricity with an eleven degree temperature difference. This engine can operate on solar heat, low temperature geothermal heat, or other low temperature heat sources. Key to engine is material used for wire loops - 55-Nitinol (nickel titanium alloy).
•

Thermo Electron Corporation

Organic Rankine Cycle System for Power Generation from Low Temperature Heat Sources

September 10, 1973
Thermo Electron Corporation
85 First Avenue
Waltham, MA 02154
For further information, contact: Jerry P. Davis
Thermo Electron Corporation
(617) 890-8700

Report on development of Rankine Cycle systems at the Thermo Electron Corporation which utilize organic fluids which have the requisite thermodynamic and physical properties for application to closed cycle power generation. The feasibility of utilizing such an engine for solar energy conversion is discussed briefly.
•

WATER HEATING

Chinnery, D.N.W.

Solar Hot Water Heating in South Africa

1971
CSIR Research Report 248
CSIR, P.O. Box 395
Pretoria, South Africa

The relative performance of nine different absorber designs are compared on the basis of a three week performance testing program, and the influence of efficiency of such factors as climatic variations, insulation, transparent cover, height of storage tank, point of connection of flow-pipe inlet and forced circulation is discussed. Full details are given for the construction of a solar water heater and on methods of combining it with other water heating systems. Other practical aspects are mentioned and attention is drawn to certain by-laws and electrical safety regulations.
•

WATER HEATING

Commonwealth Scientific and Industrial Research Organization (CSIRO)

Solar Water Heaters: Principles of Design, Construction, and Installation, Circular 2.

1964
CSIRO
Division of Mechanical Engineering
Melbourne, Australia
Available: OALS (#221)

Provides explanation on how a flat plate type solar heater works, how to install one, and what properties are required of its components. Solar water heaters are used extensively in Australia at a reasonable cost. Note: Collector orientation in Australia is north rather than south because it is located in the southern hemisphere.
•

Vincze, S.A.

A High-Speed Cylindrical Solar Water Heater

November, 1973
Solar Energy: pp. 339-344

Presents principles and formulae used in designing a cylindrical type of solar water heater.
•

MANUALS FOR DO-IT-YOURSELF SOLAR HOT WATER HEATING:

Brace Research Institute
How to Build a Solar Water Heater - Leaflet No. L-4
1973
Brace Research Institute
MacDonald College of McGill University
Montreal, Quebec, Canada

Brook, F.A.
Solar Energy and Its Use for Heating Water in California
1936
University of California
Agriculture Experiment Station Bulletin

deWinter, F.
How to Design and Build a Solar Swimming Pool Heater
1974
Copper Development Association
New York, NY 10017

WATER HEATING

MANUALS FOR DO-IT-YOURSELF SOLAR HOT WATER HEATING (continued):

Hawkins, H.M.
Domestic Solar Water Heating in Florida
1947
University of Florida
Engineering and Industrial Experiment Station Bulletin 18

Morse, R.N.
Solar Water Heaters for Domestic and Farm Use
1956
CSIRO
Melbourne, Australia

Sinson, D.A.
How to Build a Solar Water Heater
1965
Do-it-yourself Leaflet
Brace Experimental Station
St. James, Barbados

Whillier, A.
How to Heat Your Swimming Pool Using Solar Energy
1965
Do-it-yourself Leaflet No. 3
Brace Experimental Station
St. James, Barbados

SPACE HEATING

Anderson, B.

Solar Energy and Shelter Design

1973
Available: Bruce Anderson
Total Environmental Action
Box 47
Harrisville, NH 03450
$7.00

Comprehensive work on solar heating. Includes charts and graphs necessary to make all calculations to solar heat your shelter: site, orientation, heat loss and heat gain, integrated systems, separate collectors. Also includes description of several solar houses. Contains both general and technical information. This was written as a masters thesis in architecture and has been used as a text on the subject. Good bibliography.
•

SPACE HEATING

Anonymous

Prize Winning Designs for Solar Heated Residences

May, 1958
Air Conditioning, Heating, and Ventilation: Vol. 55, No. 5, pp. 76-77

Shows the top three prize winners of the 1957 International Architecture Competition conducted by the Association for Applied Solar Energy. Also enumerates the scope and design requirements of the competition. Sixty entries were selected for presentation in "Living with the Sun," published by the Association.
•

Barber, E.M./ Watson, D.

Criteria for the Preliminary Design of Solar Heated Buildings

1974
Available: Sunworks, Inc.
669 Boston Post Road
Guilford, CT 06437
$10.00

Describes the applications of solar energy for buildings (heating, hot water, etc.), components of a solar heating system, building design and construction, and factors such as financing, zoning, building codes, and taxes.
•

Bridgers, F.H./ Paxton, D.D./ Haines, R.

Performance of a Solar Heated Office Building

November, 1957
Heating, Piping, and Air Conditioning: Vol. 29, No. 11, pp. 165-170

Discusses successful operation of one of the first solar heated office buildings in the world, located in Albuquerque, NM. A description of the system is provided as well as information on its operation.
•

Curtis, E.J.W.

Solar Energy Applications in Architecture

1974
The Polytechnic of North London Medical Architecture Research Unit
Department of Environmental Design
Holloway, London N78DB, England
£1.55

Continued on next page.

SPACE HEATING

Solar Energy Applications in Architecture (continued)

Review of four solar projects located in Great Britain, including homes and swimming pools. Also provides a report on a research project being conducted by the Polytechnic of North London with water cooled flat plate collectors.

•

Holland, R.

Cheap Solar Heat

February, 1974
Alternative Sources of Energy: No. 13, pp. 22-23

A do-it-yourself article which explains how the author built a collector to heat a one-room addition to his home. Cost-$300; time-12 man days.

•

Löf, G.O.G./ Tybout, R.A.

Cost of House Heating with Solar Energy

1973
Solar Energy: Vol. 14, pp. 253-278

A computer analysis of projected solar heat costs and optimization of heating systems in eight U.S. cities under particular climatic, geographic, and resident characteristics. Heating costs as a function of various design and location factors are provided as well as the relative importance of each factor. Solar heating costs vs. conventional heating costs are provided.

•

SPACE COOLING

Chung, R./ Löf, G.O.G./ Duffie, J.A.

A Study of Solar Air Conditioners

August, 1963
Mechanical Engineering: Vol. 85, No. 8, pp. 31-35

Demonstrates the technical feasibility of solar cooling as a result of experiments done at the Solar Energy Laboratory at the University of Wisconsin. The system evaluated utilizes lithium bromide and water with flat plate collectors.

•

Farber, E.A./ Flanigan, F.M.

Operation and Performance of the University of Florida Solar Air Conditioning System

Continued

SPACE COOLING

Operation and Performance of the University of Florida Solar Air Conditioning System

April-June, 1966
Solar Energy: Vol. 10, No. 2, pp. 91-95

Operation and performance of the continuous ammonia and water absorption refrigeration system at the University of Florida. The system utilizes flat plate solar absorbers and has produced 2.4 tons of refrigeration (with peak loads of 3.7) over most of the day.

•

HEATING AND COOLING

ASHRAE

Solar Energy's Role in Heating and Cooling of Buildings

September, 1974
ASHRAE Journal: Vol. 16, No. 9

Special issue featuring several articles and news briefs. Some include: "Applying Solar Energy for Cooling and Heating Institutional Buildings" by Frank H. Bridgers, "Solar Energy Storage" by Dr. Maria Telkes, "ASHRAE Journal article leads to Georgia Solar Water Heater Experiment" by Franklin M. Cloud, "Solar Energy Will Cool and Heat Atlanta School" by Richard T. Duncan, and "Special Report on Solar Energy by NPSE for ASHRAE."

•

Denton, J.

Integrated Solar Powered Climate Conditioning Systems

July, 1974
Available: National Center for Energy Management and Power
University of Pennsylvania
Philadelphia, PA 19174
(limited supply)

This Semi-Annual Progress Report covers the period of January 1 through June 30, 1974, for work on integrated solar powered space conditioning systems. Performance comparisons were made between direct solar heating, solar powered vapor compression and gas absorption heat pumps, electric resistance heating, and combustion furnace heating; seasonal resource energy consumption for a Philadelphia single-family residence was used as a measure of comparison. A first draft of Interim Standards for Solar Collectors was also prepared. The attitudes of prospective purchasers toward using solar heating in their homes is surveyed. Financial institutions were polled to determine whether they would grant additional loans on buildings equipped with solar heating systems in view of expected operating cost savings. Government agencies were contacted to elicit plans for encouraging such loans.

•

HEATING AND COOLING

General Electric/ et al

Solar Heating and Cooling of Buildings (Phase 0) Reports

1974
General Electric, Valley Forge, PA
TRW Systems Group, Redondo, CA
Westinghouse Electric Corp., Baltimore, MD
Available: RANN Document Center
National Science Foundation
1800 G. Street, N.W.
Washington, D.C.

Reports which summarize the results of studies of the feasibility of using solar energy for heating and cooling buildings. Different reports deal with various aspects, including technical, economic, social, environmental, and institutional factors. Studies were conducted with the support of the National Science Foundation.
•

Hay, H./ Yellott, J.I.

A Naturally Air-Conditioned Building

January, 1970
Mechanical Engineering: Vol. 92, No. 1, pp. 19-25

Air conditioning system used in home in Phoenix, Arizona, is described. System reported entails a movable insulated roof which permits water bags on the roof to absorb and retain solar heat during the winter and dissipate heat to the night sky in the summer. Room temperatures were kept between 68-82°F throughout a normal year.
•

Löf, G.O.G./ Ward, D.S.

Solar Heating and Cooling: Untapped Energy Put to Use

September, 1973
Civil Engineering: Vol. 43, No. 9, pp. 88-92

Use of solar energy for heating and cooling has become increasingly economically feasible. Recent applications discussed and illustrated.
•

Moorcraft, C.

Solar Energy in Housing

Continued

HEATING AND COOLING

Solar Energy in Housing (continued)

October, 1973
Architectural Design: pp. 634-643; 652-666

Stresses the importance of bioclimatic conditions, particularly solar radiation in formulating the best design for a house. The physiological basis of comfort and its relevance to solar energy and housing is discussed. Examples of methods by which the outer skin of buildings can modulate heat in-out flows are provided, including: the Trombe-Michel solar wall, and solar architecture in Phoenix, Arizona, and Atascadero, California. Solar water heating and distillation are also mentioned. Very informative.
•

National Bureau of Standards

Interim Performance Criteria for Solar Heating/Cooling Systems and Dwellings, 2 volumes

November 15, 1974 (draft); January, 1975 (final)
Available: Solar Energy Demonstration Program
Center for Building Technology
National Bureau of Standards
Building 226
Washington, D.C. 20234

Prepared for HUD under the Solar Heating and Cooling Demonstration Act of 1974. These criteria are intended for use in the design, technical evaluation, and procurement of heating, cooling, and domestic hot water systems to be used in the heating/cooling demonstration project.
•

Thomason, H.E.

Solar House Plans

1972
Available: Edmund Scientific Co.
150 Edscorp Building
Barrington, NJ

$10.00

Contains Thomason's basic solar house plans, a system which utilizes water run over corrugated aluminum collector surface, stored in a tank surrounded by rock. Air is then forced over the hot rocks and circulated to heat the home. Discusses the advantages of the house, in which 50% or more of required heat is supplied by solar energy.
•

HEATING AND COOLING

Yellott, J.I./ MacPhee, C.W.

Solar Energy Utilization for Heating and Cooling

1974
National Science Foundation
Research Applied to National Needs (RANN)
Washington, D.C.
Available: GPO
$.70

Reprint of Chapter 59 of the 1974 ASHRAE Handbook and Product Directory, Applications Volume. An excellent report on the design and basic principles of "how to" utilize solar energy for heating and cooling.
•

SOLAR POWER GENERATION: SEE ALSO OCEAN THERMAL GRADIENT

Colorado State University/ et al

Solar Thermal Electric Power Systems: Annual Progress Report Ending December 31, 1973

January, 1974
Colorado State University
Fort Collins, CO 80521
Westinghouse Electric Corporation
Georesearch Lab
Boulder, CO 80303
Report No. NSF/RANN/SE/G-I-37815/PR/73/4
Available: NTIS (PB 231 115)
$6.75 (paper copy) $1.45 (microfiche)
•

Institute of Energy Conversion/University of Delaware

Direct Solar Energy Conversion for Large-Scale Terrestrial Use, Annual Progress Report (covering the period January 1, 1973, to December 31, 1973)

January, 1974
Institute of Energy Conversion
University of Delaware
70 South Chapel Street
Newark, DE 19711

Progress is reported on all aspects of the work on Cu2S/CdS solar cells and their application to direct solar energy conversion for large-scale terrestrial use. Research is sponsored by NSF/RANN.
•

SOLAR POWER GENERATION

Meinel, A.B./ Meinel, M.P.

Thermal Conversion Offers Potential in Solar Energy Projection

1973
Aware: No. 33, pp. 7-10
Available: OALS (#270)

The dream of large-scale use of solar energy for electric power supply for communities has recently become feasible. Diagrams of this thermal storage and conversion unit discussed. Has potential of assisting utilities meet growing power demands.
•

Seraphin, B.O.

Research Applied to Solar Thermal Power Systems - Chemical Vapor Desposition Research for Fabrication of Solar Energy Convertors, Annual Progress Report (Covering the period January 1, 1973, to December 31, 1973)

January, 1974
Optical Sciences Center
University of Arizona
Tucson, AZ 85421

Discusses research on a new approach to a selective solar energy converter that can be used to transform solar radiation into high temperature heat. This heat can be transferred and applied in a steam turbine-generator unit to produce electricity.
•

Sparrow, E.M./ Ramsey, J.W./ Wehner, G.K.

Research Applied to Solar-Thermal Power Systems, Semi-Annual Progress Report No. 3 (Covering the period July 1, 1973, to December 31, 1973)

January, 1974
University of Minnesota
Honeywell Systems and Research Center
Minneapolis, MN

One of a series of reports which illustrate the problems and solutions in the area of concentrating collectors. Materials and optimization problems in a cylindrical parabolic concentrator operating at 300°C are considered in detail. Of special interest are results of weathering tests on reflector materials and corrosion tests on receiver/heat pipe materials.
•

Sutton, G.W.

Direct Energy Conversion

Continued on next page.

SOLAR POWER GENERATION

Direct Energy Conversion (continued)

1966
McGraw-Hill
330 W. 42nd Street
New York, NY 10036

Details principles of direct energy conversion, devices for applying these principles, and the future prospects of such applications. Photovoltaics, fuel cells, thermoelectric power generation, magnetohydrodynamic power generation and thermionic energy conversion are covered.
•

Tabor, H./ Zeimer, H.

Low Cost Focussing Collector for Solar Power Units

April-June, 1962
Solar Energy: Vol. 6, No. 2, pp. 55-59

This focussing collector consists of an inflated cylinder of plastic film, 12 m long and 1.5 m in diameter. The portion exposed to the sun is clear to permit radiation to enter and fall upon rear segment, which is aluminized to act as a mirror. This system is demonstrated as being superior to a paraboloidal concentrator for overall focussing and ease, and low expense of building. Cost: $20/sq. m. Collection efficiency: 40%.
•

Stickley, R.A./ et al

Solar Power Array for the Concentration of Energy (S.P.A.C.E. on Earth) Semi-Annual Progress Report

July, 1974
Sheldahl, Inc.
Advanced Products Division
Northfield, MN 55057

Prepared in conjunction with Foster Wheeler Corp., the University of Minnesota, and Northern States Power Company. The objectives of the project study are to (a) estimate technical performance and economics, and (b) analyze and test metallized thin film reflective materials. The most significant results include: (a) development of method for predicting the theoretical limits of performance of ideal heliostat arrays in terms of a continuum field modeled in closed form, (b) definition and analysis of a turntable heliostat concept utilizing multiple edge-mounted reflectors, (c) analysis of heliostat command and control requirements, (d) selection of metallized thin film candidates and initiation of environmental exposure tests, and (e) parametric design and performance predictions.
•

SOLAR POWER GENERATION

see: NSF/RANN Abstracts and Energy Abstracts for Policy Analysis for information on several reports on solar thermal research being conducted by a variety of universities and private companies.

PROCEEDINGS

Allen, R. (ed.)

Proceedings of the Solar Heating and Cooling for Buildings Workshop, Washington, D.C., March 21-23, 1973. Part I: Technical Sessions, March 21 and 22, 1973.

1973
University of Maryland
Department of Mechanical Engineering
College Park, MD
Available: NTIS (PB-223 536/4)
$3.00 (paper) $1.45 (microfiche)

The proceedings contain thirty-six technical papers on solar energy for U.S. building applications areas; namely, solar collectors, energy storage, domestic hot water, heating, energy conservation and insulation, solar air-conditioning, and systems for solar heating and cooling. Some foreign activities are also reviewed. Each technical paper is a report on: proposed research, on-going research, proposed systems, or operating systems. Questions and answers from discussion periods are included, as is an agenda and list of attendees.
•

ASHRAE

Solar Energy Applications

1974
Available: Manager
Publication Sales Department
ASHRAE
United Engineering Center
345 E. 47th Street
New York, NY 10017
$5.00 (members) $10.00 (non-members)

Bulletin compiled by ASHRAE. Contains five papers presented at the Society's Annual Meeting in Montreal. Authors address themselves to the problems of heating and cooling institutional and commercial office buildings, the mechanics of solar energy storage, the use of solar insolation data and the status of the National Science Foundation's research activities in the solar area.
•

PROCEEDINGS

ASHRAE/University of Virginia

NSF/RANN Workshop on Solar Heating and Cooling of Buildings, Washington, D.C., June 18-19, 1974

1974
Contact: Ms. Arlene Spadafino
Technical Secretary
ASHRAE
345 E 47th Street
New York, NY 10017

Sponsored by the University of Virginia and ASHRAE with support from the National Science Foundation. As yet unpublished.
•

ASME

Proceedings, 9th Intersociety Energy Conversion Engineering Conference

1974
American Society of Mechanical Engineers
345 East 47th Street
New York, NY 10017
$50.

1343 pages comprised of over 150 papers given at the conference. Several of these papers (20) dealt with solar power. The following topics are included: photovoltaics, solar thermal power, heating and cooling of buildings, wind power, collectors, photochemical hydrogen production, and an integration of solar and wind power to support a small residence.
•

Biotechnic Press

Life Support Technics - Conference Proceedings - Ghost Ranch - Albuquerque, N.M.

1972
Biotechnic Press
P.O. Box 26091
Albuquerque, NM 87125

Includes discussion by Fred Hopman on an air-cooled flat plate collector system with rock storage, "Passive Energy Systems" by Buck Rogers, "Wind Energy" by Robert Reines, "Heating and Cooling by Controlled Radiation Balance" by Harold Hay, and more - including a presentation on Silva Mind Control.
•

PROCEEDINGS

Carnegie-Mellon University

Solar Sea Power Plant Conference and Workshop, Carnegie-Mellon University, Pittsburgh, PA

1973

see ocean thermal gradient section for annotation
•

Carpenter, E.F. (ed.)

International Conference on the Use of Solar Energy - The Scientific Basis

1958
University of Arizona Press
Box 3398
College Station
Tucson, AZ 85700
Available: OALS (#215)

Held in 1955, this conference was sponsored by the University of Arizona, The Association of Applied Solar Energy, and the Stanford Research Institute. Topics covered include: The Available Energy; Measurement of the Radiation; Thermal Process; Photochemical Processes; Electrical Processes. 5 volumes.
•

Eldridge, F. (ed.)

Japanese/United States Symposium on Solar Energy Systems - Volume II, Summary of Technical Presentations, Washington, D.C., June 3-5, 1974

1974
Available: The MITRE Corporation
1820 Dolley Madison Boulevard
McLean, VA 22101
(MTR-6284)

Summary of proceedings of conference held at the National Science Foundation, organized by the MITRE Corporation. Papers provide an overview of state-of-the art in Japan and U.S. Volume I, a summary of the final session of the conference, is also available.
•

George, K. (ed.)

Proceedings: Solar Heating, Cooling, and Energy Conservation Conference, Denver, Colorado, May 1-3, 1974

Continued on next page.

PROCEEDINGS

Proceedings: Solar Heating, Cooling and Energy Conservation Conference, May 1-3, 1974 (continued)

Environmental Action of Colorado
University of Colorado at Denver
1100 14th Street
Denver, CO 80202

Geared to an emerging solar industry, this conference hosted over 50 speakers and panelists from a wide variety of disciplines. Papers and workshops address aspects of solar energy ranging from the role of various government agencies to the problems of installing a solar energy system. To be published in early 1975. For further information, please send your name and address.

•

Institute of Electrical and Electronics Engineers(IEEE)/ et al

Solar Energy Lecture Series

1974
Sponsored by IEEE, Washington Academy of Sciences, and Washington Society of Engineers
Available: Solar Energy, Mail Stop 3608
c/o Westinghouse ATL
P.O. Box 1521
Baltimore, MD 21203
Single lecture: $3.00 (member) $5.00 (non-member)
Complete set of 8: $15.00 (member) $25.00 (non-member)

Lectures:
1) "Solar Energy as an Alternative Energy Resource" by William R. Cherry
2) "Photovoltaic Solar Power Systems" by E.L. Ralph
3) "Space Satellite Power Systems" by Peter Glaser
4) "Electric Power for Space Satellites" by Charles M. McKenzie
5) "Solar Photothermal Power Conversion" by Dr. Aden Meinel
6) "Oceanic and Atmospheric Energy Sources" by William E. Heronemous
7) "The National Solar Energy Program" by Frederick H. Morse
8) "The Energy Plantation" by Clinton Kemp

•

International Solar Energy Society

Annual Meeting, U.S. Section, International Solar Energy Society, Colorado State University, Ft. Collins, CO

August 21-23, 1974

Over 100 papers dealing primarily with technical aspects of solar energy. No compiled proceedings exist, but in all probability many of the papers will be published in future issues of Solar Energy, which is distributed to ISES members. Abstracts of presentations were handed out in booklet form at the conference. Excellent source of current information.

•

PROCEEDINGS

Rocky Mountain Science Council

Proceedings of Energy Research in the Rocky Mountain States, Albuquerque, NM, February 14 and 15, 1974

1974
Available: Technology Application Center
University of New Mexico
Albuquerque, NM 87131
$2.00

Sponsored by the Rocky Mountain Science Council, this symposium contains the texts of major addresses and summaries of the technical papers with cross reference indexes by programs, institutions, and individuals in the areas of conservation, economics, ecology, fossil fuels, geothermal studies, hydrogen, nuclear, solar and wind energies.

•

United Nations

Proceedings of the United Nations Conference on New Sources of Energy, Rome, August, 1961

Available: Sales Section
United Nations
United Nations Plaza
New York, NY 10017

Volume 1:	General Sessions	$2.50
Volume 2:	Geothermal I	$5.00
Volume 3:	Geothermal II	$5.50
Volume 4:	Solar Energy I	$7.50
Volume 5:	Solar Energy II	$4.50
Volume 6:	Solar Energy III	$5.50
Volume 7:	Wind Power	$3.50

Large quantity of information, generally very good.

•

United Nations

The Sun in the Service of Mankind

1973
Available: The Science Sector
UNESCO House
9 Place de Fontenoy
Paris Ville, France

Vast amount of information on all aspects of solar energy. Excellent reference.

•

MAJOR BIBLIOGRAPHIES AND INVENTORIES

deWinter, F.

Solar Energy and the Flat Plate Collector: An Annotated Bibliography

1974
Not yet published; will probably be available through the National Science Foundation and the Government Printing Office in late 1974/early 1975.

The literature on the flat plate collector covers a number of decades and involves many journals and books. To date there seems to be no single source in which the significance of the literature is discussed in some detail. This annotated bibliography constitutes an attempt to do this for the flat plate collector itself, and for the solar input quantities which must be used for performance calculations. The possible uses of flat plate collectors are not discussed in detail.
•

Alternative Sources of Energy Lending Library

Route 2, Box 90A
Milaca, MN 56353

A service established in 1973 whereby individuals can obtain practical and/or hard-to-get information on alternative sources of energy.

A listing of publications and their rental fees can be found in ASE's October, 1974, special access issue. Items are loaned for a period of three weeks.
•

Beecher, M. (librarian)

Solar Energy Collection

Room 257
Science/Engineering Reference Service
Arizona State University Library
Tempe, AZ 85281
(602) 965-7608

The nucleus of this large, computerized collection is the former research collection of the International Solar Energy Society, a gift to ASU. Included are books, journals, government documents, microforms, patents, extracts, reprints, pictures, and correspondence dealing with all phases of solar energy and its applications.

This is a full-time project which is updated occasionally. Materials are made available for on-site reference, through inter-library loan, or reproduction services for a fee.
•

MAJOR BIBLIOGRAPHIES AND INVENTORIES

Jensen, J.S.

Applied Solar Energy Research

1959
Stanford Research Institute
Menlo Park, CA

Review of research and development activities as of 1959, with thorough review of solar energy literature (2916 references). Excellent reference.
•

NASA

Energy: A Special Bibliography with Indexes <u>and</u> Quarterly Supplement

1974
Available: NTIS (NASA SP-7042, NASA SP-7043)
\$6.00 (initial volume)
\$4.00 (for quarterly supplement) <u>or</u>
\$15.00/year

Literature survey of special energy and energy related topics. Lists 1708 reports, articles, and other documents. Subject categories are as follows: energy systems; primary energy sources; secondary energy transport, transmission, and distribution; and energy storage. Solar listings are primarily in the field of photovoltaics and wind.
•

NTIS

Solar Energy and Wind Power
A Bibliography with Abstracts

February, 1974
NTIS (NTIS-WIN-74-016)
\$20.00

Prepared by the NTIS computer retrieval system, this bibliography contains 74 selected abstracts of research reports. Included are abstracts of reports on solar heating for buildings, solar electric power generation, solar cells, solar energy as a national energy resource, and a special section on wind power generation. Available on paper or microfiche.
•

MAJOR BIBLIOGRAPHIES AND INVENTORIES

Stanley, K. (ed.)

Special Retrospective Bibliography: Solar Energy. Arid land Abstracts, No. 6

June, 1974
Available: OALS
$3.50

This special bibliography contains 114 listings and includes short abstracts on each entry.

•

Subcommittee on Energy, Committee on Science and Astronautics

Inventory of Current Energy Research and Development, Vol. I-III

1974
Volume I: 5270-02174 $9.80
Volume II: 5270-02173 $3.10
Volume III: 5270-02174 $5.00
Available: GPO

Compiled by the staff of the Oak Ridge National Laboratory with support from the National Science Foundation/Research Applied to National Needs, this inventory provides information about more than 4,900 energy research projects underway in the United States at the present time. Projects are arranged according to the subject matter of the research, and each project has a description which provides project title, research institution, investigator(s), funding organization, duration, amount of funding by year, brief summary of the research, list of pertinent publications, and location of project.

•

REDE/R.I.S.D. Energy Library Listing

January, 1974
Contact: Professor Gerald Howes
Division of Architectural Studies
Rhode Island School of Design
2 College Street
Providence, RI 02903

Compiled by students at Rhode Island School of Design, this bibliography lists over 300 publications of REDE and R.I.S.D. that are filed at the School of Design and are available for reproduction. Indexed by major subjects (including solar) and divided into subtopics.

•

MAJOR BIBLIOGRAPHIES AND INVENTORIES

Technology Application Center

Solar Thermal Energy Utilization

December, 1974
Available: Technology Application Center
University of New Mexico
Albuquerque, NM 87131
(505) 277-3622
$37.50 (2 volumes)

Bibliography with abstracts and indexes on solar thermal energy which goes back to 1957. The initial volumes cover through 1974 and will be followed by quarterly updates. Contains subject, KWIC title, author, and corporate source indexes. Copies of the documents cited/abstracted are generally available from TAC. 1600 pages, 2100 references with abstracts.

•

Solar Energy and Agriculture: Selected Bibliographies

June 5, 1974
Congressional Record - Senate
S9755-S9760

Entered into the record by Senator McGovern, this offers the publication list of Brace Research Institute of McGill University, Quebec, Canada. Areas covered include: reprints, technical reports, miscellaneous reports, masters' theses, doctoral theses, course papers, do-it-yourself leaflets, and engineering projects.

for further information see:
section 1 - solar research (all solar technologies)
McGill University MacDonald College, Brace Research Institute

•

ON-GOING INFORMATION/ABSTRACTING AND INDEXING SERVICES

Alternative Sources of Energy

Route 2, Box 90A
Milaca, MN 56353

$5.00/year

A.S.E. is a magazine/newsletter aimed at individuals interested in the development of alternative sources of energy so that society will become more decentralized. Alternative, environmentally sound technologies in all basic fields are stressed. A grass-roots publication that provides information for the do-it-yourselfer and a communications network for all those interested in alternative energy.

•

ON-GOING INFORMATION/ABSTRACTING AND INDEXING SERVICES

Cansdale, J.H. (ed.)

ASHRAE Journal

American Society of Heating, Refrigerating, and Air-conditioning Engineers
345 E. 47th Street
New York, NY
(212) 752-6800
$9/year (members) $15/year (non-members)

One of the best on-going sources of information on solar energy and energy conservation.

•

COMPLES

Cooperation Mediterraneenne pour L'Energie Solaire, Bulletin

1961
COMPLES
32 Cours Pierre-Puget
13006 Marseilles, France

Largely in French, this publication serves as a technical journal for COMPLES members.

•

Energy Digest

Scope Publications
1067 National Press Building
Washington, D.C. 20004
(202) 347-9288
$138/year

Available by subscription only, this is an excellent newsletter which provides the latest news on energy--patents, activities in government, reports on industry, scheduled meetings, etc. Usually has information on solar.

•

Engineering Index, Inc.

Engineering Index/Energy Abstracts

Available: The Engineering Index, Inc.
United Engineering Center
345 E. 49th Street
New York, NY 10017
(212) 752-6800

Continued

ON-GOING INFORMATION/ABSTRACTING AND INDEXING SERVICES

Engineering Index, Inc. (continued)

$187.00 (entire series - 9 monthly issues)
$18.50 - Energy Sources Abstracts
$93.50 - Energy Production, Transmission, and Distribution
$37.50 - Energy Utilization Abstracts
$14.50 - Energy Conservation Abstracts
$37.50 - Energy Conversion Abstracts

Excellent reference service which often includes publications on solar energy.

•

Energy Info

Robert Morey Associates
P.O. Box 98
Dana Point, CA 92629
$30.00/year (add $5.00 for overseas airmail delivery)

Monthly newsletter which reports current meetings, progress in energy R&D, and other newsworthy items. Often features info on solar.

•

Energy Today

Trends Publishing, Inc.
National Press Building
Washington, D.C. 20004
$85/year

Bi-monthly publication in loose-leaf form which reports on all aspects of energy. Often features news from Washington on solar energy. Delves into new projects, day-to-day activities, Federal agencies, technical and economic information.

•

Environment Information Center, Inc.

The Energy Index

1974
Environment Information Center, Inc.
124 East 39th Street
New York, NY 10016
Contact: Christopher Kimball
(212) 679-0810
$50.00

Continued on next page.

ON-GOING INFORMATION/ABSTRACTING AND INDEXING SERVICES

Environment Information Center, Inc. (continued)

This reference book has 2,800 abstracts of energy documents and 15,000 cross-referenced citations of technical papers, articles, patents, etc. Also included are world-wide production statistics, energy supply and demand projections, consumption patterns, and analysis of reserves. 524 pages.

•

Guthrie, M.P. (ed.)

NSF/RANN Energy Abstracts - A Monthly Abstract Journal of Energy Research

1973-June, 1974
Available: NTIS or GPO

Sponsored by NSF/RANN, this monthly was published by Oak Ridge National Laboratory. Computer generated indexes were prepared twice a year. Purpose of bibliography is to disseminate results of NSF/RANN research and that of others as far as possible. In general, subjects covered are energy sources, electric power, and energy consumption. Publications cited include technical articles, popular and semi-technical articles, reports, symposium papers, and proceedings. Back issues are available from NTIS at approximately $3.00 per monthly issue.

•

Guthrie M./ Thompson L.
Oak Ridge National Laboratory and USAEC Technical Information Center

Energy Abstracts for Policy Analysis (EAPA)

November, 1974 - on-going

EAPA is replacing NSF/RANN Abstracts, the last issue of which was dated June 1974. EAPA provides abstracting and indexing coverage of selected publicly available non-technological literature contributing to energy-related analysis and evaluation in the following areas: policy; conservation; research and development studies; economics; supply and demand; forecasting; systems studies; and environmental effects. Specific fields of energy covered are: energy sources, including fossil fuels, nuclear fuels, hydrogen and synthetic fuels, and hydroelectric; unconventional energy sources, including solar, wind, geothermal, tidal, and waste products; electric power; energy conversion and storage; and energy consumption, including residential, commercial, industrial, agricultural, and transportation sectors and intersectorial studies; and efficient energy utilization in these sectors. Supported by the U.S. Atomic Energy Commission, National Science Foundation, and the Federal Energy Administration.

•

ON-GOING INFORMATION/ABSTRACTING AND INDEXING SERVICES

Heliotechnology (Geliotechnika)/ Applied Solar Energy

From 1965
Allerton Press
150 Fifth Avenue
New York, NY 10011
also
NTIS
$110.00/annual subscription(6 issues)
$45.00/single issue

A Russian language journal available in English. Excellent source of technical information.

•

Howe, E. (ed.)

Solar News and Views

3112 Ptarmigan Drive, #1
Walnut Creek, CA 94595

Newsletter of U.S. Section of International Solar Energy Society. Available through membership.

•

International Solar Energy Society

ISES News

P.O. Box 52
Parkville, Victoria, Australia 3052

Newsletter provided to members of ISES.

•

International Solar Energy Society

Solar Energy
The Journal of Solar Energy Science and Technology

Since 1957
International Solar Energy Society
c/o Smithsonian Radiation Laboratory
12441 Parklawn Drive
Rockville, MD 20852

Members of ISES receive this publication as part of their membership. (See ISES in section 1.) For subscription without membership, inquire to the Subscription

Continued on next page.

ON-GOING INFORMATION/ABSTRACTING AND INDEXING SERVICES

International Solar Energy Society (continued)

Manager, Headington Hill Hall, Oxford OX3 OBW, England. Current subscriptions and back files are on microfilm and are available from Microforms International Marketing Corporation, Inc., 380 Saw Mill River Road, Elmsford, New York, 10523. Excellent source of technical information on all aspects of solar energy.
•

King, L. (ed.)

Weekly Energy Report

1238 National Press Building
Washington, D.C. 20004
(202) 638-4260
$3.00/year

Newsletter emphasizing the latest energy information from Washington as well as Canada and Europe. Solar information often included.
•

Korte, R. (ed.)

Heating/Piping/Air Conditioning

Available: Circulation Department
Heating/Piping/Air Conditioning
25 Sullivan Street
Westwood, NJ 07675
(201) 262-3030
$9/year (U.S. and Canada)
$19/year (other)
$2.00 (single copy)

Articles, news, and reports for the mechanical systems engineer. Good reference for solar systems components and small parts related to HVAC systems. HPAC also publishes HPAC Info-dex-Mechanical Systems Information Index Issue, which is a directory of manufacturers, products, manufacturers product information, engineering societies and associations, sources for code requirements, engineering reference books and manuals, and recently published HPAC articles.
•

NTIS

Weekly Government Abstracts - Energy

$35.00

Weekly newsletter which provides titles and abstracts of the most recent publications on the subject of energy.
•

ON-GOING INFORMATION/ABSTRACTING AND INDEXING SERVICES

Pulford, A. (ed.)

Sun at Work in Britain

Available: United Kingdom Section of International Solar Energy Society
c/o The Royal Institution
21 Albemarle Street
London, W 1, England
£1.00 (single issue)

Recently initiated magazine which offers articles, news items, and references for a growing section of ISES.
•

Shuttleworth, J. (ed.)

The Mother Earth News

P.O. Box 70
Hendersonville, NC 28739
(704) 692-4256
$1.50 (single issue) $8.00 (one year) $14.00 (two years)

Published every two months, this magazine is aimed at individuals interested in a lifestyle in which environmentally sound practices, simple country living, and doing-it-yourself are emphasized. Written in quasi-folksy style, Mother offers instructions for basic survival skills, including how to utilize solar energy.
•

Solar Energy Digest

Available: William B. Edmondson
Solar Energy Digest
P.O. Box 17776
San Diego, CA 92117
$27.50/year $15.00/6 months

Excellent report on current direct and indirect solar energy conversion work.
•

Solar Energy Research and Information Center

Solar Energy Industry Report
Solar Energy Washington Newsletter

Available: Solar Energy Research and Information Center
1001 Connecticut Avenue, N.W.
Washington, D.C. 20036
$75/each (24 issues/year)

Newsletters covering current events in the solar energy field.
for further information see:
section 1 - information dissemination (all solar technologies)
•

ON-GOING INFORMATION/ABSTRACTING AND INDEXING SERVICES

The following are abstracting and indexing services which often contain information on publications about solar energy:

Applied Science and Technology
(1913-)
The H.W. Wilson Company
New York, NY

Electrical and Electronic Abstracts
Institute of Electrical and Electronics Engineers, Inc.
345 E. 47th St.
New York, NY 10017
$216

International Aerospace Abstracts
(1961-)
AIAA Technical Information Service
750 Third Avenue
New York, NY 10017
$110

Pandex Current Index to Scientific and Technical Literature
(1969-)
MacMillian Information
866 Third Ave.
New York, NY 10022
$360

Physics Abstracts
(1898-)
American Institute of Physics
335 E. 45th St.
New York, NY 10017
$380

Scientific and Technical Aerospace Reports (STAR)
(1963-)
NASA
Scientific and Technical Information Division
Washington, D.C. 20546
$54

a

ABSORP	absorption
ADV	advanced
AG	agriculture
AIA	American Institute of Architects
AK	Alaska
AL	Alabama
ALT	alternative
AL2	aluminum
APPLC	application
ARCH	architecture
ASHRAE	American Society of Heating Refrigeration & Airconditioning Engineers
ASSOC	association
ATMO	atmospheric
AUX	auxiliary
AV	audio visual
AVAIL	available
AZ	Arizona

b

BLDG	building(s)

c

CA	California
CHEM	chemical
CLG	college
CNCL	council
CO	Colorado
CO	company
COLL	collector
COMBO	combination
COMM	commercial
CONCEN	concentrate (or)
CONF	conference
CONSLT	consultant
CONSV	conservation
CRC	circle
CT	Connecticut
CTR	center
CU	copper

d

DC	District of Columbia
DE	Delaware
DEG	degree
DEMO	demonstration
DEPT	department
DEV	develop (ment)
DOC	document
DRWGS	drawings

e

EFF	efficiency
ELECT	electric
ENGG	engineer
ENGR	engineer (ing)
ENGY	energy
EQUIP	equipment
EST	estimate
EVAL	evaluate
EXP	experiment (al)

f

F	degrees Fahrenheit (e.g., 20F)
FL	Florida
FT	foot

g

GAL	gallon
GEN	generate (tion)
GENL	general

h

HI	Hawaii
H&C	heating and cooling
H2O	water

i

IA	Iowa
ID	Idaho
IL	Illinois
IN	Indiana
INC	incorporated
INCL	including
INFO	information
INST	institute
IRRAD	irradiation

j

JOUR	journal

k

KS	Kansas
KY	Kentucky

l

LA	Louisiana
LGR	large
LIBR	lithium bromide

m

MA	Massachusetts
MD	Maryland
ME	Maine
MED	medium
MFG	manufacture
MGT	management
MI	Michigan
MN	Minnesota
MO	Missouri
MPH	miles per hour
MS	Mississippi
MT	Montana

n

NAT	natural
NATL	national
NC	North Carolina
ND	North Dakota
NE	Nebraska
NH	New Hampshire
NJ	New Jersey
NM	New Mexico
NSF	National Science Foundation
NV	Nevada
NY	New York

o

OCC	occasionally
OH	Ohio
OK	Oklahoma
OR	Oregon

p

PA	Pennsylvania
PE	Professional Engineer
POT	potential
POTL	potential
PR	Puerto Rico
PROB	problem
PROD	produce, production
PROG	program
PROJ	project
PUB	publish
PWR	power

q

QUANT	quantity

r

RAD	radiation
R&D	research and development
RANN	Research Applied to National Needs
REFNG	refining
REP	representative
REQ	require (d) requirement (s)
RES	residential
RESCH	research
RI	Rhode Island

s

SC	South Carolina
SCI	science
SD	South Dakota
SERV	service
SOC	society
SOLR	solar
SQ	square
STRG	storage
SUPLMT	supplementary
SYS	system

t

TECH	technology
TEMP	temperature
TN	Tennessee
TNK	tank
TX	Texas

u

UT	Utah
UNIV	university
UTIL	utilization

v

VA	Virginia
VI	Virgin Islands
VOL	volume
VT	Vermont

w

WA	Washington
WI	Wisconsin
WV	West Virginia
WY	Wyoming

x

X__

y

YR	year
YRS	years

z

Z__

SOLAR/ENERGY CONSERV DISSEMINATION ORG

(BIOCONVERSION)

Rodale Press, Inc.
Organic Farming and Gardening
33 East Minor Street
Emmaus, PA 18049

Franklin Brooks
Subscription Manager

Excellent magazine for the homesteader. Contains info on how to get the most from your garden, composting, and alternative energy sources. Published monthly. Subscriptions for U.S., possessions, and Canada: $6.85/yr, $12.25/2 yrs, $16.95/3 yrs (foreign add $1.00/yr).

(INORGANIC SEMI-CONDUCTORS)

Gowen Solar Cells
Box 5337
Beverly Hills, CA 90210

Jack J. Perlmuth
Marketing Consultant
-Phone-
213 2788714

Gowen solar cells consist of thin film of silicon deposited on 10"x10" Al_2 substrates. According to Hammond E.T. Gowen, inventor, cells achieve 70% efficiency and cost $20.73/Kw.

(TOTAL SYSTEMS INTEGRATION)

Upland Hills Ecological Awareness Center
481 Lake George Road
Oxford, MI 48051

Educational facility in farm setting which will demonstrate the use of solar energy, wind power, and methane production. Goal is to show how man is dependent on nature and his relationship with the environment.

(PROVIDE INFORMATION SEARCHES TO CLIENTS)

Solar-con
Division of Space-Optics Corp.
1888 Century Park East
Suite 105
Century City, CA 90067

Al Ottum
-Phone-
213-2778884
213-8792992

Design and application of all forms of solar energy

SOLAR/ENERGY CONSERV INFO DISEMMINATION ORG

(LOBBY TO PROMOTE SOLAR LEGISLATION)

1975 Solar Energy Year
8705 Higate Road
Alexandria, VA 22308

Mrs. Stewart Homme

SOLAR COMPONENTS MANUFACTURER

(MEDIUM TEMPERATURE SOLAR COLLECTOR SYSTEMS)

Alcan Booth Industries, Ltd.
Banbury, Oxfordshire
England

Aluminum sheeting

Nuclear Technology
P.O. Box 1
Amston, CT 06231
-Phone-
202-5372387

Listed in Section 1 under Research, (Medium Temperature Solar Collector Systems). Now mfg. and markets corrosion inhibitor for fresh water systems. Inhibitor, NT-869, can be used for copper, AL_2, etc.

Tye Udylite Company
3628 E. Olympic Blvd.
Los Angeles, CA 90023
-Phone-
213-2624101

Chemical processes for blackening iron and steel

(INORGANIC SEMI-CONDUCTORS)
(HIGH TEMPERATURE SOLAR COLLECTOR SYSTEMS)

Bausch and Lomb
Vacuum Coating Division
10 Champeney Terrace
Rochester, NY 14602

David Hoagland
Technical Salesman
-Phone-
716-2321181

Metal cold mirrors in variety of shapes, filters, and solar cell cover coatings

---

SOLAR COMPONENTS MANUFACTURER

(SOLAR HEATING)

Zia Associates
P.O. Box 1466
Boulder, CO 80302

Lee Salmon
Sales Representative
-Phone-
303-4499170

Mfg. solid state differential controls for heating systems (air and H_2O flat-plate type).

(SOLAR SPACE HEATING AND COOLING)

Bell & Gossett (ITT)
8200 North Austin Ave.
Morton Grove, IL 60053

Hydronic specialists. Produce booster pumps, heat exchangers, hot water heaters, expansion tanks, controls and relief valves

---

Dilectrix Corp.
69 Allen Blvd.
Farmingdale, NY 11735
-Phone-
516-2497800

Pressure sensitive tapes and films; teflon tapes and films

Solar Control Products Corp.
217 California St
Newton, MA 02158

Solar films

Southwestern Reflective Film Co.
7341 E. Sixth Ave.
Scottsdale, AZ 85252

Mfg. reflective films

Standard Solar Control
5321 E. Washington
Phoenix, AZ

Mylar sheets for windows

SOLAR COMPONENTS MANUFACTURER

(SOLAR SPACE HEATING AND COOLING)

Sun-X
FABCO
809 E. 18th St.
Los Angeles, CA

Reflective films

West Wind
309.5 W. Boyd Drive
Farmington, NM 87401

Geoffrey Gerhard

Manufacture small parts related to solar heating and cooling and wind conversion. All parts guaranteed 2 yrs. Voltage control switch-$40. Auto load switch-$54. Differential temp. switch-$49. Thermal power switch-$47

(STORAGE)

ESB, Inc.
Wisco Division
1222 18th St.
Racine, WI 53403
-Phone-
414-6322771

Mfg. Wisco charge retaining batteries. Have the portability of dry cells, but the capabilities and reliability of storage batteries

---

SOLAR PRODUCTS DISTRIBUTORS

(WIND CONVERSION)

Budgen & Assoc.
72 Broadview Ave.
Pointe Claire, 710
Quebec, Canada

Distribute Electro and Lubing wind generator systems

Energy Alternatives
Box 223
Leverett, MA 01054

Distribute Quirks Dunlite and Winco Windcharger systems

SOLAR PRODUCTS DISTRIBUTORS

(WIND CONVERSION)

Independent Power Developers
Box 618
Noxon, MT 59853

Distribute Quirks Dunlite wind generators; also Hutter WE-10 6 kw, 10 meter

---

Low Impact Technology
73 Molesworth St.
Wadebridge
Cornwall, England

Listed in Section 1, under Solar Components Manufacturer, (Solar Space Heating and Cooling). Also distribute Electro, Quirks Dunlite, and Winco Windcharger generator systems

---

Sencenbaugh Wind Electric
Box 11174
Palo Alto, CA 94306

Distribute Electro and Quirks Dunlite wind generators . Also have plans and list available

---

Solar Energy Company
810 18th St.,N.W.
Washington, D.C. 20006

Distribute Aerowatt and Winco Windcharger wind generator systems

---

WINDLITE
Box 43
Anchorage, AK 99510

Distribute Electro, Quirks Dunlite, and Winco Windcharger systems

---

SOLAR MANUFACTURERS

(SOLAR WATER HEATING)

Amcor Export Co., Ltd 98
Giborei Israel St.
Tel Aviv, Israel

D. Henigman
Deputy Manager

SOLAR MANUFACTURERS

(SOLAR WATER HEATING)

Bell Engineering
1802 W. Grant Rd., #114
Tucson, AZ 85705
-Phone-
602-7919005

Custom manufacturer of solar water heaters for household uses and swimming pools. No off the shelf items

---

Jetel Co.
2811 N. 24th St.
Phoenix, AZ 85008
-Phone-
602-9566190

Solar products for swimming pools and home heaters

Solar Energy Applications, Inc.
2200 E. Washington St.
Phoenix, AZ 85034

Offers "Sunergy" system, which uses flat plate collectors-price depending on the surface area of the pool to be heated. One 4x12 collector is required for each 100 ft^2 of surface area. Each such panel costs $150. plus mounting

---

Solar-Ray Appliances
P.O. Box 75
Tuart Hill, W.A.
Australia

Solar H_2O heater with built -in electric booster in 20, 40, and 80 gallon models. Flat-plate collectors for (1) high volume and industrial applications and (2) swimming pool heating. Also solar stills for water desalination

---

Vincze, Stephen A.
Paragon Chambers
Lambton Quay
Wellington, C.I.
New Zealand

Solar water heaters(10 gallon system for $140.00--1973 price)

SOLAR MANUFACTURERS

(SOLAR RADIATION MEASUREMENT)

Matrix, Inc.
537 South 31st St
Mesa, AZ 85204
-Phone-
602-8321380

Meteorology Research,Inc.
Box 637
Altadena, CA 91001
-Phone-
213-7911901

Researching and mfg. meteorological sensors and systems

(MEDIUM TEMPERATURE SOLAR COLLECTOR SYSTEMS)

Rodgers & MacDonald
3003 NE 19th Drive
Gainesville, FL 32601

Rick Rodgers
-Phone-
904-3777883

Mfg. flat-plate collectors(Al_2 Roll-Bond panels)and related equipment. Sell treated solution to inhibit corrosion--estimate 6 yrs without corrosion for Al_2 panels. Also components for integrated roof systems and collectors. Catalog available in March 1975

Shelley Radiant Ceiling Co.
8110 St. Louis Ave.
Skokie, IL 60076

William Shelley
-Phone-
312-6758899

Mfg. flat-plate collectors with Al_2 sheets and Cu tubing

Solar Heaters, Inc.
3536 W. Peoria
Phoenix, AZ 85029

Offers 4x8 and 4x10 ft. panels for $72. and $90, plus mounting. Installation manual available for $4.00

SOLAR MANUFACTURERS

(MEDIUM TEMPERATURE SOLAR COLLECTOR SYSTEMS)

U.S. Solar Pillow
416 E. Oak St.
P.O. Box 1044
Tucumcari, NM 88401

Steve Langford
-Phone-
505-4612608

Newly formed company distributing collector "pillow" which consists of one sheet of clear plastic and one sheet of black plastic. Air or water can be used as a heat transfer medium

(WIND CONVERSION)

Dempster Industries, Inc.
P.O. Box 848
Beatrice, NE 68310
-Phone-
402-2234026

Wind mills, pumps, and cylinders

Quirk's Victory Light Co., Pty. Ltd.
33 Fairweather St.
Bellevue Hills, N.S.W.
Australia

Wind generators

Real Gas & Electric Co., Inc.
P.O. Box "A"
Guerneville, CA 95446

Mr. S. Kagin

Distribute Electro, Quirks Dunlite, and Winco Windcharger wind generator systems

Sidney Williams and Co., Pty. Ltd.
Williams Parade
P.O. Box 22
Dulwich, N.S.W.
Australia 2203

Wind powered pumping systems

SOLAR MANUFACTURERS

(WIND CONVERSION)

Winco
P.O. Box 3262
Sioux City, Iowa 51102

Offers 200 watt, 12 volt D.C. wind generator units

(INORGANIC SEMI-CONDUCTORS)

Globe Union-Centralab Semiconductor Division
4501 N. Arden Dr.
El Monte, CA 91734

Pessey Semiconductors
1674 McGaw Ave.
Santa Ana, CA 92705

Manufactures solar energy convertors using silicon cells in arrays

---

Solar Systems, Inc.
8124 N. Central Park
Skokie, IL 60076
-Phone-
312-6762040

Silicon solar cells and photocells

(SOLAR COOKERS)

Devidaysl Metal Industries, Ltd.
Gupta Mills Estate
Darukhana, Bombay 10
India
-Phone-
70558

(TECHNOLOGY APPLICATION, OTHER)

Aerospace Controls Corp.
13434 S. Normandie Ave.
Gardens, CA 90249

James H. Beauchene
Operations Manager
-Phone-
213-5329755

Solar simulation systems for component testing

SOLAR MANUF, EDUCATIONAL EQUIPMENT

(HIGH TEMPERATURE SOLAR COLLECTOR SYSTEMS)

Ken Fisher Sun Systems
716 Main St.
Berlin, PA 15530
-Phone-
814-2673337

Kits for making solar mirrors

Strong Electric Corp
87 City Park Ave
Toledo, OH 43602

Offers a 14" parabolic solar furnace for educational purposes for $100.00 F.O.B. Toledo

---

SOLAR RESEARCH

(ALL SOLAR TECHNOLOGIES)

Federal Trade Commission
Bureau of Competition
Energy Study Unit
Washington, D.C. 20580

Arthur S. Weissbrodt
Attorney

Conducting an analysis of solar energy industry. Focussing on factors which might impede the entry of new firms into the field. Interested in projecting growth of industry and developing better idea of its evolving structure

---

(SOLAR HEATING)

Yanda, Bill
Rt. 1 Box 107AA
Santa Fe, NM 87501
-Phone-
505-4557550

Received $15,000 grant from the Four Corners Council to integrate low tech solar heat and greenhouses into indian villages in northern NM, utilizing indigenous materials

---

SOLAR RESEARCH

(SOLAR SPACE HEATING AND COOLING)

Dept. of Housing and Urban Development
Washington, D.C.

Mr. Thomas F. Carter, Jr.
Deputy Program Manager
Solar Heating and Cooling Demonstration Program

Overall program director--handles interagency coordination and assures implementation of agency policy decisions

---

More Combs Burch
3911 E. Exposition Ave
Denver, CO 80209

Lawrence Atkinson, AIA
-Phone-
303-7443157

Designing solar heated and cooled homes

PROJECTS

Colorado
airport

Aspen, Colorado

Description: Will utilize passive drum wall collector system

Solar System Engineer: Zomeworks
Box 712
Albuquerque, NM 87103

Contractor: John McBride
Box 1940
Aspen, CO 86111
(303) 925-2102

Colorado
retrofit

Boulder, CO

Description: Retrofit of 1200 ft^2 home

solar system: 350 ft^2 flat-plate collector area. Air cooled, using black cloth absorber and rock storage

Engineer & contractor: Earth Dynamics
P.O. Box 1175
Boulder, CO 80302
(303) 447-8168

Connecticut
housing project

Description: Forty unit housing project for the elderly which will be a prototype for future state assisted projects and a demonstration of the use of solar energy. Project planned through the Department of Community Affairs of the state of Connecticut. $131,200 NSF grant for design phase

solar system: Half of project will be solar heated, the other half heated by conventional means for performance and life cycle costing experiments

---

PROJECTS

United States
various housing types

Description: Program of the AIA Research Corporation to assist HUD and NBS respond to general objectives of the Solar Heating and Cooling Demonstration Act--to develop housing design concepts incorporating the use of solar heating/cooling systems.

Subcontractors:

Community Design Associates
Cos Cob, CT

Donald Watson, AIA
Guilford, CT

Giffels Associates
Detroit, MI

Joint Venture
Denver, CO

Massdesign
Cambridge, MA

RTL, Inc.
Paramount, CA

The Architects
Taos of Taos, NM

Total Environmental Action
Harrisville, NH

School of Architecture and Environmental Studies
University of Detroit

College of Architecture
Arizona State University

For info contact:

Nancy Hallmark
Public Relations Dept.
AIA
1735 New York Ave., N.W.
Washington, D.C. 20006
(202) 785-7264

BIBLIOGRAPHY

Kreith, F./ Kreider, J.

The Practical Design and Economics of Solar Heating and Cooling Systems

To be published June, 1975
McGraw-Hill Book Co.
330 W. 42nd St.
New York, NY 10036

Compiled for a course conducted on solar heating and cooling systems, this book emphasizes practical applications and economics. Includes many tables, charts, and illustrations. 350 pages.

LATE ADDITIONS, CORRECTIONS

SOLAR/ENERGY CONSERV INFO DISSEMINATION ORG

Environmental Action of Colorado
2239 East Colfax
Denver, CO 80206
303-320-6537

Note change in address and phone.

Solar Energy Exhibition Program
University of Colorado at Denver
1100 14th Street
Denver, CO 80202
303-321-0562
303-892-1117 X414

Note change in address and phone.

SECTION 1

SOLAR PRODUCTS DISTRIBUTOR

(WIND CONVERSION)

Environmental Energies, Inc.
21243 Grand River
Detroit, MI 48219

Note change of address. Also now distribute Electro and Quirks Dunlite wind generator systems.

SOLAR MANUFACTURERS

(MEDIUM TEMPERATURE SOLAR COLLECTOR SYSTEMS)

Raypak, Inc.
3111 Agoura Road
Thousand Oaks, CA 91360

Note change in address.

Solaron Corp
4850 Olive St
Denver, CO 80022

Note correct spelling of name.

SOLAR POTENTIAL MANUFACTURERS, NEW BUSINESS

(GREENHOUSES)

Soltec
70 Adams St.
Denver, CO 80206

Note change in address.

SOLAR RESEARCH

(MEDIUM TEMPERATURE SOLAR COLLECTOR SYSTEMS)

Nuclear Technology
P.O. Box 1
Amston, CT 06231

Note state is CT not CN.

SOLAR RESEARCH

(ALL TECHNOLOGIES)

ERDA (Engy Resch & Dev Adm)
1800 G Street N W
Washington DC 20545

this agency was formed by the Energy Reorganization Act of 1974...the following agencies are now under the jurisdiction of ERDA:

- AEC (Atomic Engy Comm)
  - all depts except Nuclear Regulatory Agency
- NSF (Natl Sci Found)
  - Solar Heating & Cooling/Geothermal/Power Dev
- EPA (Enviro Protect Agency)
  - Alternative Automotive Power Resch & Dev
- DEPT OF INTERIOR
  - Bureau of Mines & Engy/Resources/Alt Engy Resch/Engy Waste Mgt

publication #15 outlines ERDA's Solar Engy Prog for further details inquire to ERDA

note above change of name and address for all departments listed

SECTION 2, SOLAR PROJECTS

JAPAN

page 87 photo caption reads: Japanese Solar Laboratory
caption should read: Yanagimachi Solar House II
Note this house has glazed collectors

SECTION 3, BIBLIOGRAPHY

MEDIUM TEMPERATURE COLLECTOR SYSTEMS

Jordon, R.C. (ed)

Low Temperature Engineering Application of Solar Energy

Price is $8.00, not $9.00.

HEATING AND COOLING OF BUILDINGS

INSTITUTION & P.I.	TITLE	AMT. & DURATION(MOS.)
AAI Corporation (Irwin Barr)	Solar energy school augmentation experiment	$ 637,282/05
Allied Chemical Corp. (Robert A. Allen)	Research on new working fluids for Rankine heat pumps for heating and cooling of buildings	42,300/13
Arkla Industries, Inc. (Philip Anderson)	Engineering, design, construction & testing of salt-water absorption unit optimized for use with solar collector heat source	114,800/12
Arthur D. Little, Inc. (Joan Berkowitz)	Technology assessment of terrestrial solar energy resources development	246,664/12
Battelle Memorial Inst. (James Eibling)	Contract for a solar powered heat pump utilizing pivoting tip-vane rotating equipment	88,743/13
Calif. Inst. of Tech./JPL (V. Truscello)	Workshop on solar cooling for bldgs.	24,822/12
Cal. Poly. St. U/ San Luis Obispo (Thomas M. Lukes)	Research on application of solar energy to food drying industry	254,000/15
Center for the Environment and Man, Inc. (Peter J. Lunde)	Studies of silica gel absorption for dehumidification & air conditioning of buildings	26,100/12
Colorado Springs, City of (J.D. Phillips)	Assessment of single family residence solar heating system in suburban development setting	125,000/25
Colorado State Univ. (John C. Ward)	Evaluation of Löf residence, Denver, Colorado	37,100/12
Colorado State Univ. (George Löf)	Design, construction, & testing of residential solar heating & cooling system	238,000/24
Colorado, University of (Jerome E. Scott)	Consumer demand analysis: solar heating and cooling of buildings	30,500/05

HEATING AND COOLING OF BUILDINGS (Continued)

INSTITUTION & P.I.	TITLE	AMT. & DURATION(MOS.)
Dow Chemical Co. (George A. Lane)	Research on solar energy storage subsystems employing isothermal heat sink materials together with foam-like encapsulating matrices for the heating & cooling of bldgs.	$ 147,159/12
Drexel University (Joseph K. Nwude)	Research Initiation--energy conservation for the food industry	17,300
General Electric Co. (A.D. Cohen)	Solar heating and cooling of bldgs. feasibility study	584,065/08
General Electric Co. (Calvin Fowler)	Solar energy school heating augmentation experiment	459,604/05
General Electric Co. (J.F. Ladd)	Solar heating and cooling of mobile homes	153,900/12
General Electric Co. (A.T. Tweedie)	Research on encapsulated two component energy storage materials for subsystems applicable to heating/cooling of bldgs.	29,800/12
Georgia Inst. of Tech. (J. Richard Williams)	Experimental solar heat supply system with fixed mirror concentrator for the heating and cooling of buildings	116,500/12
Hittman Associates, Inc. (Wm. P. Menchen)	Assessment of Rankine cycle engines for potential application to solar powered cooling of buildings	49,149/06
Honeywell, Inc. (John D. Kopecky)	Solar heating applied to public schools	465,841/05,12
Honeywell, Inc. (R. Schmidt and R. Lechevalier)	Transportable solar heating/cooling laboratory	225,000/07,12
Houston, Univ. of (J.R. Howell and R.B. Bannerot)	Evaluation of surface geometry modification to improve the directional selectivity of solar energy collectors	53,800/12
InterTechnology, Inc (George Szego)	Solar heating applied to public school	237,563/05,06

APPENDIX (C) NSF/RANN SOLAR ENERGY RESEARCH GRANTS & CONTRACTS FOR FY 1974

HEATING AND COOLING OF BUILDINGS (Continued)

INSTITUTION & P.I.	TITLE	AMT. & DURATION(MOS.)
InterTechnology, Inc. (George Szego)	Test and evaluation of Fauquier High School solar energy system	$ 55,000/06
Lawrence Berkeley Lab. (H. Paul Hernadez)	Studies of solar energy conversion systems using Nitinol engines for the cooling of buildings	113,000/12
Lawrence Berkeley Lab. (Michael Wahlig)	Studies of low-cost control system and interface problems of combined solar heating & ammonia-water absorption cooling systems	157,700/12
Lehigh University (Dale R. Simpson)	Studies of solar energy subsystems using eutectic salt phase changes for application in the heating and cooling of buildings	31,500/12
Lockheed Missile & Space Co., Inc. (M.J. O'Neill)	Improvement and extension of the Lockheed simulation program into a computational tool for the analysis, evaluation, and optimization of solar HVAC systems	32,213/13
Mass. Inst. of Tech. (B. Shawn Buckley)	Thermic controls to regulate solar heat flux into buildings	87,500/12
Mass. Inst. of Tech. (T.E. Johnson)	Exploring space conditioning with variable membranes	49,800/12
National Acad. of Sci. (Robert M. Dillon)	Studies of private sector research on solar energy for the heating & cooling of buildings	99,700/12
National Bureau of Standards (J.E. Hill)	Development of methods of evaluation and test procedures for solar collectors and storage devices	73,600/06
Pennsylvania St. U. (Stanley F. Gilman)	Evaluation of the solar building, Albuquerque, New Mexico	189,400/12
RCA Corporation (B. Shelpuk)	Exploration of the feasibilty of the vuilleumier cycle as an air conditioner for bldgs. with a solar energy heat source	151,700/12

HEATING AND COOLING OF BUILDINGS (Continued)

INSTITUTION & P.I.	TITLE	AMT. & DURATION(MOS.)
Rockwell Int'l Corp. (Guy Irwin)	Solar energy storage system using the decomposition of inorganic compounds for the heating/cooling of bldgs.	$ 98,800/12
Schlesinger, Robert J. (Individual)	Development of an instrumentation package for installation in existing dwellings to stimulate the collector and storage elements of a solar HVAC system	25,400/12
Southern Calif. Gas Co. (Samuel J. Cunningham)	Solar water heating utilization proof-of-concept effort	24,900/01
Stanford Research Inst. (Ellis E. Pickering)	Research on the design and construction of water storage methods for solar energy storage subsystems used in the heating and cooling of residential buildings	59,791/13
Syracuse University (Eugene E. Drucker)	Simulation studies of solar assisted, closed loop heat pump systems for HVAC of commercial and institutional bldgs.	123,300/12
Texas A&M Univ. (R.R. Davison)	Further development of the compressed-film floating-deck solar water heater	36,900/12
TRW, Inc., Systems Group (Dimiter I. Tchernev)	Solar heating & cooling of buildings feasibility study	511,548/08,02
United Aircraft Res. Labs. (Frank R. Biancardi)	Test and evaluation of a turbocompressor (Rankine cycle) for the air conditioning of buildings using solar energy as a heat source	161,100/13
Utah State Univ. (H. Craig Peterson)	A dynamic analysis of the economics of scale in space heating and cooling with solar energy	43,800/12
Virginia, Univ. of (F.A. Iachetta)	Workshop on solar heating and cooling of buildings	53,000/12
Westinghouse Electric Corp. (Albert Weinstein)	Conduct Phase 0 of a multiphased solar heating and cooling of bldgs. program	503,085/08,02

INSTITUTION & P.I.	TITLE	AMT. & DURATION(MOS.)
HEATING AND COOLING OF BUILDINGS (Continued)		
Westinghouse Elec. Corp.	Solar heating and cooling proof-of-concept experiment for a school in Atlanta, GA	$ 180,000/13
	TOTAL	$7,406,429
PHOTOVOLTAIC CONVERSION		
Aerospace Corporation (A.B. Greenberg)	Mission analysis of photovoltaic solar energy systems	233,900/15
Arizona State University (Charles Backus) and Spectrolab, Division of Textron, Inc.	Investigation of terrestrial photovoltaic power systems with sunlight concentration	204,200/12
Arizona, Univ. of (B.O. Seraphin)	Symposium on the material science aspects of thin-film systems for solar energy conversion	15,000/06
Arthur D. Little, Inc. (Joan Berkowitz)	Feasibility investigation of growing and characterizing gallium arsenide crystals in ribbon form	224,700/18
Boston College (P.H. Fang) and Ion Physics Corp.	Low cost polycrystalline silicon solar cells for terrestrial applications	98,400/07
Boston University (Norman N. Lichtin) and Esso Research and Engineering Co.	Photochemical conversion of solar energy	179,700/12
Brown University (Joseph Loferski)	Investigation of thin film solar cells based on Cu_2S and ternary compounds	100,400/12
Delaware, Univ. of (K.W. Boer)	Direct solar energy conversion for large-scale terrestrial use	323,300/12.01
Harvard Univ./Tyco Labs. (Bruce Chalmers)	Low-cost continuous fabrication of silicon solar cells	150,000/12
Jet Propulsion Lab./NASA (R. Lutwack)	Assessment of photovoltaic conversion of solar energy for terrestrial applications	69,100/06,06

INSTITUTION & P.I.	TITLE	AMT. & DURATION(MOS.)
PHOTOVOLTAIC CONVERSION		
Pennsylvania, Univ. of (Martin Wolf) and Dow Corning	Investigation of low-cost processes for integrated silicon solar arrays	$ 93,400/12
Solarex Corp. (Joseph Lindmayer)	Feasibility investigation of achieving 20% terrestrial conversion efficiency silicon solar cells	119,400/15
Solarex Corporation (Joseph Lindmayer)	Feasibility investigation of low-cost thin film silicon solar cells	114,100/15
Tyco Labs.,Inc. (A.I. Mlavsky)	Scale up of program on continuous silicon solar cells	321,000/15
Wisconsin, Univ. of/ Milwaukee (T.S. Jayadevaiah)	Solar energy conversion, amorphous and magnetic semiconductor interfaces	29,200/24
	TOTAL	$2,276,500
SOLAR THERMAL CONVERSION		
Aerospace Corp. (A.B. Greenberg)	Solar thermal conversion mission analysis	253,799/12
American Tech. Univ. (Alex A.J. Hoffman) Cerebonics	Analysis of the applicability of solar thermal system for military installations	122,200/09
Arizona, Univ. of (B.O. Seraphin)	Chemical vapor desposition research for fabrication of solar energy converters	175,000/12
Colorado State Univ. (G.O.G. Löf) Westinghouse Elec. Corp.	Analysis of solar-thermal electric power systems	19,700/06
Helio Associates (Aden B. Meinel)	Air-stable selective surfaces for solar energy collectors	216,800/18
Houston, Univ. of (L.L. Vant-Hull) McDonnell Douglas	Heliostat model tests	112,800/05

APPENDIX (C) NSF/RANN SOLAR ENERGY RESEARCH GRANTS & CONTRACTS FOR FY 1974

INSTITUTION & P.I.	TITLE	AMT. & DURATION(MOS.)
SOLAR THERMAL CONVERSION		
Houston, Univ. of (L.L. Vant-Hull) McDonnell Douglas	Solar thermal power system based upon optical transmission	$ 176,900/12
Itek Corp. (R.D. Cummings)	Solar power collector breadboard test	98,800/09
Lawrence Berkeley Lab. (Michael Wahlig)	Measurement of circumsolar radiation	93,600/06
McDonnell Douglas Astronautics Co.-West (G.G. Pittinato)	Elimination of control of material problems in water heat pipes	189,500/12
Martin Marietta Corp. (F. Blake) Georgia Tech.	Solar power system and component research	229,500/10
Sandia Laboratories/AEC (R.P. Stromberg)	Systems analysis computer program for solar community total energy concept	99,900/10
G.T. Schjedahl Co. (Ross A. Stickley) Northern States Power Univ. of Minnesota	Solar power array for the concentration of energy(space)	266,100/12
	TOTAL	$1,948,599
OCEAN THERMAL CONVERSION		
DSS Engineers, Inc. (W.B. Stuart)	Development of plastic tubed heat exchangers for sea power plants	59,900/08
Hawaii, University of (Karl H. Bathen)	Near-shore application for ocean thermal energy conversion pilot plant in Hawaii	48,600/10
Lockheed Missiles & Space, Inc. (Lloyd C. Trimble) Bechtel	Ocean thermal energy program--research on an engineering evaluation and test program	329,600/09

INSTITUTION & P.I.	TITLE	AMT. & DURATION (MOS.)
OCEAN THERMAL CONVERSION		
Mass., Univ. of (W.E. Heronemus)	Technical and economic feasibility of the ocean thermal differences process with potential for significant impact on the United States energy market	$ 169,800/12
Sea Solar Power, Inc. (J. Hilbert Anderson)	To design, construct and test an operating model of a sea solar power plant	31,000/05
Union Carbide Corp./ Linde Division (Frank Notaro)	Heat exchangers for ocean thermal power plants	93,200/09
	TOTAL	$ 732,100
BIOCONVERSION TO FUELS		
Calif. Inst. of Tech. (Wheeler J. North)	Evaluating oceanic farming of seaweeds as sources of organics and energy	110,600/24
Calif., Univ. of/ Berkeley (C.R. Wilke)	Special seminar on cellulose as an energy resource	7,900/08
Calif., Univ. of/ San Diego (M.D. Kamen, N.O. Kaplan)	Bioconversion of solar energy and production of hydrogen by photolysis	75,000/12
Case Western Reserve U. (L.O. Krampitz)	Hydrogen production by photosynthesis and hydrogenase activity-an energy source	107,700/12
Cornell University (William J. Jewell)	Bioconversion of agricultural wastes for energy conservation and pollution control	111,800/12
Illinois, Univ. of/ Urbana (J.T. Pfeffer)	Biological conversion of organic refuse to methane	101,000/12

INSTITUTION & P.I.	TITLE	AMT. & DURATION(MOS.)
BIOCONVERSION TO FUELS		
Indiana, Univ. of/ Bloomington (A.San Pietro)	Workshop on hydrogen production by biophotolysis	$ 13,900/12
Nebraska, Univ. of/ Lincoln (James G. Kendrick)	High protein food isolates from selected grain alcohol fermentation by-products	62,950/15
Stanford Research Inst. (Robert E. Inman)	An evaluation of the use of agricultural residues as an energy feedstock	149,300/15
Stanford Univ. (Perry L. McCarthy)	Heat treatment of refuse for increasing anaerobic biodegradability	36,200/12
	TOTAL	$ 776,350
WIND ENERGY CONVERSION		
Alaska, Univ. of (Tunis Wentink,Jr.)	Study of Alaska wind power and its possible applications	118,100/12
NASA/Lewis Res. Ctr. (Ronald L. Thomas)	Experimental 100 KW wind power and its possible applications	865,000/24
National Oceanographic and Atmospheric Administration (NOAA) (M. Changery)	Initial wind energy data assessment	14,500/09
Oregon State Univ. (Robert E. Wilson) Aerovironment, Inc.	Applied dynamics of wind power machines	19,300/06
Princeton Univ. (Thos. E. Sweeney)	Optimization and characterization of a sail-wing rotor for wind energy conversion	74,700/12
	TOTAL	$1,591,600

INSTITUTION & P.I.	TITLE	AMT. & DURATION(MOS.)
GENERAL PROJECTS		
Associated Universities, Inc. (V. Bremenkamp)	Support of U.S.-U.S.S.R. cooperation in solar and geothermal energy	$ 120,000/11
Colorado, Univ. of (Eliz. A. Kingman)	Dawning of solar energy--an exhibition program (Phase I)	8,300/06
Harbridge House, Inc. (Robert D. Crangle)	Contract for a training course in program management	7,701
Image Associates (Bastian Wimmer)	Technical briefing film on new developments in the use of solar energy in school buildings	26,646/02
Maryland, Univ. of (R. Allen)	Solar Energy Panel	15,300/12
NOAA (R. Cohen)	Solar heating and cooling for bldgs., ocean thermal energy conversion and wind conversion	46,300/12
NOAA (L. Machta)	Workshop on solar radiation monitoring for solar energy applications	19,000/06
Performance Commun., Inc. (NY) (Arthur G. Smith)	An intensive analysis of RANN communication, management of the first annual RANN symposium and preparation of copy for the first annual RANN monograph	86,393/12
State Univ. at Stony Brook (H. Richard Blieden)	Personnel Mobility Assignment	45,355/12
Virginia, Univ. of (Charles S. Chen)	Personnel Mobility Assignment	42,930/13
Washington, Univ. of (Richard Bogan)	Personnel Mobility Assignment	56,560/14
	TOTAL	$ 474,485

SOLAR ENERGY LEGISLATION ROUNDUP

Bill No.	Title	Sponsor	Status/Key Points
S 1283 law# PL 93-577	Non-nuclear Energy R&D Act	Jackson (Udall)	Establishes national R&D program; all programs to be carried out in conjunction with existing programs, including McCormack Act; no specific funding; passed Senate 12/73, House 9/74; conferees chosen, no meetings scheduled yet.
S 3234 (HR 16371) law# PL 93-473	Solar Energy Research, Development and Demonstration Act	Humphrey (McCormack)	Forms Solar Energy Coordination & Mgmt. Project to oversee solar R&D, full- and small-scale demos; establishes Solar Research Institute, Solar Information Data Bank; authorizes $75 million in FY76, $2 million for NSF program definition in FY75; conference report passed by House, Senate 10/74, awaiting President's signature.
HR 11510 (S 2744) law# PL 93-438	Energy Reorganization Act	Holifield (Ribicoff)	Sets up Energy Research & Development Administration and Nuclear Regulatory Commission; assumes R&D activities of NSF, AEC, EPA, Interior Dept.; signed into law 10/74; search is on for ERDA Administrator (see p.1).
HR 11864 (S 2650, 2658) law# PL 93-409	Solar Heating & Cooling Demonstration Act	McCormack (Cranston, Moss)	Provides for demonstration of solar heating technology in 3 years, heating and cooling in 5; $60 million authorized for FY75-79; signed into law 9/74; NASA, HUD have till Jan. to prepare demonstration program plan; NBS, HUD working on standards for solar systems, buildings.
HR 14434 law# PL 93-322	FY75 Energy Appropriations	Mahon	Enacted 7/74; provides $2.2 billion for energy; $50 million to NSF, $1.8 million to NASA and $1.2 million to AEC for solar.

SOLAR ENERGY LEGISLATION ROUNDUP (continued)

Bill No.	Title	Sponsor	Key Points/Status/Related Bills
H.Res.265	Federal Budgets	McCormack	Disapproves deferral of $5 million appropriated in 1974 to NSF for solar and geothermal R&D programs. Referred to Appropriations.
H.R. 301	Homeowners' Energy Conservation Act	Cohen	Establishes direct, low-interest loan program in HUD to assist homeowners in purchase and installation of more effective insulation and heating equipment. Pending in Subcommittee on Housing and Community Development. Related bills: H.R. 307, Conte; H.R.2647, Clausen; H.R. 2980, 3712, Cohen.
H.R. 302	Tax Credits	Cohen	Amends IRS Code to allow credit for 25% of insulation and heating equipment costs in existing homes. Pending in Ways and Means. Related bills: H.R.2981, 3713, Cohen.
H.R. 793	Tax Deductions	O'Brien	Amends IRS Code to provide tax deductions for home insulation and heating equipment costs. Pending in Ways and Means. Related bills: H.R. 933, Robinson.
H.R. 1697	Solar Tax Deductions	McClory	Amends IRS Code to permit tax deductions for purchase and installation of solar heating and cooling equipment. Pending in Ways and Means.
H.R. 1807	Tax Credits	Conte	Amends IRS Code to allow a tax credit for certain building insulation and heating improvements. Pending in Ways and Means. Related bills: H.R.3004, Goldwater; H.R.5003, Conte.
H.R.3474	ERDA Authorization Bill	Price	ERDA solar budget request: FY76 operating costs, $57.1 million; obligational authority, $70.3 million. Referred to Subcommittee on Energy Research, Development, and Demonstration. Subcommittee recommended solar budget: FY76 operating costs, $82.7 million; obligational authority, $143.7. Mark-up began Ap. 17; resumption date has not yet been scheduled.

this page brought to you through the courtesy of:
Solar Energy Washington Letter, 1001 Connecticut Ave. N.W.
Washington, D.C. 20036

SOLAR ENERGY LEGISLATION ROUNDUP

Bill No.	Title	Sponsor	Key Points/Status/Related Bills
H.R.5005	Energy Conservation and Conversion Act.	Ullman	Provides for a comprehensive national energy conservation and conversion program. Includes tax credit to homeowners for installation of solar equipment, provides money for development and demonstration of new energy technology, and extends investment credit to include insulation and solar equipment. Currently being marked up by Ways and Means; mark-up expected to continue for 2-3 weeks.
H.R.5959	Solar Tax Credits	Wylie	Amends IRS Code to provide tax credit for expenditures for solar heating and cooling equipment installed in new or existing buildings and for home insulation. Pending in Ways and Means.
S.168	Tax Incentives	Domenici	Amends IRS Code to allow tax credit or deduction for certain expenditures relating to thermal design of homes. Pending in Finance. Related bills: H.R.1505, 2066, 2067, 2068, 2482, 3064, 4728, Vanik; H.R.2648, Clausen, all pending in Ways and Means.
S.311	Excess Petroleum Profits	McClure	Amends IRS Code, taxing excess industry profits and allowing deduction for investments in R&D of new fuels, including solar. Pending in Finance.
S.489	Antitrust Law	Abourezk	Amends Clayton Act to preserve and promote competition among corporations in production of oil, natural gas, coal, oil shale, tar sands, uranium, geothermal steam, and solar energy. Pending in Judiciary. Related bills: H.R.2873, Conte; H.R.4907, Hechler (W.Va.), both pending in Subcommittee on Monopolies and Commercial Law.
S.617	Power Resources	Jackson (by request)	"Winterization Assistance Act of 1975". Pending in Banking, Housing & Urban Affairs. Executive comment requested from HUD, FEA; thus far, favorable comment received from FEA.
S.622	Standby Energy Authorities	Jackson	Establishes standby emergency authority to assure essential U.S. energy needs are met. Passed Senate 60-25, with amendment

SOLAR ENERGY LEGISLATION ROUNDUP

Bill No.	Title	Sponsor	Key Points/Status/Related Bills
			327 introduced by Fannin providing low-interest loans and loan guarantees to homeowners for installation of solar equipment.
S.875	Solar Energy	Hart (Colo.)	Authorizes HUD Secretary to make direct, low-interest loans to assist homeowners and builders in purchasing and installing solar heating and cooling equipment. Pending in Subcommittee on Housing. Executive comment requested from HUD. Related bills: H.R.3849, 4507, 4619, 5460, Gude, all pending in Banking, Currency, and Housing.
S.973	Tax Incentives	Bentsen	Amends IRS Code to provide incentives for efficient use of gas and increased use of coal and to encourage development of synthetic fuels and solar energy. Pending in Finance.
S1112	Energy Revenue and Development Act	Gravel	Establishes an energy trust fund (funded by tax on energy sources) to provide for development of domestic sources of energy and for more efficient energy use. Also amends IRS Code to allow tax credit or deduction for residential energy conservation expenditures, including solar equipment. Pending in Finance.
S.1379	Solar Tax Incentives Act	Fannin	Amends IRS Code to provide tax credit for solar equipment installed in new or existing residential or commercial buildings. Pending in Finance.
S.1392	Energy Conservation	Tunney	Establishes energy conservation demo program by using promising innovative technology to maximum extent, through retrofit of existing buildings. Pending in Public Works, Commerce, and Government Operations Committees.

this page brought to you through the courtesy of:
Solar Energy Washington Letter, 1001 Connecticut Ave. N.W.
Washington, D.C. 20036

SUMMARY OF STATE LEGISLATION

Arizona

S.B. 1230- Introduced February 12, 1974. Relating to power; providing that the Arizona Power Authority shall encourage development of solar, nuclear, and geothermal energy, and amending sections 30-121 and 30-123, Arizona revised statutes.

S.B. 1231- Relating to taxation and taxation of income; prescribing powers and duties of the department of revenue; prescribing tax benefits for installation of a solar energy device designed to produce heat or electricity; amending Title 42, Chapter 1, Article 2, Arizona revised statutes, by adding section 43-123.37. Approved by the Governor and filed on May 15, 1974.

Florida

HCR 2919- Proposes the creation of a joint select committee on solar energy to study the potential uses and practical applications of solar energy. The proposed committee to be composed of 5 members of the House of Representatives and 5 members of the Senate. Provides for employment of a director and staff.
Requires the committee to report its findings and recommendations to the next regular session of the legislature.

HB 2944- Creates a Solar Energy Center in the College of Engineering at the University of Florida to research and develop the use of solar energy for the production of electricity.
Authorized the Center to apply for and accept grants. Provides that subject to contractual rights of the federal government, all discoveries shall be the property of the state. Appropriates from the general revenue fund to the Division of Universities $1,000,000 for the establishment of the Center and $200,000 for the operation of the Center for fiscal year 1974-75.

SB 721- Solar Energy Center was recorded in Chapter 74-185--an act relating to solar energy, creating a solar energy center, authorizing research, dissemination of information and providing demonstrations, providing an effective date.

HB 3616- Authorizes the Board of Regents of the State University System to establish a solar energy center. Provides for powers and duties, authorizes research and the dissemination of information. Appropriates from the General Revenue Fund, an amount sufficient to carry out this act.

HB 3617- Exempts certain solar energy systems from sales and use tax. Defines "solar energy systems." Exempts from the corporate income tax derived from the manufacture , sale or installation of certain solar energy systems.

HB 3992- Authorizes and directs the University of Florida solar energy laboratory to develop prototype relocatable total solar energy system for the generation of hot water, heat, cooling, electricity, and ventilation for multipurpose uses and prototype systems for the several Florida school program-grade group facilities. Authorizes the laboratory to cooperate with state and federal agencies in the development and construction of prototype systems.
Appropriates $250,000 from the General Revenue Fund to the Board of Regents for the laboratory for such purposes.

SUMMARY OF STATE LEGISLATION

Florida (continued)

HB 3624- Requires the Division of Building, Construction and Maintenance of the Department of General Services to set standards for the testing of solar energy systems, set testing fees, and provide for the display of Division's approval on systems manufactured or sold within the state. Appropriates $50,000 to carry out this act.

HB 3584- Exempts from the sales and use tax the sale of solar water heaters and other appliances powered by solar energy.

SB 158- Solar Water Heating-Building Standards was recorded in Chapter 74-361--an act relating to building construction standards; creating 553.065, Florida Statutes; providing that no building be constructed without provision for future installation of solar hot water heating equipment; providing an effective date.

Hawaii

SR 114- Requesting the University of Hawaii to conduct research on solar energy.

Indiana

SB 223- A bill for an act to amend IC 1971, 6-1 by adding a new chapter concerning property tax deductions for solar energy systems.

New York

S. 7194- Introduced January 14, 1974. To amend the tax law and public service law, in relation to deductions for expenses incurred in research for improving the efficiency of existing fuel and energy sources, fuel and energy conservation, and developing alternative sources of energy and power.

S. 7264- An act to amend the environmental conservation law, in relation to authorizing and directing the commissioner of environmental conservation to establish programs and policies for the development of prototypes of solar powered heat units for residential and commercial buildings and make an appropriation therefor.

Pennsylvania

HB 2214- An act to encourage the use of solar energy systems for heating and cooling and excluding material and installation costs thereof from sales and use taxes and excluding the value added by such tangible property from ad valorem property tax.

SOLAR SPACE HEATING AND COOLING

Arizona State University
College of Architecture and College of Engineering Sciences
Tempe, AZ 85281

SOLAR UTILIZATION NOW
SOLAR ENERGY APPLICATIONS FOR BUILDINGS

January 13-17, 1975
$300
Coordinators: Dr. Byard Wood, College of Engineering Sciences, ASU, (602) 965-7298
Prof. Jeffrey Cook, College of Architecture, address and phone same
Faculty: Dr. Charles E. Barkus, Frank H. Bridgers, Prof. Jeffrey Cook, Frank E. Edlin, Dr. Donovan L. Evans, Dr. Erich Farber, Ray Fields, Dr. Leon Florchuetz, Dr. Harold Hay, Dr. Richard C. Jordon, Dr. Don V. Plantz, Dr. Charles M. Randall, Dr. Bernard Seraphin, Dr. Byard D. Wood, Prof. John I. Yellott

---

California, University of
Riverside

USING SOLAR ENERGY IN THE HOME

February 22, 1975
Faculty: Jack Shultz
Shultz Field Enterprises
6986 La Jolla Blvd.
La Jolla, CA 92037
(714) 454-1312

Oriented to plumbers and heating contractors

California, University of
San Diego

SOLAR ENERGY FOR THE CONSTRUCTION INDUSTRY

Spring, 1975
Faculty: Jack Shultz
Shultz Field Enterprises
6986 La Jolla Blvd.
La Jolla, CA 92037
(714) 454-1312

SOLAR SPACE HEATING AND COOLING

Colorado, University of
Center for Management and Technical Programs
Boulder, CO 80302
(303) 443-4865

SOLAR HEATING AND COOLING SYSTEMS

February 12-14, 1975 (Vail, CO)
April 9-11, 1975 (Boulder, CO)
June 18-20, 1975 (Colorado Springs, CO)
$280
Faculty: Richard L. Crowther, A.I.A., Dr. George O.G. Löf, Dr. Jan Kreider, Dr. Frank Kreith, Philip Tabb

---

Georgia Institute of Technology
Department of Continuing Education
Atlanta, GA 30332

SOLAR HEATING AND COOLING FOR HOMES AND BUILDINGS

February 3-5, 1975
$150
Academic Administrators: Dr. A.P. Sheppard, Professor of Electrical Engineering
Dr. J.R. Williams, Assoc. Prof. of Mechanical Engineering
Faculty: Dr. Lorin Vant Hull, Mr. P.R. Rittelmann, Mr. Glen McDonald, Mr. J.M. Akridge, Dr. S.C. Bailey, Dr. J.F. Benzel, Dr. Steve Bomar, Dr. F.J. Clarke, Dr. J.I. Craig, Dr. A.S. Debs, Dr. Robert B. Evans, Dr. D.C. Ray, Dr. J.H. Schlag, Mr. J.D. Walton

---

Total Environmental Action
Box 47
Harrisville, NH. 03450
(603) 827-3087

SOLAR ENERGY UTILIZATION

every second Saturday of each month
January 11, 1975 — April 12, 1975
February 8, 1975 — May 10, 1975
March 8, 1975 — June 14, 1975
$30 ($22 for students)
Course leader: Bruce Anderson

SOLAR SPACE HEATING AND COOLING

Wisconsin, University of
College of Engineering
432 North Lake Street
Madison, WI 53706

SOLAR THERMAL PROCESSES

January 6-10, 1975
$350
Program Director: Robert C. Lutton
Dept. of Engineering
Faculty: John A. Duffie, William A. Beckman
For program information call: (608) 262-2061
For enrollment call: (608) 262-1299

PHOTOVOLTAICS

Arizona State University
Tempe, AZ 85281

SIXTH ANNUAL DIRECT ENERGY CONVERSION SHORT COURSE

March 17-21, 1975
$325
Coordinator: Dr. C.E. Backus
College of Engineering Sciences
ASU
(602) 965-3857
Faculty: Dr. S.W. Angrist, Dr. C.E. Backus, Mr. E.G. Hampe, Dr. D.L. Jacobson
Dr. W.J. Lueckel, Dr. P. Rappaport, Dr. R.J. Rosa, Mr. J.J. Werth

Also covers thermoelectronics, magnethydroynamics, thermionics, fuel cells, and batteries, unconventional DEC

WIND CONVERSION

Total Environmental Action
Box 47
Harrisville, NH 03450
(603) 827-3087

WIND ENERGY UTILIZATION

Every second Saturday of each month
January 11, 1975 April 12, 1975
February 8, 1975 May 10, 1975
March 8, 1975 June 14, 1975
$30 ($22 for students)
Course leader: Douglas Coonley

OTHER/RELATED TO SOLAR TECHNOLOGIES

New Mexico, University of
Technology Application Center
Albuquerque, NM 87131
(505) 277-3622

HEAT PIPE TECHNOLOGY

January 6-10, 1975
$325
Coordinator and lecturer: K.T. Feldman
Director, Energy Research Center
University of New Mexico
Lecturers: Al Basuilis, Joe E. Kemme, Jack P. Kirkpatrick, Dr. Bruce Marcus, and heat pipe manufacturers

---

OTHER UNIVERSITIES WHICH PROVIDE EITHER ON-GOING OR FREQUENT COURSES ON SOLAR ENERGY

Arizona, University of , Tucson, AZ
Colorado State University, Fort Collins, CO
Florida, University of, Gainesville, FL
George Washington University, Washington, DC
Maryland, University of, College Park, MD
Minnesota, University of, Minneapolis, MN
Sagamon State University, Springfield, IL
Youngstown State University, Youngstown, OH

Environmental Action of Colorado

Founded in 1970 to coordinate activities in Denver for the April 22nd Earth Day Teach-In, EAC has developed since that time into an information clearinghouse, an energy study and advocacy group, and a catalyst for community action on environmental problems. Much of our staff is composed of dedicated volunteers.

At present, EAC is channeling much of its effort into bringing about increased recognition and development of solar energy. There exists an urgent and pressing need to secure an adequate energy supply and concurrently insure that its utilization has minimal harmful effects on our environment. Evidence indicates solar energy is an especially meritorious alternative to fossil and nuclear fuels inasmuch as it represents an inexhaustible supply that satisfies requirements of environmental quality. Therefore, we have conducted a number of projects aimed at increasing public awareness of the practicality of utilizing solar energy and are presently coordinating several more. These include:

PROCEEDINGS: SOLAR HEATING, COOLING, AND ENERGY CONSERVATION CONFERENCE - In early 1975 we will be publishing a compilation of the presentations and workshops of our Solar Heating, Cooling, and Energy Conservation Conference. This was a multidisciplinary meeting which featured over 50 noted speakers and panelists as well as technological displays. Geared to the emerging solar industry, a wide variety of topics were discussed, ranging from housing design for passive solar energy utilization to aspects of marketing solar energy equipment.

LIBRARY - EAC is compiling a library of publications and slides on the subject of solar energy. This collection provides a unique resource to the general public and the professional. At present, the facility can be utilized by appointment. Materials are available for on-site reference or reproduction for a fee.

ENVIRONMENTAL ACTION REPRINT SERVICE (EARS) - A service that reproduces and distributes information on solar energy, nuclear energy, and other energy/environment issues.

COMMUNITY SERVICES - EAC strives to provide information and services to the academic community and the community at large. As much as possible, referrals and consultation on energy/environment issues are provided. In addition, speakers, seminars, and other educational programs are often sponsored by EAC for university students and/or the community.

Solar Energy Exhibition Program

The purpose of the Solar Energy Exhibition Program is to increase the public's awareness of the viability of solar energy as an abundant, clean, and practical energy source.

It is a four-phase project that entails creating a major traveling museum exhibition that will open in Denver, and then visit museums and other educational institutions across the country, as well as in Colorado.

The exhibition is designed by Joseph A. Wetzel Associates, a leading exhibit design firm, widely recognized for its innovative approach. The concept Wetzel Associates has developed for the solar exhibition centers around a learning and communications environment - prepared for involvement and random discovery. Comprised of modular units in a variety of sizes, the exhibition is designed for ease of installation and transportation.

There will be two versions of the exhibition. The large version, between 2,500 and 3,000 square feet with a maximum height of 12 feet, will be shown in major museums (mainly Science and Technology Centers) across the nation. The smaller version, about 500 square feet, with a maximum height of 8 feet, will travel to smaller museums and educational institutions around the country.

In addition to the exhibition itself, a program will be offered that will provide an opportunity for continued public interest and investigation of solar energy and related issues, such as energy conservation. It will give people access to more comprehensive information, and will encourage public participation in discussions, workshops, and other activities. The program will add greatly to the effectiveness of the exhibition as a whole.

The exhibition program has received support from Environic Foundation International at Notre Dame University, and it is currently funded by the National Science Foundation. NSF funds are channeled through the University of Colorado.

Funds thus far provided have enabled the first two phases to be completed. Phase 1 entailed research, planning, and organizing the general content of the exhibition. Phase 2 entailed designing the exhibit and initiating development of the program that will accompany the exhibition.

Continued funding from the National Science Foundation will provide partial support to Phase 3: Production and Installation. Additional funds are required to complete this phase of the project, and subsequently for Phase 4: Circulation.

Current efforts are directed at obtaining these needed funds. The more groups and individuals that contribute, the more successful the exhibition program will be. If you would like to know more about this project, and would like to make a contribution, please contact:

Elizabeth Kingman - Director
Solar Energy Exhibition Program
University of Colorado at Denver
1100 14th Street
Denver, Colorado 80202
303-321-0562
303-892-1117 X414

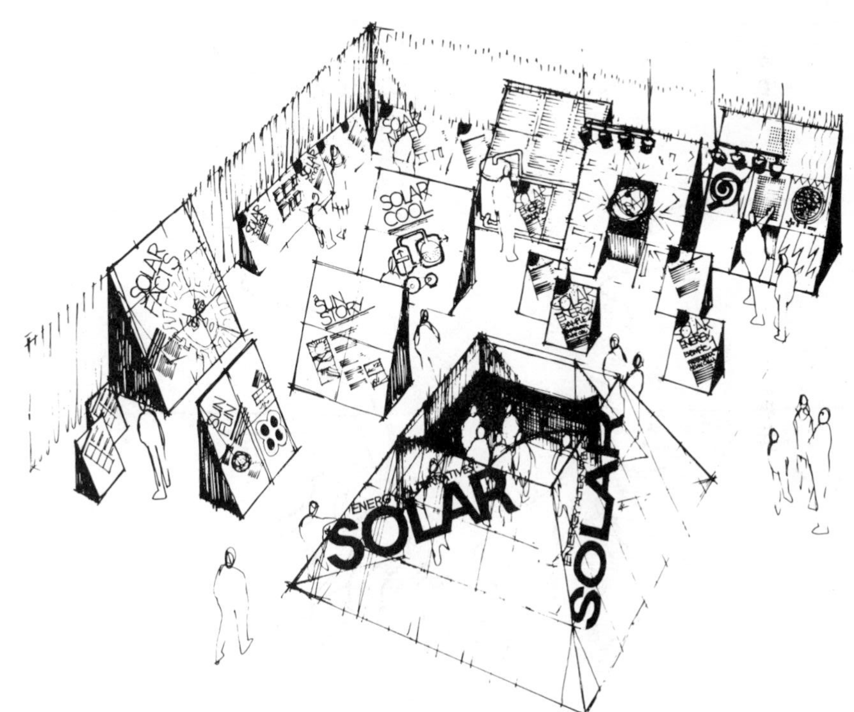

Artist's rendering of exhibition concept - designed by Joseph Wetzel Associates.

In 1972, NSF and NASA organized a solar energy panel, consisting of scientists, engineers, economists, environmentalists, and sociologists, to assess the potential of solar energy as a national energy resource. In its findings the panel concluded:

SOLAR ENERGY 'IS AN INEXHAUSTIBLE SOURCE OF ENORMOUS AMOUNTS OF CLEAN ENERGY.'

'IN PRINCIPLE, SOLAR ENERGY CAN BE USED FOR ANY ENERGY NEED NOW BEING MET BY CONVENTIONAL FUELS.'

'THERE ARE NUMEROUS CONVERSION METHODS BY WHICH SOLAR ENERGY CAN BE UTILIZED FOR HEAT AND POWER, E.G., THERMAL, PHOTOSYNTHESIS, BIOCONVERSION, PHOTOVOLTAICS, WINDS AND OCEAN TEMPERATURE DIFFERENCES.'

FROM SEA-THERMAL ENERGY ALONE, 'THE TOTAL ANNUAL PRODUCTION COULD EXCEED THE YEAR 2000 PROJECTED TOTAL ENERGY DEMANDS.'

'IN 1969, THE TOTAL ELECTRIC ENERGY CONSUMED IN THE U.S. COULD HAVE BEEN SUPPLIED BY THE SOLAR ENERGY INCIDENT ON 0.14% OF THE U.S. LAND AREA,' A STATEMENT BASED ON U.S. AVERAGE SOLAR INCIDENCE AND THE ASSUMPTION OF A 10% CONVERSION EFFICIENCY.

'THERE ARE NO TECHNICAL BARRIERS TO WIDE APPLICATION OF SOLAR ENERGY TO MEET U.S. NEEDS.'

'FOR MOST APPLICATIONS, THE COST OF CONVERTING SOLAR ENERGY TO USEFUL FORMS OF ENERGY IS NOW HIGHER THAN CONVENTIONAL SOURCES, BUT DUE TO INCREASING PRICES OF CONVENTIONAL FUELS AND INCREASING CONSTRAINTS ON THEIR USE, IT WILL BECOME COMPETITIVE IN THE NEAR FUTURE.'

WITH SUPPORT BEHIND SOLAR ENERGY PROGRAMS, 'BUILDING HEATING COULD REACH PUBLIC USE WITHIN 5 YEARS, BUILDING COOLING IN 6 TO 10 YEARS, SYNTHETIC FUELS FROM ORGANIC MATERIALS IN 5 TO 8 YEARS, AND ELECTRICITY PRODUCTION IN 10 TO 15 YEARS.'

'SOLAR ENERGY UTILIZATION ON A LARGE SCALE COULD HAVE MINIMAL IMPACT ON THE ENVIRONMENT IF PROPERLY PLANNED.'

'IT APPEARS THAT AN OBJECTIVE ALLOCATION OF R&D FUNDS WOULD CALL FOR SUBSTANTIALLY INCREASED R&D SUPPORT FOR A NUMBER OF SOLAR ENERGY OPPORTUNITIES. THERE ARE ALSO INTERNATIONAL BENEFITS IN MAKING A VIABLE SOLAR ENERGY TECHNOLOGY AVAILABLE TO THE WORLD, AS WELL AS BALANCE OF PAYMENTS AND NATIONAL SECURITY BENEFITS IN LIMITING OUR ALMOST INEVITABLE DEPENDENCE ON FOREIGN ENERGY SOURCES.'*

It is true that solar energy offers tremendous potential and opportunities, but if widespread utilization of the sun's energy is to be achieved, a major implementation program is called for. This program requires a cooperative effort between government, industries, universities, communities, and people. Crucial to its success is communication and education. We in solar energy have directed our efforts toward this goal.

---

*Solar Energy as a National Energy Resource, Prepared by NSF/NASA Solar Energy Panel, December 1972.

Reversed print of a solar eruption. In addition to giving off light and heat, the sun, during solar flares, billows forth clouds of particles and other emissions of varying intensity. When these particles are strong enough and collide with the earth some hours later, they can dissipate more energy in the earth's high atmosphere than the most destructive hurricanes on record. The resulting electrical and magnetic disturbances can black out long range radio communications, cause airplane and ship compasses to shift erratically and trigger brilliant auroral flashes. Photo courtesy of NASA.

A A I CORP
WEAPONS&AERO SYS DIV
PO BOX 6767
BALTIMORE MD 21204
IRWIN R BARR
-PHONE-
301 6661400 X232
SOLAR MANUFACTURERS
SOLAR SPACE HEATING & COOLING

A F E D E S
28 RUE DE LA SOURCE
75016 PARIS FRANCE
PAUL GIRARD
GEN DELEGATE AFEDES
-PHONE-
224 59 35
SOLAR/ENERGY CONSERV INFO DISSEMINATION ORG
ALL SOLAR TECHNOLOGIES

ACORN GLAS-TINT INTERNATIONAL
1123 W CENTURY BLVD
LOS ANGELES CA 90044
CHRIS T DUNKLE
PRESIDENT
-PHONE-
213 7561471
SOLAR PRODUCTS DISTRIBUTORS
OFFICE/BANK BUILDINGS

ADELAIDE, UNIV OF
ARCH & TOWN PLANNING
ADELAIDE, S A AUSTRALIA
J D KENDRICK
SR LECTURER BLDG SCI
-PHONE-
223 4333 X2444
SOLAR RESEARCH
ALL SOLAR TECHNOLOGIES

ADRIAN MANUF & RESEARCH
619 19TH STREET
GOLDEN CO 80401
FREDERICK T VARANI

-PHONE-
303 2794536
SOLAR RESEARCH
SOLAR SPACE HEATING

ADV SOLAR ENGY TECH NEWSLETTER
1609 WEST WINDROSE
PHOENIX AZ 85029
CARL M LANGDON
EDITOR-PUBLISHER
-PHONE-
602 9423796
SOLAR/ENERGY CONSERV INFO DISSEMINATION ORG
PUBLISH LAY + TECHNICAL PERIODICALS

AEROSPACE CORP
ADVANCED TECH APPLIC
P O BOX 92957
LOS ANGELES CA 90009
DR A B GREENBERG

SOLAR RESEARCH
ALL SOLAR TECHNOLOGIES

AEROWATT
37 RUE CHANZY
PARIS 75 FRANCE

SOLAR MANUFACTURERS
WIND CONVERSION

AGRO-CITY PLANNING CONSULTANTS
3410 GARROTT RD
HOUSTON TX 77006
GEOFFREY STANFORD

-PHONE-
713 5297266
SOLAR RESEARCH
BIO-CONVERSION

AJAX MAGNETHERMIC
1745 OVERLAND AVE NE
WARREN OH 44482
THEODORE BURKE JR

-PHONE-
216 3722529
SOLAR RESEARCH
SOLAR SPACE HEATING

ALABAMA, UNIVERSITY OF
CTR ENVIRONMENT STUD
P O BOX 1247
HUNTSVILLE AL 35807

SOLAR RESEARCH
SOLAR POWERED TRANSPORTATION

ALABAMA,UNIV OF/IN HUNTSVILLE
CTR ENVIRONMENT STUD
P O BOX 1247
HUNTSVILLE AL 35807
D L CHRISTENSEN
-PHONE-
205 8956257
SOLAR RESEARCH
TOTAL SYSTEMS INTEGRATION

ALASKA,UNIV OF
GEOPHYSICAL INSTITUT
FAIRBANKS AK 99701
DR KEITH MATHER
DIRECTOR
-PHONE-
907 4797607
907 4797558
SOLAR RESEARCH
WIND CONVERSION

ALBERTA, UNIV OF
DEPT OF MECH ENGR
EDMONTON 7 ALBERTA CANADA
GERALD W SADLER

-PHONE-
403 4323450

SOLAR RESEARCH
SOLAR RADIATION MEASUREMENTS

ALBRIGHT COLLEGE
DEPT OF PHYSICS
READING PA 19604
DR THURMAN R KREMSER

-PHONE-
215 9212381 X266

SOLAR INTEREST
SOLAR SPACE HEATING

ALBUQUERQUE WESTERN INDUS INC
ALBUQUERQUE NM 87107

-PHONE-
505 3447224
SOLAR MANUFACTURERS
HOMES

ALCOA ALUMINUM
1501 ALCOA BLDG
PITTSBURGH PA 15219
JOHN E WRIGHT *
SUPERVISOR PUB REL
-PHONE-
412 5534751
SOLAR COMPONENTS MANUFACTURER
MEDIUM TEMPERATURE SOLAR COLLECTOR SYSTEMS

ALCOA ALUMINUM
1501 ALCOA BLDG
PITTSBURGH PA 15219
WILLIAM S LEWIS
MARKET DEV MANAGER
-PHONE-
412 5532748
SOLAR COMPONENTS MANUFACTURER
MEDIUM TEMPERATURE SOLAR COLLECTOR SYSTEMS

ALLIED CHEMICAL CORP
MATERIAL RESERCH CTR
P O BOX 1021R
MORRISTOWN NJ 07960
A J NOZIK

SOLAR RESEARCH
SOLAR ENERGY STORAGE

ALTERNATIVE SOURCES OF ENERGY
RT 2, BOX 90A
MILACA MN 56353
DONALD MARINER

-PHONE-
612 9836892
SOLAR/ENERGY CONSERV INFO DISSEMINATION ORG
PUBLISH LAY PERIODICALS

ALTERNATIVE SOURCES OF ENGY
928 2ND ST SW
ROANOKE VA 24016
EUGENE E COLI
CO-EDITOR
-PHONE-
703 3448412
SOLAR RESEARCH
SOLAR SPACE HEATING & COOLING

AM SOC HEAT REFRIG AIRCON ENGR
PUBLICATION SALES
345 EAST 47TH ST
NEW YORK NY 10017

-PHONE-
212 7526800
SOLAR/ENERGY CONSERV INFO DISSEMINATION ORG
PUBLISH TECHNICAL PERIODICALS

AM SOC HEAT REFRIG AIRCON ENGR
345 EAST 47TH ST
NEW YORK NY 10017
JOSEPH F CUBA
DIR OF RESEARCH
-PHONE-
212 7526800 X367
SOLAR RESEARCH
SOLAR SPACE HEATING & COOLING

AMELCO WINDOW CORP
77 RT 17 BOX 333
HASBROUCK HTS NJ 07604
R C CURRENT JR

-PHONE-
212 2440610
SOLAR COMPONENTS MANUFACTURER
SOLAR SPACE HEATING & COOLING

AMERICAN CYANAMID CO
CHEM RESEARCH DIV
1937 WEST MAIN ST
STAMFORD CT 06904
DR GOTTFRIED HAACKE
-PHONE-
203 3487331 X329
SOLAR RESEARCH
INORGANIC SEMI-CONDUCTORS

AMERICAN ENERGY ALTERNATVS INC
P O BOX 905
BOULDER CO 80302
J P SAYLER
-PHONE-
303 4471865
SOLAR PRODUCTS DISTRIBUTORS
WIND CONVERSION

AMERICAN GAS ASSOCIATION
ENGY&ENVIROMENT SYS
1515 WILSON BLVD
ARLINGTON VA 22209
PETER E SUSEY
MANAGER
-PHONE-
703 5242000
SOLAR RESEARCH
SOLAR SPACE HEATING & COOLING

AMERICAN INST AERONAUT&ASTRONT
MILWAUKEE STUDENT BR
3228 N 26TH STREET
MILWAUKEE WI 53206
DANIEL M LENTZ
-PHONE-
414 8730133
414 9635168
SOLAR/ENERGY CONSERV INFO DISSEMINATION ORG
INORGANIC SEMI-CONDUCTORS

AMERICAN INST OF ARCHITECTS
1735 NEW YORK AVE NW
WASHINGTON DC 20006
-PHONE-
202 7857300
SOLAR/ENERGY CONSERV INFO DISSEMINATION ORG
PUBLISH LAY + TECHNICAL PERIODICALS

AMERICAN SCIENCE&ENGINEER INC
SOLAR CLIMATE CONTRL
995 MASS AVE
CAMBRIDGE MA 02139
DAVID A BOYD
MANAGER
-PHONE-
617 8681600
SOLAR RESEARCH
SOLAR SPACE HEATING & COOLING

AMERICAN SCIENCE&ENGINEER INC
995 MASS AVE
CAMBRIDGE MA 02139
RYSZARD GAJEWSKI
DIRECTOR OF RESCH
-PHONE-
617 8681600
SOLAR RESEARCH
HIGH TEMPERATURE SOLAR COLLECTOR SYSTEMS

AMHERST COLLEGE
PHYSICS DEPT
AMHERST MA 01002
ROBERT H ROMER
-PHONE-
413 5422258
413 5422251
SOLAR RESEARCH
SOLAR SPACE HEATING

ARCHITECTS COLLABORATIVE INC
46 BRATTLE ST
CAMBRIDGE MA 02138
GERALD L FOSTER
-PHONE-
617 8684200
SOLAR INTEREST
MEDIUM TEMPERATURE SOLAR COLLECTOR SYSTEMS

ARCHITECTS COLLABORATIVE INC
46 BRATTLE STREET
CAMBRIDGE MA 02138
SARAH P HARKNESS
-PHONE-
617 8684200 X333
SOLAR RESEARCH
ALL SOLAR TECHNOLOGIES

ARCHITECTURE PLUS
1345 SIXTH AVE
NEW YORK NY 10019
MARGOT VILLECCO
SENIOR EDITOR
-PHONE-
212 4898697
SOLAR/ENERGY CONSERV INFO DISSEMINATION ORG
PUBLISH LAY + TECHNICAL PERIODICALS

ARGONNE NATIONAL LAB
ENERGY & ENVIRONMENT
9500 SOUTH CASS AVE
ARGONNE IL 60439
RALPH C NIEMANN
MECH ENGR
-PHONE-
312 7397711 X5007
SOLAR RESEARCH
SOLAR HEAT,COOL,&ELECTRIC POWER GENERATION

ARGONNE NATIONAL LABORATORY
ACCELERATOR FACILITY
9700 SOUTH CASS AVE
ARGONNE IL 60439
JOHN H MARTIN
SENIOR PHYSICIST
-PHONE-
312 7397711 X4881
312 8529714
SOLAR/ENERGY CONSERV INFO DISSEMINATION ORG
SOLAR HEAT,COOL,&ELECTRIC POWER GENERATION

ARGONNE NATIONAL LABORATORY
ARF
9700 SOUTH CASS AVE
ARGONNE IL 60439
VACLAV J SEVCIK
MECH ENGR
-PHONE-
312 7397711 X4182
SOLAR RESEARCH
SOLAR SPACE HEATING & COOLING

ARGONNE NATIONAL LABORATORY
9700 SOUTH CASS AVE
ARGONNE IL 60439
PROF ROLAND WINSTON

SOLAR RESEARCH
HIGH TEMPERATURE SOLAR COLLECTOR SYSTEMS

ARIES CONSULTING ENGRS INC
SUITE D
1810 N W 6TH STREET
GAINESVILLE FL 32601
ROBLEY S PULLER

-PHONE-
904 3726687

SOLAR RESEARCH
SOLAR SPACE HEATING & COOLING

ARIZONA STATE UNIV
LIBRARY-SCI-ENGR SER
ROOM 257
TEMPE AZ 85281
MARY BEECHER
LIBRARIAN
-PHONE-
602 9657608

SOLAR/ENERGY CONSERV INFO DISSEMINATION ORG
ALL SOLAR TECHNOLOGIES

ARIZONA STATE UNIVERSITY
COLL OF ARCHITECTURE
TEMPE AZ 85281
JOHN I YELLOTT

-PHONE-
602 9655562
602 9653216
SOLAR RESEARCH
SOLAR SPACE HEATING & COOLING

ARIZONA, UNIV OF
OPTICAL SCIENCES CTR
TUCSON AZ 85721
ADEN B MEINEL

-PHONE-
602 7493322
602 8843138
SOLAR RESEARCH
HIGH TEMPERATURE SOLAR COLLECTOR SYSTEMS

ARIZONA, UNIV OF
OPTICAL SCIENCES CTR
TUCSON AZ 85721
MARJORIE P MEINEL

-PHONE-
602 7493322
602 8843138
SOLAR RESEARCH
HIGH TEMPERATURE SOLAR COLLECTOR SYSTEMS

ARIZONA, UNIV OF
OPTICAL SCIENCE CTR
TUCSON AZ 85721
BERNHARD O SERAPHIN

-PHONE-
602 8842263

SOLAR RESEARCH
INORGANIC SEMI-CONDUCTORS

ARIZONA, UNIVERSITY OF
ENVIRONMENT RES LAB
TUCSON INT'L AIRPORT
TUCSON AZ 85706
DR JOHN F PECK
-PHONE-
602 8842931
SOLAR RESEARCH
HIGH TEMPERATURE SOLAR COLLECTOR SYSTEMS

ARKLA INDUST INC
509 MARSHALL ST
SHREVEPORT LA 71151
PHILIP ANDERSON

-PHONE-
318 4251271
SOLAR COMPONENTS MANUFACTURER
SOLAR SPACE COOLING

ARKLA INDUSTRIES INC
BOX 751
LITTLE ROCK AR 72203
ROBERT J DESTICHE
VP SALES
-PHONE-
501 3726241
SOLAR POTENTIAL MFG,ESTABLISHED BUSINESS
SOLAR SPACE HEATING & COOLING

ARMSTRONG CORK CO. R&D CENTER
BUILDING SYSTEMS
2500 COLUMBIA AVE
LANCASTER PA 17603
ALAN S GLASSMAN
RESEARCH ARCHITECT
-PHONE-
717 3970611 X7128

SOLAR RESEARCH
SOLAR SPACE HEATING & COOLING

ARTHUR D LITTLE, INC.
3601 S WADSWORTH
DENVER CO 80235
JOHN K REYNOLDS
CONSULTANT

-PHONE-
303 9854505
303 9854741
SOLAR RESEARCH
ALL SOLAR TECHNOLOGIES

ARTHUR D LITTLE,INC
PUBLIC REALTIONS
20 ACORN PARK
CAMBRIDGE MA 01240
ELLEN MCCAULEY *

-PHONE-
617 8645770

SOLAR RESEARCH
SOLAR SPACE HEATING & COOLING

ARTHUR D LITTLE,INC
ENGINEERING SCIENCES
20 ACORN PARK
CAMBRIDGE MA 01240
DR.PETER E GLASER
VICE PRESIDENT
-PHONE-
617 8645770

SOLAR RESEARCH
SOLAR SPACE HEATING & COOLING

ASE INFO SERVICES
1448 E 7TH ST
TUCSON AZ 85719
HARRY D BURRIS
PERSON

-PHONE-
602 8826571
703 2814564
SOLAR/ENERGY CONSERV INFO DISSEMINATION ORG
PROVIDE INFORMATION SEARCHES TO CLIENTS

ASG INDUSTRIES, INC
BOX 929
KINGSPORT TN 37662
JAMES S HERBERT *
DIR CUST TECH SERV

-PHONE-
615 2450211 X325
SOLAR COMPONENTS MANUFACTURER
SOLAR SPACE HEATING & COOLING

ASG INDUSTRIES,INC
1B BELMAR ROAD
CRANBURY NJ 08512
CLARENCE W CLARKSON
DIR LTG PROD TECH SV
-PHONE-
609 6550058
609 6550057
SOLAR COMPONENTS MANUFACTURER
SOLAR SPACE HEATING & COOLING

ASHLAND OIL INC
P O 391
ASHLAND KY 41101
HARRY T WILEY *

-PHONE-
606 3293333

SOLAR RESEARCH
SOLAR SPACE HEATING & COOLING

ASHLAND OIL INC
RESEARCH & DEVELOP
P O BOX 391
ASHLAND KY 41101
NORMAN W HALL
-PHONE-
606 3293333 X8374

SOLAR RESEARCH
SOLAR SPACE HEATING & COOLING

ASKELON METAL PRODUCTS LTD
TEL AVIV ISRAEL

SOLAR MANUFACTURERS
SOLAR WATER HEATING

ATLANTIC RESEARCH CORPORATION
5390 CHEROKEE AVENUE
ALEXANDRIA VA 22314
GEORGE D SUMMERS
DIRECTOR
-PHONE-
703 3543400 X315
703 3543400 X312
SOLAR POTENTIAL MFG,ESTABLISHED BUSINESS
ALL SOLAR TECHNOLOGIES

ATOMICS INTERNATIONAL
P O BOX 309
CANOGA PARK CA 91304
W V BOTTS

-PHONE-
213 3411000 X2062

SOLAR RESEARCH
SOLAR ELECTRIC POWER GENERATION

AUBURN UNIVERSITY
DEPT OF INDUST ENGR
AUBURN AL 36830
CHARLES R WHITE

-PHONE-
205 8264340
SOLAR RESEARCH
MEDIUM TEMPERATURE SOLAR COLLECTOR SYSTEMS

AUST INVENT & DEVELOPMENTS PLD
EVELOPMENTS
P O BOX 227
FILLMORE UT 84631
GORDON D GRIFFIN
MANAGING DIRECTOR
-PHONE-
801 7436817
SOLAR RESEARCH
SOLAR ENERGY STORAGE

AVCO EVERETT RESEARCH LAB INC
2385 REVERE BEACH PK
EVERETT MA 02149
RAYMOND JANNEY *
ASSIST TO CHAIRMAN

-PHONE-
617 3893000
SOLAR RESEARCH
WIND CONVERSION

AVCO EVERETT RESEARCH LAB INC
2385 REVERE BEACH PK
EVERETT MA 02149
ARTHUR KANTROWITZ
CHAIRMAN

-PHONE-
617 3893000
SOLAR RESEARCH
WIND CONVERSION

AYRES&HAYAKAWA ENERGY MGT
CONSULTING ENGINEERS
1190 SOUTH BEVERLY
LOS ANGELES CA 90035
JEROME WEINGART
SOLAR CONSULTANT
-PHONE-
213 8794477
SOLAR RESEARCH
SOLAR SPACE HEATING & COOLING

B F SALES ENGR CO
4770 FOX ST NO 9
DENVER CO 80216
GLENN SELCH

SOLAR PRODUCTS DISTRIBUTORS
MEDIUM TEMPERATURE SOLAR COLLECTOR SYSTEMS

BABCOCK & WILCOX CO
RESEARCH CENTER
1562 BEESON
ALLIANCE OH 44601
LARRY LANDIS *
PERSONNEL&PUBLIC REL
-PHONE-
216 8219110 X387
SOLAR RESEARCH
SOLAR ELECTRIC POWER GENERATION

BABCOCK & WILCOX CO
RESEARCH CENTER
1562 BEESON
ALLIANCE OH 44601
ROBERT A LEE
STEAM GEN TECH SECT
-PHONE-
216 8219110 X376
SOLAR RESEARCH
SOLAR ELECTRIC POWER GENERATION

BABCOCK & WILCOX CO
RESEARCH CENTER
1562 BEESON STREET
ALLIANCE OH 44601
KARL H SCHULZE
RESEARCH PHYSICIST
-PHONE-
216 8219110 X227
SOLAR RESEARCH
SOLAR ELECTRIC POWER GENERATION

BALL BROTHERS RESEARCH CORP
AEROSPACE DIV
P O BOX 1062
BOULDER CO 80302
M W FRANK
PROG MGR
-PHONE-
303 4414627
SOLAR RESEARCH
TOTAL SYSTEMS INTEGRATION

BARBER-NICHOLS ENGR CO
6325 W 55TH STREET
ARVADA CO 80002
ROBERT BARBER

-PHONE-
303 4218111
SOLAR RESEARCH
SOLAR SPACE COOLING

BARI, UNIV OF
ENERGY SOURCES LAB
LARGO FRACCACRETA 1
1 70122 BARY ITALY
GIORGIO NEBBIA

-PHONE-
080 214495
SOLAR RESEARCH
SOLAR WATER DISTILLATION

BATTELLE COLUMBUS LABS
UNCONVENTIONALENERGY
505 KING AVENUE
COLUMBUS OH 43201
JAMES A EIBLING
PROGRAM MANAGER
-PHONE-
614 2993151

SOLAR RESEARCH
ALL SOLAR TECHNOLOGIES

BATTELLE COLUMBUS LABS
505 KING AVENUE
COLUMBUS OH 43201
RICHARD A NATHAN
PRINCIPAL CHEMIST

-PHONE-
614 2993151 X2049

SOLAR RESEARCH
SOLAR ENERGY STORAGE

BATTELLE COLUMBUS LABS
ORGANIC CHEMISTRY
505 KING AVENUE
COLUMBUS OH 43201
DR ROBERT E SCHWERZEL
RESEARCH CHEMIST
-PHONE-
614 2993151 X3457
614 2993151 X3154
SOLAR RESEARCH
ALL SOLAR TECHNOLOGIES

BATTELLE COLUMBUS LABS
THERMOMECH ENGY SYS
505 KING AVENUE
COLUMBUS OH 43201
SHERWOOD G TALBERT
SR MECH ENGR
-PHONE-
614 2993151
SOLAR RESEARCH
MEDIUM TEMPERATURE SOLAR COLLECTOR SYSTEMS

BATTELLE NORTHWEST LABS
BATTELLE BLVD
RICHLAND WA 99352
KIRK DRUMHELLER

-PHONE-
509 9462349
SOLAR RESEARCH
SOLAR SPACE HEATING & COOLING

BEASLEY INDUSTRIES PTY LTD
BOLTON AVENUE
DEVON PARK, S A AUSTRALIA
E L BEASLEY
SALES MANAGER

-PHONE-
082 464871
SOLAR MANUFACTURERS
SOLAR WATER HEATING

BECHTEL CORPORATION
ENERGY INFO CENTER
P O BOX 3965
SAN FRANCISCO CA 43201
BARBARA CLEMENTS
-PHONE-
614 2993151
SOLAR/ENERGY CONSERV INFO DISSEMINATION ORG
PROVIDE INFORMATION SEARCHES TO CLIENTS

BENBOW RESEARCH
650 CALIFORNIA ST
SAN FRANCISCO CA 99108
J BENBOW BULLOCK
PRESIDENT
-PHONE-
415 9812516
SOLAR RESEARCH
SOLAR SPACE HEATING

BEUTELS SOLAR HEATER INC
1527 N MIAMI AVE
MIAMI FL 33136

-PHONE-
305 3711426
SOLAR MANUFACTURERS
SOLAR WATER HEATING

BIG OUTDOORS PEOPLE
P O BOX 936
MINNEAPOLIS MN 55440
DENNIS O JOHNSON

-PHONE-
612 3315344
612 3315430
SOLAR POTENTIAL MFG,ESTABLISHED BUSINESS
MEDIUM TEMPERATURE SOLAR COLLECTOR SYSTEMS

BILL FLANNIGANS PLANS
20032 23RD ST
ASTORIA NY 11100
BILL FLANNIGAN

-PHONE-
212 7282083

SOLAR/ENERGY CONSERV INFO DISSEMINATION ORG
WIND CONVERSION

BIOMASS ENERGY INST INC
PO BOX 129 P STA C
WINNIPEG MANITOBA CANADA
JOHN MORRISS

-PHONE-
204 2840472

SOLAR RESEARCH
BIO-CONVERSION

BIOTECHNIC PRESS
P O BOX 26091
ALBUQUERQUE NM 87125

SOLAR/ENERGY CONSERV INFO DISSEMINATION ORG
PUBLISH LAY BOOKS

BLACK & VEATCH
1500 MEADOW LAKE PKY
KANSAS CITY MO 64114
J C GROSSKREUTZ

-PHONE-
816 3617000
SOLAR RESEARCH
SOLAR ELECTRIC POWER GENERATION

BOARD OF COUNTY COMMISSIONERS
BROWARD COUNTY
CO COURTHOUSE RM 270
FT LAUDERDALE FL 33301
THOMAS HUGHES

SOLAR INTEREST
SOLAR WATER HEATING

BOB SCHMITT HOMES INC
13079 FALLING WATER
STRONGSVILLE OH 44136
D J DIRKSE

-PHONE-
216 2386915
SOLAR MANUFACTURERS
HOMES

BOEING AEROSPACE CO
MAIL STOP 8C-64
P O BOX 3999
SEATTLE WA 98124
RONALD B CAIRNS
RESEARCH & ENGR DEPT
-PHONE-
206 7730464
SOLAR RESEARCH
SOLAR SPACE HEATING & COOLING

BORG WARNER CORP.
RESEARCH CENTER
ALGONQUIN&WOLF RDS
DES PLAINES IL 60018
NEAL F CARDNER
RESEARCH PHYSICIST
-PHONE-
312 8273131 X266
ENERGY CONSERVATION RESEARCH
NEW CONSTRUCTION, RESEARCHING

BORG WARNER CORP.
RESEARCH CENTER
ALGONQUIN&WOLF RDS
DES PLAINES IL 60018
RICHARD L KUEHNER
STAFF SCIENTIST
-PHONE-
312 8273131
SOLAR RESEARCH
SOLAR SPACE HEATING & COOLING

BOSTON COLLEGE
PHYSICS DEPT
140 COMMONWEALTH
CHESTNUT HILL MA 02167
P H FANG

SOLAR RESEARCH
INORGANIC SEMI-CONDUCTORS

BOSTON UNIVERSITY
CHEMISTRY DEPT
685 COMMONWEALTH AVE
BOSTON MA 02215
N N LICHTIN
CHAIRMAN
-PHONE-
617 3532493
SOLAR RESEARCH
INORGANIC SEMI-CONDUCTORS

BOULDER SOLAR ENERGY SOCIETY
P O BOX 3431
BOULDER CO 80303

SOLAR/ENERGY CONSERV INFO DISSEMINATION ORG
ALL SOLAR TECHNOLOGIES

BRADFORD,UNIV OF
BRADFORD YORKSHIRE ENGLAND
TOM T STONIER
PROF OF PEACE STUD

-PHONE-
33466
SOLAR POLITICAL CONSIDERATIONS
LOBBY TO PROMOTE SOLAR LEGISLATION

BRADLEY UNIV
MECH ENGR DEPT
PEORIA IL 61606
Y B SAFDARI

-PHONE-
309 6767611
SOLAR RESEARCH
SOLAR SPACE HEATING & COOLING

BRAEMAR ENGINEERING PTY LTD
BOX 63 PO
BROOKLY PARK, S A AUSTRALIA

-PHONE-
56 6644
SOLAR MANUFACTURERS
SOLAR WATER HEATING

BRAEMAR ENGR PTY LTD
P O BOX 58
GEEBUNG AUSTRALIA

SOLAR MANUFACTURERS
SOLAR WATER HEATING

BROOKHAVEN NATIONAL LAB
PUBLIC RELATIONS
40 BROOKHAVEN AVE
UPTON NY 11973
CARL R THIEN *
INFORMATION OFFICER
-PHONE-
516 3452345
SOLAR RESEARCH
BIO-CONVERSION

BROOKHAVEN NATIONAL LABORATORY
BIOLOGY DEPT
50 BELL AVENUE
UPTON NY 11973
JOHN M OLSEN
BIOPHYSICIST
-PHONE-
516 3453382
SOLAR RESEARCH
WIND CONVERSION

BROWN MANUFACTURING CO
P O BOX 14546
OKLAHOMA CITY OK 73114
RUSSELL BROWN
PRESIDENT

-PHONE-
405 7511323
SOLAR COMPONENTS MANUFACTURER
MEDIUM TEMPERATURE SOLAR COLLECTOR SYSTEMS

BROWN UNIVERSITY
ENGINEERING DEPT
UNIV HALL BOX 1867
PROVIDENCE RI 02912
J J LOFERSKI

SOLAR RESEARCH
INORGANIC SEMI-CONDUCTORS

BSIC/EFL
3000 SAND HILL RD
MENLO PARK CA 94025
JOSHUA A BURNS
ASSOCIATE DIRECTOR
-PHONE-
415 8542300
SOLAR/ENERGY CONSERV INFO DISSEMINATION ORG
PUBLISH ORIGINAL RESEARCHED INFORMATION

BUILDING RESCH STATION
GARSTON WATFORD
WD2 7JR UNITED KINGDOM
P PERTHERBRIDGE

-PHONE-
092 7 3 740 X40
ENERGY CONSERVATION RESEARCH
SOLAR RADIATION MEASUREMENTS

BUILDING RESEARCH STATION
ADVISORY SERVICE
GARSTON WATFORD
WD2 7JR UNITED KINGDOM
*
-PHONE-
092 7 3 766 X12
SOLAR/ENERGY CONSERV INFO DISSEMINATION ORG
PUBLISH ORIGINAL RESEARCHED INFORMATION

BULLOCK ENGINEERING & DEV CO
70 WEST 6TH AVE
DENVER CO 80222
HARRY C BULLOCK
PRESIDENT

-PHONE-
303 8926416
303 8926483
SOLAR INTEREST
ALL SOLAR TECHNOLOGIES

C I R S O
DIV OF MECH ENGR
P O BOX 26
HIGHETT, VICTORIA AUSTRALIA
F G HOGG *
CHIEF
-PHONE-
03 950333

SOLAR RESEARCH
MEDIUM TEMPERATURE SOLAR COLLECTOR SYSTEMS

C I R S O
DIV OF MECH ENGR
P O BOX 26
HIGHETT, VICTORIA AUSTRALIA
ROBERT V DUNKLE
CHIEF RESEARCH SCI
-PHONE-
950333

SOLAR RESEARCH
MEDIUM TEMPERATURE SOLAR COLLECTOR SYSTEMS

C N E S
91 BRETIGNY BOX 04
FRANCE
WOLFGANG PALZ

-PHONE-
4909220
SOLAR RESEARCH
INORGANIC SEMI-CONDUCTORS

CALIF INSTITUT OF TECHNOLOGY
JET PROPULSION LAB
4800 OAK GROVE DRIVE
PASADENA CA 91103
PAUL A BERMAN
-PHONE-
213 3542816
SOLAR RESEARCH
INORGANIC SEMI-CONDUCTORS

CALIF, UNIV OF--LOS ALAMOS
LOS ALAMOS SCI LAB
P O BOX 1663
LOS ALAMOS NM 87544
WILLIAM A RANKEN
GROUP LEADER Q-25
-PHONE-
505 6676578
SOLAR RESEARCH
SOLAR SPACE HEATING & COOLING

CALIF, UNIV OF-BERKELEY
LAWRENCE BERKELEY LE
BERKELEY CA 94720
MICHAEL WAHIG
ENERGY&ENVIRO PROG
-PHONE-
415 8432740 X5787
415 8432740 X5001
SOLAR RESEARCH
INORGANIC SEMI-CONDUCTORS

CALIF DEPT PARKS & RECREATION
SPECIAL STUDIES SECT
1416 9TH ST
SACRAMENTO CA 95811
DR ROBERT B DEERING

-PHONE-
916 4453130
ENERGY CONSERVATION RESEARCH
BIO-CONVERSION

CALIF POLYTECHNIC STATE UNIV
ARVH&ENVIRO DESGN&EN
SAN LUIS OBISPO CA 93401
KEN HAGGARD

-PHONE-
805 5462573
SOLAR RESEARCH
SOLAR SPACE HEATING & COOLING

CALIF, UNIV OF-BERKELEY
BIOCHEM DEPT
BERKELEY CA 94720
J B NEILANDS

-PHONE-
415 6427460
SOLAR RESEARCH
BIO-CONVERSION

CALIF, UNIV OF-DAVIS
DEPT ENVIRO HORTCULT
DAVIS CA 95616
JONATHAN HAMMOND

-PHONE-
916 7523697

ENERGY CONSERVATION RESEARCH
SOLAR SPACE HEATING & COOLING

CALIF INST OF THE ARTS
SCHOOL OF DESIGN
24700 MCBEAN PKWY
VALENCIA CA 91355
STEPHEN SELKOWITZ
INSTRUCTOR
-PHONE-
805 2551050 X328
SOLAR RESEARCH
SOLAR WATER HEATING

CALIF, UNIV OF--L A
SCHOOL OF ENGR
ENGY & KINETICS DEPT
LOS ANGELES CA 90024
D K EDWARDS
-PHONE-
213 8255313
SOLAR RESEARCH
MEDIUM TEMPERATURE SOLAR COLLECTOR SYSTEMS

CALIF, UNIV OF-BERKELEY
SCH OF PUBLIC HEALTH
BERKELEY CA 94720
W J OSWALD

SOLAR RESEARCH
BIO-CONVERSION

CALIF, UNIV OF-DAVIS
AGR ENGR DEPT
DAVIS CA 95616
DR L W NEUBAUER

-PHONE-
916 7520102

ENERGY CONSERVATION RESEARCH
SOLAR SPACE HEATING & COOLING

CALIF, UNIV OF-L A
ENGR & APPLIED SCIEN
ENGY&KINETICS 5531BH
LOS ANGELES CA 90024
HARRY BUCHBERG

-PHONE-
213 8255313
SOLAR RESEARCH
SOLAR SPACE HEATING & COOLING

CALIF, UNIV OF-LA JOLLA
CHEM DEPT
LA JOLLA CA 92037
JOHN BENEMAN

-PHONE-
714 4532000 X1639
SOLAR RESEARCH
BIO-CONVERSION

CALIF, UNIV OF-LIVERMORE
LAWRENCE LIVER LAB
P O BOX 808
LIVERMORE CA 94550
S SZYBALSKI
ENERGY CONSV SECT
-PHONE-
415 4471100 X7721
SOLAR RESEARCH
SOLAR HEAT,COOL,&ELECTRIC POWER GENERATION

CALIF, UNIV OF-LOS ALAMOS
LOS ALAMOS SCI LAB
Q-DIVISON (Q-DOT)
LOS ALAMOS NM 87544
DR J D BALCOMB
ASST ENERGY DIV-LDR
-PHONE-
505 6676441

SOLAR RESEARCH
SOLAR SPACE HEATING & COOLING

CALIF, UNIV OF-RICHMOND FLD ST
RICHMOND FIELD STA
1301 SOUTH 46TH ST
RICHMOND CA 94704
DON D HORNING

-PHONE-
415 2356000 X211

SOLAR RESEARCH
HIGH TEMPERATURE SOLAR COLLECTOR SYSTEMS

CALIF, UNIV OF-RICHMOND FLD ST
177 RICHMOND FLD STA
1301 SOUTH 46TH ST
RICHMOND CA 94704
EVERETT D HOWE

-PHONE-
415 6426352
415 5254990 X237
SOLAR RESEARCH
SOLAR WATER DISTILLATION

CALIF, UNIV OF-SEA H2O CON LAB
SEA WATER CONV LAB
1301 SOUTH 46TH ST
RICHMOND CA 94804
BADAWI W TLEIMAT
-PHONE-
415 2356000 X231
SOLAR RESEARCH
SOLAR SPACE HEATING & COOLING

CALIF,UNIV OF--BERKELEY
MATERIALS SCI DEPT
HEARST MINING BLDG
BERKELEY CA 94720
MARSHAL F MERRIAM
-PHONE-
415 6423815
SOLAR RESEARCH
MEDIUM TEMPERATURE SOLAR COLLECTOR SYSTEMS

CARNEGIE MELLON UNIV
6123 SCIENCE HALL
PITTSBURG PA 15213
CLARENCE ZENER

-PHONE-
412 6212600 X229
SOLAR RESEARCH
OCEAN THERMAL GRADIENT SYSTEMS

CARNEGIE MELLON UNIVERSITY
ELECTRICAL ENGR DEPT
PITTSBURG PA 15213
PROF A G MILNES

-PHONE-
412 6212600
SOLAR RESEARCH
INORGANIC SEMI-CONDUCTORS

CARRIER CORP
RESEARCH DIVISION
CARRIER PKWY
SYRACUSE NY 13201
DR WENDELL J BIERMAN
-PHONE-
315 4638411 X3481
SOLAR RESEARCH
SOLAR SPACE HEATING & COOLING

CARSON ASTRONOMICAL INSTRUMENT
P O BOX 5566
VALENCIA CA 91355
CURTIS L OLSON
DIR OF MARKETING
-PHONE-
805 2551234
SOLAR POTENTIAL MFG,ESTABLISHED BUSINESS
HIGH TEMPERATURE SOLAR COLLECTOR SYSTEMS

CASSEL SOLAR ENGINEERING
ROUTE 10 BOX 17
ANNAPOLIS MD 21401
DAVID E CASSEL

-PHONE-
301 9740897

SOLAR/ENERGY CONSERV INFO DISSEMINATION ORG
SOLAR SPACE HEATING & COOLING

CATHOLIC UNIVERSITY OF AMERICA
DEPT OF ELECT ENGR
4 ST ON MICH AVE N E
WASHINGTON DC 20017
CHARLES F PULVARI
-PHONE-
202 6355193
202 5262400
SOLAR RESEARCH
INORGANIC SEMI-CONDUCTORS

CAUDILL ROWLETT SCOTT ARCH/ENG
111 W LOOP SOUTH
HOUSTON TX 77027
MICHEL BEZMAN

-PHONE-
713 6219600

SOLAR RESEARCH
SOLAR SPACE HEATING

CCR-EURATOM
SCIENTIF DIRECTORATE
ISPRA (VARESE) ITALY
JOACHIM GRETZ

-PHONE-
003 9332780 X131
SOLAR RESEARCH
TOTAL SYSTEMS INTEGRATION

CENTER FOR ENVIRONMENT & MAN
275 WINDSOR STREET
HARTFORD CT 06120
RAYMOND J DOLAN

-PHONE-
203 5494400 X348
ENERGY CONSERVATION RESEARCH
SOLAR RADIATION MEASUREMENTS

CENTRAL RESEARCH ORGANIZATION
PHYSICS & ENGR RESCH
KANBE
YANKIN P O RANGOON BURMA
U MAUNG MAUNG
-PHONE-
50544
SOLAR RESEARCH
ALL SOLAR TECHNOLOGIES

CERTAIN-TEED PRODUCTS CORP
P O BOX 860
VALLEY FORGE PA 19482
PHILIP W BACON
SOLAR ENGY COORD
-PHONE-
215 6875000 X7989
SOLAR POTENTIAL MFG,ESTABLISHED BUSINESS
MEDIUM TEMPERATURE SOLAR COLLECTOR SYSTEMS

CHAFFEY COLLEGE
ALTA LOMA CA 91701
HOMER W DAVIS

-PHONE-
714 9871737 X240
SOLAR RESEARCH
HOMES

CHAMBER OF MINES OF S AFRICA
RESEARCH LABS
PO BOX 61809
JOHANNESBURG SOUTH AFRICA
AUSTIN WHILLIER
-PHONE-
JOH 31161
SOLAR RESEARCH
MEDIUM TEMPERATURE SOLAR COLLECTOR SYSTEMS

CHARLES F KETTERING RESRCH LAB
150 E SO COLLEGE ST
YELLOW SPRINGS OH 45387
DR DONALD L KEISTER

-PHONE-
513 7677271
SOLAR RESEARCH
BIO-CONVERSION

CHARLES W MOORE ASSOCIATES
ESSEX CT 06426
WILLIAM H GROVER
MANAGER

-PHONE-
203 7670101
SOLAR RESEARCH
HOMES

CHEMALLOY-ELECTRONICS CORP
7947 TOWERS-PO DR 10
SANTEE CA 92071
MR FRIEDMAN

-PHONE-
714 4485715
SOLAR MANUFACTURERS
SOLAR ELECTRIC POWER GENERATION

CHICAGO, UNIV OF
BIOLOGY DEPT
5630 S INGLESIDE AVE
CHICAGO IL 60637
RONALD G ALDERFER
-PHONE-
312 7532666
SOLAR RESEARCH
BIO-CONVERSION

CLARK UNIVERSITY
PHYSICS DEPT
WORCESTER MA 01610
ROGER P KOHIN

-PHONE-
617 7937711

SOLAR RESEARCH
SOLAR SPACE HEATING

COGSWELL/HAUSLER ASSOC
P O BOX 2214
CHAPEL HILL NC 27514
EDWARD HOSKINS

-PHONE-
919 9425197
SOLAR/ENERGY CONSERV INFO DISSEMINATION ORG
SOLAR SPACE HEATING & COOLING

COLORADO ENERGY RESEARCH INST
COLO SCHOOL OF MINES
GOLDEN CO 80401
THOMAS VOGENTHALER
DIRECTOR
-PHONE-
303 2790300
SOLAR RESEARCH
GOVERNMENT AGENCIES PROVIDING GRANTS

CHURCHILL FILMS
CUSTOMER RELATIONS
662 NO ROBERTSON
LOS ANGELES CA 90069
PRISCILLA FLORENCE *
-PHONE-
213 6575110
SOLAR/ENERGY CONSERV INFO DISSEMINATION ORG
PUBLISH AND/OR DISTRIBUTE FILMS

CNTRL SALT&MARNE CHEM RES INST
BHAVNAGAR (GUJARAT) INDIA
DR R L DATTA

-PHONE-
3960

SOLAR RESEARCH
INORGANIC SEMI-CONDUCTORS

COLEMAN SOLAR SERVICE
8900 NW 34 AVE RD
MIAMI FL 33147
JAMES J GAY

-PHONE-
305 6914126
SOLAR MANUFACTURERS
SOLAR WATER HEATING

COLORADO NATIONAL BANK
P O BOX 5168TA
DENVER CO 80217
ROGER WHITE *
ASST TO PRESIDENT
-PHONE-
303 8931862
SOLAR FINANCIAL CONSIDERATIONS
BANKERS

CHURCHILL FILMS
662 NO ROBERTSON
LOS ANGELES CA 90069
GEORGE H MCQUILKIN
PRODUCER-DIRECTOR
-PHONE-
213 6575110
SOLAR/ENERGY CONSERV INFO DISSEMINATION ORG
PUBLISH AND/OR DISTRIBUTE FILMS

COBALT ENGINEERING CO
164 PENINGTN-HPOURTN
PENNINGTON NJ 08534
BRAD OWEN,PE

-PHONE-
609 7371376
609 8821414 X360
SOLAR POTENTIAL MFG,ESTABLISHED BUSINESS
SOLAR SPACE HEATING

COLORADO COLLEGE
PHYSICS
COLORADO SPRINGS CO 80903
DAVID F KERN
STUDENT
-PHONE-
303 6328168
SOLAR RESEARCH
SOLAR SPACE HEATING

COLORADO STATE UNIV
SOLAR ENGY APPL LAB
FORT COLLINS CO 80521
DAN S WARD

-PHONE-
303 4918211
SOLAR RESEARCH
SOLAR SPACE HEATING & COOLING

COLORADO STATE UNIVERSITY
PROJECT SURGE
FT COLLINS CO 80521
-PHONE-
303 4911101
SOLAR/ENERGY CONSERV INFO DISSEMINATION ORG
SOLAR SPACE HEATING & COOLING

COLORADO STATE UNIVERSITY
SOLAR ENGY APPL LAB
ENGR RESEARCH CENTER
FT COLLINS CO 80521
SUSUMU KARAKI
ASSOC DIRECTOR
-PHONE-
303 4911101
SOLAR RESEARCH
SOLAR SPACE HEATING & COOLING

COLORADO STATE UNIVERSITY
SOLAR ENGY APPL LAB
ENGR RESEARCH CENTER
FORT COLLINS CO 80521
DR GEORGE LOF
DIRECTOR
-PHONE-
303 4918632
303 4918325
SOLAR RESEARCH
SOLAR SPACE HEATING & COOLING

COLORADO UNIV OF-BOULDER
CTR FOR MGT&TEC PROG
BOULDER CO 80302
-PHONE-
303 4434865
SOLAR/ENERGY CONSERV INFO DISSEMINATION ORG
SOLAR HEAT,COOL,&ELECTRIC POWER GENERATION

COLORADO UNIV OF-BOULDER
ELECT ENGR
BOULDER CO 80302
FRANK BARNES
-PHONE-
303 4927327
SOLAR RESEARCH
INORGANIC SEMI-CONDUCTORS

COLORADO UNIV OF-BOULDER
GEOGRAPHY DEPT
BOULDER CO 80302
R G BARRY
-PHONE-
303 4926387
SOLAR RESEARCH
SOLAR RADIATION MEASUREMENTS

COLORADO UNIV OF-BOULDER
DIV OF CONTINUING ED
BOULDER CO 80302
MARGARET EVANS
-PHONE-
303 4927416
SOLAR/ENERGY CONSERV INFO DISSEMINATION ORG
PUBLISH AND/OR DISTRIBUTE FILMS

COLORADO UNIV OF-BOULDER
CIVIL & ENVIR ENGR
BOULDER CO 80302
DAVID FENG
-PHONE-
303 4927112
ENERGY CONSERVATION RESEARCH
NEW CONSTRUCTION, PASSIVE SYSTEMS

COLORADO UNIV OF-BOULDER
CIRES
BOULDER CO 80302
HARTMUT SPETZER
-PHONE-
303 4928028
SOLAR RESEARCH
HIGH TEMPERATURE SOLAR COLLECTOR SYSTEMS

COLORADO UNIV OF-BOULDER
AEROSPACE ENG SCI
BOULDER CO 80302
MAHINDER UBEROI
-PHONE-
303 4926612
SOLAR RESEARCH
HIGH TEMPERATURE SOLAR COLLECTOR SYSTEMS

COLORADO, UNIV OF
MEDICAL SCHOOL
DENVER CO 80220
JOHN C COBB MD
-PHONE-
303 3947150
SOLAR RESEARCH
HOMES

COLORADO, UNIV OF-BOULDER
ATIP
RM 331A UMC
BOULDER CO 80302
MILTON LOEB
-PHONE-
303 4926659
SOLAR/ENERGY CONSERV INFO DISSEMINATION ORG
HAVE&MAINTAIN ENERGY DATA BANK

ENVIRONMTL ACTION-CO

DENVER CO 80206
BEN BILLINGS
ENVIRO ACT REPRINT S
-PHONE-
303 5341602
SEE PAGE 4.009 + 4.021
SOLAR/ENERGY CONSERV INFO DISSEMINATION ORG
PUBLISH INFORMATIONAL REPRINTS

COLORADO, UNIV OF-DENVER

1100 14TH STREET
DENVER CO 80202
ELIZABETH A KINGMAN
SOLAR EXHIBIT PROG
-PHONE-
SEE PAGE 4.009 + 4.022
303 8921117 X471
SOLAR/ENERGY CONSERV INFO DISSEMINATION ORG
PROVIDE AUDIO & VISUAL SERVICES

ENVIRONMTL ACTION-CO

DENVER CO 80206
ALBERT NUNEZ JR
SOLAR PROJECT COORD
-PHONE-
303 5341602
SEE PAGE 4.009
SOLAR/ENERGY CONSERV INFO DISSEMINATION ORG
PROVIDE INFORMATION SEARCHES TO CLIENTS

COLORADO, UNIV OF-DENVER

1100 14TH STREET
DENVER CO 80202
CAROLYN M PESKO
SOLAR DIRECT PROJ DR
-PHONE-
SEE PAGE 4.009
303 8921117 X471
SOLAR/ENERGY CONSERV INFO DISSEMINATION ORG
HAVE&MAINTAIN ENERGY DATA BANK

ENVIRONMTL ACTION-CO

DENVER CO 80206
MOREY WOLFSON
DIRECTOR
-PHONE-
303 5341602
SEE PAGE 4.009 + 4.021
SOLAR/ENERGY CONSERV INFO DISSEMINATION ORG
ALL SOLAR TECHNOLOGIES

COLSPAN ENVIRONMENTAL SYS INC
P O BOX 3467
BOULDER CO 80303
ALAN M FRANK

-PHONE-
303 4494411
SOLAR POTENTIAL MFG,NEW BUSINESS
MEDIUM TEMPERATURE SOLAR COLLECTOR SYSTEMS

COLUMBIA UNIVERSITY
MECHANICAL ENGR
214 MUDD
NEW YORK NY 10027
PROF ROBERT TAUSSIG

-PHONE-
212 2802961
SOLAR RESEARCH
MEDIUM TEMPERATURE SOLAR COLLECTOR SYSTEMS

COMFORT TRAIN AIR COND CO
4060 S KALAMATH
ENGLEWOOD CO 80110
WILLIAM BROWNING

-PHONE-
303 7894497
SOLAR PRODUCTS DISTRIBUTORS
SOLAR SPACE HEATING & COOLING

COMSTAT LABORATORIES
APPLIED SCI DIV
P O BOX 115
CLARKSBURG MD 20734
DR EDMUND S RITTNER
DIRECTOR
-PHONE-
301 4284541
SOLAR RESEARCH
INORGANIC SEMI-CONDUCTORS

CON EDISON CO OF NY INC
RESEARCH & DEVELOP
H IRVING PLACE
NEW YORK NY 10003
DR ROBERT A BELL
DIRECTOR
-PHONE-
212 4603882
SOLAR RESEARCH
MEDIUM TEMPERATURE SOLAR COLLECTOR SYSTEMS

CONNECTICUT, UNIV OF
ENVIRONMENTAL ECON.
BOX U-21 UNIV. OF CT
STORRS CT 06268
CARLOS D STERN
ASSISTANT PROFESSOR
-PHONE-
203 4862740
ENERGY CONSERVATION RESEARCH
RETROFITTING, TECHNOLOGIES

CONNECTICUT, UNIVERSITY OF
CHEMICAL ENGR
STORRS CT 06268
DR MICHAEL HOWARD *
ASST PROF

-PHONE-
203 4864020
ENERGY CONSERVATION RESEARCH
RETROFITTING, TECHNOLOGIES

CONSOLIDATED EDISON
4 IRVING PLACE
NEW YORK NY 10003
R A BELL

-PHONE-
212 4604600
SOLAR RESEARCH
SOLAR SPACE HEATING & COOLING

CONTINENTAL CAN CO INC
PROCUREMENT PLANNING
633 THIRD AVE
NEW YORK NY 10017
R W DIEHL
CORP DIRECTOR
-PHONE-
212 5517851
SOLAR RESEARCH
ALL SOLAR TECHNOLOGIES

CONTINENTAL CAN COMPANY INC
CORP R&D DEPT
7622 S RACINE AVE
CHICAGO IL 60620
DR J G BUCK

-PHONE-
312 8463800
SOLAR RESEARCH
ALL SOLAR TECHNOLOGIES

CONTOUR CO INC
9300 WHITMORE ST
EL MONTE CA 91731
DAN MAROVSH

-PHONE-
212 2834101
SOLAR POTENTIAL MFG,ESTABLISHED BUSINESS
SOLAR SPACE HEATING & COOLING

CORNING GLASS WORKS
LIGHTING PROD DIV
CORNING NY 14830
DONALD A URQUHART *
MANAGER SPECIAL PROJ
-PHONE-
607 9747306
SOLAR POTENTIAL MFG,ESTABLISHED BUSINESS
MEDIUM TEMPERATURE SOLAR COLLECTOR SYSTEMS

CORNING GLASS WORKS
PHYSICAL RESEARCH
CORNING NY 14830
UGUR ORTABASI
SENIOR PHYSIST
-PHONE-
607 9743370
SOLAR RESEARCH
MEDIUM TEMPERATURE SOLAR COLLECTOR SYSTEMS

COSANTI FOUNDATION
6433 DOUBLETREE ROAD
SCOTTSDALE AZ 85253
PAOLO SOLERI
PRESIDENT
-PHONE-
602 9486145

SOLAR RESEARCH
ALL SOLAR TECHNOLOGIES

CRISS-CROSS
BOX 173
BLACKHAWK CO 80422

-PHONE-
303 5829000
303 4437035
SOLAR RESEARCH
SOLAR SPACE HEATING

CROWER CAMS & EQUIPMENT CO
3333 MAIN STREET
CHULA VISTA CA 92011
BRUCE CROWER
PRESIDENT

SOLAR RESEARCH
SOLAR ENGINES

CTR FOR RESEARCH-ACTS OF MAN
4025 CHESTNUT STREET
PHILADELPHIA PA 19104
MADELEINE S KLAUSNER
DIR OF ADMIN
-PHONE-
215 5946241

SOLAR SOCIAL CONSIDERATIONS
ALL SOLAR TECHNOLOGIES

CURTIS&DAVIS ARCHITECTS/PLANER
126 EAST 38TH STREET
NEW YORK NY 10016
CARL T GRIMM

-PHONE-
212 6899590

SOLAR/ENERGY CONSERV INFO DISSEMINATION ORG
PROVIDE INFORMATION SEARCHES TO CLIENTS

CUTTER LABS
808 PARKER
BERKELEY CA 94710
EDGAR M FRIED, PE
STAFF ENG
-PHONE-
415 8410123 X278
415 2842858
SOLAR INTEREST
HOMES

D&J SHEET METAL CO
10555 NW 7TH AVE
MIAMI FL 33150
JAKE STICHER

-PHONE-
305 7577033
SOLAR MANUFACTURERS
SOLAR WATER HEATING

DARWIN TEAGUE INC
375 SYLVAN AVE
ENGLEWOOD CLIFFS NJ 07632
W D TEAGUE
PRES

-PHONE-
201 5685124
SOLAR RESEARCH
SOLAR SPACE HEATING & COOLING

DAYLIN INC
SUN SOURCE
9606 SANTA MONICA
BEVERLY HILLS CA 90035
DAVID L COLLINS
GENERAL MANAGER
-PHONE-
213 8783211
SOLAR MANUFACTURERS
SOLAR WATER HEATING

DAYTON UNIV OF
DEPT OF MECH ENGR
COLLEGE PK AV
DAYTON OH 45469
JOHN E MINARDI
-PHONE-
513 2290123
SOLAR RESEARCH
MEDIUM TEMPERATURE SOLAR COLLECTOR SYSTEMS

DEETER HOMES
246 PALM AVE
AUBURN CA 95603
DAN DEETER

-PHONE-
916 8858448
SOLAR INTEREST
SOLAR SPACE HEATING & COOLING

DEKO LABS
BOX 12841 UNIV STATN
GAINESVILLE FL 32604
DONALD F DEKOLD

-PHONE-
904 3726009
SOLAR COMPONENTS MANUFACTURER
MEDIUM TEMPERATURE SOLAR COLLECTOR SYSTEMS

DELAWARE, UNIV OF
INST OF ENGY CONVSN
NEWARK DE 19711
DR KARL W BOER *

-PHONE-
302 7388481

SOLAR RESEARCH
HOMES

DELAWARE, UNIV OF
INST OF ENGY CONVSN
NEWARK DE 19711
L R CRITTENDON

-PHONE-
302 7388481

SOLAR RESEARCH
HOMES

DELAWARE, UNIV OF
CHEMISTRY DEPT
NEWARK DE 19711
MARJORIE P KRAUS

-PHONE-
302 7382458
215 2748114
SOLAR RESEARCH
BIO-CONVERSION

DELAWARE, UNIV OF
ENERGY CONVERSION
NEWARK DE 19711
MARIA TELKES
ADJUNCT PROFESSOR
-PHONE-
302 7388481
SOLAR RESEARCH
SOLAR ENERGY STORAGE

DENVER MUSEUM OF NATURAL HIST
GATES PLANETARIUM
CITY PARK
DENVER CO 80205
MARK PETERSON
-PHONE-
303 3884201
SOLAR/ENERGY CONSERV INFO DISSEMINATION ORG
PROVIDE AUDIO & VISUAL SERVICES

DENVER PLANNING OFFICE
1445 CLEVELAND PLACE
DENVER CO 80202
WAYLAND WALKER *
CITY PLANNER
-PHONE-
303 2972736
SOLAR INTEREST
PLANNING OFFICES, LOCAL

DENVER PLANNING OFFICE
1445 CLEVELAND PLACE
DENVER CO 80202
ALAN L CANTER
DIRECTOR OF PLANNING
-PHONE-
303 2972736
303 2972737
SOLAR INTEREST
PLANNING OFFICES, LOCAL

DENVER, UNIV OF
DENVER RESEARCH INST
UNIVERSITY OF DENVER
DENVER CO 80210
FRED P VENDITTI
SR RES ENGR
-PHONE-
303 7532241
SOLAR RESEARCH
SOLAR ELECTRIC POWER GENERATION

DEPT OF HOUSING & URBAN DEVEL
BUILDING TECH DIV
451 7TH STREET S W
WASHINGTON DC 20410
ORVILLE LEE
SOLAR ENGY PROG MGR
-PHONE-
202 7575356
SOLAR RESEARCH
SOLAR SPACE HEATING & COOLING

DEPT OF HOUSING & URBAN DEVEL
ENERGY AFF RM 10230
451 7TH STREET S W
WASHINGTON DC 20410
GEORGE TAPPERT
-PHONE-
202 7556480
SOLAR FINANCIAL CONSIDERATIONS
ALL SOLAR TECHNOLOGIES

DESERT RESEARCH INSTITUTE
OFFICE OF PRESIDENT
RENO NV 89507
JOHN R DOHERTY
-PHONE-
702 7846131
SOLAR RESEARCH
ACADEMIC BUILDINGS

DESERT SUNSHINE EXPOSURE TESTS
CLIENT SERVICES
BOX185 BLACKCAN STGE
PHOENIX AZ 85020
JOHN L SCOTT *
-PHONE-
602 4657525
602 4657521
SOLAR ENVIRONMENTAL CONSIDERATIONS
SOLAR,MATERIAL WEATHERABILITY TESTS

DESERT SUNSHINE EXPOSURE TESTS
BOX185 BLACKCAN STGE
PHOENIX AZ 85020
GENE A ZERLAUT
PRESIDENT
-PHONE-
602 4657525
602 4657521
SOLAR RESEARCH
SOLAR,MATERIAL WEATHERABILITY TESTS

DO LITTLE INDUSTRIES
23 TERRACE PLACE
DANBURY CT 06810
SAM SAHLY
SOLAR RESEARCH
HOMES

DOME EAST CORPORATION
325 DUFFY AVENUE
HICKSVILLE NY 11801
DAVID A ROBINSON
DESIGN ENGINEER
-PHONE-
516 9380545
SOLAR MANUFACTURERS
MEDIUM TEMPERATURE SOLAR COLLECTOR SYSTEMS

DOMENICO SPERANDO
VIA CIMAROSA 13-21
FOLLONICA (GR) ITALY
SOLAR MANUFACTURERS
WIND CONVERSION

DOW CHEMICAL USA
ROCKY FLATS DIV
P O BOX 888
GOLDEN CO 80401
DEVELOPMENT PROJECTS
-PHONE-
303 4942951
SOLAR RESEARCH
SOLAR HEAT,COOL,&ELECTRIC POWER GENERATION

DOW CORNING
SEMICONDUCTOR PROD
12334 GEDDES RD
HEMLOCK MI 48626
LEON D CROSSMAN
MGR SOLID STATE RES
-PHONE-
517 6425201 X344
SOLAR RESEARCH
INORGANIC SEMI-CONDUCTORS

DOW CORNING CORP
RESEARCH DEPT
S SAGINAW RD
MIDLAND MI 48640
H A CLARK

-PHONE-
517 6368676
SOLAR RESEARCH
INORGANIC SEMI-CONDUCTORS

DOWNING-LEACH ASSOC ARCH&PLANN
2305 BROADWAY
BOULDER CO 80302
R B DOWNING

-PHONE-
303 4437533
SOLAR RESEARCH
CONDOMINIUMS

DUBIN-MINDELL-BLOOME ASSOC PE
CONSULTING ENGINEERS
42 W39 ST
NEW YORK NY 10018
FRED S DUBIN
PRESIDENT
-PHONE-
212 8689700
SOLAR/ENERGY CONSERV INFO DISSEMINATION ORG
TOTAL SYSTEMS INTEGRATION

DUNLITE ELECTRICAL CO LTD
FROME ST
ADELAIDE SOUTH AUSTRALIA

SOLAR MANUFACTURERS
WIND CONVERSION

DUPONT E I DE NEMOURS CO
DEVELOPMENT DEPT
1007 MARKET ST
WILMINGTON DE 19898

SOLAR COMPONENTS MANUFACTURER
ALL SOLAR TECHNOLOGIES

DYNA TECHNOLOGY INC
P O BOX 3263
SIOUX CITY IA 51102

SOLAR MANUFACTURERS
WIND CONVERSION

DYNATHERM CORP
MARELE CT OFF IND LN
COCKEYSVILLE MD 21030
J A STREB
V.P.MARKETING
-PHONE-
301 6669151
SOLAR POTENTIAL MFG,ESTABLISHED BUSINESS
SOLAR SPACE HEATING & COOLING

EARTHMIND
JOSEL SAUGUS CA 91350
DAVID HOUSE

-PHONE-
805 2513053
SOLAR/ENERGY CONSERV INFO DISSEMINATION ORG
WIND CONVERSION

ECO-SOLUTIONS,INC.
P O BOX 4117
BOULDER CO 80302
ROBERT HESS *
PRESIDENT
-PHONE-
303 4436798
SOLAR POTENTIAL MFG,NEW BUSINESS
SOLAR SPACE HEATING

ECO-SOLUTIONS,INC.
PO BOX 4117
BOULDER CO 80302
DAVID A SCHOEN
EXECUTIVE SECRETARY
-PHONE-
303 4436798
SOLAR POTENTIAL MFG,NEW BUSINESS
SOLAR SPACE HEATING

ECOLOGY ACTION EDUCATION INST
BOX 3895
MODESTO CA 95352
CLIFFORD C HUMPHREY

-PHONE-
209 5291964
SOLAR/ENERGY CONSERV INFO DISSEMINATION ORG
PUBLISH ORIGINAL RESEARCHED INFORMATION

ECOTEC FOUNDATION INC
2923 WOR
LD AVENUE
CINCINNATI OH 45206
PETER SEICLEL
-PHONE-
513 8618324
SOLAR RESEARCH
HOMES

EDMUND SCIENTIFIC CO
ADV & MKT
101 E GLOUCESTER PK
BARRINGTON NJ 08007
JACK SCHARFF
DIR MKT & ADV
-PHONE-
609-547-3488
SOLAR MANUF, EDUCATIONAL EQUIPMENT
SOLAR SPACE HEATING & COOLING

EDWARDS ENGINEERING CORP
101 ALEXANDER AVE
POMPTON NJ 07444
RAY C EDWARDS
PRESIDENT
-PHONE-
202 8352808 X49
SOLAR POTENTIAL MFG,ESTABLISHED BUSINESS
SOLAR SPACE HEATING & COOLING

ENAG S A
DE PONT L'ABBE 29S
QUIMPER FRANCE

SOLAR MANUFACTURERS
WIND CONVERSION

ENERGY DEVELOPMENT ASSOC
1100 W WHITCOMB AVE
MADISON HTS MI 48071
NEVILLE MAPHAM
GENERAL MANAGER
-PHONE-
313 5664000
313 5839434
SOLAR COMPONENTS MANUFACTURER
SOLAR ENERGY STORAGE

EDUCATIONAL MATERIALS&EQUIP CO
BOX 17
PELHAM NY 10803
THOMAS J MCMAHON

-PHONE-
914 5761121
SOLAR MANUF, EDUCATIONAL EQUIPMENT
ALL SOLAR TECHNOLOGIES

ELECTRICAL RESEARCH ASSOC
CLEEVE RD
LEATHERHEAD SURREF ENGLAND

SOLAR RESEARCH
ALL SOLAR TECHNOLOGIES

ENERGEX CORP
418 TROPICANA
LAS VEGAS NV 89109
ALFRED JENKINS
PRESIDENT
-PHONE-
702 7362994
SOLAR MANUFACTURERS
ALL SOLAR TECHNOLOGIES

ENERGY MANAGEMENT CO
8900 MELROSE AVE
LOS ANGELES CA 90069

-PHONE-
213 2746881

SOLAR INTEREST
ALL SOLAR TECHNOLOGIES

EDWARDS ENGINEERING CORP
101 ALEXANDER AVE
POMPTON NJ 07444
EDWARD BOGUCZ *

-PHONE-
201 8352808
SOLAR POTENTIAL MFG,ESTABLISHED BUSINESS
SOLAR SPACE HEATING & COOLING

ELECTROTECHNICAL LAB
ENGY SYSTEM SECTION
TANASHI
TOYKO JAPAN
TATSU TANI

SOLAR RESEARCH
HIGH TEMPERATURE SOLAR COLLECTOR SYSTEMS

ENERGY CONVERSION DEVICES INC
TROY MI 48084
STANFORD R OVSHINSKY

-PHONE-
313 5497300
SOLAR MANUFACTURERS
INORGANIC SEMI-CONDUCTORS

ENERGY POLICY PROJECT
1776 MASS AV NW
WASHINGTON DC 20036
DAVID S FREEMAN
DIRECTOR
-PHONE-
202 8721064

SOLAR RESEARCH
PUBLISH ORIGINAL RESEARCHED INFORMATION

ENERGY RESEARCH CORP
ENERGY REVIEW
6 EAST VALERIO
SANTA BARBARA CA 93101
RONALD A ZUCKERMAN
EDITOR
-PHONE-
805 9631388
SOLAR/ENERGY CONSERV INFO DISSEMINATION ORG
HAVE&MAINTAIN ENERGY DATA BANK

ENERGY SYSTEMS INC
634 CREST DR
EL CAJON CA 92021
TERRANCE R CASTER
PRESIDENT
-PHONE-
714 4471000
SOLAR MANUFACTURERS
SOLAR SPACE HEATING

ENGINEERS COLLABORATIVE LTD
101 E ONTARIO
CHICAGO IL 60611
L A BIHLER
EXEC V P
-PHONE-
312 6449600
SOLAR RESEARCH
TOTAL SYSTEMS INTEGRATION

ENVIRONMENTAL ACTION COALITION
EDUC MATERIALS PROG
235 EAST 49TH ST
NEW YORK NY 10017
JOAN EDWARDS
PROGRAM OFFICER
-PHONE-
212 4869550
SOLAR/ENERGY CONSERV INFO DISSEMINATION ORG
HAVE&MAINTAIN ENERGY DATA BANK

ENVIRONMENTAL CONSULT SERV INC
1485 SIERRA DRIVE
BOULDER CO 80302
DR JAN F KREIDER
CHIEF ENGR
-PHONE-
303 4472218
303 4431235
SOLAR RESEARCH
SOLAR HEAT,COOL,&ELECTRIC POWER GENERATION

ENVIRONMENTAL DESIGN
PETTIGREW AR 72752
JOEL DAVIDSON
SOLAR RESEARCH
SOLAR SPACE HEATING & COOLING

ENVIRONMENTAL EDUCATION GROUP
67 LURLINE AVENUE
CANOGO CA 91306
ALLAN TRATNER
SOLAR/ENERGY CONSERV INFO DISSEMINATION ORG
PUBLISH ORIGINAL RESEARCH + REPRINTS

ENVIRONMENTAL ENERGIES INC
11350 SCHAEFER ST
DETROIT MI 48227
SOLAR POTENTIAL MFG,ESTABLISHED BUSINESS
WIND CONVERSION

ENVIRONMENTAL ENERGY FOUNDATN
ROOM 1001
12000 EDGEWATER DR
LAKEWOOD OH 44107
G GILBERT OUTTERSON
DIRECTOR
-PHONE-
216 2218130
SOLAR RESEARCH
SOLAR HEAT,COOL,&ELECTRIC POWER GENERATION

ENVIRONMENTAL INFORMATION CTR
124 EAST 39TH STREET
NEW YORK NY 10016
CHRISTOPHER KIMBALL
-PHONE-
212 6790810
SOLAR/ENERGY CONSERV INFO DISSEMINATION ORG
PUBLISH LAY + TECHNICAL BOOKS

ENVIRONMENTAL SYS FOR PEOPLE
213 SUN 2030 E SPDWY
TUCSON AZ 85719
ROBERT G WASON
CONSULTING ENGR
-PHONE-
602 7955437
SOLAR INTEREST
SOLAR SPACE HEATING & COOLING

EPPLEY LABORATORY INC
12 SHEFFIELD AVE
NEWPORT RI 02840
GEORGE L KIRK
PRESIDENT
-PHONE-
4 0184710 X20
SOLAR MANUFACTURERS
SOLAR RADIATION MEASUREMENTS

ESB INCORPORATED
19 WEST COLLEGE AVE
PA 19067
EDOUARD JORDI

-PHONE-
215 4933659
SOLAR RESEARCH
SOLAR WATER DISTILLATION

ESCHER TECHNOLOGY ASSOC
P O BOX 189
ST JOHNS MI 48879
WILLIAM ESCHER

-PHONE-
517 2246726
SOLAR RESEARCH
OCEAN THERMAL GRADIENT SYSTEMS

EXXON RESCH&ENGR CO
SOLAR ENGY CONV UNIT
LINDEN NJ 07036
J A ECKERT
PROJECT MANAGER
-PHONE-
201 4740100
SOLAR RESEARCH
INORGANIC SEMI-CONDUCTORS

FAFCO INC
2860 SPRING ST
REDWOOD CITY CA 94065
RALPH SCHNEIDER
MARKETING MGR
-PHONE-
415 3646772
SOLAR MANUFACTURERS
SOLAR WATER HEATING

FEDERAL ENGY OFFICE REGION II
26 FEDERAL PLAZA
NEW YORK NY 10017
KENNETH A PORTER

-PHONE-
212 2641041
SOLAR RESEARCH
ALL SOLAR TECHNOLOGIES

FEDERAL POWER COMMISSION
OFFICE CHIEF ENGR
825 N CAPITOL ST NE
WASHINGTON DC 20426
WALTER E CUSHEN
-PHONE-
202 3765270
SOLAR RESEARCH
SOLAR WATER HEATING

FEUER CORPORATION
5401 MCCONNELL AVE
LOS ANGELES CA 90066
STANLEY FEUER ASHRAE
PRESIDENT

-PHONE-
213 3904046
213 4562894
SOLAR INTEREST
SOLAR SPACE HEATING & COOLING

FEUER CORPORATION
ENGINEERING
5401 MCCONNELL AVE
LOS ANGELES CA 90066
RICHARD D MILLER ASHRAE ASME
CHIEF ENGINEER
-PHONE-
213 3904046
213 8217209
SOLAR INTEREST
SOLAR SPACE HEATING & COOLING

FIRST NATIONAL CITY BANK
AREA 3 OPERATING GP
PO BOX 939 CHURCH ST
NEW YORK NY 10008
HENRY DEFORD
SENIOR VP

SOLAR FINANCIAL CONSIDERATIONS
BANKERS

FLINDERS UNIV OF S AUSTRALIA
BEDFORD PARK
SOUTH AUSTRALIA AUSTRALIA
J O'M BOCKRIS

-PHONE-
76 0511

SOLAR RESEARCH
INORGANIC SEMI-CONDUCTORS

FLORIDA ENERGY COMMISSION
SENATE OFFICE BLDG
TALLAHASSEE FL 32304
MARVIN YAROSH

-PHONE-
904 4881078
904 4881167
SOLAR POLITICAL CONSIDERATIONS
ALL SOLAR TECHNOLOGIES

FLORIDA, UNIV OF
SOLAR ENGY/ENGY CONV
GAINESVILLE FL 92611
ERICH H FARBER *
DIRECTOR
-PHONE-
904 3920820

SOLAR RESEARCH
SOLAR ENGINES

FLORIDA, UNIV OF
ELECTRIC ENGR DEPT
GAINESVILLE FL 32601
ROBERT L BAILEY

-PHONE-
904 3920916
904 3920921
SOLAR RESEARCH
INORGANIC SEMI-CONDUCTORS

FLORIDA, UNIV OF
126 MECH ENGR BLDG
GAINESVILLE FL 32611
JOHN C PEED
PROF EMERITUS
-PHONE-
904 3920805

SOLAR RESEARCH
ALL SOLAR TECHNOLOGIES

FLORIDA, UNIV OF
DEPT MECH ENGR
GAINESVILLE FL 32601
DR HERBERT SCHAEPER

-PHONE-
904 3920809

SOLAR RESEARCH
WIND CONVERSION

FLORIDA,UNIV OF/MECH ENGR
1222 N W 34 TERRACE
GAINESVILLE FL 32605
JOHN ATKINS

-PHONE-
904 3723117
SOLAR RESEARCH
SOLAR SPACE COOLING

FORD MOTOR CO
SCIENTIFIC RES STAFF
PO BOX 2053
DEARBORN MI 48121
ROY L GEALER
PRINCIPAL STAFF ENGR
-PHONE-
313 3231219
ENERGY CONSERVATION RESEARCH
RETROFITTING, TECHNOLOGIES

FORD MOTOR COMPANY
SCIENTIFIC RES STAFF
ADVANCED TECH APP/RA
DEARBORN MI 48121
J V PETROCELLI

SOLAR RESEARCH
SOLAR ENERGY STORAGE

FOSTER WHEELER CORP
APP THERMODYNAMICS
110 S ORANGE AVE
LIVINGSTON NJ 07039
R J ZOSCHAK
MANAGER
-PHONE-
201 5331100
SOLAR RESEARCH
SOLAR ELECTRIC POWER GENERATION

FRANKLIN INST RESEARCH LABS
MECH&NUCLEAR ENGR DP
20TH & RACE STREETS
PHILADELPHIA PA 19103
W H STEIGLEMANN MANAGER
ENERGY SYS LABS
-PHONE-
215 4481138
SOLAR RESEARCH
SOLAR SPACE HEATING & COOLING

FRANKLIN INST RESEARCH LABS
SOLAR APPLC INFO CTR
20TH & RACE STREETS
PHILADELPHIA PA 19103
FRANK WEINSTEIN

-PHONE-
215 4481500
SOLAR/ENERGY CONSERV INFO DISSEMINATION ORG
HAVE&MAINTAIN ENERGY DATA BANK

FRED RICE PRODUCTIONS INC
6313 PEACH AVE
VAN NUYS CA 91401
FRED RICE
PRESIDENT
-PHONE-
213 7863860
SOLAR MANUFACTURERS
SOLAR WATER HEATING

FRIENDS RESEARCH CENTER
ENVIRONMENTAL STUDYS
308 HILTON AVE
CATONSVILLE MD 21228
BRUCE L WELCH
-PHONE-
301 7472773
SOLAR POLITICAL CONSIDERATIONS
ALL SOLAR TECHNOLOGIES

FULTON FEDERAL SAVINGS & LOAN
P O BOX 1077
ATLANTA GA 30301
ROBERT FREDERIC

-PHONE-
404 5222300
SOLAR FINANCIAL CONSIDERATIONS
BANKERS

FUN & FROLIC INC
P O BOX 277
MADISON HEIGHTS MI 48071
EDWARD J KONOPKA
PRESIDENT

SOLAR MANUFACTURERS
SOLAR WATER HEATING

GENERAL ATOMIC COMPANY
P O BOX 81608
SAN DIEGO CA 92138
DAVE WALCK *

-PHONE-
714 4531000

SOLAR RESEARCH
SOLAR ELECTRIC POWER GENERATION

GENERAL DYNAMICS CORP
CONVAIR AEROSPACE
P O BOX 748
FT WORTH TX 76101
DONALD W GOODWIN
DESIGN SPECIALIST
-PHONE-
817 7324811 X3113
817 7324811 X3292
SOLAR POTENTIAL MFG,NEW BUSINESS
MEDIUM TEMPERATURE SOLAR COLLECTOR SYSTEMS

GENERAL RESEARCH CORPORATION
SYSTEMS RESEARCH DIV
1501 WILSON BLVD
ARLINGTON VA 22209
NANCY C LESTER

-PHONE-
703 5347206

SOLAR/ENERGY CONSERV INFO DISSEMINATION ORG
PUBLISH ORIGINAL RESEARCHED INFORMATION

GARDEN WAY LABORATORIES
P O BOX 66
CHARLOTTE VT 05445
DR DOUGLAS C TAFF

-PHONE-
802 8626501
SOLAR MANUFACTURERS
WIND CONVERSION

GENERAL ATOMIC COMPANY
P O BOX 81608
SAN DIEGO CA 92138
DR JOHN RUSSELL
HEAD,SOLAR RESEARCH

-PHONE-
714 4531000

SOLAR RESEARCH
SOLAR ELECTRIC POWER GENERATION

GENERAL ELECTRIC COMPANY
TPD
MOUNTAIN VIEW ROAD
LYNCHBURG VA 24502
OTWARD M MUELLER
ELECTRONICS ENGINEER
-PHONE-
804 8467311 X2765
804 2371161
SOLAR INTEREST
SOLAR HEAT,COOL,&ELECTRIC POWER GENERATION

GEOCENTRIC SYSTEMS INC
ARCHITECTURAL
PO BOX 49
TAOS NM 87571
WILLIAM A MINGENBACH
PRESIDENT
-PHONE-
505 7584111
505 7582864
SOLAR RESEARCH
NEW CONSTRUCTION, PRACTICING

GAYDARDT INDUSTRIES INC
8542 EDGEWORTH DR
CAPITAL HEIGHTS MD 20028
GEORGE R GAYDOS

-PHONE-
301 3500614
SOLAR MANUFACTURERS
MEDIUM TEMPERATURE SOLAR COLLECTOR SYSTEMS

GENERAL DYNAMICS CORP
CONVAIR AEROSPACE
P O BOX 748
FORT WORTH TX 76101
DR RALPH O DOUGHTY
GROUP ENGINEER
-PHONE-
817 7324811 X3113
817 7324811 X3292
SOLAR POTENTIAL MFG,NEW BUSINESS
MEDIUM TEMPERATURE SOLAR COLLECTOR SYSTEMS

GENERAL MOTORS CORP
HARRISON RADIATOR
UPPER MOUNTAIN ROAD
LOCKPORT NY 14094
VERNON L ERIKSEN
DIRECTOR OF R AND D
-PHONE-
716 4393362

SOLAR POTENTIAL MFG,NEW BUSINESS
MEDIUM TEMPERATURE SOLAR COLLECTOR SYSTEMS

GEORGE O G LOF CONS ENGR
158 FILLMORE ST R204
DENVER CO 80206
GEORGE O LOF
CONSULTING ENGINEER

-PHONE-
303 3220446
303 4918632
SOLAR RESEARCH
SOLAR SPACE HEATING & COOLING

GEORGIA INST OF TECHNOLOGY
ENGINEERING EXP STAT
ATLANTA GA 30332
STEVE H BOMAR

-PHONE-
404 8943661
SOLAR RESEARCH
SOLAR FURNACE

GEORGIA INST OF TECHNOLOGY
SCHOOL OF ARCH
ATLANTA GA 30332
FRANK J CLARKE

-PHONE-
404 8944885
ENERGY CONSERVATION RESEARCH
SOLAR SPACE HEATING & COOLING

GEORGIA INST OF TECHNOLOGY
ENGINEERING EXP STAT
ATLANTA GA 30332
ATIP DEBS

-PHONE-
404 8943661
SOLAR RESEARCH
SOLAR SPACE HEATING

GEORGIA INST OF TECHNOLOGY
ENGINEERING EXP STAT
ATLANTA GA 30332
J F KINNEY

-PHONE-
404 8943661

SOLAR RESEARCH
MEDIUM TEMPERATURE SOLAR COLLECTOR SYSTEMS

GEORGIA INST OF TECHNOLOGY
ENGINEERING EXP STAT
ATLANTA GA 30332
JESSE D WALTON JR

-PHONE-
404 8943661

SOLAR RESEARCH
SOLAR ELECTRIC POWER GENERATION

GLEN-KAYE FILMS
100 EAST 21ST STREET
BROOKLYN NY 11226
DOUGLAS R KAYE
PRODUCER
-PHONE-
212 2872929
212 6939618
SOLAR/ENERGY CONSERV INFO DISSEMINATION ORG
PUBLISH AND/OR DISTRIBUTE FILMS

GOBAR GAS RESEARCH STATION
AJITMAL ETAWAH U P INDIA
RAM BUX SINGH
INCHARGE

SOLAR/ENERGY CONSERV INFO DISSEMINATION ORG
BIO-CONVERSION

GOVT INDUST RESCH INST NAGOYA
SOLAR RESEARCH LAB
1 HIRATE-MACHI
KITA-KU NAGOYA JAPAN
TETSUO NOGUCHI

-PHONE-
052 9112111
SOLAR RESEARCH
SOLAR FURNACE

GOVT OF INDIA YOJANA BHAVAN
POWER & ENERGY SECT
PARLIAMENT STREET
NEW DELHI 1 INDIA
MANAS KUMAR CHATTERJEE
DIR ENGY PLANN COMM
-PHONE-
381109
SOLAR RESEARCH
INORGANIC SEMI-CONDUCTORS

GRAY COMPANY ENTERPRISES
SUITE 245
7701 N STEMMONS
DALLAS TX 75247
DAN R FOSTER
DIR-MEDIA RELATIONS
-PHONE-
214 6303841

SOLAR/ENERGY CONSERV INFO DISSEMINATION ORG
PROVIDE AUDIO SERVICES

GROUP SEVEN CHEMICALS CO
1210 DONA DRIVE
CORPUS CHRISTI TX 78407
GEORGE R ULVILD, FAIC
OWNER

-PHONE-
512 8886191
512 8835809
SOLAR RESEARCH
HIGH TEMPERATURE SOLAR COLLECTOR SYSTEMS

GRUMMAN AEROSPACE CORPORATION
PLANT 30
BETHPAGE NY 11714
JOHN MOCKOVCIAK, JR *
MGR ENERGY SYS DEPT

-PHONE-
516 5753785

SOLAR POTENTIAL MFG,ESTABLISHED BUSINESS
MEDIUM TEMPERATURE SOLAR COLLECTOR SYSTEMS

GRUMMAN AEROSPACE CORPORATION
ENERGY SYSTEMS DEPT
PLANT 30
BETHPAGE NY 11714
DR ROBERT MADLEY
MGR RESEARCH&DEVEL
-PHONE-
516 5751573
SOLAR POTENTIAL MFG,ESTABLISHED BUSINESS
MEDIUM TEMPERATURE SOLAR COLLECTOR SYSTEMS

H G MULTITRADE INC
P O BOX 01 4781
MIAMI FL 33130
-PHONE-
305 8569603
SOLAR INTEREST
MEDIUM TEMPERATURE SOLAR COLLECTOR SYSTEMS

HARVARD UNIV
GRADUATE SCH DESIGN
DEPT OF ARCHITECTURE
CAMBRIDGE MA 02138
DAVID LORD
ASST PROF
-PHONE-
617 4952570
SOLAR RESEARCH
ALL SOLAR TECHNOLOGIES

HARVARD UNIVERSITY
GRADUATE SCHOOL
CAMBRIDGE STATION
CAMBRIDGE MA 02138
B CHALMERS
SOLAR RESEARCH
INORGANIC SEMI-CONDUCTORS

HAWAII, UNIV OF
HI INST OF GEOPHYSIC
2525 CORREA ROAD
HONOLULU HI 96822
MARTIN VITOUSEK
-PHONE-
808 9487668
SOLAR RESEARCH
OCEAN THERMAL GRADIENT SYSTEMS

HEIZER CORPORATION
20 NORTH WACKER DR.
CHICAGO IL 60606
PETER J GILLESPIE
VICE PRESIDENT
-PHONE-
312 6412200
SOLAR FINANCIAL CONSIDERATIONS
PROVIDE VENTURE CAPITOL

HELIO ASSOCIATES INC
8230 E BROADWAY
TUCSON AZ 85710
ROBERT E GERLACH
VICE PRESIDENT
-PHONE-
602 8865376
SOLAR RESEARCH
SOLAR HEAT,COOL,&ELECTRIC POWER GENERATION

HELIO-DYNAMICS INC
518 SOUTH VAN NESS A
LOS ANGELES CA 90020
TRUMAN D TEMPLE
PRES
-PHONE-
213 3849853
SOLAR MANUFACTURERS
MEDIUM TEMPERATURE SOLAR COLLECTOR SYSTEMS

HELIOTEC INC
33 EDINBORO ST
BOSTON MA 02160
-PHONE-
617 4821245
SOLAR MANUFACTURERS
MEDIUM TEMPERATURE SOLAR COLLECTOR SYSTEMS

HELTER-SKELTER ENGINEERING
BOX 479
FOREST KNOLLS CA 94933
WILLIAM WARD-SWANN
-PHONE-
415 4572986
SOLAR MANUFACTURERS
SOLAR SPACE HEATING

HITACHI
437 MADISON AVE
NEW YORK NY 10022
-PHONE-
212 8384804
SOLAR MANUFACTURERS
SOLAR WATER HEATING

HITTMAN ASSOCIATES INC
ENGY & ENVIR SYS
9190 REED BRANCH RD
COLUMBIA MD 21043
H M CURRAN
DIRECTOR OF ENGR
-PHONE-
301 7307800
SOLAR RESEARCH
SOLAR SPACE COOLING

HOFLAR INDUSTRIES
5511 128TH ST
SURREY B C CANADA
ERICH W HOFFMANN

-PHONE-
604 5962665
SOLAR MANUFACTURERS
SOLAR SPACE HEATING

HONEYWELL INC
SYSTEMS&RESEARCH CTR
2600 RIDGWAY PARKWAY
MINNEAPOLIS MN 55413
ROGER N SCHMIDT
-PHONE-
612 3314141 X4078
SOLAR RESEARCH
SOLAR ELECTRIC POWER GENERATION

HONEYWELL INC
SYSTEMS&RESEARCH CTR
2600 RIDGWAY PARKWAY
MINNEAPOLIS MN 55413
DR M K SELCUK
-PHONE-
612 3314141
SOLAR RESEARCH
SOLAR SPACE HEATING & COOLING

HONEYWELL INCORP
URBAN & ENVIRO SYS
2700 RIDGWAY PARKWAY
MINNEAPOLIS MN 55413
NEIL C SHER
-PHONE-
612 3314141 X4033
SOLAR RESEARCH
SOLAR HEAT,COOL,&ELECTRIC POWER GENERATION

HOUSTON, UNIV OF
PHYSICS DEPT
HOUSTON TX 77004
ALVIN HILDEBRANDT
PROFESSOR
-PHONE-
713 7492834
SOLAR RESEARCH
SOLAR ELECTRIC POWER GENERATION

HOUSTON, UNIV OF
PHYSICS DEPT
HOUSTON TX 77004
DR LORIN L VANT-HULL

-PHONE-
713 7493809
SOLAR RESEARCH
SOLAR ELECTRIC POWER GENERATION

I B M
T WATSON RESERCH CTR
PO BOX 218
YORKTOWN HEIGHTS NY 10598
ROBERT W KEYES
-PHONE-
914 9453000
SOLAR RESEARCH
INORGANIC SEMI-CONDUCTORS

IBM
ELECTRONICS SYSTEMS
HUNTSVILLE AL 35804
LOLLAR MANDT

SOLAR MANUFACTURERS
SOLAR RADIATION MEASUREMENTS

ILLINOIS, STATE OF
CAPITAL DEVELOP BORD
1030 S LAGRANGE RD
LAGRANGE IL 60525
ROBERT A KEGEL PE
-PHONE-
312 3528134
SOLAR/ENERGY CONSERV INFO DISSEMINATION ORG
NEW CONSTRUCTION, TECHNOLOGIES

ILLINOIS, UNIV OF
ADVANCED COMP CENTER
URBANA IL 61801
ROBERT A HERENDEEN

SOLAR/ENERGY CONSERV INFO DISSEMINATION ORG
HAVE&MAINTAIN ENERGY DATA BANK

ILLINOIS, UNIV OF
SCHOOL OF ENGINEERNG
URBANA IL 61801
J T PFEFFER

SOLAR RESEARCH
BIO-CONVERSION

ILLINOIS, UNIV OF-CHICAGO
ARCHITECTURE&ART CLG
BOX 4348
CHICAGO IL 60680
ROGER G WHITMER AIA
ASSOC PROF ARCH DEPT
-PHONE-
312 9963335
SOLAR RESEARCH
NEW CONSTRUCTION, TECHNOLOGIES

ILLINOIS, UNIVERSITY OF
SMALL HOMES COUNCIL
ONE EAST ST MARYS RD
CHAMPAIGN IL 61820
WAYNE L SHICK
-PHONE-
217 3331801 X457
SOLAR/ENERGY CONSERV INFO DISSEMINATION ORG
PROVIDE INFORMATION SEARCHES TO CLIENTS

INNOTECH CORP
181 MAIN ST
NORWALK CN 06851
RAY PENNOYER
V P
-PHONE-
203 8462041
SOLAR RESEARCH
INORGANIC SEMI-CONDUCTORS

INSTITUTE D'ENERGIE SOLAIRE
OBSERVATOIRE
BOUZAREAH ALGERIA
D M DAMERDJI

-PHONE-
784229
SOLAR RESEARCH
SOLAR WATER DISTILLATION

INT'L RESEARCH&TECHNOLOGY CORP
1225 CONN AVENUE NW
WASHINGTON DC 20036
MARTIN O STERN

SOLAR RESEARCH
SOLAR ENGINES

INDIAN INSTITUTE OF TECHNOLOGY
MADRAS 600036 INDIA
DR MAKAM C GUPTA
PROF MECH ENGR

-PHONE-
802742 X235
SOLAR RESEARCH
SOLAR SPACE HEATING & COOLING

INST FOR FARM ELECTRIFICATION
1U VESHNIAKOVSKI
DOM 2 MOSCOW USSR

SOLAR MANUFACTURERS
WIND CONVERSION

INSTITUTE OF GAS TECHNOLOGY
3424 S STATE
CHICAGO IL 60616
WILLIAM F RUSH

-PHONE-
312 2259600 X685
SOLAR RESEARCH
SOLAR SPACE HEATING & COOLING

INTERNATIONAL RECTIFIER
SEMICONDUCTOR DIV
233 KANSAS ST
EL SEGUNDO CA 90245

-PHONE-
213 6786281
SOLAR MANUFACTURERS
INORGANIC SEMI-CONDUCTORS

INFORM
25 BROAD ST
NEW YORK NY 10004
STEWART W HERMAN

-PHONE-
212 4253550
SOLAR/ENERGY CONSERV INFO DISSEMINATION ORG
PUBLISH ORIGINAL RESEARCHED INFORMATION

INSTITUT ROYAL METEOROLOGIQUE
RADIOMETRY SECTION
3 AVENUE CIRCULAIRE
B 1180 BRUSSELS BELGIUM
R DOGNIAUX
-PHONE-
02 746788
SOLAR RESEARCH
SOLAR RADIATION MEASUREMENTS

INT'L RESEARCH&TECH CORP
1501 WILSON BLVD
ARLINGTON VA 22209
THEODORE B TAYLOR

-PHONE-
703 5245834
SOLAR RESEARCH
HIGH TEMPERATURE SOLAR COLLECTOR SYSTEMS

INTERTECHNOLOGY CORPORATION
MARKETING OPERATIONS
PO BOX 340
WARRENTON VA 22186
NORRIS L BEARD *
DIRECTOR
-PHONE-
703 3477900
SOLAR RESEARCH
SOLAR SPACE HEATING & COOLING

INTERTECHNOLOGY CORPORATION
PO BOX 340
WARRENTON VA 22186
GEORGE C SZEGO
PRESIDENT
-PHONE-
703 3477900
703 3471113 X4121
SOLAR RESEARCH
SOLAR SPACE HEATING & COOLING

INTL SOLAR ENERGY SOCIETY
PO BOX 52
PARKVILLE VIC AUSTRALIA
THE SECRETARY

SOLAR/ENERGY CONSERV INFO DISSEMINATION ORG
ALL SOLAR TECHNOLOGIES

INTL SOLAR ENGY SOC/UK SECT
21 ALBEMARLE ST
LONDON ENGLAND
J K PAGE
CHAIRMAN

SOLAR/ENERGY CONSERV INFO DISSEMINATION ORG
ALL SOLAR TECHNOLOGIES

INTL SOLAR ENGY SOC/US SECT
C/O SMITHSONIAN REL
12441 PARKLAWN DR
ROCKVILLE MD 20852
DR WILLIAM H KLEIN
SECRETARY/TREASURER

SOLAR/ENERGY CONSERV INFO DISSEMINATION ORG
ALL SOLAR TECHNOLOGIES

ION PHYSICS CORPORATION
SOUTH BEDFORD ST
BURLINGTON MA 01803
ROBERT BERMAN
MANAGER

-PHONE-
617 2721313 X462
SOLAR POTENTIAL MFG,ESTABLISHED BUSINESS
INORGANIC SEMI-CONDUCTORS

IOWA STATE UNIV
DEPT OF ARCHITECTURE
AMES IA 50010
RAY D CRITES FAIA

-PHONE-
515 2948460
SOLAR RESEARCH
SOLAR SPACE HEATING & COOLING

IRVIN INDUSTRIES INC
STRUCTURES GROUP
555 S BROADWAY
LEXINGTON KY 40508
LLOYD RAIN
MANAGER-PROD DEVEL
-PHONE-
606 2546406
SOLAR POTENTIAL MFG,ESTABLISHED BUSINESS
SOLAR SPACE HEATING

ISOTHERMICS
P O BOX 86
AUGUSTA NJ 07822

-PHONE-
201 3833500
SOLAR COMPONENTS MANUFACTURER
SOLAR SPACE HEATING

ISRAEL INSTITUTE OF TECHNOLOGY
TECHNION
HAIFA ISRAEL
M S ERLICKI

-PHONE-
04 235151
SOLAR RESEARCH
INORGANIC SEMI-CONDUCTORS

ISRAEL INSTITUTE OF TECHNOLOGY
TECHNION
HAIFA ISRAEL
S GAVRIL

-PHONE-
04 235151
SOLAR RESEARCH
INORGANIC SEMI-CONDUCTORS

ISRAEL INSTITUTE OF TECHNOLOGY
TECHNION
HAIFA ISRAEL
D SCHIEBER

-PHONE-
04 235151
SOLAR RESEARCH
INORGANIC SEMI-CONDUCTORS

ITEK CORPORATION
OPTICAL SYSTEMS DIV
10 MAGUIRE ROAD
LEXINGTON MA 02173
GARY NELSON
-PHONE-
617 2763411
SOLAR POTENTIAL MFG,ESTABLISHED BUSINESS
SOLAR SPACE HEATING & COOLING

JADAVPUR UNIVERSITY
ELEC & TELECOMM ENGR
CALCUTTA 700032 INDIA
DR J S CHATTERJEE *
PROF IN CHARGE

SOLAR RESEARCH
INORGANIC SEMI-CONDUCTORS

JOHN SIMKINS SCHOOL
MAIN ST
SOUTH YARMOUTH MA 02664
JOHN SILVER

-PHONE-
617 3982412
SOLAR RESEARCH
MEDIUM TEMPERATURE SOLAR COLLECTOR SYSTEMS

JOHNS-MANVILLE
RES AND DEV
PO BOX 5108
DENVER CO 80217
JOHN O COLLINS, JR
MGR GOVT RES LIASON
-PHONE-
303 7701000 X404

ENERGY CONSERVATION RESEARCH
NEW CONSTRUCTION, PRACTICING

JOHNSON DIVERSIFIED INC
2340 QUEEN ANN ST
MERRITT ISLAND FL 32952
STAN JOHNSON

-PHONE-
305 4525545

SOLAR MANUFACTURERS
SOLAR WATER HEATING

JADAVPUR UNIVERSITY
ELEC & TELECOMM ENGR
CALCUTTA 700032 INDIA
MANISH KR MUKHERJEE
PROFESSOR

SOLAR RESEARCH
INORGANIC SEMI-CONDUCTORS

JOHNS HOPKINS UNIV
APPLIED PHYSICS LAB
8621 GEORGIA AVE
SILVER SPRINGS MD 20910
W H AVERY
-PHONE-
301 9537100 X2411
SOLAR RESEARCH
OCEAN THERMAL GRADIENT SYSTEMS

JOHNS-MANVILLE CORP
PRODUCT INFO CENTER
GREENWOOD PLAZA
DENVER CO 80217
ANN PADOVANI *
CURRENT PRODUCTS
-PHONE-
303 7701000

ENERGY CONSERVATION RESEARCH
NEW CONSTRUCTION, PRACTICING

JOHNSON SERVICE COMPANY
RESEARCH DIVISION
507 E MICHIGAN ST
MILWAUKEE WI 53201
S R BUCHANAN
ASST DIR OF RESEARCH
-PHONE-
414 2769200 X677

SOLAR POTENTIAL MFG,ESTABLISHED BUSINESS
SOLAR SPACE HEATING & COOLING

JEFFERSON CO PUBLIC SCHOOLS
ENVIRONMENTAL EDUC
2192 COORS DRIVE
GOLDEN CO 80401
BETTY JANE MEADOWS
ENV EDUC SPECIALIST
-PHONE-
303 2793346
303 2793347
SOLAR ENVIRONMENTAL CONSIDERATIONS
INCENTIVES FOR ENERGY CONSERVATION

JOHNS HOPKINS UNIV
APPLIED PHYSICS LAB
8621 GEORGIA AVE
SILVER SPRING MD 20910
CHARLES J SWET
-PHONE-
301 9537100
SOLAR RESEARCH
ALL SOLAR TECHNOLOGIES

JOHNS-MANVILLE CORP
GROWTH PLAN&DEVELOP
GREENWOOD PLAZA
DENVER CO 80217
JACK D VERSCHOOR
MANAGER-PLANNING
-PHONE-
303 7701000 X2879
303 9865757
ENERGY CONSERVATION RESEARCH
NEW CONSTRUCTION, MECHANICAL SYSTEMS

KALWALL CORPORATION
P O BOX 237
MANCHESTER NH 03105
KEITH HARRISON
DIVISION MGR

-PHONE-
603 6273861
603 6277887
SOLAR MANUFACTURERS
SOLAR SPACE HEATING

KALWALL CORPORATION
KAL-LITE DIVISION
P O BOX 237
MANCHESTER NH 03105
RONALD K HERMSDORF

-PHONE-
603 6273861
SOLAR COMPONENTS MANUFACTURER
MEDIUM TEMPERATURE SOLAR COLLECTOR SYSTEMS

KALYANI, UNIV OF
DEPT OF PHYSICS
KALYANI NADIA
WEST BENGAL INDIA
PROF D C SARKAR *
HEAD

SOLAR RESEARCH
INORGANIC SEMI-CONDUCTORS

KALYANI, UNIV OF
DEPT OF PHYSICS
KALYANI NADIA
WEST BENGAL INDIA
HIRANMAY SAHA

SOLAR RESEARCH
INORGANIC SEMI-CONDUCTORS

KANSAS STATE UNIV
CHEM ENGR DEPT
MANHATTAN KS 66506
ROBERT H SHIPMAN

-PHONE-
913 5325584
SOLAR RESEARCH
BIO-CONVERSION

KANSAS,UNIV OF
DEPT OF GEOGRAPHY
LAWRENCE KS 66044
STEVE BLAKE

-PHONE-
913 8644276
SOLAR RESEARCH
WIND CONVERSION

KAPUR SOLAR FARMS
BIJWASAN-NAJAFGARH R
P O KAPAS HERA
NEW DELHI 110037 INDIA
J C KAPUR
-PHONE-
391747
SOLAR RESEARCH
MEDIUM TEMPERATURE SOLAR COLLECTOR SYSTEMS

KNME CHANNEL 5
1130 UNIV BLVD NE
ALBUQUERQUE NM
FRED MANFREDI
PRODUCER
-PHONE-
505 2772121
SOLAR/ENERGY CONSERV INFO DISSEMINATION ORG
PUBLISH AND/OR DISTRIBUTE FILMS

LAB SCIENCES INC
PO BOX 1236
BOCA RATON FA 33432
JOSEPH A CARISEO
PRESIDENT
-PHONE-
305 3951917
SOLAR MANUF, EDUCATIONAL EQUIPMENT
MEDIUM TEMPERATURE SOLAR COLLECTOR SYSTEMS

LAMA FOUNDATION
PO BOX 444
SAN CRISTOBAL NM 87564
HANS VON BREIEN JR
COORDINATOR

SOLAR RESEARCH
MEDIUM TEMPERATURE SOLAR COLLECTOR SYSTEMS

LAWRENCE LIVERMORE LAB
SOLAR ENERGY GROUP
P O BOX 808
LIVERMORE CA 94550
ARNOLD F CLARK
PHYSICIST
-PHONE-
415 4471100 X8693
SOLAR RESEARCH
SOLAR ELECTRIC POWER GENERATION

LEAR SIEGLER INC
MAMMOTH DIVISION
13120 B COUNTY RD 6
MINNEAPOLIS MN 55441
FLOYD H SCHENEEBERG
VICE PRES
-PHONE-
612 5442711
SOLAR RESEARCH
SOLAR SPACE HEATING

LEGISLATIVE ENERGY COMMISSION
NEW YORK STATE
ROOM 718 L.O.B.
ALBANY NY 12207
JAMES I MONROE
SCIENTIFIC ADVISOR
-PHONE-
518 4722700
SOLAR RESEARCH
LEGISLATIVE, STATE

LEHR ASSOC,CONSULTING ENGRS
ONE PENN PLAZA
NEW YORK NY 10001
VALENTINE A LEHR

-PHONE-
212 9478050
SOLAR/ENERGY CONSERV INFO DISSEMINATION ORG
SOLAR WATER HEATING

LENNOX INDUSTRIES INC
HEATING EQUIP ENGR
PO BOX 250
MARSHALLTOWN IA 50158
GARRY D HANSON
MANAGER
-PHONE-
515 7544011
SOLAR INTEREST
SOLAR SPACE HEATING & COOLING

LEROY S TROYER & ASSOC
112 1/2 LINCOLNWAY E
MISHAWAKA IN 46544
LEROY S TROYER

-PHONE-
219 2599976
SOLAR INTEREST
ALL SOLAR TECHNOLOGIES

LIBBEY OWENS FORD CO
811 MADISON AVE
TOLEDO OH 43695
DAVID R GROVE
SALES MGR
-PHONE-
419 2425781
SOLAR COMPONENTS MANUFACTURER
HIGH TEMPERATURE SOLAR COLLECTOR SYSTEMS

LIBBEY OWENS FORD CO
811 MADISON AVE
TOLEDO OH 43695
Y K PEI DR

-PHONE-
419 2473731
SOLAR POTENTIAL MFG,ESTABLISHED BUSINESS
MEDIUM TEMPERATURE SOLAR COLLECTOR SYSTEMS

LIBRARY OF CONGRESS
SCIENCE POLICY DIV
WASHINGTON DC 20540
JAMES G MOORE

-PHONE-
202 4266040
SOLAR/ENERGY CONSERV INFO DISSEMINATION ORG
PROVIDE INFORMATION SEARCHES TO CLIENTS

LISTON & ASSOC,MECH ENGRS
890 SARATOGA AVE
SAN JOSE CA 95129
THOMAS L LISTON

-PHONE-
408 2466281
SOLAR RESEARCH
SOLAR SPACE HEATING

LOCKHEED PALO ALTO RESRCH LAB
DEPT 52-01 BLDG 201
3251 HANOVER ST
PALO ALTO CA 94304
WAYNE SHANNON *
-PHONE-
415 4934411
SOLAR RESEARCH
SOLAR SPACE HEATING & COOLING

LOCKHEED PALO ALTO RESRCH LAB
DEPT 52/21 BLDG 205
3251 HANOVER ST
PALO ALTO CA 94304
ELMER R STREED
-PHONE-
415 4934411
SOLAR RESEARCH
SOLAR SPACE HEATING & COOLING

LOF BROS SOLAR APPLIANCES
P O BOX 10594
DENVER CO 80210
L LOF

-PHONE-
303 7446852
303 7582585
SOLAR MANUFACTURERS
SOLAR WATER HEATING

LOS ALAMOS SCIENTIFIC LAB
ENGINEERING
ENG-9 MS 636
LOS ALAMOS NM 87544
RICHARD A HEMPHILL
ENGINEER
-PHONE-
505 6676261
505 6721965
SOLAR INTEREST
NEW CONSTRUCTION, MECHANICAL SYSTEMS

LOUISIANA TECH UNIVERSITY
ELECTRICAL ENGINEER
PO BOX 4967,TECH STA
RUSTON LA 71270
DEAN L SMITH
ASSISTANT PROFESSOR
-PHONE-
318 2572262

SOLAR RESEARCH
INORGANIC SEMI-CONDUCTORS

LOW IMPACT TECHNOLOGY LTD
73 MOLESWORTH ST
CORNWALL ENGLAND
ANDREW MACKILLOP
DIRECTOR

SOLAR COMPONENTS MANUFACTURER
SOLAR SPACE HEATING & COOLING

M I T
ARCHITECTURE
3-407 MIT 77MASSAVE
CAMBRIDGE MA 02139
DAY CHAHROUDI
RESEARCH AFILLIATE
-PHONE-
617 2537647
SOLAR RESEARCH
SOLAR SPACE HEATING & COOLING

MAGRUDER MIDDLE SCHOOL
TORRANCE CA 90510
ALLEN GOLDSMITH

SOLAR RESEARCH
SOLAR FURNACE

MANITOBA
4IO- 352 DONALD ST
WINNIPEG MANITOBA CANADA
JOHN M GLASS CO
ENVIRONMENTAL DES

-PHONE-
204 9424735

SOLAR RESEARCH
TOTAL SYSTEMS INTEGRATION

MARTIN MARIETTA AEROSPACE
MAIL I661
PO BOX 179
DENVER CO 80201
DAVID H BACHMAN
SR FACILITIES ENGR
-PHONE-
303 7945211 X4051
303 7945211 X3427
SOLAR INTEREST
SOLAR ENGINES

MARTIN MARIETTA AEROSPACE
MAIL # E0455
PO BOX 179
DENVER CO 80201
FLOYD A BLAKE

-PHONE-
303 7945211 X3677

SOLAR RESEARCH
SOLAR ELECTRIC POWER GENERATION

MARYLAND DEPT OF NAT RESOURCES
PWR PLT SITING PROG
TAWES ST OFFICE BLDG
ANNAPOLIS MD 21401
DR PAUL MASSICOT

-PHONE-
301 2675384
SOLAR RESEARCH
SOLAR ELECTRIC POWER GENERATION

MARYLAND, UNIV OF
CAMPUS PLANNING DEPT
SERVICE BLDG 2ND FL
COLLEGE PARK MD 20740
WILLIAM E JONES
CAMPUS ARCHITECT
-PHONE-
301 4545770
SOLAR RESEARCH
SOLAR SPACE HEATING

MARYLAND, UNIVERSITY OF
MECHANICAL ENGR
COLLEGE PARK MD 20740
REDFIELD W ALLEN

SOLAR RESEARCH
SOLAR SPACE COOLING

MASCHINEFABRIK LUBING BENING
2847 BARNSTORF PO110
GERMANY

-PHONE-
054 42625
SOLAR MANUFACTURERS
WIND CONVERSION

MASSACHUSETTS, UNIV OF
CIVIL ENGINEERING
AMHERST MA 01002
W E HERONEMUS

SOLAR RESEARCH
OCEAN THERMAL GRADIENT SYSTEMS

MASSACHUSETTS, UNIV OF
PHYSICS & ASTRONOMY
AMHERST MA 01002
ALLAN R HOFFMAN

-PHONE-
413 5450933
SOLAR RESEARCH
SOLAR SPACE HEATING

MASSACHUSETTS, UNIV OF
DEPT CHEM ENGR
AMHERST MA 01002
W LEIGH SHORT
PROF

-PHONE-
413 5451588
413 5452507
ENERGY CONSERVATION RESEARCH
BIO-CONVERSION

MASSACHUSETTS, UNIVERSITY OF
PHYSICS DEPT
HASBROUCK LABORATORY
AMHERST MA 01002
ROBERT A JOHNSON
RESEARCH ASSOCIATE
-PHONE-
413 5450346

SOLAR INTEREST
NEW CONSTRUCTION, RESEARCHING

MCDONNELL DOUGLAS ASTRONAUTICS
ADV SOL&NVC ENGY SYS
5301 BOLSA AVE
HUNTINGTON BEACH CA 92646
RAYMON W HALLET JR
DIRECTOR
-PHONE-
714 8963664

SOLAR RESEARCH
SOLAR ELECTRIC POWER GENERATION

MCFALL AND KONKEL CONSULT ENGR
2160 S CLERMONT ST
DENVER CO 80222
TEMPLE LOONEY

-PHONE-
303 7574976
SOLAR RESEARCH
SOLAR SPACE HEATING & COOLING

MCGILL UNIV MACDONALD COLLEGE
BRACE RESEARCH INST
ST ANN DE BELEVUE800
QUEBEC H9X 3MI CANADA
T A LAWAND *
DIR FIELD OPERATIONS
-PHONE-
514 4576580 X341
SOLAR RESEARCH
ALL SOLAR TECHNOLOGIES

MCGILL UNIV MACDONALD COLLEGE
BRACE RESEARCH INST
ST ANN DE BELEVUE800
QUEBEC H9X 3MI CANADA
RON ALWARD

-PHONE-
514 4576580 X341
SOLAR/ENERGY CONSERV INFO DISSEMINATION ORG
ALL SOLAR TECHNOLOGIES

MCGILL UNIVERSITY
CIVIL ENGR DEPT
MONTREAL P Q CANADA
NICOLAS CHEPURNIY

-PHONE-
514 8437821

SOLAR RESEARCH
SOLAR PONDS

MECHANICAL ENGINEER
3881 S GRAPE ST
DENVER CO 80237
FRED L TRAUTMAN PE
OWNER
-PHONE-
303 7570416

SOLAR RESEARCH
SOLAR SPACE HEATING & COOLING

MEEK'S
6520 E 25 PLACE
TULSA OK 74129
IVAN R MEEK
PRES
-PHONE-
918 8356081
918 8356638
SOLAR POTENTIAL MFG,NEW BUSINESS
SOLAR SPACE HEATING & COOLING

MEGATHERM CORP
TAUNTON AVE
EAST PROVIDENCE RI 02914
ROBERT H STEVENSON

-PHONE-
401 4383800
SOLAR COMPONENTS MANUFACTURER
SOLAR SPACE HEATING

MELBOURNE, UNIV OF
MECH ENGR DEPT
MELBOURNE
PARKVILLE, VICTORIA AUSTRALIA
W W CHARTRES
-PHONE-
345 1844 X6744
SOLAR RESEARCH
MEDIUM TEMPERATURE SOLAR COLLECTOR SYSTEMS

METALCO
NICOSIA CYPRUS GREECE
MARIUS IONNIDES
PROD MGR

SOLAR MANUFACTURERS
SOLAR WATER HEATING

METROPOLITAN ECOLOGY WORKSHOP
74 JOY STREET
BOSTON MA 02114
ROBERT POLLUCK

-PHONE-
617 7234699
SOLAR/ENERGY CONSERV INFO DISSEMINATION ORG
ALL SOLAR TECHNOLOGIES

MICHIGAN, UNIV OF
DEPT OF ARCH&DESIGN
TAPPAN & MONROE STS
ANN ARBOR MI 48104
EDWARD J KELLY JR
INSTRUCTOR ENV TECH
-PHONE-
313 7641340
SOLAR RESEARCH
SOLAR SPACE COOLING

MINISTRY OF INT'L TRADE
ENERGY TRNSPT SECT
ELECTROTECHNICAL LAB
TOKYO JAPAN
TAKASHI HORIGOME
ENERGY DIVISION
-PHONE-
042 6612141
SOLAR RESEARCH
SOLAR ELECTRIC POWER GENERATION

MINN, UNIV OF
DEPT OF MECH ENGR
MINNEAPOLIS MN 55455
E M SPARROW

-PHONE-
612 3733034
SOLAR RESEARCH
SOLAR ELECTRIC POWER GENERATION

MIAMI, UNIV OF
ARCHITECTE PLANNING
COPAL GABLES FL 33124
RALPH WARBURTON
ASSOCIATE DEAN
-PHONE-
305 2843438
SOLAR RESEARCH
SOLAR SPACE HEATING & COOLING

MICHIGAN, UNIVERSITY OF
DEPT OF ELECT ENGR
ANN ARBOR MI 48104
DALE M GRIMES

-PHONE-
313 7649546
SOLAR RESEARCH
BIO-CONVERSION

MINN, UNIV OF
DEPT OF MECH ENGR
MINNEAPOLIS MN 55455
R C JORDAN *

-PHONE-
612 3733302
SOLAR RESEARCH
SOLAR ELECTRIC POWER GENERATION

MINN, UNIV OF
ELECT ENGINEERING
MINNEAPOLIS MN 55455
G K WEHNER

-PHONE-
612 3737831
SOLAR RESEARCH
SOLAR ELECTRIC POWER GENERATION

MICHIGAN STATE UNIV
211 AG ENGINEERING
EAST LANSING MI 48824
ERNEST H KIDDER

-PHONE-
517 3554720
SOLAR RESEARCH
SOLAR RADIATION MEASUREMENTS

MIDWEST RESEARCH INST
ENGR SCIENCES DIV
425 VOULKER BLVD
KANSAS CITY MO 64110
MICHAEL C NOLAND

-PHONE-
816 5610202
SOLAR RESEARCH
ALL SOLAR TECHNOLOGIES

MINN, UNIV OF
DEPT MECH ENGR
MINNEAPOLIS MN 55455
E R ECKERT

-PHONE-
612 3733315
SOLAR RESEARCH
SOLAR ELECTRIC POWER GENERATION

MINNESOTA STATE HEALTH DEPT
ENVIRO HEALTH DIV
5900 PENN AVE SOUTH
MINNEAPOLIS MN 55419
PAUL T PANAGOS
PUBLIC HEALTH ENGR
-PHONE-
614 9229225
SOLAR RESEARCH
MEDIUM TEMPERATURE SOLAR COLLECTOR SYSTEMS

MINNESOTA, UNIVERSITY OF
ARCH&LANDSCAPE ARCH
110 ARCHITECTURE BLG
MINNEAPOLIS MN 55455
DENNIS R HOLLOWAY
ASSISTANT PROFESSOR
-PHONE-
612 3732198

SOLAR RESEARCH
SOLAR SPACE HEATING

MINNESOTA, UNV OF
MECHANICAL ENGR
MINNEAPOLIS MN 55455
JOHN ILSE
RESEARCH ASSISTANT

-PHONE-
612 3764916
612 4233701
SOLAR RESEARCH
SOLAR SPACE HEATING

MIROMIT LTD
44 MONTEFORT ST
TEL AVIV ISRAEL

-PHONE-
611481

SOLAR MANUFACTURERS
SOLAR WATER HEATING

MISSISSIPPI, UNIVERSITY OF
CHEMICAL ENGINEERING
UNIVERSITY MS 38677
FRANK A ANDERSON

-PHONE-
601 2327023
SOLAR RESEARCH
MEDIUM TEMPERATURE SOLAR COLLECTOR SYSTEMS

MITRE CORPORATION, THE
SOLAR ENERGY LAB
WESTGATE RESEARCH PK
MCLEAN VA 22101
JAMES S BURTON
DIRECTOR
-PHONE-
703 8933500
SOLAR RESEARCH
INORGANIC SEMI-CONDUCTORS

MITRE CORPORATION, THE
SYSTEMS DEVELOP DIV
WESTGATE RESEARCH PK
MCLEAN VA 22101
FRANK R ELDRIDGE

-PHONE-
703 5368103
SOLAR RESEARCH
INORGANIC SEMI-CONDUCTORS

MITRE CORPORATION, THE
WESTGATE RESEARCH PK
MCLEAN VA 22101
RICHARD S GREELEY

-PHONE-
703 8933500
SOLAR RESEARCH
ALL SOLAR TECHNOLOGIES

MONSANTO RESEARCH CORP
1515 NICHOLAS ROAD
DAYTON OH 45407
FRANK WINSLOW

-PHONE-
513 2683411
SOLAR POTENTIAL MFG,ESTABLISHED BUSINESS
INORGANIC SEMI-CONDUCTORS

MONTANA STATE UNIVESITY
SCHOOL OF ENGINEERG
BOZEMAN MT 59715
R E POWE

SOLAR RESEARCH
WIND CONVERSION

MOTHER EARTH NEWS INC
CORRESPONDENCE SECT
P O BOX 70
HENDERSONVILLE NC 28739
*

-PHONE-
704 6924256
SOLAR/ENERGY CONSERV INFO DISSEMINATION ORG
TOTAL SYSTEMS INTEGRATION

MOTHER EARTH NEWS INC
RESEARCH DEPT
P O BOX 70
HENDERSONVILLE NC 28739
JOHN A BOLL
DIRECTOR R&D
-PHONE-
704 6924256
SOLAR/ENERGY CONSERV INFO DISSEMINATION ORG
TOTAL SYSTEMS INTEGRATION

MOTOROLA INC
SOLAR ENERGY GROUP
8201 E MCDOWELL ROAD
SCOTSDALE AZ 85252
DR ARNOLD LESK
HEAD
-PHONE-
602 9493355
SOLAR RESEARCH
SOLAR ELECTRIC POWER GENERATION

MOTOROLA INC
NEW VENTURES LABS
8201 E MCDOWELL ROAD
SCOTSDALE AZ 85252
STEVE LEVY
MANAGER
-PHONE-
602 9493355
SOLAR RESEARCH
SOLAR ELECTRIC POWER GENERATION

MTN AREA DOMESTIC SOLAR ENGY
TECH CONSORTIUM
RR 2 BOX 274
EVERGREEN CO 80439
MALCOLM A LILLYWHITE
COORDINATOR
-PHONE-
303 6746633
SOLAR/ENERGY CONSERV INFO DISSEMINATION ORG
ALL SOLAR TECHNOLOGIES

N Y SCI COMM FOR PUBLIC INFO
30 EAST 68TH ST
NEW YORK NY 10021
CAROLYN S KONHFIM
EXECUTIVE DIR

-PHONE-
212 7377302
SOLAR/ENERGY CONSERV INFO DISSEMINATION ORG
ALL SOLAR TECHNOLOGIES

N Y STATE COUNCIL ON ARCHITECT
810 SEVENTH AVE
NEW YORK NY 10019
BARBARA J SCHNEPP

-PHONE-
212 7657630
SOLAR/ENERGY CONSERV INFO DISSEMINATION ORG
NEW CONSTRUCTION, TECHNOLOGIES

NAHB RESEARCH FOUNDATION INC
PO BOX 1627
ROCKVILLE MD 20850
RALPH J JOHNSON
STAFF VICE PRESIDENT
-PHONE-
301 7624200
ENERGY CONSERVATION RESEARCH
PUBLISH TECHNICAL BOOKS

NALLE PLASTICS INC
203 COLORADO STREET
AUSTIN TX 78701
GEORGE S NALLE JR

-PHONE-
512 4776168
SOLAR COMPONENTS MANUFACTURER
SOLAR WATER DISTILLATION

NASA
CODE R-2
WASHINGTON DC 20546
PUBLIC AFFAIRS OFFICER *

SOLAR RESEARCH
ALL SOLAR TECHNOLOGIES

NASA
CODE RPP
WASHINGTON DC 20546
ERNST M COHN

-PHONE-
202 7552400
SOLAR RESEARCH
ALL SOLAR TECHNOLOGIES

NASA G C MARSHALL SPACE FLT CT
PUBLIC AFFAIRS
MARSHAL SPACE F C AL 35812
JOSEPH M JONES
DIRECTOR

SOLAR RESEARCH
MEDIUM TEMPERATURE SOLAR COLLECTOR SYSTEMS

NASA GODDARD SPACE FLIGHT CTR
CODE 912
GREENBELT MD 20770
DR MATTHEW P THEKAEKARA

-PHONE-
301 9825034
SOLAR RESEARCH
SOLAR RADIATION MEASUREMENTS

NASA JOHNSON SPACE CRAFT CTR
URBAN SYS PROJ OFF
HOUSTON TX 77053
MARTIN B KEOUGH
MAIL CODE FZ
-PHONE-
713 4835557
SOLAR RESEARCH
SOLAR SPACE HEATING & COOLING

NASA LANGLEY RESEARCH CTR
MAIL STOP 231A
HAMPTON VA 23665
DR EDMUND J CONWAY

-PHONE-
804 8273127
SOLAR RESEARCH
INORGANIC SEMI-CONDUCTORS

NASA LEWIS RESCH CTR
POWER APPLICATIONS
MAIL DROP 500 201
CLEVELAND OH 44142
ROBERT RAGSDALE
ASST TO CHIEF
-PHONE-
216 4334000 X6943
SOLAR RESEARCH
MEDIUM TEMPERATURE SOLAR COLLECTOR SYSTEMS

NASA LEWIS RESEARCH CTR
PHYSICAL SCIENCE DIV
21000 BROOKPARK ROAD
CLEVELAND OH 44135
YIH-YUN HSU
MAIL STOP 301-1
-PHONE-
216 4334000 X205
SOLAR RESEARCH
BIO-CONVERSION

NASA LEWIS RESEARCH CTR
POWER SYSTEMS DIV
21000 BROOKPARK ROAD
CLEVELAND OH 44135
GEORGE M KAPLAN

-PHONE-
216 4334000 X6845
SOLAR RESEARCH
SOLAR ELECTRIC POWER GENERATION

NASA LEWIS RESEARCH CTR
PLUM BROOK STATION
21000 BROOKPARK ROAD
CLEVELAND OH 44135
RICHARD L PUTHOFF
-PHONE-
216 4334000
SOLAR RESEARCH
WIND CONVERSION

NASA LEWIS RESEARCH CTR
POWER ELECTRONICS BR
21000 BROOKPARK RD
CLEVELAND OH 44135
THOMAS J RILEY
-PHONE-
216 4334000 X288
SOLAR RESEARCH
INORGANIC SEMI-CONDUCTORS

NASA LEWIS RESEARCH CTR
M S 54-4
21000 BROOKPARK ROAD
CLEVELAND OH 44135
BERNARD L SATER
-PHONE-
216 4334000 X220
SOLAR RESEARCH
INORGANIC SEMI-CONDUCTORS

NASA LEWIS RESEARCH CTR
SOLAR SYS SECTION
21000 BROOKPARK ROAD
CLEVELAND OH 44135
FREDERICK SIMON
M S 500-201
-PHONE-
216 4334000 X726
SOLAR RESEARCH
SOLAR SPACE HEATING & COOLING

NASA LEWIS RESEARCH CTR
SOLAR SYS SECTION
2100 BROOKPARK RD
CLEVELAND OH 44135
RONALD THOMAS
500-201
-PHONE-
216 4334000
SOLAR RESEARCH
SOLAR SPACE HEATING & COOLING

NASA LEWIS RESEARCH CTR
SOLAR SYS SECTION
2100 BROOKPARK ROAD
CLEVELAND OH 44135
RICHARD W VERNON

-PHONE-
216 4334000 X6857
SOLAR RESEARCH
SOLAR SPACE HEATING & COOLING

NATIONAL ACADEMY OF SCIENCES
BLDG RES ADVISORY BD
2101 CONSTITUT AV NW
WASHINGTON DC 20418
BEN H EVANS AIA

-PHONE-
202 3896515
SOLAR RESEARCH
SOLAR SPACE HEATING & COOLING

NATIONAL BLDG RESCH INST CSIR
ENVIRONMENT ENGR DIV
PO BOX 395
PRETORIA SOUTH AFRICA
C S GROBBELAAS
DIRECTOR
-PHONE-
PRT 746011
SOLAR RESEARCH
SOLAR WATER DISTILLATION

NATIONAL BUREAU OF STANDARDS
CTR FOR BLDG TECH
BLDG 226 RM B104
WASHINGTON DC 20234
JAMES E HILL
THERMAL ENGR SYS SEC
-PHONE-
301 9213503
SOLAR RESEARCH
SOLAR SPACE HEATING & COOLING

NATIONAL BUREAU OF STANDRDS
OFFICE OF HOUS TECH
BLDG 226 ROOM B146
WASHINGTON DC 20234
ROBERT D DIKKERS
MGR SOL ENG DEM PROG
-PHONE-
301 9213285

SOLAR RESEARCH
SOLAR SPACE HEATING & COOLING

NATIONAL BUREAU OF STNDRDS
CTR FOR BLDG TECH
ROOM 100F BLDG 226
WASHINGTON DC 20234
T KUSUDA

-PHONE-
301 9213522
301 6566556
SOLAR RESEARCH
SOLAR SPACE HEATING & COOLING

NATIONAL CAPITAL PARKS
KLINGLE URBAN ENV CT
1100 OHIO DRIVE S W
WASHINGTON DC 20242
NANCY STRADER

-PHONE-
212 2827020

SOLAR/ENERGY CONSERV INFO DISSEMINATION ORG
PROVIDE AUDIO & VISUAL SERVICES

NATIONAL PHYSICAL LAB ISRAEL
HEBREW UNIV CAMPUS
DANZIGER BLDG A
JERUSALEM ISRAEL
DR HARRY TABOR
-PHONE-
30211 X475
SOLAR RESEARCH
MEDIUM TEMPERATURE SOLAR COLLECTOR SYSTEMS

NATIONAL RESEARCH COUNCIL
OTTAWA CANADA
PETER SOUTH
ASSOC RESCH OFFICER

SOLAR RESEARCH
WIND CONVERSION

NATIONAL SCIENCE FOUNDATION
ADVNCED ENGY RES&TEC
WASHINGTON DC 20550
LLOYD O HERWIG *
DIRECTOR
-PHONE-
202 6327230
SOLAR RESEARCH
ALL SOLAR TECHNOLOGIES

NATIONAL SCIENCE FOUNDATION
ADVNCED ENGY RES&TEC
WASHINGTON DC 20550
DONALD A BEATTIE
ACTING DIVISION DIR
-PHONE-
202 6325726
SOLAR RESEARCH
ALL SOLAR TECHNOLOGIES

NATIONAL SCIENCE FOUNDATION
ADVNCED ENGY RES&TEC
WASHINGTON DC 20550
H RICHARD BLIEDEN
PROG MGR PHOTOVOLTAC
-PHONE-
202 6327364
SOLAR RESEARCH
INORGANIC SEMI-CONDUCTORS

NATIONAL SCIENCE FOUNDATION
ADVNCED ENGY RES&TEC
WASHINGTON DC 20550
RICHARD BOGAN
PROG MGR BIOCONVERSN
-PHONE-
202 6327364
SOLAR RESEARCH
BIO-CONVERSION

NATIONAL SCIENCE FOUNDATION
ADVNCED ENGY RES&TEC
WASHINGTON DC 20550
ROBERT COHEN
PROG MGR OCEAN THERM
-PHONE-
202 6327364
SOLAR RESEARCH
OCEAN THERMAL GRADIENT SYSTEMS

NATIONAL SCIENCE FOUNDATION
ADVNCED ENGY RES&TEC
WASHINGTON DC 20550
LOUIS DIVONE
PROG MGR WIND ENGY
-PHONE-
202 6327364
SOLAR RESEARCH
WIND CONVERSION

NATIONAL SCIENCE FOUNDATION
ADVNCED ENGY RES&TEC
WASHINGTON DC 20550
HAROLD HOROWITZ
PROG MGR HEAT&COOL B
-PHONE-
202 6327364
SOLAR RESEARCH
SOLAR SPACE HEATING & COOLING

NATIONAL SCIENCE FOUNDATION
ADVNCED ENGY RES&TEC
WASHINGTON DC 20550
GEORGE JAMES
INFORMATION PROG MGR
-PHONE-
202 6327398
SOLAR RESEARCH
ALL SOLAR TECHNOLOGIES

NATIONAL SCIENCE FOUNDATN
PUBLIC TECH PROJECTS
WASHINGTON DC 20550
CHARLES CHEN
PROG MGR HEAT&COOL B
-PHONE-
202 6324175
SOLAR RESEARCH
SOLAR SPACE HEATING & COOLING

NATIONAL SUN CONTROL
PO BOX 330205
MIAMI FL 33133
JACK TRENT
OWNER

-PHONE-
305 4447494
SOLAR COMPONENTS MANUFACTURER
SOLAR SPACE COOLING

NATL CONCRETE MASONRY ASSOC
PO BOX 9185 ROSLYN S
ARLINGTON VA 22209
THOMAS E REDMOND
MGR RESEARCH & DEVEL

-PHONE-
703 5240815
SOLAR RESEARCH
SOLAR SPACE HEATING & COOLING

NATIONAL SCIENCE FOUNDATION
ADVNCED ENGY RES&TEC
WASHINGTON DC 20550
GEORGE M KAPLAN
PROG MGR SOLAR THERM
-PHONE-
202 6327364
SOLAR RESEARCH
MEDIUM TEMPERATURE SOLAR COLLECTOR SYSTEMS

NATIONAL SCIENCE FOUNDATN
PUBLIC TECH PROJECTS
WASHINGTON DC 20550
JOHN DEL GOBBO
PROG MGR HEAT&COOL B
-PHONE-
202 6324175
SOLAR RESEARCH
SOLAR SPACE HEATING & COOLING

NATIONAL TECHNICAL INFO SER
5285 PORT ROYAL ROAD
SPRINGFIELD VA 22151
SALES OFFICE

-PHONE-
703 3218500
SOLAR/ENERGY CONSERV INFO DISSEMINATION ORG
PROVIDE INFORMATION SEARCHES TO CLIENTS

NATL CTR FOR ATMOS RESEARCH
BUDGET PLAN & INFO
BOX 3000
BOULDER CO 80303
HENRY H LANSFORD *
INFORMATION OFFICER
-PHONE-
303 4945151
SOLAR RESEARCH
SOLAR SPACE HEATING & COOLING

NATIONAL SCIENCE FOUNDATN
EXPLOR RESCH&PROB AS
WASHINGTON DC 20550
CHARLES R HAUER *
DIRECTOR
-PHONE-
202 6324175
SOLAR RESEARCH
SOLAR SPACE HEATING & COOLING

NATIONAL SCIENCE FOUNDATN
PUBLIC TECH PROJECTS
WASHINGTON DC 20550
RAYMOND FIELDS
DEPUTY DIRECTOR
-PHONE-
202 6324175
SOLAR RESEARCH
SOLAR SPACE HEATING & COOLING

NATL CNTR FOR ATMOS RESEARCH
PLANT FACILITIES
BOX 3000
BOULDER CO 80303
L PAUL MOORE
PLANT ENGINEER
-PHONE-
303 4945151 X547
SOLAR RESEARCH
SOLAR SPACE HEATING & COOLING

NATL CTR FOR ATMOS RESEARCH
BOX 3000
BOULDER CO 80303
PHILIP C BENEDICT

-PHONE-
303 4945151
SOLAR RESEARCH
SOLAR SPACE HEATING

NAVAL RESEARCH LAB
UPPER AIR PHYSICS BR
CODE 7120
WASHINGTON DC 20375
TALBOT CHUBB
-PHONE-
202 7673580
SOLAR RESEARCH
HIGH TEMPERATURE SOLAR COLLECTOR SYSTEMS

NAVAL UNDERSEAS CENTER
SAN DIEGO CA 92132
HOWARD WILCOX
-PHONE-
714 2257957
SOLAR RESEARCH
BIO-CONVERSION

NEGEV, UNIV OF THE
R & D AUTHORITY
P O B 1025
BEER SHEVA ISRAEL
J SCHECHTER
DIRECTOR
-PHONE-
0573382
SOLAR RESEARCH
SOLAR PONDS

NETHERS COMMUNITY SCHOOL
BOX 41
WOODVILLE VA 22749
-PHONE-
703 9879011
703 9879041
SOLAR RESEARCH
SOLAR SPACE HEATING

NEVADA SYSTEM, UNIV OF
DESERT RESEARCH INST
RENO NV 89507
JOHN R DOHERTY *
SOLAR RESEARCH
ACADEMIC BUILDINGS

NEVADA,STATE OF
HIGHWAY DEPARTMENT
1263 SO STEWART ST
CARSON CITY NV 89701
JAMES O HUBBARD
ARCHITECT
-PHONE-
702 8827485
SOLAR INTEREST
SOLAR SPACE HEATING

NEW AGE DOMES
EVENING STAR RANCH
RR 4 BOX 90
GOLDEN CO 80422
STEVEN COFFEL
-PHONE-
303 5825719
SOLAR POTENTIAL MFG,NEW BUSINESS
GREENHOUSES

NEW ALCHEMY INSTITUTE EAST
P O BOX 432
WOODS HOLE MA 02543
JOHN TODD
PRES
-PHONE-
617 5632655
SOLAR RESEARCH
TOTAL SYSTEMS INTEGRATION

NEW ALCHEMY INSTITUTE WEST
BOX 376
PESCADERO CA 94040
RICHARD MERRILL
DIRECTOR
-PHONE-
805 9690101
SOLAR RESEARCH
MEDIUM TEMPERATURE SOLAR COLLECTOR SYSTEMS

NEW ALCHEMY INSTITUTE WEST
BOX 376
PESCADERO CA 94040
YEDIDA MERRILL
-PHONE-
805 9690101
SOLAR/ENERGY CONSERV INFO DISSEMINATION ORG
PUBLISH LAY PERIODICALS

NEW ENGLAND SIERRA CLUB
14 BEACON ST
BOSTON MA 02108
RICHARD ELY
-PHONE-
617 2275339
SOLAR/ENERGY CONSERV INFO DISSEMINATION ORG
ALL SOLAR TECHNOLOGIES

NEW LIFE ENVIRO DESIGNS INST
P O BOX 648
KALAMAZOO MI 49005
RICHARD TILMANN
-PHONE-
616 3452948
SOLAR RESEARCH
TOTAL SYSTEMS INTEGRATION

NEW LIFE FARM
DRURY MO 65638
TED LANDERS
FARMER
-PHONE-
417 2612553
SOLAR RESEARCH
TOTAL SYSTEMS INTEGRATION

NEW MEXICO SOLAR ENERGY ASSOC
602 1/2 CANYON RD
SANTA FE NM 87501
KEITH HAGGART
PRESIDENT
-PHONE-
505 9832861
SOLAR/ENERGY CONSERV INFO DISSEMINATION ORG
ALL SOLAR TECHNOLOGIES

NEW MEXICO STATE UNIV
MECHANICAL ENGR DEPT
LAS CRUCES NM 88003
DR ROBERT L SAN MARTIN
BOX 3450
-PHONE-
505 6463501
SOLAR RESEARCH
SOLAR SPACE HEATING & COOLING

NEW SOUTH WALES, UNIV OF
SCHOOL OF PHYSICS
BOX 1
KENSINGTON, N S W AUSTRALIA
J GIUTRONICH
-PHONE-
663 0351
SOLAR RESEARCH
HIGH TEMPERATURE SOLAR COLLECTOR SYSTEMS

NEWCASTLE UNIV OF
DEPT OF ELECT ENGR
NEW SOUTH WALES AUSTRALIA
JOHN P MOORE
SOLAR RESEARCH
SOLAR SPACE HEATING & COOLING

NIAGARA FLOWER CO
405 LEXINGTON AVE
NEW YORK NY 10017
ERWIN LODWIG
-PHONE-
212 6976151
SOLAR COMPONENTS MANUFACTURER
SOLAR SPACE COOLING

NORTH MOUNTAIN COMMUNITY
RR 2 BOX 207
LEXINGTON VA 24450
WYNN SOLOMON
-PHONE-
703 4637095
SOLAR RESEARCH
TOTAL SYSTEMS INTEGRATION

NORTHERN ARIZ UNIV
MECH ENGR
NAU BOX 15600
FLAGSTAFF AZ 86001
WILLIAM W DAVIS
CHAIRMAN
-PHONE-
602 5233698
SOLAR RESEARCH
MEDIUM TEMPERATURE SOLAR COLLECTOR SYSTEMS

NORTHERN PLAINS RESOURCE CONSL
421 STAPLETON BLDG
BILLINGS MT 59101
KYE COCHRAN
-PHONE-
406 2596114
SOLAR/ENERGY CONSERV INFO DISSEMINATION ORG
ALL SOLAR TECHNOLOGIES

NORTHROP RESEARCH&TECH CENTER
3401 WEST BROADWAY
HAWTHORNE CA 90250
WALTER E CRANDALL
-PHONE-
213 6754611 X1526
SOLAR POTENTIAL MFG,ESTABLISHED BUSINESS
SOLAR SPACE HEATING

NORTHRUP INC
2208 CANTON ST
DALLAS TX 75201
LYNN L NORTHRUP
-PHONE-
214 7415631
SOLAR MANUFACTURERS
MEDIUM TEMPERATURE SOLAR COLLECTOR SYSTEMS

NOTRE DAME, UNIV OF
ELECT ENGR DEPT
NOTRE DAME IN 46556
WILLIAM B BERRY
-PHONE-
219 2837193
SOLAR RESEARCH
INORGANIC SEMI-CONDUCTORS

NUCLEAR TECHNOLOGY
P O BOX 1
AMSTON CN 06231
FRANK KERNOZEK

-PHONE-
203 5372387
SOLAR RESEARCH
MEDIUM TEMPEPATURE SOLAR COLLECTOR SYSTEMS

OAK RIDGE NATIONAL LAB
REACTOR DIVISION
PO BOX Y
OAK RIDGE TN 37830
SAM E BEALL JR
-PHONE-
615 4838611
SOLAR RESEARCH
SOLAR HEAT,COOL,&ELECTRIC POWER GENERATION

OAK RIDGE NATIONAL LABORATORY
PO BOX X
OAK RIDGE TN 37830
R CARLSMITH

ENERGY CONSERVATION RESEARCH
NEW CONSTRUCTION, TECHNOLOGIES

OBC INTERNATIONAL MARKETING
P O BOX 205
GLEN IRIS VICTORIA AUSTRALIA

-PHONE-
203 063
208 321
SOLAR PRODUCTS DISTRIBUTORS
SOLAR WATER HEATING

OBELITZ INDUSTRIES INC
BOX 2788
SEAL BEACH CA 90740

-PHONE-
213 5985425

SOLAR/ENERGY CONSERV INFO DISSEMINATION ORG
PUBLISH ORIGINAL RESEARCHED INFORMATION

OFFICE OF TECH ASSESSMENT
119 'D' STREET NE
WASHINGTON DC 20510
RONAL W LARSON

-PHONE-
202 2258996

SOLAR LEGAL CONSIDERATIONS
LEGISLATIVE, FEDERAL

OHIO ST UNIV
DEPT OF PHYSICS
COLUMBUS OH 43210
ARI RAHL

-PHONE-
614 4226446

SOLAR RESEARCH
SOLAR PONDS

OHIO STATE UNIV
ELECT ENGR LAB
2015 NEIL RM300 CL
COLUMBUS OH 43210
ARTHUR E MIDDLETON DIRECTOR
ELECT MAT&DEVICE LAB
-PHONE-
614 4226210
614 4869561
ENERGY CONSERVATION RESEARCH
INORGANIC SEMI-CONDUCTORS

OHIO STATE UNIVERSITY
NORTH HIGH ST
COLUMBUS OH 43210
CHARLES F SEPSY

-PHONE-
614 4226898

SOLAR RESEARCH
SOLAR SPACE HEATING & COOLING

OKLAHOMA STATE UNIVERSITY
ELECTRICAL ENGG
ENGINEERING SOUTH
STILLWATER OK 74074
H JACK ALLISON
ASSOC PROF ELEC ENGG
-PHONE-
405 3726211 X6476
SOLAR RESEARCH
SOLAR ENERGY STORAGE

OKLAHOMA STATE UNIVERSITY
ELECTRICAL ENGG
ENGINEERING SOUTH
STILLWATER OK 74074
WILLIAM L HUGHES
HEAD SCHOOL OF E E
-PHONE-
405 3726211 X7581
SOLAR RESEARCH
WIND CONVERSION

OKLAHOMA STATE UNIVERSITY
ELECTRICAL ENGG
ENGINEERING SOUTH
STILLWATER OK 74074
RAMACHANDRA G RAMAKUMAR
ASSOC PROF ELEC ENGG
-PHONE-
405 3726211 X6476
SOLAR RESEARCH
WIND CONVERSION

OLIN CORPORATION,BRASS GROUP
ROLL-BOND PRODUCTS
EAST ALTON IL 62024
JOHN I BARTON
MARKETING MANAGER
-PHONE-
618 2582000
618 2582443
SOLAR COMPONENTS MANUFACTURER
MEDIUM TEMPERATURE SOLAR COLLECTOR SYSTEMS

OPEN NORTHWEST INFO NETWORK
608 19TH AVE E
SEATTLE WA 98112
-PHONE-
206 3238506
SOLAR/ENERGY CONSERV INFO DISSEMINATION ORG
PUBLISH LAY PERIODICALS

OPTICAL COATING LAB INC
P O BOX 1599
SANTA ROSA CA 95403
JOSEPH H APFEL
DIRECTOR RESEARCH
-PHONE-
707 5456440
SOLAR COMPONENTS MANUFACTURER
MEDIUM TEMPERATURE SOLAR COLLECTOR SYSTEMS

OPTICAL COATING LABORATORY INC
2789 GIFFEN AVENUE
SANTA ROSA CA 95403
DANFORTH JOSLYN
-PHONE-
707 5456440
SOLAR COMPONENTS MANUFACTURER
MEDIUM TEMPERATURE SOLAR COLLECTOR SYSTEMS

OWENS-CORNING FIBERGLAS CORP
TECHNICAL CENTER
PO BOX 415
GRANVILLE OH 43023
DR GEORGE W GRIMM
MGR CORP RESEARCH
-PHONE-
614 5870610
ENERGY CONSERVATION RESEARCH
NEW CONSTRUCTION, INTEREST

OZARK ACCESS CENTER INC
55 SPRING ST BOX 506
EUREKA SPRINGS AR 72632
EDD JEFFORDS
-PHONE-
501 2539616
501 2539601
SOLAR/ENERGY CONSERV INFO DISSEMINATION ORG
PUBLISH LAY PERIODICALS

PACIFIC POWER & LIGHT
CUST TECH SVCS
920 SW 6TH AVE
PORTLAND OR 97204
LAWTHAN M AUSTIN JR
TECH SVCS DIR
-PHONE-
503 2267411 X1054
503 6469894
SOLAR INTEREST
RETROFITTING, MECHANICAL SYSTEMS

PAUL MUELLER CO
P O BOX 828
SPRINGFIELD MO 65801
-PHONE-
417 8652831
SOLAR COMPONENTS MANUFACTURER
MEDIUM TEMPERATURE SOLAR COLLECTOR SYSTEMS

PELCASP SOLAR DESIGNERS
240 LANGDON ST
MADISON WI 53703
MICHAEL B PELLETT
PRES
-PHONE-
608 2519967
SOLAR INTEREST
SOLAR SPACE HEATING

PENN, UNIV OF
N CTR ENGY MGT&POWER
260 TOWNE BLDG D3
PHILADELPHIA PA 19174
NOAM LIOR
ASSISTANT PROFESSOR
-PHONE-
215 5944803
SOLAR RESEARCH
SOLAR SPACE HEATING & COOLING

PENN, UNIV OF
N CTR ENGY MGT&POWER
260 TOWNE BLDG D3
PHILADELPHIA PA 19174
HAROLD G LORSCH
-PHONE-
215 5947181
SOLAR RESEARCH
SOLAR SPACE HEATING & COOLING

PENN, UNIVERSITY OF
N CTR ENGY MGT&POWER
TOWNE BUILDING
PHILADELPHIA PA 19104
JESSE C DENTON *
-PHONE-
215 5945122
SOLAR RESEARCH
HIGH TEMPERATURE SOLAR COLLECTOR SYSTEMS

PENN, UNIVERSITY OF
N CTR ENGY MGT&POWER
TOWNE BUILDING
PHILADELPHIA PA 19104
GEORGE L SCHRENK

-PHONE-
215 5945121
SOLAR RESEARCH
HIGH TEMPERATURE SOLAR COLLECTOR SYSTEMS

PENNSYLVANIA STATE UNIV
MATERIAL SCIENCE DPT
320 M I BLDG
UNIVERSITY PARK PA 16802
HOWARD F PALMER
PROF OF FUEL SCIENCE
-PHONE-
814 8656512
SOLAR RESEARCH
HIGH TEMPERATURE SOLAR COLLECTOR SYSTEMS

PENNWALT CORPORATION
AUTOMATIC POWER DIV
P O BOX 18738
HOUSTON TX 77023
ROBERT J DODGE

-PHONE-
713 2285208
SOLAR PRODUCTS DISTRIBUTORS
WIND CONVERSION

PERCY WILSON MORTGAGE&FINANCE
COMM & INDUST LOAN
221 N LASALLE ST
CHICAGO IL 60601
TODD J BROMS
ASST LOAN OFFICER
-PHONE-
312 8556726
SOLAR FINANCIAL CONSIDERATIONS
MORTGAGE INSTITUTIONS

PERRY-DEAN-STEWART ARCHITECTS
955 PARK SQ BLDG
BOSTON MA 02116
CHARLES F ROGERS
PARTNER

-PHONE-
617 4829160
SOLAR INTEREST
SOLAR HEAT,COOL,&ELECTRIC POWER GENERATION

PGA ENGINEERS INC
10 BROADWAY
ST LOUIS MO 63102
WILLIAM EVERS JR
VICE PRESIDENT

-PHONE-
314 2317318
ENERGY CONSERVATION RESEARCH
NEW CONSTRUCTION, MECHANICAL SYSTEMS

PHILADELPHIA ELECTRIC CO
ENGY CONVERSON RESCH
2301 MARKET ST S10-1
PHILADELPHIA PA 19101
OLIVER D GILDERSLEEVE JR
-PHONE-
215 8414856
SOLAR RESEARCH
SOLAR ELECTRIC POWER GENERATION

PHILCO-FORD CORP
3939 FABIAN WAY
PALO ALTO CA 94303
HOWARD E POLLARD
SR ENG SPECIALIST
-PHONE-
415 3264350 X6329
SOLAR RESEARCH
INORGANIC SEMI-CONDUCTORS

PHILIPS FORSCHUNGSLABORATORIUM
AACHEN GMBH
D-51 AACHEN
WEISSHAUSSTRASSE WEST GERMANY
DR HORST HORSTER
-PHONE-
024 1 62071
SOLAR RESEARCH
MEDIUM TEMPERATURE SOLAR COLLECTOR SYSTEMS

PHYSICAL INDUSTRIES CORP
SOLAR DIVISION
P O BOX 357
LAKESIDE CA 92040
JOHN H HEDGER
PRESIDENT
-PHONE-
714 5611266
SOLAR MANUFACTURERS
SOLAR HEAT,COOL,&ELECTRIC POWER GENERATION

PHYSICS & ENGR LAB D S I R
PRIVATE BAG
LOWER HUTT NEW ZEALAND
D P MCAULIFFE

-PHONE-
WEL 699199 X818
SOLAR RESEARCH
SOLAR WATER HEATING

PILKINGTON BROTHERS LTD
R&D LABS
LATHOM ORMSKIRE
LANCASHIRE ENGLAND
A PULFORD
HEAD NEW VENTURES

SOLAR RESEARCH
MEDIUM TEMPERATURE SOLAR COLLECTOR SYSTEMS

INDEX OF ACADEMIC INSTITUTIONS/GOVERNMENT AGENCY /INDUSTRIES

PORTABLE ALUMINUM IRRIG CO INC
PO BOX 878
VISTA CA 92083
WILL C KINNEY

-PHONE-
714 7242163

SOLAR POTENTIAL MFG,ESTABLISHED BUSINESS
SOLAR WATER DISTILLATION

PROF ENG CONSULTANTS
MECHANICAL
1440 E ENGLISH
WICHITA KS 67211
ERVIN E HYSOM
CONSULTING MECH ENG
-PHONE-
316 2622691

SOLAR INTEREST
SOLAR SPACE HEATING & COOLING

PUERTO RICO,UNIV OF
CENTRO DE INFO TECNI
RECINTO UNIV MAYAGUZ
MAYAGUEZ PUERTO RICO
ERNESTO Q MARTIN
DIRECTOR

SOLAR/ENERGY CONSERV INFO DISSEMINATION ORG
PROVIDE INFORMATION SEARCHES TO CLIENTS

R&D CONSULTANTS
8713 EASELINE ROAD
LAFAYETTE CO 80026
ROGER D BENNETT

-PHONE-
303 4946131
SOLAR POTENTIAL MFG,NEW BUSINESS
SOLAR SPACE HEATING

PORTOLA INSTITUTE
ENERGY PRIMER
558 SANTA CRUZ AVE
MENLO PARK CA 94025
CHARLES F MISSAR
EDITOR
-PHONE-
415 3235535
415 3230313
SOLAR/ENERGY CONSERV INFO DISSEMINATION ORG
PUBLISH LAY + TECHNICAL BOOKS

PRUDENTIAL INS CO OF AMERICA
REAL ESTATE INVESTNT
1180 RAYMOND BLVD
NEWARK NJ 07102
CHARLES F REYNOLDS JR
VICE PRES
-PHONE-
201 3364578

ENERGY CONSERVAT'N,FINANCIAL CONSIDERATIONS
NEW CONSTRUCTION, RESOURCES

PURDUE UNIV
DEPT OF AG ENGR
W LAFAYETTE IN 47907
BRIAN HORSFIELD

-PHONE-
317 7492971
SOLAR RESEARCH
SOLAR AGRICULTURAL DRYING

RADISOL S A
31 BOULEVARD DANTON
CASABLANCO MOROCCO

SOLAR MANUFACTURERS
SOLAR WATER HEATING

PPG INDUSTRIES INC
NEW PROD DEVELOPMENT
ONE GATEWAY CENTER
PITTSBURGH PA 15222
R R LEWCHUK
PROJECT MGR
-PHONE-
412 4342645
SOLAR MANUFACTURERS
MEDIUM TEMPERATURE SOLAR COLLECTOR SYSTEMS

PUERTO RICO, UNIV OF
GEN ENGR DEPT
MAYAGUEZ PUERTO RICO
DR MANUEL RODRIGUEZ PERAZZA

-PHONE-
809 8324040 X209
809 8324040 X455
SOLAR RESEARCH
MEDIUM TEMPERATURE SOLAR COLLECTOR SYSTEMS

QUEENSLAND, UNIV OF
DEPT OF MECH ENGR
ST LUCIA, QUEENSLAND AUSTRALIA
NORMAN R SHERIDAN

-PHONE-
072 700111
SOLAR RESEARCH
MEDIUM TEMPERATURE SOLAR COLLECTOR SYSTEMS

RAM PRODUCTS INC
BOX 340
STURGIS MI 49091
PAUL SORKORAM

-PHONE-
616 6519351
SOLAR POTENTIAL MFG,ESTABLISHED BUSINESS
HIGH TEMPERATURE SOLAR COLLECTOR SYSTEMS

RAND CORP
1700 MAIN ST
SANTA MONICA CA 90406
D N MORRIS
-PHONE-
213 3930411
ENERGY CONSERVATION RESEARCH
INCENTIVES FOR ENERGY CONSERVATION

RAYPAK INC
31111 AGOURA RD
THOUSAND OAKS CA 91360
ALLEN BONIFACE
-PHONE-
213 8891500
SOLAR MANUFACTURERS
MEDIUM TEMPERATURE SOLAR COLLECTOR SYSTEMS

RCA CORPORATION
D SARNOFF RESCH CTR
PRINCETON NJ 08540
PAUL RAPPAPORT
DIRECTOR M&P LAB
-PHONE-
609 4522700
SOLAR RESEARCH
INORGANIC SEMI-CONDUCTORS

READING UNIV OF
CHEMISTRY
READING ENGLAND
D BRYCE-SMITH
SOLAR RESEARCH
INORGANIC SEMI-CONDUCTORS

REDE RESEARCH&DESIGN INSTITUTE
P O BOX 307
PROVIDENCE RI 02901
RALPH BECKMAN
-PHONE-
401 8615390
SOLAR RESEARCH
SOLAR HEAT,COOL,&ELECTRIC POWER GENERATION

REDE RESEARCH&DESIGN INSTITUTE
P O BOX 307
PROVIDENCE RI 02901
RONALD BECKMAN
DIRECTOR
-PHONE-
401 8615390
SOLAR RESEARCH
SOLAR ELECTRIC POWER GENERATION

RENSSELAER POLYTECHNIC INST
ARCHITECT RESERCH CT
TROY NY 12181
WALTER KRONER
-PHONE-
518 2706461
SOLAR RESEARCH
ALL SOLAR TECHNOLOGIES

RESOURCES FOR THE FUTURE INC
ENERGY RESEARCH&TECH
1755 MASS AVE NW
WASHINGTON DC 20036
SAM H SCHURR
ENERGY CONSERVATION RESEARCH
FOUNDATIONS PROVIDING GRANTS/FUNDS

REVERE COPPER & BRASS INC
BLDG PROD DEPT
P O BOX 151
ROME NY 13440
WILLIAM J HEIDRICH
-PHONE-
315 3382022
SOLAR MANUFACTURERS
SOLAR SPACE HEATING & COOLING

REYNOLDS METALS
EXTRUSION DIV
6601 W BROAD ST
RICHMOND VA 23230
ADAM J DRESCHER
-PHONE-
804 2822331 X2197
SOLAR COMPONENTS MANUFACTURER
MEDIUM TEMPERATURE SOLAR COLLECTOR SYSTEMS

RHO SIGMA UNLIMITED
5108 MELVIN AVE
TARZANA CA 91356
ROBERT J SCHLESINGER P E
-PHONE-
213 3424376
SOLAR COMPONENTS MANUFACTURER
SOLAR SPACE HEATING

RHODE ISLAND, UNIV OF
MECHANICAL ENGR DEPT
KINGSTON RI 02881
DR RICHARD C LESSMANN
SOLAR RESEARCH
TOTAL SYSTEMS INTEGRATION

RICE MAREK HARRAL HOLTZ INC
PO BOX 12037
DENVER CO 80212
CLIFT EPPS

-PHONE-
303 4204455

SOLAR RESEARCH
SOLAR SPACE HEATING & COOLING

RICHARD P MUELLER & ASSOC INC
ENVTL ENGRG DEPT
1900 SULPHUR SPRG RD
BALTIMORE MD 21227
ANDREW J PARKER
MANAGER
-PHONE-
301 2745666
202 7839159
SOLAR RESEARCH
SOLAR SPACE HEATING & COOLING

ROARING FORK RESOURCE CENTER
P O BOX 9950
ASPEN CO 81611
GREGORY E FRANTA

-PHONE-
303 9253481
303 9255125
SOLAR/ENERGY CONSERV INFO DISSEMINATION ORG
PROVIDE INFORMATION SEARCHES TO CLIENTS

ROBERT G WERDEN&ASSOCIATES INC
PO BOX 414
JENKINTOWN PA 19046
ROBERT G WERDEN

-PHONE-
215 8852500
SOLAR RESEARCH
SOLAR SPACE HEATING & COOLING

ROBERT MOREY ASSOCIATES
ENERGY INFO
PO BOX 98
DANA POINT CA 92629

SOLAR/ENERGY CONSERV INFO DISSEMINATION ORG
PUBLISH LAY PERIODICALS

ROCHESTER INST OF TECH
50 WEST MAIN
ROCHESTER NY 14614
PAUL IVOJCIECHOIVSKI

-PHONE-
716 4642411
SOLAR RESEARCH
SOLAR SPACE HEATING

ROCKWELL INTERNATIONAL
SPACE DIVISION
12214 LAKEWOOD BLVD
DOWNEY CA 90241
CHARLES L GOULD SF26

-PHONE-
213 5943921
SOLAR POTENTIAL MFG,ESTABLISHED BUSINESS
SOLAR SPACE HEATING & COOLING

ROHM AND HAAS
PLASTICS DEPT
INDEPENDENCE MALL W
PHILADELPHIA PA 19105
ROBERT F BITLER
TECH COORD
-PHONE-
215 5926799
SOLAR COMPONENTS MANUFACTURER
SOLAR SPACE HEATING & COOLING

ROORKEE, UNIV OF
DEPT OF MECH ENGR
ROORKEE 247667 INDIA
R K MEHROTRA

SOLAR RESEARCH
SOLAR SPACE HEATING & COOLING

ROORKEE, UNIV OF
DEPT OF MECH ENGR
ROORKEE 247667 INDIA
RAJENDRA PRAKASH

-PHONE-
289
SOLAR RESEARCH
MEDIUM TEMPERATURE SOLAR COLLECTOR SYSTEMS

RUDKIN-WILEY CORPORATION
POLARCH
760 HONEYSPOT
STRATFORD CT 06497
SAM HILL
MGR SALES&MARKETING
-PHONE-
202 3755966
SOLAR INTEREST
SOLAR SPACE HEATING & COOLING

RUTGERS UNIVERSITY
ELECTRIC ENGR DEPT
NEW BRUNSWICK NJ 08903
WAYNE ANDERSON

-PHONE-
201 9323871
SOLAR RESEARCH
INORGANIC SEMI-CONDUCTORS

SAMUEL PAUL ARCHITECT, OFF OF
107-40 QUEENS BLVD
FOREST HILLS NY 11375
SAMUEL PAUL

-PHONE-
212 2615100
SOLAR INTEREST
SOLAR SPACE HEATING

SANDHU AND ASSOCIATES
309 S RANDOLPH
CHAMPAIGN IL 61820
MARTIN D IGNAZITO

-PHONE-
217 3522999
SOLAR RESEARCH
SOLAR SPACE HEATING & COOLING

SANDIA LABORATORIES
DIVISION 8184
LIVERMORE CA 94550
THOMAS D BRUMLEVE
STAFF MEMBER
-PHONE-
415 4552941
SOLAR RESEARCH
HIGH TEMPERATURE SOLAR COLLECTOR SYSTEMS

SANDIA LABORATORIES
DIVISION 8184
LIVERMORE CA 94550
ALAN C SKINROOD
SUPERVISOR
-PHONE-
415 4552501
SOLAR RESEARCH
TOTAL SYSTEMS INTEGRATION

SANDIA LABORATORIES
P O BOX 969
LIVERMORE CA 94550
CLIFFORD T YOKOMIZO
TECHNICAL STAFF
-PHONE-
415 4557011 X2668
SOLAR RESEARCH
HIGH TEMPERATURE SOLAR COLLECTOR SYSTEMS

SANDIA LABS
ORGANIZATION 5717
ALBUQUERQUE NM 87110
RAYMOND HARRISON

-PHONE-
505 2648669
SOLAR RESEARCH
TOTAL SYSTEMS INTEGRATION

SANDIA LABS
DIV 4736 BOX 5800
ALBUQUERQUE NM 87115
R P STROMBERG

-PHONE-
505 2648170

SOLAR RESEARCH
TOTAL SYSTEMS INTEGRATION

SANGAMON STATE UNIV
PHYSICAL SCI PROG
SPRINGFIELD IL 62708
AL CASELLA
ASST PROF
-PHONE-
217 7866600

SOLAR RESEARCH
SOLAR SPACE HEATING

SCHOOL OF LIVING
RT 1 BOX 129
FREELAND MD 21053
MILDRED J LOOMIS

-PHONE-
301 3575619
301 3296549
SOLAR/ENERGY CONSERV INFO DISSEMINATION ORG
PUBLISH LAY PERIODICALS

SCIENTIFIC RESEARCH FOUNDATION
P O B 3745
JERUSALEM ISRAEL
DR H TABOR

-PHONE-
02 534515
SOLAR RESEARCH
MEDIUM TEMPERATURE SOLAR COLLECTOR SYSTEMS

SEA PINES COMPANY
ENVIRONMENTAL SERV
PO BOX 5608
HILTON HEAD ISLAND SC 29928
RICHARD F WILDERMAN
-PHONE-
803 7853333 X3144
SOLAR RESEARCH
SOLAR SPACE HEATING & COOLING

SEA SOLAR POWER INC
1615 HILLOCK LANE
YORK PA 17403
JAMES H ANDERSON *
VICE PRES
-PHONE-
717 8438594
SOLAR POTENTIAL MFG,NEW BUSINESS
OCEAN THERMAL GRADIENT SYSTEMS

SEA SOLAR POWER INC
1615 HILLOCK LANE
YORK PA 17403
J HILBERT ANDERSON
PRESIDENT
-PHONE-
717 8438594
SOLAR POTENTIAL MFG,NEW BUSINESS
OCEAN THERMAL GRADIENT SYSTEMS

SEKISUI KAYAKU KOGIYO CO
2 KINUGASA MACHI
OSAKA JAPAN
SOLAR MANUFACTURERS
SOLAR WATER HEATING

SELECT RAD SURFACES PLATING CO
P O BOX 8364
SAN JOSE CA 95155
RICHARD T AVALON
PRES
-PHONE-
408 2959040
SOLAR COMPONENTS MANUFACTURER
ALL SOLAR TECHNOLOGIES

SENCENBAUGH WIND ELECTRIC
P O BOX 11174
PALO ALTO CA 94306
SOLAR MANUFACTURERS
WIND CONVERSION

SENSENICH CORPORATION
PO BOX 1168
LANCASTER PA 17604
A C WEDGE
EXECUTIVE VP
SOLAR COMPONENTS MANUFACTURER
WIND CONVERSION

SERVICE UNLIMITED INC
SECOND & WALNUT STS
WILMINGTON DE 19801
RICHARD W WOLF
-PHONE-
302 6551568
SOLAR POTENTIAL MFG,ESTABLISHED BUSINESS
SOLAR SPACE HEATING

SHAW PUMP COMPANY
9660 EAST RUSH ST
SOUTH EL MONTE CA 91733
-PHONE-
213 4431784
SOLAR MANUFACTURERS
WIND CONVERSION

SHELDAHL INC
CORP R & D
HWY 3
NORTHFIELD MN 55057
DONALD E ANDERSON
VICE PRES
-PHONE-
507 6455633
SOLAR POTENTIAL MFG,ESTABLISHED BUSINESS
HIGH TEMPERATURE SOLAR COLLECTOR SYSTEMS

SIDNEY WILLIAMS CO PTY LTD
WM PARADE P O BOX 22
DULWICH HILL NSW AUSTRALIA
SOLAR MANUFACTURERS
WIND CONVERSION

SILVER INSTITUTE INC
COLO REPRESENTATIVE
1001 CONNECTICUT, NW
WASHINGTON DC 20036
ROBERT R DAVIES
LITERATURE CONSULT
-PHONE-
303 4430105 XCOLO
202 3932285 XDC
SOLAR/ENERGY CONSERV INFO DISSEMINATION ORG
PROVIDE INFORMATION SEARCHES TO CLIENTS

SISTEMAS DE ENERGIA SOLAR
AVENIDA MAGDALA 1151
CONDADO
SANTURCE PUERTO RICO
DAVID L GUTHRIE
-PHONE-
809 7257825
SOLAR/ENERGY CONSERV INFO DISSEMINATION ORG
ALL SOLAR TECHNOLOGIES

SKY THERM PROCESS& ENGINEERING
945 WILSHIRE BLVD
LOS ANGELES CA 90017
HAROLD R HAY
-PHONE-
213 6247261 X306
SOLAR POTENTIAL MFG,ESTABLISHED BUSINESS
SOLAR SPACE HEATING & COOLING

SMALLS SOLA HEETA CO
10 GOONGARRIE ST
BAYSWATER AUSTRALIA

SOLAR MANUFACTURERS
SOLAR WATER HEATING

SOLAR DESIGNS
203 HOLLY LANE
ORINDA CA 94563
ROBERT M GRAVEN

-PHONE-
415 2543938
SOLAR/ENERGY CONSERV INFO DISSEMINATION ORG
PROVIDE INFORMATION SEARCHES TO CLIENTS

SOLAR ENERGY DIGEST
PO BOX 17776
SAN DIEGO CA 92117
WILLIAM B EDMONDSON CWO-4
EDITOR & PUBLISHER
-PHONE-
714 2272980
SOLAR/ENERGY CONSERV INFO DISSEMINATION ORG
PUBLISH LAY PERIODICALS

SOLAR ENERGY RESEARCH CORP
RT 4 BOX 26B
LONGMONT CO 80501
JAMES B WIEGAND
PRESIDENT
-PHONE-
303 7724522
SOLAR MANUFACTURERS
SOLAR SPACE HEATING & COOLING

SMITHSONIAN RADIATION BIO LAB
12441 PARKLAWN DR
ROCKVILLE MD 20852
WALTER SHROPSHIRE JR

-PHONE-
301 4432333
SOLAR RESEARCH
BIO-CONVERSION

SOLAR ENERGY CO
DEERWOOD DR
MERRIMACK NH 03054
ROGER PAPINEAU
PRESIDENT
-PHONE-
603 4245168
SOLAR MANUFACTURERS
SOLAR WATER HEATING

SOLAR ENERGY DIGEST
P O BOX 17776
SAN DIEGO CA 92117
SOLAR EQUIP SALES DIVISION

-PHONE-
714 2772980
SOLAR PRODUCTS DISTRIBUTORS
SOLAR WATER HEATING

SOLAR ENERGY RESEARCH&INFO CTR
1001 CONN AVE NW
WASHINGTON DC 20036
LUCILLE ALFIERI
ACTING DIRECTOR

SOLAR/ENERGY CONSERV INFO DISSEMINATION ORG
PUBLISH LAY PERIODICALS

SOL-THERM CORP
7 WEST 14TH STREET
NEW YORK NY 10011
ITAMAR SITTENFELD
VICE PRESIDENT
-PHONE-
212 6914632
SOLAR MANUFACTURERS
SOLAR WATER HEATING

SOLAR ENERGY DEVELOPMENT INC
1457 ALAMEDA AVE
LAKEWOOD OH 44107
NICHOLAS S MACRON

-PHONE-
216 2213500
SOLAR RESEARCH
MEDIUM TEMPERATURE SOLAR COLLECTOR SYSTEMS

SOLAR ENERGY INST OF TURKEY
POBOX 37 BAKANLIKLAR
ANKARA TURKEY
MUAMMER CETINCELIK
PRESIDENT
-PHONE-
234 255378
SOLAR MANUFACTURERS
MEDIUM TEMPERATURE SOLAR COLLECTOR SYSTEMS

SOLAR ENERGY RESEARCH&INFO CTR
1001 CONN AVE NW
WASHINGTON DC 20036
SOLAR ENERGY INDUSTRIES ASSOC

SOLAR/ENERGY CONSERV INFO DISSEMINATION ORG
ALL SOLAR TECHNOLOGIES

SOLAR ENERGY RESEARCH&INFO CTR
1001 CONN AVE NW
WASHINGTON DC 20036
RACHEL SYNDER
WRITER
-PHONE-
202 2931000
SOLAR/ENERGY CONSERV INFO DISSEMINATION ORG
PUBLISH LAY PERIODICALS

SOLAR ENERGY SOC OF AMERICA
2780 SEPULA BLVD
TORRANCE CA 90510
JUDY STERNBERG *
DIRECTOR INFO EXCHNG
-PHONE-
213 3263283
SOLAR/ENERGY CONSERV INFO DISSEMINATION ORG
PUBLISH LAY + TECHNICAL PERIODICALS

SOLAR ENERGY SOC OF AMERICA
2780 SEPULA BLVD
TORRANCE CA 90510
DAVID SATCHWELL
PRESIDENT
-PHONE-
213 3263283
SOLAR/ENERGY CONSERV INFO DISSEMINATION ORG
PUBLISH LAY + TECHNICAL PERIODICALS

SOLAR ENERGY SYSTEMS INC.
70 S. CHAPEL ST.
NEWARK, DE 19711
DR. KARL W. BOER
-PHONE-
302 7310990
SOLAR MANUFACTURERS
INORGANIC SEMI-CONDUCTORS

SOLAR ENVIRONMENTAL ENGR CO
PO BOX 1914
FORT COLLINS CO 80521
THOMAS E CORDER
VICE PRESIDENT
-PHONE-
303 4934480
303 4846386
SOLAR ECONOMIC CONSIDERATIONS
TOTAL SYSTEMS INTEGRATION

SOLAR HEAT LTD
99 MIDDLETON HALL RD
KINGS NORTON BIRMGHM ENGLAND
A A BLANCO
DIRECTOR
SOLAR MANUFACTURERS
SOLAR WATER HEATING

SOLAR HOME SYSTEMS
ROOM 202
38518 OAKHILL CIRCLE
WILLOUGHBY OH 44094
JOSEPH BARBISH
-PHONE-
216 9511119
SOLAR POTENTIAL MFG,ESTABLISHED BUSINESS
SOLAR SPACE HEATING & COOLING

SOLAR HOMES INC
2 NARRAGANSETT AVE
JAMESTOWN RI 02835
STENCER DICKENSON
-PHONE-
401 4231025
SOLAR MANUFACTURERS
HOMES

SOLAR HYDRONICS
P O BOX 4484
BOULDER CO 80302
TEAGUE VAN BUREN
-PHONE-
303 4479828
SOLAR RESEARCH
SOLAR SPACE HEATING

SOLAR INC
4706 FAIRVIEW AVE
BOISE ID 83704
JAMES H BALLANTYNE JR
SOLAR MANUFACTURERS
SOLAR SPACE HEATING

SOLAR POWER CORP
186 FORBES ROAD
BRAINTREE MA 02184
ROBERT W WILLIS
VICE PRESIDENT
-PHONE-
617 8486877
SOLAR MANUFACTURERS
INORGANIC SEMI-CONDUCTORS

SOLAR POWER SUPPLY
RT 3 BOX A10
EVERGREEN CO 80439
DOUG DONNER *
V P MARKETING
-PHONE-
303 6744734
SOLAR POTENTIAL MFG,NEW BUSINESS
SOLAR HEAT,COOL,&ELECTRIC POWER GENERATION

SOLAR POWER SUPPLY
RT 3 BOX A10
EVERGREEN CO 80439
MALCOM LILLYWHITE
V P RESEARCH&DEVELOP
-PHONE-
303 6744734
SOLAR POTENTIAL MFG,NEW BUSINESS
SOLAR HEAT,COOL,&ELECTRIC POWER GENERATION

SOLAR PRODUCTS CORP
3500 EAST 1ST AVE
DENVER CO 80206

SOLAR MANUFACTURERS
SOLAR COOKING

SOLAR SUNSTILL INC
SETAUKET NY 11733
CHAD J RASEMAN

-PHONE-
516 9414078
SOLAR COMPONENTS MANUFACTURER
GREENHOUSES

SOLAR SYS INC
323 COUNTRY CLUB DR
REHOBOTH BEACH DE 19971
DANE DE REIMER

-PHONE-
302 2272323
SOLAR MANUFACTURERS
MEDIUM TEMPERATURE SOLAR COLLECTOR SYSTEMS

SOLAR WIND
PO BOX 7
EAST HOLDEN ME 04429
HENRY M CLEWS
PRES
-PHONE-
207 8435168
SOLAR MANUFACTURERS
WIND CONVERSION

SOLAR-X CONTROL PRODUCTS CORP
217 CALF ST
NEWTON MA 02158

-PHONE-
617 2448686
SOLAR COMPONENTS MANUFACTURER
SOLAR SPACE COOLING

SOLAREX CORP
1335 PICCARD DR
ROCKVILLE MD 20850
PETER F VARADI
EXEC V P
-PHONE-
301 9480202

SOLAR MANUFACTURERS
INORGANIC SEMI-CONDUCTORS

SOLARSYSTEMS INC
1515 WSW LOOP 323
TYLER TX 75701
CONE RICE *
V P SALES
-PHONE-
214 5920945
214 5975534
SOLAR MANUFACTURERS
SOLAR SPACE HEATING & COOLING

SOLARSYSTEMS INC
1515 WSW LOOP 323
TYLER TX 75701
JOHN L DECKER
PRESIDENT
-PHONE-
214 5920945
214 5975537
SOLAR MANUFACTURERS
SOLAR SPACE HEATING & COOLING

SOLERG ASSOCIATES
PO BOX 90691
LOS ANGELES CA 90009
EGAN J RATTIN

-PHONE-
213 8364200
SOLAR MANUFACTURERS
MEDIUM TEMPERATURE SOLAR COLLECTOR SYSTEMS

SOLORON
4850 OLIVE ST
DENVER CO 80022
JOHN BAYLESS
PRES & DIRECTOR
-PHONE-
303 2892288
SOLAR MANUFACTURERS
MEDIUM TEMPERATURE SOLAR COLLECTOR SYSTEMS

SOLTEC
PO BOX 6844
DENVER CO 80206
RICHARD S SPEED
PRESIDENT
-PHONE-
303 3338869
SOLAR POTENTIAL MFG,NEW BUSINESS
GREENHOUSES

SONNEWALD SERVICE
RD 1 BOX 457
SPRING GROVE PA 17362
H R LEFEVER

-PHONE-
717 2253456

SOLAR RESEARCH
SOLAR SPACE HEATING

SOPHIA UNIVERSITY
DEPT OF PHYSICS
7 KIOI-CHO CHIYODA-K
TOKYO JAPAN
ISAO OSHIDA
FACULTY OF SCI & TEC
-PHONE-
03 2659211 X567

SOLAR RESEARCH
HIGH TEMPERATURE SOLAR COLLECTOR SYSTEMS

SOUTHERN CALIF, UNIV OF
WESRAC
809 WEST 34TH ST
LOS ANGELES CA 90007
WILLIAM SKINNER *
MARKETING MGR
-PHONE-
213 7466132
213 7463675
SOLAR/ENERGY CONSERV INFO DISSEMINATION ORG
PROVIDE INFORMATION SEARCHES TO CLIENTS

SOUTHERN CALIF, UNIV OF
WESRAC
809 WEST 34TH ST
LOS ANGELES CA 90007
HERBERT O ASBURY
MGR SPECIAL PROJECTS
-PHONE-
213 7466132
213 7463675
SOLAR/ENERGY CONSERV INFO DISSEMINATION ORG
PROVIDE INFORMATION SEARCHES TO CLIENTS

SOUTHERN METHODIST UNIVERSITY
INSTITUTE OF TECHN
DALLAS TX 75275
TING L CHU
PROFESSOR

-PHONE-
214 6923014
214 6923017
SOLAR RESEARCH
INORGANIC SEMI-CONDUCTORS

SPECTROLAB DIV OF TEXTRON
SOLAR POWER SYSTEMS
12484 GLADSTONE AVE
SYLMAR CA 91342
JERRY W RAVIN *
MARKETING MGR
-PHONE-
213 3654611

SOLAR MANUFACTURERS
INORGANIC SEMI-CONDUCTORS

SPECTROLAB DIV OF TEXTRON
SOLAR POWER SYSTEMS
12484 GLADSTONE AVE
SYLMAR CA 91342
DICK DONNELLY
-PHONE-
213 3654611
SOLAR MANUFACTURERS
INORGANIC SEMI-CONDUCTORS

SPENSLEY HORN LUBITZ AND JUBAS
PATENT LAW
1880 CENTURY PK EAST
LOS ANGELES CA 90066
DANIEL L DAWES
-PHONE-
213 5535050
SOLAR RESEARCH
ZONING ORDINANCES FOR SOLAR UTILIZATION

ST JEROME UNIVERSITY
DEPT D'HELIOPHYSIQUE
13013 MARSEILLE FRANCE
GEORGES PERI
FACULTY OF SCIENCE
-PHONE-
93 98 0901
SOLAR RESEARCH
SOLAR RADIATION MEASUREMENTS

STANFIELD AIR SYSTEMS
110 TARA WAY
ATHENS GA 30601
LYNN M STANFIELD
PRESIDENT

-PHONE-
404 5494767
404 5434202
SOLAR POTENTIAL MFG,ESTABLISHED BUSINESS
RETROFITTING, INTEREST

STANFORD RESEARCH INSTITUTE
333 RAVENSWOOD AVE
MENLO PARK CA 94025
RONALD I DEUTSCH *
GEN MGR OFF PUB REL

-PHONE-
415 3266200 X3754

SOLAR RESEARCH
PUBLISH ORIGINAL RESEARCHED INFORMATION

STANFORD RESEARCH INSTITUTE
ENERGY TECHNOLOGY
333 RAVENSWOOD AVE
MENLO PARK CA 94025
JOHN A ALICH JR
ENGINEER ECONOMIST
-PHONE-
415 3266200 X2343
415 3266200 X2704
SOLAR RESEARCH
PUBLISH ORIGINAL RESEARCHED INFORMATION

STANFORD RESEARCH INSTITUTE
333 RAVENSWOOD AVE
MENLO PARK CA 94025
RICHARD L GOEN

-PHONE-
415 3266200 X3198

SOLAR RESEARCH
MEDIUM TEMPERATURE SOLAR COLLECTOR SYSTEMS

STANFORD RESEARCH INSTITUTE
ECONOMICS DIVISION
333 RAVENSWOOD AVE
MENLO PARK CA 94025
JOHN P HENRY JR
DIR ENERGY TECH DEPT
-PHONE-
415 3266200 X2358
415 3266200 X2704
SOLAR RESEARCH
PUBLISH ORIGINAL RESEARCHED INFORMATION

STANFORD RESEARCH INSTITUTE
FOOD&PLANT SCIENCES
333 RAVENSWOOD AVE
MENLO PARK CA 94025
ROBERT E INMAN
PLANT BIOLOGIST
-PHONE-
415 3266200 X2155
408 2558853
SOLAR RESEARCH
BIO-CONVERSION

STANFORD UNIVERSITY
NEWS SERVICE
STANFORD CA 94305
ROBERT W BEYERS *
DIR NEWS SERVICE
-PHONE-
415 3212300 X2558

SOLAR RESEARCH
INORGANIC SEMI-CONDUCTORS

STANFORD UNIVERSITY
MATERIALS SCIENCE
STANFORD CA 94305
RICHARD H BUBE
PROF MAT SCI AND EE
-PHONE-
415 3212300 X2535
415 3215796
SOLAR RESEARCH
INORGANIC SEMI-CONDUCTORS

STANFORD UNIVERSITY
DEPT OF MECH ENGR
STANFORD CA 94305
JOEL H FERZEGER

-PHONE-
415 3212300 X3615

SOLAR RESEARCH
MEDIUM TEMPERATURE SOLAR COLLECTOR SYSTEMS

STEELCRAFT CORP
ENVIRONMENTAL DESIGN
PO BOX 12408
MEMPHIS TN 38112
GARY FORD
V P
-PHONE-
901 3242151
901 4581885
SOLAR MANUFACTURERS
MEDIUM TEMPERATURE SOLAR COLLECTOR SYSTEMS

STOLLE CORP
1501 MICHIGAN ST
SIDNEY OH 45365
EDWARD G BECK JR

SOLAR MANUFACTURERS
MEDIUM TEMPERATURE SOLAR COLLECTOR SYSTEMS

STONEWALLEN ENGINEERING
RT 2 BOX 381A
DELTA CO 81416
R TODD MAC DOUGALL
ENGINEER

SOLAR RESEARCH
SOLAR ENERGY STORAGE

SUHAY ENTERPRISES
1505 E WINDSOR RD
GLENDALE CA 91204
FRANK L SUHAY

-PHONE-
213 2469352
SOLAR MANUFACTURERS
SOLAR WATER HEATING

SUN MOUNTAIN DESIGN
960 CAMINO SANTANDER
SANTA FE NM 87501
HERMAN G BARKMANN

-PHONE-
505 9831680
SOLAR RESEARCH
SOLAR SPACE HEATING & COOLING

SUN PUBLISHING CO
P O BOX 4383
ALBUQUERQUE NM 87106
SKIP WATSON *
EDITOR
-PHONE-
505 2556550
SOLAR/ENERGY CONSERV INFO DISSEMINATION ORG
PUBLISH LAY + TECHNICAL PERIODICALS

SUN PUBLISHING CO
P O BOX 4383
ALBUQUERQUE NM 87106
THE SUN
SOLAR ENERGY SERIES
-PHONE-
505 2556550
SOLAR/ENERGY CONSERV INFO DISSEMINATION ORG
PUBLISH LAY + TECHNICAL PERIODICALS

SUNDSTRAND HYDRAULICS
2210 HARRISON AVE
ROCKFORD IL 61101
JAMES F NELSON *
PRODUCT MANAGER
-PHONE-
815 2267212
SOLAR POTENTIAL MFG,ESTABLISHED BUSINESS
SOLAR SPACE HEATING & COOLING

SUNDSTRAND HYDRAULICS
2210 HARRISON AVE
ROCKFORD IL 61101
JAMES H MEYER
SR PROJ ENGR
-PHONE-
815 2267246
SOLAR POTENTIAL MFG,ESTABLISHED BUSINESS
SOLAR SPACE HEATING & COOLING

SUNDU COMPANY
3319 KEYES LANE
ANAHEIM CA 92804
ANTHONY MEAGHER

-PHONE-
714 8282873
SOLAR MANUFACTURERS
SOLAR WATER HEATING

SUNEARTH CONSTRUCTION CO INC
BOX 99
MILFORD SQUARE PA 18935

-PHONE-
215 5368555
SOLAR MANUFACTURERS
SOLAR SPACE HEATING & COOLING

SUNPOWER
RT 4 BOX 275
ATHENS OH 45701
WILLIAM T BEALE
PRESIDENT
-PHONE-
614 5936934
SOLAR POTENTIAL MFG,NEW BUSINESS
SOLAR ENGINES

SUNSHINE ENERGY CORP
ROUTE 25
BROOKFIELD CENTER CT 06805
BRUCE GRIFFEN

-PHONE-
203 7753369
SOLAR POTENTIAL MFG,NEW BUSINESS
SOLAR SPACE HEATING

SUNSTRAND AVIATION
RESEARCH DEPT
4747 HARRISON AVE
ROCKFORD IL 61101
R G MOKADAM
-PHONE-
815 2266780
SOLAR POTENTIAL MFG,ESTABLISHED BUSINESS
SOLAR HEAT,COOL,&ELECTRIC POWER GENERATION

SUNWATER CO
1112 PIONEER WAY
EL CAJON CA 92020
HORACE MCCRACKEN

-PHONE-
714 2830454
SOLAR MANUFACTURERS
SOLAR WATER HEATING

SUNWORKS INC
669 BOSTON POST ROAD
GUILFORD CT 06437
EVERETT M BARBER JR
PRESIDENT
-PHONE-
203 4536191
SOLAR MANUFACTURERS
MEDIUM TEMPERATURE SOLAR COLLECTOR SYSTEMS

SYNERGY PEOPLES' PAGES
P O BOX AH
STANFORD CA 94305
ALLEN BORNING
TEXT EDITOR

SOLAR/ENERGY CONSERV INFO DISSEMINATION ORG
PUBLISH LAY BOOKS

SYSTEM PLANNING CORP
1500 WILSON BLVD
ARLINGTON VA 22209
BRUCE W MACDONALD

-PHONE-
703 5254904
SOLAR RESEARCH
PROVIDE ECONOMIC ANALYSIS OF SOLAR TECH

SYSTEMS RESEARCH LABS INC
2800 INDIAN RIPPLE
DAYTON OH 45440
KEITH KLOPFENSTEIN

-PHONE-
513 2528758
SOLAR RESEARCH
SOLAR LASERS

T-K SOLAR DISTRIBUTORS
5 TH FLOOR
9 EAST 16TH STREET
NEW YORK NY 10003
THOMAS P KAY
PRESIDENT
-PHONE-
212 9894422
SOLAR PRODUCTS DISTRIBUTORS
SOLAR WATER HEATING

TECHNION-ISRAEL INST OF TECH
BLDG RESEARCH STATN
TECHNION CITY
HAIFA ISRAEL
BARUCH GIVONI
HD BLDG CLIMATOLOGY
-PHONE-
225111 X285
SOLAR RESEARCH
SOLAR SPACE HEATING & COOLING

TECHNION-ISRAEL INST OF TECH
MECHANICAL ENG DEPT
TECHNION CITY
HAIFA ISRAEL
GERSHON GROSSMAN
SENIOR LECTURER
-PHONE-
04 235105
04 235106
SOLAR RESEARCH
SOLAR SPACE HEATING & COOLING

TECHNION-ISRAEL INST OF TECH
DEPT MECH ENGR
TECHNION CITY
HAIFA ISRAEL
AVRAHAM SHITZER

-PHONE-
04 235105

SOLAR RESEARCH
SOLAR SPACE HEATING & COOLING

TECHNION-ISRAEL INST OF TECH
MECHANICAL ENGINEERG
TECHNION CITY
HAIFA ISRAEL
ALEXANDER SOLAN
ASSOCIATE PROFESSOR

ENERGY CONSERVATION RESEARCH
MEDIUM TEMPERATURE SOLAR COLLECTOR SYSTEMS

TENNESSEE, UNIV OF
MECH&AERO ENGR DEPT
KNOXVILLE TN 37916
DR EDWARD LUMSDAINE
ASSOCIATE PROFESSOR
-PHONE-
615 9743035
SOLAR RESEARCH
SOLAR SPACE HEATING & COOLING

TERRUM SUN&ELECT BOILER INDUST
14 HAYOTZER HOLDN
TEL AVIV ISRAEL 611481

SOLAR MANUFACTURERS
SOLAR WATER HEATING

TEXAS A&M UNIV
CHEMICAL ENGRG DEPT
COLLEGE STATION TX 77843
RICHARD R DAVISON

-PHONE-
713 8453361
SOLAR RESEARCH
MEDIUM TEMPERATURE SOLAR COLLECTOR SYSTEMS

TEXAS A&M UNIV
CHEMICAL ENGR DEPT
COLLEGE STATION TX 77843
W B HARRIS

-PHONE-
713 8453361
SOLAR RESEARCH
MEDIUM TEMPERATURE SOLAR COLLECTOR SYSTEMS

TEXAS A&M UNIV
CHEMICAL ENGRG DEPT
COLLEGE STATION TX 77843
J H MARTIN

-PHONE-
713 8453361
SOLAR ECONOMIC CONSIDERATIONS
SOLAR SPACE HEATING & COOLING

TEXAS INSTRUMENTS
34 FOREST ST
ATTLEBORO MA 02703
TUEVO SANTALA

-PHONE-
217 2222800 X6150
SOLAR RESEARCH
MEDIUM TEMPERATURE SOLAR COLLECTOR SYSTEMS

TEXAS INSTRUMENTS INC
GOVT PROD MS10-10
ATTLEBORO MA 02703
WILLIAM C PAYNTON
MANAGER
-PHONE-
617 2222800 X6654
617 2222800 X484
SOLAR POTENTIAL MFG,ESTABLISHED BUSINESS
INORGANIC SEMI-CONDUCTORS

TEXAS INSTRUMENTS INC
MS10-16
ATTLEBORO MA 02703
UNTO SAVOLAINEN
-PHONE-
617 2222800 X272
SOLAR RESEARCH
MEDIUM TEMPERATURE SOLAR COLLECTOR SYSTEMS

TEXAS TECH UNIVERSITY
DEPT OF CIVIL ENGRG
PO BOX 4089
LUBBOCK TX 79409
GEORGE A WHETSTONE
-PHONE-
806 7421234
SOLAR RESEARCH
SOLAR ELECTRIC POWER GENERATION

TEXAS, UNIV OF
ZOOLOGY DEPT
AUSTIN TX 78712
JACK M YERS
-PHONE-
512 4711686
SOLAR RESEARCH
BIO-CONVERSION

THE ROYAL INSTITUTION
M D ARCHER
SOLAR RESEARCH
INORGANIC SEMI-CONDUCTORS

THERMAX ELECTRIC WATER HEATERS
P O BOX 173
HAMILTON CENTRAL AUSTRALIA
SOLAR MANUFACTURERS
SOLAR WATER HEATING

THERMO ELECTRON CORPORATION
85 FIRST AVENUE
WALTHAM MA 02154
JERRY P DAVIS
-PHONE-
674 8908700 X266
SOLAR POTENTIAL MFG,ESTABLISHED BUSINESS
SOLAR ENGINES

THOMASON SOLAR HOMES INC
6802 WALKER MILL SE
WASHINGTON DC 20027
DR HARRY E THOMASON
-PHONE-
301 3364042
301 3365329
SOLAR MANUFACTURERS
SOLAR SPACE HEATING & COOLING

TOKIWA OPTIC&ELECT MANUFACTORY
2-9 1CHOME KOYAMADAI
SHINAGAWA-KU TOKYO JAPAN
HIROSHI MIZOJIRI
PRESIDENT
-PHONE-
03 7126212
03 7149947
SOLAR MANUFACTURERS
HIGH TEMPERATURE SOLAR COLLECTOR SYSTEMS

TOKYO UNIVERSITY
FAC OF ENGINEERING
BUNKYO-KU
TOKYO JAPAN
TAKASHI HIRAYAMA
-PHONE-
03 8122111
SOLAR RESEARCH
SOLAR SPACE HEATING

TOTAL ENVIRONMENTAL ACTION
DESIGN
BOX 47
HARRISVILLE NH 03450
DOUGLAS E MAHONE *
-PHONE-
603 8273087
SOLAR/ENERGY CONSERV INFO DISSEMINATION ORG
ALL SOLAR TECHNOLOGIES

TOTAL ENVIRONMENTAL ACTION
BOX 47
HARRISVILLE NH 03450
BRUCE ANDERSON
PRINCIPAL
-PHONE-
603 8273087
603 8273097
SOLAR/ENERGY CONSERV INFO DISSEMINATION ORG
ALL SOLAR TECHNOLOGIES

TR PRODUCTIONS
SOLAR ENERGY PROJECT
1031 COMMONWEALTH AV
BOSTON MA 02215
CHAS SCHWARTZ
DIRECTOR
-PHONE-
617 7830200
SOLAR/ENERGY CONSERV INFO DISSEMINATION ORG
PUBLISH AND/OR DISTRIBUTE FILMS

TRANTER INC
SPECIAL PROJECTS
735 EAST HAZEL ST
LANSING MI 48909
ROBERT S BOWLAND
ASST TO VP
-PHONE-
517 3728410
SOLAR MANUFACTURERS
MEDIUM TEMPERATURE SOLAR COLLECTOR SYSTEMS

TRW SYSTEMS
ONE SPACE PARK
REDONDO BEACH CA 90278
KEN MORITZ *
-PHONE-
213 5362253
SOLAR POTENTIAL MFG,ESTABLISHED BUSINESS
SOLAR SPACE HEATING & COOLING

TRW SYSTEMS
BLDG 42
ONE SPACE PARK
REDONDO BEACH CA 90278
JACK M CHERNE
MGR SOLAR ENGY PROG
-PHONE-
213 5352871
SOLAR POTENTIAL MFG,ESTABLISHED BUSINESS
SOLAR SPACE HEATING & COOLING

TURBON ENGINEERING PTY LTD
PIRUDI STREET
COORPAROO AUSTRALIA
SOLAR MANUFACTURERS
SOLAR WATER HEATING

TYCO LABORATORIES INC
SILICON DEPARTMENT
16 HICKORY DRIVE
WALTHAM MA 02154
DAVID N JEWETT
MANAGER
-PHONE-
617 8902400
SOLAR RESEARCH
INORGANIC SEMI-CONDUCTORS

TYCO LABORATORIES INC
16 HICKORY DRIVE
WALTHAM MA 02154
A I MLAVSKY
-PHONE-
617 8902400
SOLAR RESEARCH
INORGANIC SEMI-CONDUCTORS

U N E S C O
ELHARRACH B P 26
ALGER ALGERIA
DR Y ELMAHGARY
SOLAR RESEARCH
SOLAR WATER DISTILLATION

U S DEPT HOUSING & URBAN DEV
OFF POLICY DEV&RESCH
451 7TH ST S W
WASHINGTON DC 20410
JOSEPH SHERMAN
DIR BLDG TECH&SAFETY
-PHONE-
202 7555364
SOLAR ECONOMIC CONSIDERATIONS
GOVERNMENT AGENCIES PROVIDING GRANTS

U S GOVT GSA
CONSTRUCTION MGMT
BLDG#41 DFC
DENVER CO 80225
CARL SWENSON *
CONST ENGR SUPERVISR
-PHONE-
303 2342641
SOLAR INTEREST
PLANNING OFFICES, FEDERAL

U S GOVT GSA
CONSTRUCTION MGMT
BLDG#41 DFC
DENVER CO 80225
JAMES H OTTMER
MECHANICAL ENGINEER
-PHONE-
303 2342641
SOLAR INTEREST
PLANNING OFFICES, FEDERAL

U S HOME CORP
RESOURCE SAVING HOME
1437 S BELCHER RD
CLEARWATER FL 33516
ALAN C BOMSTEIN
DIRECTOR
-PHONE-
813 5310441
SOLAR POTENTIAL MFG,ESTABLISHED BUSINESS
HOMES

U S NAVAL AMMUNITION DEPOT
HAWTHORNE NV 89415
ERNERST J KIRSCHKE CAPT

-PHONE-
702 9452451 X590
SOLAR RESEARCH
SOLAR SPACE HEATING

U S WEATHER BUREAU
NATL WEATH RECOD CTR
FEDERAL BLDG
ASHEVILLE NC 28801

SOLAR/ENERGY CONSERV INFO DISSEMINATION ORG
SOLAR RADIATION MEASUREMENTS

UNITED AUTO WORKERS
CONSERVATION DEPT
UAW FAMILY ED CENTER
ONAWAY MI 49765
PHILIP J PERKINS
ENVIRONMENTAL ED
-PHONE-
517 7338521
SOLAR/ENERGY CONSERV INFO DISSEMINATION ORG
PROVIDE AUDIO & VISUAL SERVICES

UNITED NATIONS
OFFICE FOR SCI &TECH
NEW YORK NY 10017
GERTRAND H CHATEL
SCI APPL SEC RM3146E
-PHONE-
212 7541234 X2591
SOLAR RESEARCH
ALL SOLAR TECHNOLOGIES

U S NAVY
CODE L 80 CEL, NCBC
PORT HUENEME CA 93043
EARL BECK

-PHONE-
805 9824623
SOLAR RESEARCH
SOLAR WATER HEATING

UK ATOMIC ENERGY AUTHORITY
REACTOR GROUP
RISLEY
WARRINGTON LANCASHIR ENGLAND
N KIRKER
TECH OPERATIONS DIR
-PHONE-
WAR 31244 X2290
SOLAR RESEARCH
INORGANIC SEMI-CONDUCTORS

UNITED BANK OF BOULDER
1300 WALNUT ST
BOULDER CO 80302
JACK MINNEMAN
V P

-PHONE-
303 4423734
SOLAR FINANCIAL CONSIDERATIONS
BANKERS

UNITED NATIONS
RESOURCES& TRANSPORT
NEW YORK NY 10017
THE SECTION HEAD
ENERGY SECTION ESA
-PHONE-
212 7541234 X4339
SOLAR RESEARCH
ALL SOLAR TECHNOLOGIES

U S PATENT OFFICE
TECH ASSESMNT&FORCST
WASHINGTON DC 20231
ALFRED C MARMOR
DIRECTOR
-PHONE-
703 5573051
SOLAR/ENERGY CONSERV INFO DISSEMINATION ORG
PROVIDE INFORMATION SEARCHES TO CLIENTS

UNITED AIRCRAFT RESEARCH LABS
ENERGY RESEARCH DEPT
SILVER LANE
EAST HARTFORD CT 06108
DR SIMION C KUO
PRINCIPAL SYS ENGR
-PHONE-
203 5658758
SOLAR RESEARCH
SOLAR SPACE HEATING & COOLING

UNITED ENGRS&CONSTRUCTORS INC
1401 ARCH
PHILADELPHIA PA

SOLAR RESEARCH
OCEAN THERMAL GRADIENT SYSTEMS

UNITSPAN ARCH SYS INC
9419 MASON AVE
CHATSWORTH CA 91311
AL OTTUM

-PHONE-
213 9981131
SOLAR MANUFACTURERS
MEDIUM TEMPERATURE SOLAR COLLECTOR SYSTEMS

UNIV TECNICA FED SANTA MARIA
LAB DE ENERGIA SOLAR
CASILLA 110 V
VALPARAISO CHILE
JULIO HIRSCHMANN R

-PHONE-
61268
SOLAR RESEARCH
MEDIUM TEMPERATURE SOLAR COLLECTOR SYSTEMS

UNIVERSAL OIL PRODUCTS CO
WOLVERINE TUBE DIV
P O BOX 2202
DECATUR AL 35601
JOHN K THORNE
MGR PRODUCT DEVELOP
-PHONE-
205 3531310 X290
SOLAR POTENTIAL MFG,ESTABLISHED BUSINESS
MEDIUM TEMPERATURE SOLAR COLLECTOR SYSTEMS

UNIVERSAL OIL PRODUCTS COMPANY
10 UOP PLAZA
DES PLAINES IL 60016
WILLIAM C HOLT JR

-PHONE-
312 3913330
SOLAR POTENTIAL MFG,ESTABLISHED BUSINESS
MEDIUM TEMPERATURE SOLAR COLLECTOR SYSTEMS

UNIVERSIDADE FED DE PARAIBA
ESCOLA DE ENGENHARIA
58.000 JOAO PESSOA
PARAIBA BRASIL
CLEANTHO DA CAMARA TORRES
LAB DE ENERGIA SOLAR

SOLAR RESEARCH
ALL SOLAR TECHNOLOGIES

UNIVERSITY CITY SCIENCE INST
POWER INFO CENTER
3401 MARKET-RM 2107
PHILADELPHIA PA 19104

-PHONE-
215 3828683
SOLAR/ENERGY CONSERV INFO DISSEMINATION ORG
SOLAR ELECTRIC POWER GENERATION

UNIVERSITY OF TECHNOLOGY
LOUGHBOROUGH
LEIES ENGLAND
D J CROOME

-PHONE-
LOU 63171
SOLAR RESEARCH
MEDIUM TEMPERATURE SOLAR COLLECTOR SYSTEMS

US AFCRL-PHF
HANSCOM FIELD
BEDFORD MA 01730
NICHOLAS F YANNONI

-PHONE-
617 8612265
SOLAR RESEARCH
INORGANIC SEMI-CONDUCTORS

US DEPT OF AGRICULTURE
INDEPENDENCE AVE SW
WASHINGTON DC 20250
CHARLES BEER
DIR EXTENSION SER

SOLAR RESEARCH
SOLAR WATER HEATING

US NAVAL WEAPONS CENTER
CHINA LAKE CA 93555
RICHARD D ULRICH
CODE 45701

-PHONE-
714 9397384
SOLAR RESEARCH
TOTAL SYSTEMS INTEGRATION

UTAH STATE UNIVERSITY
DEPT OF ECONOMICS
LOGAN UT 84322
CRAIG H PETERSON
ASST PROF ECONOMICS

-PHONE-
801 7524100 X7307
SOLAR ECONOMIC CONSIDERATIONS
PROVIDE ECONOMIC ANALYSIS OF SOLAR TECH

UTAH, UNIV OF
MECH ENGR DEPT
SALT LAKE CITY UT 84112
ROBERT F BOEHM

-PHONE-
801 5816441
SOLAR RESEARCH
MEDIUM TEMPERATURE SOLAR COLLECTOR SYSTEMS

VANDERBILT UNIV
ELECT ENGR
21ST AV S
NASHVILLE TN 37203
ROBERT NASH
ASSOC PROF
-PHONE-
615 3227311
SOLAR RESEARCH
SOLAR SPACE HEATING

VANTAGE HEATING&AIR COND INC
5611 KENDALL CT
ARVADA CO 80002
RICHARD PENCE

-PHONE-
303 4204410
SOLAR PRODUCTS DISTRIBUTORS
SOLAR SPACE HEATING & COOLING

VIRGINIA, UNIV OF
DEPT OF MECH ENGR
CHARLOTTESVILLE VA 22901
DR JAMES J KAUZLARICH

-PHONE-
804 9243661
SOLAR RESEARCH
SOLAR SPACE HEATING & COOLING

WALTER V STERLING INC
TECH&MANAGMT CONSULT
16262 E WHITTIER B
WHITTIER CA 90603
CHESTER W YOUNG
-PHONE-
213 9438703
SOLAR RESEARCH
SOLAR ELECTRIC POWER GENERATION

WASHINGTON UNIVERSITY
CTR BIOL OF NAT SYS
BOX 1126
ST LOUIS MO 63130
BARRY COMMONER
DIRECTOR
-PHONE-
314 8630100
314 8636768
SOLAR ENVIRONMENTAL CONSIDERATIONS
ALL SOLAR TECHNOLOGIES

VERMONT, UNIV OF
ELECT ENGR DEPT
BURLINGTON VT 05401
LLOYD M LAMBERT

-PHONE-
802 6563332
SOLAR RESEARCH
INORGANIC SEMI-CONDUCTORS

VOLUNTEERS IN INTL TECH ASSIST
3706 RHODE ISLAND AV
MT RANIER MD 20822
THOMAS H FOX

-PHONE-
301 2777000
SOLAR RESEARCH
SOLAR COOKING

WARNER BURNS TOAN LUNDE
724 FIFTH AVE
NEW YORK NY 10019
JEAN DUFFY

-PHONE-
212 7578000
SOLAR INTEREST
SOLAR HEAT,COOL,&ELECTRIC POWER GENERATION

WATT ENGR LTD
8395 BASELINE RD
BOULDER CO 80303
ARTHUR D WATT

-PHONE-
303 4946734

SOLAR/ENERGY CONSERV INFO DISSEMINATION ORG
SOLAR SPACE HEATING & COOLING

VEXOCAVE COMPANY
STAR ROUTE BOX 400
N TURKEY CREEK
MORRISON CO 80465
RICHARD KITHIL
-PHONE-
303 6978246
SOLAR RESEARCH
SOLAR SPACE HEATING

WALLACE AND TIERMAN LTD
925 WARDEN AVE
SCARBOROUGH ONTARIO CANADA

SOLAR PRODUCTS DISTRIBUTORS
WIND CONVERSION

WASEDA UNIVERSITY
DEPT OF ARCHITECTURE
4-170 NISHIOKUBO
SHINJUKU-KU TOKYO JAPAN
KEN-ICHI KIMURA
-PHONE-
03 2093211 X420
SOLAR RESEARCH
SOLAR SPACE HEATING & COOLING

WEATHER ENGINEERING & MFG INC
4703 N PARK DR
COLORADO SPRINGS CO 80907
GEORGE R PEACORE
PRESIDENT

-PHONE-
303 5988228

SOLAR PRODUCTS DISTRIBUTORS
SOLAR HEAT,COOL,&ELECTRIC POWER GENERATION

WEATHER ENGINEERING & MFG INC
4703 N PARK DR
COLORADO SPRINGS CO 80907
LANE A PINNOW
MGR ENGR & SALES
-PHONE-
303 5988228
SOLAR PRODUCTS DISTRIBUTORS
SOLAR HEAT,COOL,&ELECTRIC POWER GENERATION

WEST TEXAS STATE UNIV
BOX 248
CANYON TX 79016
DR VAUGHN NELSON

-PHONE-
806 6563904
SOLAR RESEARCH
WIND CONVERSION

WESTERN ONTARIO, UNIV OF
FAC OF ENGINEERING
LONDON N6A3K7 CANADA
ROBERT K SWARTMAN

-PHONE-
519 6793332
SOLAR RESEARCH
SOLAR REFRIGERATION

WESTINGHOUSE ELECT CORP
8401 BASELINE RD
BOULDER CO 80303
DR CHARLES D BEACH

-PHONE-
303 4946363

SOLAR RESEARCH
SOLAR ELECTRIC POWER GENERATION

WESTINGHOUSE ELECTRIC CORP
MFG DEVEL. LAB
R & D CENTER
PITTSBURGH PA 15235
FREDERICK J MICHEL
MGR INT'L SERVICES
-PHONE-
412 2563548
412 2564724
SOLAR POTENTIAL MFG,ESTABLISHED BUSINESS
SOLAR HEAT,COOL,&ELECTRIC POWER GENERATION

WESTINGHOUSE ELECTRIC CORP.
SPECIAL ENERGY SYST
PO BOX 1693 MS-999
BALTIMORE MD 21203
ALBERT WEINSTEIN
MANAGER
-PHONE-
301 7653454

SOLAR RESEARCH
SOLAR SPACE HEATING & COOLING

WESTINGHOUSE RESEARCH LAB
CHURCHILL BORO
PITTSBURGH PA 15235
ROGER W WARREN
MANAGER
-PHONE-
412 2565187
SOLAR RESEARCH
HIGH TEMPERATURE SOLAR COLLECTOR SYSTEMS

WESTINGHOUSE RESEARCH LABS
BEULAH RD
PITTSBURG PA 15235
R K RIEL *

-PHONE-
412 2563614
SOLAR RESEARCH
INORGANIC SEMI-CONDUCTORS

WESTINGHOUSE RESEARCH LABS
SOLID STATE RES
BEULAH RD
PITTSBURGH PA 15235
F A SHIRLAND
-PHONE-
412 2563613
SOLAR RESEARCH
INORGANIC SEMI-CONDUCTORS

WHIRLPOOL CORP
R&E
MONTE ROAD
BENTON HARBOR MI 49022
DR EUGENE R WHITLOW
-PHONE-
616 9265341
SOLAR RESEARCH
MEDIUM TEMPERATURE SOLAR COLLECTOR SYSTEMS

WILLIAM M FROBECK & ASSOCIATES
1011 GILMAN STREET
BERKELEY CA 94710
WARREN W EUKEL

-PHONE-
415 5248664
SOLAR RESEARCH
ALL SOLAR TECHNOLOGIES

WILLIAMS COLLEGE
PHYSICS DEPT
WILLIAMSTOWN MA 01267
JAY SHELTON

-PHONE-
413 5972123
SOLAR RESEARCH
MEDIUM TEMPERATURE SOLAR COLLECTOR SYSTEMS

WINDWORKS
BOX 329 ROUTE 3
MUKWONAGO WI 53149

SOLAR/ENERGY CONSERV INFO DISSEMINATION ORG
WIND CONVERSION

WINDWORKS
BOX 329 RTE 3
MUKWONAGO WI 53149
HANS MEYER

SOLAR COMPONENTS MANUFACTURER
WIND CONVERSION

WISCONSIN, STATE OF
FACLTY MGMT BUREAU
1 WEST WILSON ST
MADISON WI 53702
PAUL L BROWN *
DIRECTOR RM 180
-PHONE-
608 2661031
SOLAR RESEARCH
SOLAR SPACE HEATING

WISCONSIN, STATE OF
FACLTY MGMT BUREAU
1 WEST WILSON ST
MADISON WI 53702
GORDON E HARMAN PE
ENVIROMT AFF OFFICER
-PHONE-
608 2662885
SOLAR RESEARCH
SOLAR SPACE HEATING

WISCONSIN, UNIV OF
ENGINEER EXP STATION
1500 JOHNSON DRIVE
MADISON WI 53706
J A DUFFIE

SOLAR RESEARCH
SOLAR SPACE HEATING & COOLING

WISCONSIN, UNIV OF
MARINE STUDIES CENTE
1225 WEST DAYTON ST
MADISON WI 53706
THOMAS W SMITH
RESEARCH SPECIALIST
-PHONE-
608 2634578
ENERGY CONSERVATION RESEARCH
NEW CONSTRUCTION, RESEARCHING

WM PATTERSON COLLEGE
SECONDARY EDUCATION
300 POMPTON RD
WAYNE NJ 07470
ZWEIG JONAS
ASSOCIATE PROFESSOR
-PHONE-
201 8812119

ENERGY CONSERVATION RESEARCH
SOLAR HEAT,COOL,&ELECTRIC POWER GENERATION

WOODS HOLE OCEANOGRAPHIC INST
CHEMISTRY DEPT
WOODS HOLE MA 02543
OLIVER C ZAFIRIOU

-PHONE-
617 5481400 X302

SOLAR RESEARCH
BIO-CONVERSION

WORMSER SCIENTIFIC CORP
88 FOXWOOD RD
STAMFORD CT 06903
ERIC M WORMSER
PRESIDENT

-PHONE-
203 3221981
203 3226200
SOLAR/ENERGY CONSERV INFO DISSEMINATION ORG
SOLAR SPACE HEATING & COOLING

WRIGHT-INGRAHAM INSTITUTE
1228 TERRACE ROAD
COLORADO SPRINGS CO 80904
ELIZABETH W INGRAHAM
DIRECTOR
-PHONE-
303 6337011
SOLAR RESEARCH
SOLAR HEAT,COOL,&ELECTRIC POWER GENERATION

XONICS INC
6837 HAYVENHURST AVE
VAN NUYS CA 91406
PAUL F SCOTT
PRINCIPAL SCIENTIST
-PHONE-
213 7877380
SOLAR RESEARCH
SOLAR SPACE HEATING

YALE UNIV
FORESTRY&ENVIRO STDY
360 PROSPECT ST
NEW HAVEN CT 06511
WILLIAM E REIFSNYDER
-PHONE-
203 4360020
SOLAR RESEARCH
SOLAR RADIATION MEASUREMENTS

YOUNGSTOWN STATE UNIV
ELECT ENGR DEPT
YOUNGSTOWN OH 44503
CHARLES K ALEXANDER, JR, PHD
DIR SOLAR ENGY TSK G
-PHONE-
216 7461851 X425
216 7592698
SOLAR RESEARCH
TOTAL SYSTEMS INTEGRATION

ZEPHER WIND DYNAMO
P O BOX 241
BRUNSWICK MA 04011
ALLEN LISHNESS

-PHONE-
207 7256534
SOLAR POTENTIAL MFG,NEW BUSINESS
WIND CONVERSION

3 M CO
MARKETING DEPT
3 M CENTER
ST PAUL MN 55101
JOHN J MUELLER
MARKETING COORD
-PHONE-
612 7337201
SOLAR POTENTIAL MFG,NEW BUSINESS
WIND CONVERSION

YOUNGSTOWN STATE UNIV
ELECT ENGR DEPT
YOUNGSTOWN OH 44503
DR D F ROST
DIR SOLAR ENGY TSK G
-PHONE-
216 7461851 X425
216 5336128
SOLAR RESEARCH
TOTAL SYSTEMS INTEGRATION

ZOMEWORKS CORPORATION
BOX 712
ALBUQUERQUE NM 87103
STEVE BAER
PRESIDENT
-PHONE-
505 2425354
SOLAR MANUFACTURERS
SOLAR SPACE HEATING

YUP COMPANY LTD
1-10-3 JUJONAKAHARA
KITA-KU TOKYO JAPAN
YOSHIO YATABE

-PHONE-
03 9008987

SOLAR POTENTIAL MFG,ESTABLISHED BUSINESS
HIGH TEMPERATURE SOLAR COLLECTOR SYSTEMS

ZURN INDUSTRIES INC
5533 PERRY HWY
ERIE PA 16509
HARRY LINDSAY

-PHONE-
814 8644041
SOLAR PRODUCTS DISTRIBUTORS
INORGANIC SEMI-CONDUCTORS

AFFAIRS OFFICER * PUBLIC
NASA
CODE R-2
WASHINGTON DC 20546

SOLAR RESEARCH
ALL SOLAR TECHNOLOGIES

ALEXANDER, JR, PHD CHARLES K
DIR SOLAR ENGY TSK G
YOUNGSTOWN STATE UNIV
ELECT ENGR DEPT
YOUNGSTOWN OH 44503

-PHONE-
216 7461851 X425
216 7592698
SOLAR RESEARCH
TOTAL SYSTEMS INTEGRATION

ALLEN REDFIELD W
MARYLAND, UNIVERSITY OF
MECHANICAL ENGR
COLLEGE PARK MD 20740

SOLAR RESEARCH
SOLAR SPACE COOLING

ANDERSON BRUCE
PRINCIPAL
TOTAL ENVIRONMENTAL ACTION
BOX 47
HARRISVILLE NH 03450

-PHONE-
603 8273087
603 8273097
SOLAR/ENERGY CONSERV INFO DISSEMINATION ORG
ALL SOLAR TECHNOLOGIES

AIMONE MICHAEL A
GRADUATE STUDENT-U F
3461 SW 2ND AVE #425
GAINESVILLE FL 32607

SOLAR RESEARCH
SOLAR ELECTRIC POWER GENERATION

ALFIERI LUCILLE
ACTING DIRECTOR
SOLAR ENERGY RESEARCH&INFO CTR
1001 CONN AVE NW
WASHINGTON DC 20036

SOLAR/ENERGY CONSERV INFO DISSEMINATION ORG
PUBLISH LAY PERIODICALS

ALLISON H JACK
ASSOC PROF ELEC ENGG
OKLAHOMA STATE UNIVERSITY
ELECTRICAL ENGG
ENGINEERING SOUTH
STILLWATER OK 74074
-PHONE-
405 3726211 X6476
SOLAR RESEARCH
SOLAR ENERGY STORAGE

ANDERSON DONALD E
VICE PRES
SHELDAHL INC
CORP R & D
HWY 3
NORTHFIELD MN 55057
-PHONE-
507 6455633
SOLAR POTENTIAL MFG,ESTABLISHED BUSINESS
HIGH TEMPERATURE SOLAR COLLECTOR SYSTEMS

ALDERFER RONALD G
CHICAGO, UNIV OF
BIOLOGY DEPT
5630 S INGLESIDE AVE
CHICAGO IL 60637
-PHONE-
312 7532666
SOLAR RESEARCH
BIO-CONVERSION

ALICH JR JOHN A
ENGINEER ECONOMIST
STANFORD RESEARCH INSTITUTE
ENERGY TECHNOLOGY
333 RAVENSWOOD AVE
MENLO PARK CA 94025
-PHONE-
415 3266200 X2343
415 3266200 X2704
SOLAR RESEARCH
PUBLISH ORIGINAL RESEARCHED INFORMATION

ALWARD RON
MCGILL UNIV MACDONALD COLLEGE
BRACE RESEARCH INST
ST ANN DE BELLEVUE800
QUEBEC H9X 3MI CANADA

-PHONE-
514 4576580 X341
SOLAR/ENERGY CONSERV INFO DISSEMINATION ORG
ALL SOLAR TECHNOLOGIES

ANDERSON FRANK A
MISSISSIPPI, UNIVERSITY OF
CHEMICAL ENGINEERING
UNIVERSITY MS 38677

-PHONE-
601 2327023

SOLAR RESEARCH
MEDIUM TEMPERATURE SOLAR COLLECTOR SYSTEMS

a

INDEX OF INDIVIDUALS

ANDERSON J HILBERT
PRESIDENT
SEA SOLAR POWER INC
1615 HILLOCK LANE
YORK PA 17403
-PHONE-
717 8438594
SOLAR POTENTIAL MFG,NEW BUSINESS
OCEAN THERMAL GRADIENT SYSTEMS

ANDERSON * JAMES H
VICE PRES
SEA SOLAR POWER INC
1615 HILLOCK LANE
YORK PA 17403
-PHONE-
717 8438594
SOLAR POTENTIAL MFG,NEW BUSINESS
OCEAN THERMAL GRADIENT SYSTEMS

ANDERSON PHILIP
ARKLA INDUST INC
509 MARSHALL ST
SHREVEPORT LA 71151
-PHONE-
318 4251271
SOLAR COMPONENTS MANUFACTURER
SOLAR SPACE COOLING

ANDERSON WAYNE
RUTGERS UNIVERSITY
ELECTRIC ENGR DEPT
NEW BRUNSWICK NJ 08903
-PHONE-
201 9323871
SOLAR RESEARCH
INORGANIC SEMI-CONDUCTORS

APFEL JOSEPH H
DIRECTOR RESEARCH
OPTICAL COATING LAB INC
P O BOX 1599
SANTA ROSA CA 95403
-PHONE-
707 5456440
SOLAR COMPONENTS MANUFACTURER
MEDIUM TEMPERATURE SOLAR COLLECTOR SYSTEMS

ARCHER M D
THE ROYAL INSTITUTION
LONDON ENGLAND
SOLAR RESEARCH
INORGANIC SEMI-CONDUCTORS

ARONIN,AIA,FRIA JEFFREY E
P O BOX 570
NEW YORK NY 10017
-PHONE-
516 3742394
SOLAR RESEARCH
SOLAR SPACE HEATING & COOLING

ASBURY HERBERT O
MGR SPECIAL PROJECTS
SOUTHERN CALIF, UNIV OF
WESRAC
809 WEST 34TH ST
LOS ANGELES CA 90007
-PHONE-
213 7466132
213 7463675
SOLAR/ENERGY CONSERV INFO DISSEMINATION ORG
PROVIDE INFORMATION SEARCHES TO CLIENTS

ATKINS JOHN
FLORIDA,UNIV OF/MECH ENGR
1222 N W 34 TERRACE
GAINESVILLE FL 32605
-PHONE-
904 3723117
SOLAR RESEARCH
SOLAR SPACE COOLING

AUSTIN JR LAWTHAN M
TECH SVCS DIR
PACIFIC POWER & LIGHT
CUST TECH SVCS
920 SW 6TH AVE
PORTLAND OR 97204
-PHONE-
503 2267411 X1054
503 6469894
SOLAR INTEREST
RETROFITTING, MECHANICAL SYSTEMS

AVALON RICHARD T
PRES
SELECT RAD SURFACES PLATING CO
P O BOX 8364
SAN JOSE CA 95155
-PHONE-
408 2959040
SOLAR COMPONENTS MANUFACTURER
ALL SOLAR TECHNOLOGIES

AVERY W H
JOHNS HOPKINS UNIV
APPLIED PHYSICS LAB
8621 GEORGIA AVE
SILVER SPRINGS MD 20910
-PHONE-
301 9537100 X2411
SOLAR RESEARCH
OCEAN THERMAL GRADIENT SYSTEMS

AYER JOHN B
P ENG
2013 DEERHURST CRT
OTTAWA ONT K1J842 CANADA

SOLAR RESEARCH
INORGANIC SEMI-CONDUCTORS

BAER STEVE
PRESIDENT
ZOMEWORKS CORPORATION
BOX 712
ALBUQUERQUE NM 87103

-PHONE-
505 2425354

SOLAR MANUFACTURERS
SOLAR SPACE HEATING

BALLANTYNE JR JAMES H
SOLAR INC
4706 FAIRVIEW AVE
BOISE ID 83704

SOLAR MANUFACTURERS
SOLAR SPACE HEATING

BARBISH JOSEPH
SOLAR HOME SYSTEMS
ROOM 202
38518 OAKHILL CIRCLE
WILLOUGHBY OH 44094
-PHONE-
216 9511119
SOLAR POTENTIAL MFG,ESTABLISHED BUSINESS
SOLAR SPACE HEATING & COOLING

BACHMAN DAVID H
SR FACILITIES ENGR
MARTIN MARIETTA AEROSPACE
MAIL I661
PO BOX 179
DENVER CO 80201
-PHONE-
303 7945211 X4051
303 7945211 X3427
SOLAR INTEREST
SOLAR ENGINES

BAILEY ROBERT L
FLORIDA, UNIV OF
ELECTRIC ENGR DEPT
GAINESVILLE FL 32601

-PHONE-
904 3920916
904 3920921
SOLAR RESEARCH
INORGANIC SEMI-CONDUCTORS

BARBER ROBERT
BARBER-NICHOLS ENGR CO
6325 W 55TH STREET
ARVADA CO 80002

-PHONE-
303 4218111
SOLAR RESEARCH
SOLAR SPACE COOLING

BARKMANN HERMAN G
SUN MOUNTAIN DESIGN
960 CAMINO SANTANDER
SANTA FE NM 87501

-PHONE-
505 9831680
SOLAR RESEARCH
SOLAR SPACE HEATING & COOLING

BACON PHILIP W
SOLAR ENGY COORD
CERTAIN-TEED PRODUCTS CORP
P O BOX 860
VALLEY FORGE PA 19482

-PHONE-
215 6875000 X7989

SOLAR POTENTIAL MFG,ESTABLISHED BUSINESS
MEDIUM TEMPERATURE SOLAR COLLECTOR SYSTEMS

BALCOMB DR J D
ASST ENERGY DIV-LDR
CALIF, UNIV OF-LOS ALAMOS
LOS ALAMOS SCI LAB
Q-DIVISON (Q-DOT)
LOS ALAMOS NM 87544
-PHONE-
505 6676441

SOLAR RESEARCH
SOLAR SPACE HEATING & COOLING

BARBER JR EVERETT M
PRESIDENT
SUNWORKS INC
669 BOSTON POST ROAD
GUILFORD CT 06437
-PHONE-
203 4536191
SOLAR MANUFACTURERS
MEDIUM TEMPERATURE SOLAR COLLECTOR SYSTEMS

BARNES FRANK
COLORADO UNIV OF-BOULDER
ELECT ENGR
BOULDER CO 80302

-PHONE-
303 4927327
SOLAR RESEARCH
INORGANIC SEMI-CONDUCTORS

BARR IRWIN R
A A I CORP
WEAPONS&AERO SYS DIV
PO BOX 6767
BALTIMORE MD 21204
-PHONE-
301 6661400 X232

SOLAR MANUFACTURERS
SOLAR SPACE HEATING & COOLING

BAYLESS JOHN
PRES & DIRECTOR
SOLORON
4850 OLIVE ST
DENVER CO 80022
-PHONE-
303 2892288
SOLAR MANUFACTURERS
MEDIUM TEMPERATURE SOLAR COLLECTOR SYSTEMS

BEALL JR SAM E
OAK RIDGE NATIONAL LAB
REACTOR DIVISION
PO BOX Y
OAK RIDGE TN 37830

-PHONE-
615 4838611
SOLAR RESEARCH
SOLAR HEAT,COOL,&ELECTRIC POWER GENERATION

BEATTIE DONALD A
ACTING DIVISION DIR
NATIONAL SCIENCE FOUNDATION
ADVNCED ENGY RES&TEC
WASHINGTON DC 20550
-PHONE-
202 6325726
SOLAR RESEARCH
ALL SOLAR TECHNOLOGIES

BARRY R G
COLORADO UNIV OF-BOULDER
GEOGRAPHY DEPT
BOULDER CO 80302

-PHONE-
303 4926387

SOLAR RESEARCH
SOLAR RADIATION MEASUREMENTS

BEACH DR CHARLES D
WESTINGHOUSE ELECT CORP
8401 BASELINE RD
BOULDER CO 80303

-PHONE-
303 4946363
SOLAR RESEARCH
SOLAR ELECTRIC POWER GENERATION

BEARD * NORRIS L
DIRECTOR
INTERTECHNOLOGY CORPORATION
MARKETING OPERATIONS
PO BOX 340
WARRENTON VA 22186
-PHONE-
703 3477900
SOLAR RESEARCH
SOLAR SPACE HEATING & COOLING

BECK EARL
U S NAVY
CODE L 80 CEL, NCBC
PORT HUENEME CA 93043

-PHONE-
805 9824623
SOLAR RESEARCH
SOLAR WATER HEATING

BARTON JOHN I
MARKETING MANAGER
OLIN CORPORATION,BRASS GROUP
ROLL-BOND PRODUCTS
EAST ALTON IL 62024
-PHONE-
618 2582000
618 2582443
SOLAR COMPONENTS MANUFACTURER
MEDIUM TEMPERATURE SOLAR COLLECTOR SYSTEMS

BEALE WILLIAM T
PRESIDENT
SUNPOWER
RT 4 BOX 275
ATHENS OH 45701
-PHONE-
614 5936934
SOLAR POTENTIAL MFG,NEW BUSINESS
SOLAR ENGINES

BEASLEY E L
SALES MANAGER
BEASLEY INDUSTRIES PTY LTD
BOLTON AVENUE
DEVON PARK, S A AUSTRALIA

-PHONE-
082 464871
SOLAR MANUFACTURERS
SOLAR WATER HEATING

BECK JR EDWARD G
STOLLE CORP
1501 MICHIGAN ST
SIDNEY OH 45365

SOLAR MANUFACTURERS
MEDIUM TEMPERATURE SOLAR COLLECTOR SYSTEMS

INDEX OF INDIVIDUALS

BECKMAN RALPH
REDE RESEARCH&DESIGN INSTITUTE
P O BOX 307
PROVIDENCE RI 02901

-PHONE-
401 8615390
SOLAR RESEARCH
SOLAR HEAT,COOL,&ELECTRIC POWER GENERATION

BECKMAN RONALD
DIRECTOR
REDE RESEARCH&DESIGN INSTITUTE
P O BOX 307
PROVIDENCE RI 02901

-PHONE-
401 8615390
SOLAR RESEARCH
SOLAR ELECTRIC POWER GENERATION

BEECHER MARY
LIBRARIAN
ARIZONA STATE UNIV
LIBRARY-SCI-ENGR SER
ROOM 257
TEMPE AZ 85281
-PHONE-
602 9657608
SOLAR/ENERGY CONSERV INFO DISSEMINATION ORG
ALL SOLAR TECHNOLOGIES

BEER CHARLES
DIR EXTENSION SER
US DEPT OF AGRICULTURE
INDEPENDENCE AVE SW
WASHINGTON DC 20250

SOLAR RESEARCH
SOLAR WATER HEATING

BELL DR ROBERT A
DIRECTOR
CON EDISON CO OF NY INC
RESEARCH & DEVELOP
H IRVING PLACE
NEW YORK NY 10003
-PHONE-
212 4603882
SOLAR RESEARCH
MEDIUM TEMPERATURE SOLAR COLLECTOR SYSTEMS

BELL R A
CONSOLIDATED EDISON
4 IRVING PLACE
NEW YORK NY 10003

-PHONE-
212 4604600
SOLAR RESEARCH
SOLAR SPACE HEATING & COOLING

BENEDICT PHILIP C
NATL CTR FOR ATMOS RESEARCH
BOX 3000
BOULDER CO 80303

-PHONE-
303 4945151
SOLAR RESEARCH
SOLAR SPACE HEATING

BENEMAN JOHN
CALIF, UNIV OF-LA JOLLA
CHEM DEPT
LA JOLLA CA 92037

-PHONE-
714 4532000 X1639
SOLAR RESEARCH
BIO-CONVERSION

BENGTSON WILLIAM L
406 OAK BROOK ROAD
OAK BROOK IL 60521

-PHONE-
312 6542492
SOLAR RESEARCH
HOMES

BERMAN PAUL A
CALIF INSTITUT OF TECHNOLOGY
JET PROPULSION LAB
4800 OAK GROVE DRIVE
PASADENA CA 91103
-PHONE-
213 3542816
SOLAR RESEARCH
INORGANIC SEMI-CONDUCTORS

BERMAN ROBERT
MANAGER
ION PHYSICS CORPORATION
SOUTH BEDFORD ST
BURLINGTON MA 01803
-PHONE-
617 2721313 X462
SOLAR POTENTIAL MFG,ESTABLISHED BUSINESS
INORGANIC SEMI-CONDUCTORS

BERRY WILLIAM B
NOTRE DAME, UNIV OF
ELECT ENGR DEPT
NOTRE DAME IN 46556

-PHONE-
219 2837193
SOLAR RESEARCH
INORGANIC SEMI-CONDUCTORS

BEYERS * ROBERT W
DIR NEWS SERVICE
STANFORD UNIVERSITY
NEWS SERVICE
STANFORD CA 94305
-PHONE-
415 3212300 X2558
SOLAR RESEARCH
INORGANIC SEMI-CONDUCTORS

BIHLER L A
EXEC V P
ENGINEERS COLLABORATIVE LTD
101 E ONTARIO
CHICAGO IL 60611

-PHONE-
312 6449600

SOLAR RESEARCH
TOTAL SYSTEMS INTEGRATION

BLAKE FLOYD A
MARTIN MARIETTA AEROSPACE
MAIL # E0455
PO BOX 179
DENVER CO 80201
-PHONE-
303 7945211 X3677
SOLAR RESEARCH
SOLAR ELECTRIC POWER GENERATION

BLELLOCH PE ANDREW
272 CLARKSVILLE RD
PRINCETON JCT NJ 08550

-PHONE-
609 7991475
SOLAR RESEARCH
SOLAR SPACE HEATING

BEZMAN MICHEL
CAUDILL ROWLETT SCOTT ARCH/ENG
111 W LOOP SOUTH
HOUSTON TX 77027

-PHONE-
713 6219600
SOLAR RESEARCH
SOLAR SPACE HEATING

BILLINGS BEN
ENVIRO ACT REPRINT S

ENVIRONMTL ACTION-CO

DENVER CO 80206
-PHONE-
303 5341602
SEE PAGE 4.009 +4.021
SOLAR/ENERGY CONSERV INFO DISSEMINATION ORG
PUBLISH INFORMATIONAL REPRINTS

BLAKE STEVE
KANSAS,UNIV OF
DEPT OF GEOGRAPHY
LAWRENCE KS 66044

-PHONE-
913 8644276
SOLAR RESEARCH
WIND CONVERSION

BLIEDEN H RICHARD
PROG MGR PHOTOVOLTAC
NATIONAL SCIENCE FOUNDATION
ADVNCED ENGY RES&TEC
WASHINGTON DC 20550
-PHONE-
202 6327364
SOLAR RESEARCH
INORGANIC SEMI-CONDUCTORS

BIERMAN DR WENDELL J
CARRIER CORP
RESEARCH DIVISION
CARRIER PKWY
SYRACUSE NY 13201
-PHONE-
315 4638411 X3481
SOLAR RESEARCH
SOLAR SPACE HEATING & COOLING

BITLER ROBERT F
TECH COORD
ROHM AND HAAS
PLASTICS DEPT
INDEPENDENCE MALL W
PHILADELPHIA PA 19105
-PHONE-
215 5926799
SOLAR COMPONENTS MANUFACTURER
SOLAR SPACE HEATING & COOLING

BLANCO A A
DIRECTOR
SOLAR HEAT LTD
99 MIDDLETON HALL RD
KINGS NORTON BIRMGHM ENGLAND

SOLAR MANUFACTURERS
SOLAR WATER HEATING

BOCKRIS J O'M
FLINDERS UNIV OF S AUSTRALIA
BEDFORD PARK
SOUTH AUSTRALIA AUSTRALIA

-PHONE-
76 0511
SOLAR RESEARCH
INORGANIC SEMI-CONDUCTORS

BOEHM ROBERT F
UTAH, UNIV OF
MECH ENGR DEPT
SALT LAKE CITY UT 84112

-PHONE-
801 5816441
SOLAR RESEARCH
MEDIUM TEMPERATURE SOLAR COLLECTOR SYSTEMS

BOGAN RICHARD
PROG MGR BIOCONVERSN
NATIONAL SCIENCE FOUNDATION
ADVNCED ENGY RES&TEC
WASHINGTON DC 20550

-PHONE-
202 6327364
SOLAR RESEARCH
BIO-CONVERSION

BOMAR STEVE H
GEORGIA INST OF TECHNOLOGY
ENGINEERING EXP STAT
ATLANTA GA 30332

-PHONE-
404 8943661
SOLAR RESEARCH
SOLAR FURNACE

BORNING ALLEN
TEXT EDITOR
SYNERGY PEOPLES' PAGES
P O BOX AH
STANFORD CA 94305

SOLAR/ENERGY CONSERV INFO DISSEMINATION ORG
PUBLISH LAY BOOKS

BOER DR KARL W
SOLAR ENERGY SYSTEMS INC
70 S. CHAPEL ST.
NEWARK, DE 19711

-PHONE-
302 7310990
SOLAR MANUFACTURERS
INORGANIC SEMI-CONDUCTORS

BOGUCZ * EDWARD
EDWARDS ENGINEERING CORP
101 ALEXANDER AVE
POMPTON NJ 07444

-PHONE-
201 8352808
SOLAR POTENTIAL MFG,ESTABLISHED BUSINESS
SOLAR SPACE HEATING & COOLING

BOMSTEIN ALAN C
DIRECTOR
U S HOME CORP
RESOURCE SAVING HOME
1437 S BELCHER RD
CLEARWATER FL 33516
-PHONE-
813 5310441
SOLAR POTENTIAL MFG,ESTABLISHED BUSINESS
HOMES

BOTTS W V
ATOMICS INTERNATIONAL
P O BOX 309
CANOGA PARK CA 91304

-PHONE-
213 3411000 X2062
SOLAR RESEARCH
SOLAR ELECTRIC POWER GENERATION

BOER * DR KARL W
DELAWARE, UNIV OF
INST OF ENGY CONVSN
NEWARK DE 19711

-PHONE-
302 7388481
SOLAR RESEARCH
HOMES

BOLL JOHN A
DIRECTOR R&D
MOTHER EARTH NEWS INC
RESEARCH DEPT
P O BOX 70
HENDERSONVILLE NC 28739
-PHONE-
704 6924256
SOLAR/ENERGY CONSERV INFO DISSEMINATION ORG
TOTAL SYSTEMS INTEGRATION

BONIFACE ALLEN
RAYPAK INC
31111 AGOURA RD
THOUSAND OAKS CA 91360

-PHONE-
213 8891500
SOLAR MANUFACTURERS
MEDIUM TEMPERATURE SOLAR COLLECTOR SYSTEMS

BOWLAND ROBERT S
ASST TO VP
TRANTER INC
SPECIAL PROJECTS
735 EAST HAZEL ST
LANSING MI 48909
-PHONE-
517 3728410
SOLAR MANUFACTURERS
MEDIUM TEMPERATURE SOLAR COLLECTOR SYSTEMS

BOYD DAVID A
MANAGER
AMERICAN SCIENCE&ENGINEER INC
SOLAR CLIMATE CONTRL
995 MASS AVE
CAMBRIDGE MA 02139
-PHONE-
617 8681600
SOLAR RESEARCH
SOLAR SPACE HEATING & COOLING

BROWN RUSSELL
PRESIDENT
BROWN MANUFACTURING CO
P O BOX 14546
OKLAHOMA CITY OK 73114
-PHONE-
405 7511323
SOLAR COMPONENTS MANUFACTURER
MEDIUM TEMPERATURE SOLAR COLLECTOR SYSTEMS

BRYCE-SMITH D
READING UNIV OF
CHEMISTRY
READING ENGLAND

SOLAR RESEARCH
INORGANIC SEMI-CONDUCTORS

BUCHBERG HARRY
CALIF, UNIV OF-L A
ENGR & APPLIED SCIEN
ENGY&KINETICS 553IBH
LOS ANGELES CA 90024
-PHONE-
213 8255313

SOLAR RESEARCH
SOLAR SPACE HEATING & COOLING

BROMS TODD J
ASST LOAN OFFICER
PERCY WILSON MORTGAGE&FINANCE
COMM & INDUST LOAN
221 N LASALLE ST
CHICAGO IL 60601
-PHONE-
312 8556726
SOLAR FINANCIAL CONSIDERATIONS
MORTGAGE INSTITUTIONS

BROWNING WILLIAM
COMFORT TRAIN AIR COND CO
4060 S KALAMATH
ENGLEWOOD CO 80110

-PHONE-
303 7894497
SOLAR PRODUCTS DISTRIBUTORS
SOLAR SPACE HEATING & COOLING

BUBE RICHARD H
PROF MAT SCI AND EE
STANFORD UNIVERSITY
MATERIALS SCIENCE
STANFORD CA 94305

-PHONE-
415 3212300 X2535
415 3215796
SOLAR RESEARCH
INORGANIC SEMI-CONDUCTORS

BUCK DR J G
CONTINENTAL CAN COMPANY INC
CORP RED DEPT
7622 S RACINE AVE
CHICAGO IL 60620
-PHONE-
312 8463800

SOLAR RESEARCH
ALL SOLAR TECHNOLOGIES

BROWN * PAUL L
DIRECTOR RM 180
WISCONSIN, STATE OF
FACLTY MGMT BUREAU
1 WEST WILSON ST
MADISON WI 53702
-PHONE-
608 2661031
SOLAR RESEARCH
SOLAR SPACE HEATING

BRUMLEVE THOMAS D
STAFF MEMBER
SANDIA LABORATORIES
DIVISION 8184
LIVERMORE CA 94550
-PHONE-
415 4552941
SOLAR RESEARCH
HIGH TEMPERATURE SOLAR COLLECTOR SYSTEMS

BUCHANAN S R
ASST DIR OF RESEARCH
JOHNSON SERVICE COMPANY
RESEARCH DIVISION
507 E MICHIGAN ST
MILWAUKEE WI 53201
-PHONE-
414 2769200 X677

SOLAR POTENTIAL MFG,ESTABLISHED BUSINESS
SOLAR SPACE HEATING & COOLING

BULLOCK HARRY C
PRESIDENT
BULLOCK ENGINEERING & DEV CO
70 WEST 6TH AVE
DENVER CO 80222
-PHONE-
303 8926416
303 8926483
SOLAR INTEREST
ALL SOLAR TECHNOLOGIES

BULLOCK J PENEOW
PRESIDENT
BENBOW RESEARCH
650 CALIFORNIA ST
SAN FRANCISCO CA 94108
-PHONE-
415 9812516
SOLAR RESEARCH
SOLAR SPACE HEATING

BURKE VERNON
BOX 2557
ASPEN CO 81611
SOLAR RESEARCH
SOLAR SPACE HEATING & COOLING

BURKE JR THEODORE
AJAX MAGNETHERMIC
1745 OVERLAND AVE NE
WARREN OH 44482
-PHONE-
216 3722529
SOLAR RESEARCH
SOLAR SPACE HEATING

BURNS JOSHUA A
ASSOCIATE DIRECTOR
BSIC/EFL
3000 SAND HILL RD
MENLO PARK CA 94025
-PHONE-
415 8542300
SOLAR/ENERGY CONSERV INFO DISSEMINATION ORG
PUBLISH ORIGINAL RESEARCHED INFORMATION

BURRIS HARRY D
PERSON
ASE INFO SERVICES
1448 E 7TH ST
TUCSON AZ 85719
-PHONE-
602 8826571
703 2814564
SOLAR/ENERGY CONSERV INFO DISSEMINATION ORG
PROVIDE INFORMATION SEARCHES TO CLIENTS

BURTON JAMES S
DIRECTOR
MITRE CORPORATION, THE
SOLAR ENERGY LAB
WESTGATE RESEARCH PK
MCLEAN VA 22101
-PHONE-
703 8933500
SOLAR RESEARCH
INORGANIC SEMI-CONDUCTORS

BUSHNELL ROBERT H
502 ORD DRIVE
BOULDER CO 80303
-PHONE-
303 4947421
SOLAR/ENERGY CONSERV INFO DISSEMINATION ORG
PROVIDE INFORMATION SEARCHES TO CLIENTS

CAIRNS RONALD E
RESEARCH & ENGR DEPT
BOEING AEROSPACE CO
MAIL STOP 8C-64
P O BOX 3999
SEATTLE WA 98124
-PHONE-
206 7730464
SOLAR RESEARCH
SOLAR SPACE HEATING & COOLING

CANTER ALAN L
DIRECTOR OF PLANNING
DENVER PLANNING OFFICE
1445 CLEVELAND PLACE
DENVER CO 80202
-PHONE-
303 2972736
303 2972737
SOLAR INTEREST
PLANNING OFFICES, LOCAL

CARDNER NEAL F
RESEARCH PHYSICIST
BORG WARNER CORP.
RESEARCH CENTER
ALGONQUIN&WOLF RDS
DES PLAINES IL 60018
-PHONE-
312 8273131 X266
ENERGY CONSERVATION RESEARCH
NEW CONSTRUCTION, RESEARCHING

CARISEO JOSEPH A
PRESIDENT
LAB SCIENCES INC
PO BOX 1236
BOCA RATON FA 33432
-PHONE-
305 3951917
SOLAR MANUF, EDUCATIONAL EQUIPMENT
MEDIUM TEMPERATURE SOLAR COLLECTOR SYSTEMS

CARLSMITH R
OAK RIDGE NATIONAL LABORATORY
PO BOX X
OAK RIDGE TN 37830
ENERGY CONSERVATION RESEARCH
NEW CONSTRUCTION, TECHNOLOGIES

CASELLA AL
ASST PROF
SANGAMON STATE UNIV
PHYSICAL SCI PROG
SPRINGFIELD IL 62708
-PHONE-
217 7866600
SOLAR RESEARCH
SOLAR SPACE HEATING

CASSEL DAVID E
CASSEL SOLAR ENGINEERING
ROUTE 10 BOX 17
ANNAPOLIS MD 21401
-PHONE-
301 9740897
SOLAR/ENERGY CONSERV INFO DISSEMINATION ORG
SOLAR SPACE HEATING & COOLING

CASTER TERRANCE R
PRESIDENT
ENERGY SYSTEMS INC
634 CREST DR
EL CAJON CA 92021
-PHONE-
714 4471000
SOLAR MANUFACTURERS
SOLAR SPACE HEATING

CCLI EUGENE E
CO-EDITOR
ALTERNATIVE SOURCES OF ENGY
928 2ND ST SW
ROANOKE VA 24016
-PHONE-
703 3448412
SOLAR RESEARCH
SOLAR SPACE HEATING & COOLING

CETINCELIK MUAMMER
PRESIDENT
SOLAR ENERGY INST OF TURKEY
POBOX 37 BAKANLIKLAR
ANKARA TURKEY
-PHONE-
234 255378
SOLAR MANUFACTURERS
MEDIUM TEMPERATURE SOLAR COLLECTOR SYSTEMS

CHAHROUDI DAY
RESEARCH AFILLIATE
M I T
ARCHITECTURE
3-407 MIT 77MASSAVE
CAMBRIDGE MA 02139
-PHONE-
617 2537647
SOLAR RESEARCH
SOLAR SPACE HEATING & COOLING

CHALMERS E
HARVARD UNIVERSITY
GRADUATE SCHOOL
CAMBRIDGE STATION
CAMBRIDGE MA 02138

SOLAR RESEARCH
INORGANIC SEMI-CONDUCTORS

CHARTER * RICH
BOX 211
CAZADERO CA 95421

SOLAR INTEREST
ALL SOLAR TECHNOLOGIES

CHARTRES W W
MELBOURNE, UNIV OF
MECH ENGR DEPT
MELBOURNE
PARKVILLE, VICTORIA AUSTRALIA
-PHONE-
345 1844 X6744
SOLAR RESEARCH
MEDIUM TEMPERATURE SOLAR COLLECTOR SYSTEMS

CHATEL BERTRAND H
SCI APPL SEC RM3146E
UNITED NATIONS
OFFICE FOR SCI &TECH
NEW YORK NY 10017
-PHONE-
212 7541234 X2591
SOLAR RESEARCH
ALL SOLAR TECHNOLOGIES

CHATTERJEE * DR J S
PROF IN CHARGE
JADAVPUR UNIVERSITY
ELEC & TELECOMM ENGR
CALCUTTA 700032 INDIA

SOLAR RESEARCH
INORGANIC SEMI-CONDUCTORS

CHATTERJEE MANAS KUMAR
DIR ENGY PLANN COMM
GOVT OF INDIA YOJANA BHAVAN
POWER & ENERGY SECT
PARLIAMENT STREET
NEW DELHI 1 INDIA
-PHONE-
381109
SOLAR RESEARCH
INORGANIC SEMI-CONDUCTORS

CHEN CHARLES
PROG MGR HEAT&COOL B
NATIONAL SCIENCE FOUNDATN
PUBLIC TECH PROJECTS
WASHINGTON DC 20550
-PHONE-
202 6324175
SOLAR RESEARCH
SOLAR SPACE HEATING & COOLING

CHEPURNIY NICOLAS
MCGILL UNIVERSITY
CIVIL ENGR DEPT
MONTREAL P Q CANADA
-PHONE-
514 8437821
SOLAR RESEARCH
SOLAR PONDS

CHERNE JACK M
MGR SOLAR ENGY PROG
TRW SYSTEMS
BLDG 42
ONE SPACE PARK
REDONDO BEACH CA 90278
-PHONE-
213 5352871
SOLAR POTENTIAL MFG,ESTABLISHED BUSINESS
SOLAR SPACE HEATING & COOLING

CHERRY JR * NED D
533 EAST 87TH STREET
NEW YORK NY 10028
-PHONE-
212 8792503
212 4896666
SOLAR RESEARCH
HOMES

CHRISTENSEN D L
ALABAMA,UNIV OF/IN HUNTSVILLE
CTR ENVIRONMENT STUD
P O BOX 1247
HUNTSVILLE AL 35807
-PHONE-
205 8956257
SOLAR RESEARCH
TOTAL SYSTEMS INTEGRATION

CHU TING L
PROFESSOR
SOUTHERN METHODIST UNIVERSITY
INSTITUTE OF TECHN
DALLAS TX 75275
-PHONE-
214 6923014
214 6923017
SOLAR RESEARCH
INORGANIC SEMI-CONDUCTORS

CHUBB TALBOT
NAVAL RESEARCH LAB
UPPER AIR PHYSICS BR
CODE 7120
WASHINGTON DC 20375
-PHONE-
202 7673580
SOLAR RESEARCH
HIGH TEMPERATURE SOLAR COLLECTOR SYSTEMS

CLARK ARNOLD F
PHYSICIST
LAWRENCE LIVERMORE LAB
SOLAR ENERGY GROUP
P O BOX 808
LIVERMORE CA 94550
-PHONE-
415 4471100 X8693
SOLAR RESEARCH
SOLAR ELECTRIC POWER GENERATION

CLARK H A
DOW CORNING CORP
RESEARCH DEPT
S SAGINAW RD
MIDLAND MI 48640
-PHONE-
517 6368676
SOLAR RESEARCH
INORGANIC SEMI-CONDUCTORS

CLARK,IV * WILLIAM A
SUGARLOAF STAR RTE
BOULDER CO 80302
-PHONE-
303 4477533
SOLAR RESEARCH
SOLAR SPACE HEATING & COOLING

CLARKE FRANK J
GEORGIA INST OF TECHNOLOGY
SCHOOL OF ARCH
ATLANTA GA 30332
-PHONE-
404 8944885
ENERGY CONSERVATION RESEARCH
SOLAR SPACE HEATING & COOLING

CLARKSON CLARENCE W
DIR LTG PROD TECH SV
ASG INDUSTRIES,INC
1B BELMAR ROAD
CRANBURY NJ 08512
-PHONE-
609 6550058
609 6550057
SOLAR COMPONENTS MANUFACTURER
SOLAR SPACE HEATING & COOLING

CLEMENTS BARBARA
BECHTEL CORPORATION
ENERGY INFO CENTER
P O BOX 3965
SAN FRANCISCO CA 43201
-PHONE-
614 2993151
SOLAR/ENERGY CONSERV INFO DISSEMINATION ORG
PROVIDE INFORMATION SEARCHES TO CLIENTS

COCHRAN * KJERSTI A
1326 PERALTA AVE
BERKELEY CA 94702

SOLAR RESEARCH
ALL SOLAR TECHNOLOGIES

COHEN ROBERT
PROG MGR OCEAN THERM
NATIONAL SCIENCE FOUNDATION
ADVNCED ENGY RES&TEC
WASHINGTON DC 20550

-PHONE-
202 6327364
SOLAR RESEARCH
OCEAN THERMAL GRADIENT SYSTEMS

COLLINS, JR JOHN O
MGR GOVT RES LIASON
JOHNS-MANVILLE
RES AND DEV
PO BOX 5108
DENVER CO 80217
-PHONE-
303 7701000 X404

ENERGY CONSERVATION RESEARCH
NEW CONSTRUCTION, PRACTICING

CLEWS HENRY M
PRES
SOLAR WIND
PO BOX 7
EAST HOLDEN ME 04429
-PHONE-
207 8435168
SOLAR MANUFACTURERS
WIND CONVERSION

COCHRAN KYE
NORTHERN PLAINS RESOURCE CONSL
421 STAPLETON BLDG
BILLINGS MT 59101

-PHONE-
406 2596114
SOLAR/ENERGY CONSERV INFO DISSEMINATION ORG
ALL SOLAR TECHNOLOGIES

COHN ERNST M
NASA
CODE RPP
WASHINGTON DC 20546

-PHONE-
202 7552400
SOLAR RESEARCH
ALL SOLAR TECHNOLOGIES

COMMONER BARRY
DIRECTOR
WASHINGTON UNIVERSITY
CTR BIOL OF NAT SYS
BOX 1126
ST LOUIS MO 63130
-PHONE-
314 8630100
314 8636768
SOLAR ENVIRONMENTAL CONSIDERATIONS
ALL SOLAR TECHNOLOGIES

COBB MD JOHN C
COLORADO, UNIV OF
MEDICAL SCHOOL
DENVER CO 80220

-PHONE-
303 3947150
SOLAR RESEARCH
HOMES

COFFEL STEVEN
NEW AGE DOMES
EVENING STAR RANCH
RR 4 BOX 90
GOLDEN CO 80422
-PHONE-
303 5825719
SOLAR POTENTIAL MFG,NEW BUSINESS
GREENHOUSES

COLLINS DAVID L
GENERAL MANAGER
DAYLIN INC
SUN SOURCE
9606 SANTA MONICA
BEVERLY HILLS CA 90035
-PHONE-
213 8783211
SOLAR MANUFACTURERS
SOLAR WATER HEATING

CONWAY DR EDMUND J
NASA LANGLEY RESEARCH CTR
MAIL STOP 231A
HAMPTON VA 23665

-PHONE-
804 8273127

SOLAR RESEARCH
INORGANIC SEMI-CONDUCTORS

INDEX OF INDIVIDUALS

CORDER THOMAS E
VICE PRESIDENT
SOLAR ENVIRONMENTAL ENGR CO
PO BOX 1914
FORT COLLINS CO 80521
-PHONE-
303 4934480
303 4846386
SOLAR ECONOMIC CONSIDERATIONS
TOTAL SYSTEMS INTEGRATION

CRITTENDON L R
DELAWARE, UNIV OF
INST OF ENGY CONVSN
NEWARK DE 19711

-PHONE-
302 7388481
SOLAR RESEARCH
HOMES

CROWER BRUCE
PRESIDENT
CROWER CAMS & EQUIPMENT CO
3333 MAIN STREET
CHULA VISTA CA 92011

SOLAR RESEARCH
SOLAR ENGINES

CURD * MICHAEL
BOX 865
FRISCO CO 80443

-PHONE-
303 4686004
SOLAR RESEARCH
HOMES

CRANDALL WALTER E
NORTHROP RESEARCH&TECH CENTER
3401 WEST BROADWAY
HAWTHORNE CA 90250

-PHONE-
213 6754611 X1526

SOLAR POTENTIAL MFG,ESTABLISHED BUSINESS
SOLAR SPACE HEATING

CROOME D J
UNIVERSITY OF TECHNOLOGY
LOUGHBOROUGH
LEICS ENGLAND

-PHONE-
LOU 63171
SOLAR RESEARCH
MEDIUM TEMPERATURE SOLAR COLLECTOR SYSTEMS

CUBA JOSEPH F
DIR OF RESEARCH
AM SOC HEAT REFRIG AIRCON ENGR
345 EAST 47TH ST
NEW YORK NY 10017
-PHONE-
212 7526800 X367
SOLAR RESEARCH
SOLAR SPACE HEATING & COOLING

CURRAN H M
DIRECTOR OF ENGR
HITTMAN ASSOCIATES INC
ENGY & ENVIR SYS
9190 REED BRANCH RD
COLUMBIA MD 21043
-PHONE-
301 7307800
SOLAR RESEARCH
SOLAR SPACE COOLING

CRITES FAIA RAY D
IOWA STATE UNIV
DEPT OF ARCHITECTURE
AMES IA 50010

-PHONE-
515 2948460

SOLAR RESEARCH
SOLAR SPACE HEATING & COOLING

CROSSMAN LEON D
MGR SOLID STATE RES
DOW CORNING
SEMICONDUCTOR PROD
12334 GEDDES RD
HEMLOCK MI 48626
-PHONE-
517 6425201 X344
SOLAR RESEARCH
INORGANIC SEMI-CONDUCTORS

CUDDY L M
722 CRESTFOREST RD
KNOXVILLE TN 37919

SOLAR INTEREST
ALL SOLAR TECHNOLOGIES

CURRENT JR R C
AMELCO WINDOW CORP
77 RT 17 BOX 333
HASBROUCK HTS NJ 07604

-PHONE-
212 2440610
SOLAR COMPONENTS MANUFACTURER
SOLAR SPACE HEATING & COOLING

CUSHEN WALTER F
FEDERAL POWER COMMISSION
OFFICE CHIEF ENGR
825 N CAPITOL ST NE
WASHINGTON DC 20426
-PHONE-
202 3765270
SOLAR RESEARCH
SOLAR WATER HEATING

DA CAMARA TORRES CLEANTHO
LAB DE ENERGIA SOLAR
UNIVERSIDADE FED DE PARAIBA
ESCOLA DE ENGENHARIA
58.000 JOAO PESSOA
PARAIBA BRASIL
SOLAR RESEARCH
ALL SOLAR TECHNOLOGIES

DAMERDJI D M
INSTITUTE D'ENERGIE SOLAIRE
OBSERVATOIRE
BOUZAREAH ALGERIA
-PHONE-
784229
SOLAR RESEARCH
SOLAR WATER DISTILLATION

DATTA DR R L
CNTRL SALT&MARNE CHEM RES INST
BHAVNAGAR (GUJARAT) INDIA
-PHONE-
3960
SOLAR RESEARCH
INORGANIC SEMI-CONDUCTORS

DAVIDSON JOEL
ENVIRONMENTAL DESIGN
PETTIGREW AR 72752
SOLAR RESEARCH
SOLAR SPACE HEATING & COOLING

DAVIES ROBERT R
LITERATURE CONSULT
SILVER INSTITUTE INC
COLO REPRESENTATIVE
1001 CONNECTICUT, NW
WASHINGTON DC 20036
-PHONE-
303 4430105 XCOLO
202 3932285 XDC
SOLAR/ENERGY CONSERV INFO DISSEMINATION ORG
PROVIDE INFORMATION SEARCHES TO CLIENTS

DAVIS * ARIEL R
3476 FLEETWOOD DR
SALT LAKE CITY UT 84109
-PHONE-
801 2774140
801 2724225
SOLAR RESEARCH
HOMES

DAVIS HOMER W
CHAFFEY COLLEGE
ALTA LOMA CA 91701
-PHONE-
714 9871737 X240
SOLAR RESEARCH
HOMES

DAVIS JERRY P
THERMO ELECTRON CORPORATION
85 FIRST AVENUE
WALTHAM MA 02154
-PHONE-
674 8908700 X266
SOLAR POTENTIAL MFG,ESTABLISHED BUSINESS
SOLAR ENGINES

DAVIS WILLIAM W
CHAIRMAN
NORTHERN ARIZ UNIV
MECH ENGR
NAU BOX 15600
FLAGSTAFF AZ 86001
-PHONE-
602 5233698
SOLAR RESEARCH
MEDIUM TEMPERATURE SOLAR COLLECTOR SYSTEMS

DAVISON FRED F
HIGHWOOD MT 59450
-PHONE-
406 7332013
SOLAR RESEARCH
WIND CONVERSION

DAVISON RICHARD R
TEXAS A&M UNIV
CHEMICAL ENGRG DEPT
COLLEGE STATION TX 77843
-PHONE-
713 8453361
SOLAR RESEARCH
MEDIUM TEMPERATURE SOLAR COLLECTOR SYSTEMS

DAWES DANIEL L
SPENSLEY HORN LUBITZ AND JUBAS
PATENT LAW
1880 CENTURY PK EAST
LOS ANGELES CA 90066
-PHONE-
213 5535050
SOLAR RESEARCH
ZONING ORDINANCES FOR SOLAR UTILIZATION

DE REIMER DANE
SOLAR SYS INC
323 COUNTRY CLUB DR
REHOBOTH BEACH DE 19971

-PHONE-
302 2272323
SOLAR MANUFACTURERS
MEDIUM TEMPERATURE SOLAR COLLECTOR SYSTEMS

DEBS ATIP
GEORGIA INST OF TECHNOLOGY
ENGINEERING EXP STAT
ATLANTA GA 30332

-PHONE-
404 8943661
SOLAR RESEARCH
SOLAR SPACE HEATING

DECKER JOHN L
PRESIDENT
SOLARSYSTEMS INC
1515 WSW LOOP 323
TYLER TX 75701
-PHONE-
214 5920945
214 5975537
SOLAR MANUFACTURERS
SOLAR SPACE HEATING & COOLING

DEERING DR ROBERT P
CALIF DEPT PARKS & RECREATION
SPECIAL STUDIES SECT
1416 9TH ST
SACRAMENTO CA 95811
-PHONE-
916 4453130

ENERGY CONSERVATION RESEARCH
BIO-CONVERSION

DEETER DAN
DEETER HOMES
246 PALM AVE
AUBURN CA 95603

-PHONE-
916 8858448

SOLAR INTEREST
SOLAR SPACE HEATING & COOLING

DEFORD HENRY
SENIOR VP
FIRST NATIONAL CITY BANK
AREA 3 OPERATING GP
PO BOX 939 CHURCH ST
NEW YORK NY 10008

SOLAR FINANCIAL CONSIDERATIONS
BANKERS

DEKOLD DONALD F
DEKO LABS
BOX 12841 UNIV STATN
GAINESVILLE FL 32604

-PHONE-
904 3726009
SOLAR COMPONENTS MANUFACTURER
MEDIUM TEMPERATURE SOLAR COLLECTOR SYSTEMS

DEL GOBBO JOHN
PROG MGR HEAT&COOL B
NATIONAL SCIENCE FOUNDATN
PUBLIC TECH PROJECTS
WASHINGTON DC 20550

-PHONE-
202 6324175
SOLAR RESEARCH
SOLAR SPACE HEATING & COOLING

DENNETT ROGER D
R&D CONSULTANTS
8713 BASELINE ROAD
LAFAYETTE CO 80026

-PHONE-
303 4946131
SOLAR POTENTIAL MFG,NEW BUSINESS
SOLAR SPACE HEATING

DENTON * JESSE C
PENN, UNIVERSITY OF
N CTR ENGY MGT&POWER
TOWNE BUILDING
PHILADELPHIA PA 19104
-PHONE-
215 5945122
SOLAR RESEARCH
HIGH TEMPERATURE SOLAR COLLECTOR SYSTEMS

DESTICHE ROBERT J
VP SALES
ARKLA INDUSTRIES INC
BOX 751
LITTLE ROCK AR 72203
-PHONE-
501 3726241
SOLAR POTENTIAL MFG,ESTABLISHED BUSINESS
SOLAR SPACE HEATING & COOLING

DEUTSCH * RONALD I
GEN MGR OFF PUB REL
STANFORD RESEARCH INSTITUTE
333 RAVENSWOOD AVE
MENLO PARK CA 94025
-PHONE-
415 3266200 X3754
SOLAR RESEARCH
PUBLISH ORIGINAL RESEARCHED INFORMATION

DICKENSON STENCER
SOLAR HOMES INC
2 NARRAGANSETT AVE
JAMESTOWN RI 02835
-PHONE-
401 4231025
SOLAR MANUFACTURERS
HOMES

DIEGES * PAUL
1051 DAVIS RD
PERRIS CA 92370
-PHONE-
714 6572822
SOLAR RESEARCH
HIGH TEMPERATURE SOLAR COLLECTOR SYSTEMS

DIEHL R W
CORP DIRECTOR
CONTINENTAL CAN CO INC
PROCUREMENT PLANNING
633 THIRD AVE
NEW YORK NY 10017
-PHONE-
212 5517851
SOLAR RESEARCH
ALL SOLAR TECHNOLOGIES

DIKKERS ROBERT D
MGR SOL ENG DEM PROG
NATIONAL BUREAU OF STANDRDS
OFFICE OF HOUS TECH
BLDG 226 ROOM B146
WASHINGTON DC 20234
-PHONE-
301 9213285
SOLAR RESEARCH
SOLAR SPACE HEATING & COOLING

DIRKSE D J
BOB SCHMITT HOMES INC
13079 FALLING WATER
STRONGSVILLE OH 44136
-PHONE-
216 2386915
SOLAR MANUFACTURERS
HOMES

DIVONE LOUIS
PROG MGR WIND ENGY
NATIONAL SCIENCE FOUNDATION
ADVNCED ENGY RES&TEC
WASHINGTON DC 20550
-PHONE-
202 6327364
SOLAR RESEARCH
WIND CONVERSION

DIXON * DANIEL W
ARCHITECT
PO BOX 797
GRANBY CO 80446
-PHONE-
303 8872200
SOLAR RESEARCH
SOLAR SPACE HEATING

DOBSON LARRY
RT 1 BOX 459
CLINTON WA 98236
-PHONE-
204 3211361
SOLAR RESEARCH
HIGH TEMPERATURE SOLAR COLLECTOR SYSTEMS

DODGE ROBERT J
PENNWALT CORPORATION
AUTOMATIC POWER DIV
P O BOX 18738
HOUSTON TX 77023
-PHONE-
713 2285208
SOLAR PRODUCTS DISTRIBUTORS
WIND CONVERSION

DOGNIAUX R
INSTITUT ROYAL METEOROLOGIQUE
RADIOMETRY SECTION
3 AVENUE CIRCULAIRE
B 1180 BRUSSELS BELGIUM
-PHONE-
02 746788
SOLAR RESEARCH
SOLAR RADIATION MEASUREMENTS

DOHERTY JOHN R
DESERT RESEARCH INSTITUTE
OFFICE OF PRESIDENT
RENO NV 89507
-PHONE-
702 7846131
SOLAR RESEARCH
ACADEMIC BUILDINGS

DOHERTY * JOHN F
NEVADA SYSTEM, UNIV OF
DESERT RESEARCH INST
RENO NV 89507

SOLAR RESEARCH
ACADEMIC BUILDINGS

DOLAN RAYMOND J
CENTER FOR ENVIRONMENT & MAN
275 WINDSOR STREET
HARTFORD CT 06120
-PHONE-
203 5494400 X348
ENERGY CONSERVATION RESEARCH
SOLAR RADIATION MEASUREMENTS

DONNELLY DICK
SPECTROLAB DIV OF TEXTRON
SOLAR POWER SYSTEMS
12484 GLADSTONE AVE
SYLMAR CA 91342
-PHONE-
213 3654611
SOLAR MANUFACTURERS
INORGANIC SEMI-CONDUCTORS

DONNER * DOUG
V P MARKETING
SOLAR POWER SUPPLY
RT 3 BOX A10
EVERGREEN CO 80439
-PHONE-
303 6744734
SOLAR POTENTIAL MFG,NEW BUSINESS
SOLAR HEAT,COOL,&ELECTRIC POWER GENERATION

DORRINGTON * RICHARD F
11231 N 34TH DRIVE
PHOENIX AZ 85029

SOLAR RESEARCH
SOLAR SPACE HEATING & COOLING

DOUGHTY DR RALPH O
GROUP ENGINEER
GENERAL DYNAMICS CORP
CONVAIR AEROSPACE
P O BOX 748
FORT WORTH TX 76101
-PHONE-
817 7324811 X3113
817 7324811 X3292
SOLAR POTENTIAL MFG,NEW BUSINESS
MEDIUM TEMPERATURE SOLAR COLLECTOR SYSTEMS

DOWNING R B
DOWNING-LEACH ASSOC ARCH&PLANN
2305 BROADWAY
BOULDER CO 80302
-PHONE-
303 4437533
SOLAR RESEARCH
CONDOMINIUMS

DRESCHER ADAM J
REYNOLDS METALS
EXTRUSION DIV
6601 W BROAD ST
RICHMOND VA 23230
-PHONE-
804 2822331 X2197
SOLAR COMPONENTS MANUFACTURER
MEDIUM TEMPERATURE SOLAR COLLECTOR SYSTEMS

DRUMHELLER KIRK
BATTELLE NORTHWEST LABS
BATTELLE BLVD
RICHLAND WA 99352
-PHONE-
509 9462349
SOLAR RESEARCH
SOLAR SPACE HEATING & COOLING

DUBIN FRED S
PRESIDENT
DUBIN-MINDELL-BLOOME ASSOC PE
CONSULTING ENGINEERS
42 W39 ST
NEW YORK NY 10018
-PHONE-
212 8689700
SOLAR/ENERGY CONSERV INFO DISSEMINATION ORG
TOTAL SYSTEMS INTEGRATION

DUBIN-VAUGHN SARAH K
ONE SEASIDE PL
E NORWALK CT 06855
-PHONE-
203 8534953
SOLAR INTEREST
ALL SOLAR TECHNOLOGIES

DUFFIE J A
WISCONSIN, UNIV OF
ENGINEER EXP STATION
1500 JOHNSON DRIVE
MADISON WI 53706

SOLAR RESEARCH
SOLAR SPACE HEATING & COOLING

DUFFY JEAN
WARNER BURNS TOAN LUNDE
724 FIFTH AVE
NEW YORK NY 10019

-PHONE-
212 7578900
SOLAR INTEREST
SOLAR HEAT,COOL,&ELECTRIC POWER GENERATION

DUNKLE CHRIS T
PRESIDENT
ACORN GLAS-TINT INTERNATIONAL
1123 W CENTURY BLVD
LOS ANGELES CA 90044

-PHONE-
213 7561471
SOLAR PRODUCTS DISTRIBUTORS
OFFICE/BANK BUILDINGS

DUNKLE ROBERT V
CHIEF RESEARCH SCI
C I R S O
DIV OF MECH ENGR
P O BOX 26
HIGHETT, VICTORIA AUSTRALIA
-PHONE-
950333
SOLAR RESEARCH
MEDIUM TEMPERATURE SOLAR COLLECTOR SYSTEMS

ECKERT E R
MINN, UNIV OF
DEPT MECH ENGR
MINNEAPOLIS MN 55455

-PHONE-
612 3733315
SOLAR RESEARCH
SOLAR ELECTRIC POWER GENERATION

ECKERT J A
PROJECT MANAGER
EXXON RESCH&ENGR CO
SOLAR ENGY CONV UNIT
LINDEN NJ 07036
-PHONE-
201 4740100
SOLAR RESEARCH
INORGANIC SEMI-CONDUCTORS

EDMONDSON CWO-4 WILLIAM B
EDITOR & PUBLISHER
SOLAR ENERGY DIGEST
PO BOX 17776
SAN DIEGO CA 92117
-PHONE-
714 2272980
SOLAR/ENERGY CONSERV INFO DISSEMINATION ORG
PUBLISH LAY PERIODICALS

EDWARDS D K
CALIF, UNIV OF--L A
SCHOOL OF ENGR
ENGY & KINETICS DEPT
LOS ANGELES CA 90024

-PHONE-
213 8255313
SOLAR RESEARCH
MEDIUM TEMPERATURE SOLAR COLLECTOR SYSTEMS

EDWARDS JOAN
PROGRAM OFFICER
ENVIRONMENTAL ACTION COALITION
EDUC MATERIALS PROG
235 EAST 49TH ST
NEW YORK NY 10017
-PHONE-
212 4869550
SOLAR/ENERGY CONSERV INFO DISSEMINATION ORG
HAVE&MAINTAIN ENERGY DATA BANK

EDWARDS RAY C
PRESIDENT
EDWARDS ENGINEERING CORP
101 ALEXANDER AVE
POMPTON NJ 07444

-PHONE-
202 8352808 X49
SOLAR POTENTIAL MFG,ESTABLISHED BUSINESS
SOLAR SPACE HEATING & COOLING

EIBLING JAMES A
PROGRAM MANAGER
BATTELLE COLUMBUS LABS
UNCONVENTIONALENERGY
505 KING AVENUE
COLUMBUS OH 43201
-PHONE-
614 2993151
SOLAR RESEARCH
ALL SOLAR TECHNOLOGIES

ELDRIDGE FRANK R
MITRE CORPORATION, THE
SYSTEMS DEVELOP DIV
WESTGATE RESEARCH PK
MCLEAN VA 22101

-PHONE-
703 5368103
SOLAR RESEARCH
INORGANIC SEMI-CONDUCTORS

ELIASON * JON T
6501 ROBINHOOD LANE
HUNTSVILLE AL 35806

-PHONE-
205 8374782
SOLAR RESEARCH
INORGANIC SEMI-CONDUCTORS

ELLIS * EDWARD E
6126 VRAIN ST
ARVADA CO 80003

-PHONE-
303 5717561

SOLAR RESEARCH
MEDIUM TEMPERATURE SOLAR COLLECTOR SYSTEMS

ELMAHGARY DR Y
U N E S C O
ELHARRACH B P 26
ALGER ALGERIA

SOLAR RESEARCH
SOLAR WATER DISTILLATION

ELTER * DR JOHN F
130 LABURN CRESCENT
ROCHESTER NY 14620

-PHONE-
716 4735314
716 8722000
SOLAR RESEARCH
SOLAR SPACE HEATING

ELY RICHARD
NEW ENGLAND SIERRA CLUB
14 BEACON ST
BOSTON MA 02108

-PHONE-
617 2275339
SOLAR/ENERGY CONSERV INFO DISSEMINATION ORG
ALL SOLAR TECHNOLOGIES

EPPS CLIFT
RICE MAREK HARRAL HOLTZ INC
PO BOX 12037
DENVER CO 80212

-PHONE-
303 4204455
SOLAR RESEARCH
SOLAR SPACE HEATING & COOLING

ERIKSEN VERNON L
DIRECTOR OF R AND D
GENERAL MOTORS CORP
HARRISON RADIATOR
UPPER MOUNTAIN ROAD
LOCKPORT NY 14094
-PHONE-
716 4393362
SOLAR POTENTIAL MFG,NEW BUSINESS
MEDIUM TEMPERATURE SOLAR COLLECTOR SYSTEMS

ERLICKI M S
ISRAEL INSTITUTE OF TECHNOLOGY
TECHNION
HAIFA ISRAEL

-PHONE-
04 235151
SOLAR RESEARCH
INORGANIC SEMI-CONDUCTORS

ESCHER WILLIAM
ESCHER TECHNOLOGY ASSOC
P O BOX 189
ST JOHNS MI 48879

-PHONE-
517 2246726
SOLAR RESEARCH
OCEAN THERMAL GRADIENT SYSTEMS

EUKEL WARREN W
WILLIAM M BROBECK & ASSOCIATES
1011 GILMAN STREET
BERKELEY CA 94710

-PHONE-
415 5248664
SOLAR RESEARCH
ALL SOLAR TECHNOLOGIES

EVANS MARGARET
COLORADO UNIV OF-BOULDER
DIV OF CONTINUING ED
BOULDER CO 80302

-PHONE-
303 4927416
SOLAR/ENERGY CONSERV INFO DISSEMINATION ORG
PUBLISH AND/OR DISTRIBUTE FILMS

EVANS AIA BEN H
NATIONAL ACADEMY OF SCIENCES
BLDG RES ADVISORY BD
2101 CONSTITUT AV NW
WASHINGTON DC 20418
-PHONE-
202 3896515
SOLAR RESEARCH
SOLAR SPACE HEATING & COOLING

EVERS JR WILLIAM
VICE PRESIDENT
PGA ENGINEERS INC
10 BROADWAY
ST LOUIS MO 63102
-PHONE-
314 2317318
ENERGY CONSERVATION RESEARCH
NEW CONSTRUCTION, MECHANICAL SYSTEMS

FANG P H
BOSTON COLLEGE
PHYSICS DEPT
140 COMMONWEALTH
CHESTNUT HILL MA 02167

SOLAR RESEARCH
INORGANIC SEMI-CONDUCTORS

FARBER * ERICH H
DIRECTOR
FLORIDA, UNIV OF
SOLAR ENGY/ENGY CONV
GAINESVILLE FL 92611
-PHONE-
904 3920820
SOLAR RESEARCH
SOLAR ENGINES

FEEMSTER * JONH R
20089 PIERCE RD
SARATOGA CA 95070

-PHONE-
408 7394880 X2733
SOLAR POTENTIAL MFG,NEW BUSINESS
SOLAR SPACE HEATING & COOLING

FENG DAVID
COLORADO UNIV OF-BOULDER
CIVIL & ENVIR ENGR
BOULDER CO 80302

-PHONE-
303 4927112

ENERGY CONSERVATION RESEARCH
NEW CONSTRUCTION, PASSIVE SYSTEMS

FERZEGER JOEL H
STANFORD UNIVERSITY
DEPT OF MECH ENGR
STANFORD CA 94305

-PHONE-
415 3212300 X3615

SOLAR RESEARCH
MEDIUM TEMPERATURE SOLAR COLLECTOR SYSTEMS

FEUER ASHRAE STANLEY
PRESIDENT
FEUER CORPORATION
5401 MCCONNELL AVE
LOS ANGELES CA 90066
-PHONE-
213 3904046
213 4562894
SOLAR INTEREST
SOLAR SPACE HEATING & COOLING

FIELDS RAYMOND
DEPUTY DIRECTOR
NATIONAL SCIENCE FOUNDATN
PUBLIC TECH PROJECTS
WASHINGTON DC 20550
-PHONE-
202 6324175
SOLAR RESEARCH
SOLAR SPACE HEATING & COOLING

FLANNIGAN BILL
BILL FLANNIGANS PLANS
20032 23RD ST
ASTORIA NY 11100

-PHONE-
212 7282083
SOLAR/ENERGY CONSERV INFO DISSEMINATION ORG
WIND CONVERSION

FLORENCE * PRISCILLA
CHURCHILL FILMS
CUSTOMER RELATIONS
662 NO ROBERTSON
LOS ANGELES CA 90069
-PHONE-
213 6575110
SOLAR/ENERGY CONSERV INFO DISSEMINATION ORG
PUBLISH AND/OR DISTRIBUTE FILMS

FORD GARY
V P
STEELCRAFT CORP
ENVIRONMENTAL DESIGN
PO BOX 12408
MEMPHIS TN 38112
-PHONE-
901 3242151
901 4581885
SOLAR MANUFACTURERS
MEDIUM TEMPERATURE SOLAR COLLECTOR SYSTEMS

FOSTER DAN R
DIR-MEDIA RELATIONS
GRAY COMPANY ENTERPRISES
SUITE 245
7701 N STEMMONS
DALLAS TX 7524,
-PHONE-
214 6303841

SOLAR/ENERGY CONSERV INFO DISSEMINATION ORG
PROVIDE AUDIO SERVICES

FOSTER GERALD L
ARCHITECTS COLLABORATIVE INC
46 BRATTLE ST
CAMBRIDGE MA 02138

-PHONE-
617 8684200

SOLAR INTEREST
MEDIUM TEMPERATURE SOLAR COLLECTOR SYSTEMS

FOX THOMAS H
VOLUNTEERS IN INTL TECH ASSIST
3706 RHODE ISLAND AV
MT RANIER MD 20822

-PHONE-
301 2777000
SOLAR RESEARCH
SOLAR COOKING

FRANK ALAN M
COLSPAN ENVIRONMENTAL SYS INC
P O BOX 3467
BOULDER CO 80303

-PHONE-
303 4494411
SOLAR POTENTIAL MFG,NEW BUSINESS
MEDIUM TEMPERATURE SOLAR COLLECTOR SYSTEMS

FRANK M W
PROG MGR
BALL BROTHERS RESEARCH CORP
AEROSPACE DIV
P O BOX 1062
BOULDER CO 80302
-PHONE-
303 4414627
SOLAR RESEARCH
TOTAL SYSTEMS INTEGRATION

FRANTA GREGORY E
ROARING FORK RESOURCE CENTER
P O BOX 9950
ASPEN CO 81611

-PHONE-
303 9253481
303 9255125
SOLAR/ENERGY CONSERV INFO DISSEMINATION ORG
PROVIDE INFORMATION SEARCHES TO CLIENTS

FREDERIC ROBERT
FULTON FEDERAL SAVINGS & LOAN
P O BOX 1077
ATLANTA GA 30301

-PHONE-
404 5272300
SOLAR FINANCIAL CONSIDERATIONS
BANKERS

FREEMAN DAVID S
DIRECTOR
ENERGY POLICY PROJECT
1776 MASS AV NW
WASHINGTON DC 20036
-PHONE-
202 8721064

SOLAR RESEARCH
PUBLISH ORIGINAL RESEARCHED INFORMATION

FRIED, PE EDGAR M
STAFF ENG
CUTTER LABS
808 PARKER
BERKELEY CA 94710
-PHONE-
415 8410123 X278
415 2842858
SOLAR INTEREST
HOMES

FRIEDMAN MR
CHEMALLOY-ELECTRONICS CORP
7947 TOWERS-PO DR 10
SANTEE CA 92071

-PHONE-
714 4485715

SOLAR MANUFACTURERS
SOLAR ELECTRIC POWER GENERATION

GAJEWSKI RYSZARD
DIRECTOR OF RESCH
AMERICAN SCIENCE&ENGINEER INC
995 MASS AVE
CAMBRIDGE MA 02139
-PHONE-
617 8681600

SOLAR RESEARCH
HIGH TEMPERATURE SOLAR COLLECTOR SYSTEMS

GAUCHER * LEON P
CONSULTANT
R D 1
FISHKILL NY 12524

-PHONE-
914 6312465
SOLAR RESEARCH
ALL SOLAR TECHNOLOGIES

GAVRIL S
ISRAEL INSTITUTE OF TECHNOLOGY
TECHNION
HAIFA ISRAEL

-PHONE-
04 235151
SOLAR RESEARCH
INORGANIC SEMI-CONDUCTORS

GAY JAMES J
COLEMAN SOLAR SERVICE
8900 NW 34 AVE RD
MIAMI FL 33147

-PHONE-
305 6914126
SOLAR MANUFACTURERS
SOLAR WATER HEATING

GAYDOS GEORGE R
GAYDARDT INDUSTRIES INC
8542 EDGEWORTH DR
CAPITAL HEIGHTS MD 20028
-PHONE-
301 3500614
SOLAR MANUFACTURERS
MEDIUM TEMPERATURE SOLAR COLLECTOR SYSTEMS

GEALER ROY L
PRINCIPAL STAFF ENGR
FORD MOTOR CO
SCIENTIFIC RES STAFF
PO BOX 2053
DEARBORN MI 48121
-PHONE-
313 3231219
ENERGY CONSERVATION RESEARCH
RETROFITTING, TECHNOLOGIES

GERLACH ROBERT E
VICE PRESIDENT
HELIO ASSOCIATES INC
8230 E BROADWAY
TUCSON AZ 85710
-PHONE-
602 8865376
SOLAR RESEARCH
SOLAR HEAT,COOL,&ELECTRIC POWER GENERATION

GILDERSLEEVE JR OLIVER D
PHILADELPHIA ELECTRIC CO
ENGY CONVERSON RESCH
2301 MARKET ST S10-1
PHILADELPHIA PA 19101
-PHONE-
215 8414856
SOLAR RESEARCH
SOLAR ELECTRIC POWER GENERATION

GILLESPIE PETER J
VICE PRESIDENT
HEIZER CORPORATION
20 NORTH WACKER DR.
CHICAGO IL 60606
-PHONE-
312 6412200
SOLAR FINANCIAL CONSIDERATIONS
PROVIDE VENTURE CAPITOL

GIRARD PAUL
GEN DELEGATE AFEDES
A F E D E S
28 RUE DE LA SOURCE
75016 PARIS FRANCE
-PHONE-
224 59 35
SOLAR/ENERGY CONSERV INFO DISSEMINATION ORG
ALL SOLAR TECHNOLOGIES

GIRVAN ROBERT F
STANHOPE IA 50246
-PHONE-
515 8263287
SOLAR RESEARCH
MEDIUM TEMPERATURE SOLAR COLLECTOR SYSTEMS

GIUTRONICH J
NEW SOUTH WALES, UNIV OF
SCHOOL OF PHYSICS
BOX 1
KENSINGTON, N S W AUSTRALIA
-PHONE-
663 0351
SOLAR RESEARCH
HIGH TEMPERATURE SOLAR COLLECTOR SYSTEMS

GIVONI BARUCH
HD BLDG CLIMATOLOGY
TECHNION-ISRAEL INST OF TECH
BLDG RESEARCH STATN
TECHNION CITY
HAIFA ISRAEL
-PHONE-
225111 X285
SOLAR RESEARCH
SOLAR SPACE HEATING & COOLING

GLASER DR.PETER E
VICE PRESIDENT
ARTHUR D LITTLE,INC
ENGINEERING SCIENCES
20 ACORN PARK
CAMBRIDGE MA 01240
-PHONE-
617 8645770
SOLAR RESEARCH
SOLAR SPACE HEATING & COOLING

GLASS CO JOHN M
ENVIRONMENTAL DES
MANITOBA
410- 352 DONALD ST
WINNIPEG MANITOBA CANADA
-PHONE-
204 9424735
SOLAR RESEARCH
TOTAL SYSTEMS INTEGRATION

GLASSMAN ALAN S
RESEARCH ARCHITECT
ARMSTRONG CORK CO. R&D CENTER
BUILDING SYSTEMS
2500 COLUMBIA AVE
LANCASTER PA 17603
-PHONE-
717 3970611 X7128
SOLAR RESEARCH
SOLAR SPACE HEATING & COOLING

INDEX OF INDIVIDUALS

GOEN RICHARD L
STANFORD RESEARCH INSTITUTE
333 RAVENSWOOD AVE
MENLO PARK CA 94025

-PHONE-
415 3266200 X3198

SOLAR RESEARCH
MEDIUM TEMPERATURE SOLAR COLLECTOR SYSTEMS

GOULD SF26 CHARLES L
ROCKWELL INTERNATIONAL
SPACE DIVISION
12214 LAKEWOOD BLVD
DOWNEY CA 90241
-PHONE-
213 5943921
SOLAR POTENTIAL MFG,ESTABLISHED BUSINESS
SOLAR SPACE HEATING & COOLING

GREENBERG DR A B
AEROSPACE CORP
ADVANCED TECH APPLIC
P O BOX 92957
LOS ANGELES CA 90009

SOLAR RESEARCH
ALL SOLAR TECHNOLOGIES

GRIFFIN GORDON D
MANAGING DIRECTOR
AUST INVENT & DEVELOPMENTS PLD
EVELOPMENTS
P O BOX 227
FILLMORE UT 84631
-PHONE-
801 7436817
SOLAR RESEARCH
SOLAR ENERGY STORAGE

GOLDSMITH ALLEN
MAGRUDER MIDDLE SCHOOL
TORRANCE CA 90510

SOLAR RESEARCH
SOLAR FURNACE

GRAVEN ROBERT M
SOLAR DESIGNS
203 HOLLY LANE
ORINDA CA 94563

-PHONE-
415 2543938
SOLAR/ENERGY CONSERV INFO DISSEMINATION ORG
PROVIDE INFORMATION SEARCHES TO CLIENTS

GRETZ JOACHIM
CCR-EURATOM
SCIENTIF DIRECTORATE
ISPRA (VARESE) ITALY

-PHONE-
003 9332780 X131
SOLAR RESEARCH
TOTAL SYSTEMS INTEGRATION

GRIMES DALE M
MICHIGAN, UNIVERSITY OF
DEPT OF ELECT ENGR
ANN ARBOR MI 48104

-PHONE-
313 7649546
SOLAR RESEARCH
BIO-CONVERSION

GOODWIN DONALD W
DESIGN SPECIALIST
GENERAL DYNAMICS CORP
CONVAIR AEROSPACE
P O BOX 748
FT WORTH TX 76101
-PHONE-
817 7324811 X3113
817 7324811 X3292
SOLAR POTENTIAL MFG,NEW BUSINESS
MEDIUM TEMPERATURE SOLAR COLLECTOR SYSTEMS

GREELEY RICHARD S
MITRE CORPORATION, THE
WESTGATE RESEARCH PK
MCLEAN VA 22101

-PHONE-
703 8933500
SOLAR RESEARCH
ALL SOLAR TECHNOLOGIES

GRIFFEN BRUCE
SUNSHINE ENERGY CORP
ROUTE 25
BROOKFIELD CENTER CT 06805

-PHONE-
203 7753369
SOLAR POTENTIAL MFG,NEW BUSINESS
SOLAR SPACE HEATING

GRIMM CARL T
ARCHITECT
5 COOK LANE
CROTON-ON-HUDSON NY 10520

-PHONE-
914 2715471
SOLAR RESEARCH
SOLAR SPACE HEATING & COOLING

GRIMM CARL T
CURTIS&DAVIS ARCHITECTS/PLANER
126 EAST 38TH STREET
NEW YORK NY 10016

-PHONE-
212 6899590
SOLAR/ENERGY CONSERV INFO DISSEMINATION ORG
PROVIDE INFORMATION SEARCHES TO CLIENTS

GRIMM DR GEORGE W
MGR CORP RESEARCH
OWENS-CORNING FIBERGLAS CORP
TECHNICAL CENTER
PO BOX 415
GRANVILLE OH 43023
-PHONE-
614 5870610
ENERGY CONSERVATION RESEARCH
NEW CONSTRUCTION, INTEREST

GROBBELAAS C S
DIRECTOR
NATIONAL BLDG RESCH INST CSIR
ENVIRONMENT ENGR DIV
PO BOX 395
PRETORIA SOUTH AFRICA
-PHONE-
PRT 746011
SOLAR RESEARCH
SOLAR WATER DISTILLATION

GROSSKREUTZ J C
BLACK & VEATCH
1500 MEADOW LAKE PKY
KANSAS CITY MO 64114

-PHONE-
816 3617000

SOLAR RESEARCH
SOLAR ELECTRIC POWER GENERATION

GROSSMAN GERSHON
SENIOR LECTURER
TECHNION-ISRAEL INST OF TECH
MECHANICAL ENG DEPT
TECHNION CITY
HAIFA ISRAEL
-PHONE-
04 235105
04 235106
SOLAR RESEARCH
SOLAR SPACE HEATING & COOLING

GROVE DAVID R
SALES MGR
LIBBEY OWENS FORD CO
811 MADISON AVE
TOLEDO OH 43695

-PHONE-
419 2425781

SOLAR COMPONENTS MANUFACTURER
HIGH TEMPERATURE SOLAR COLLECTOR SYSTEMS

GROVER WILLIAM H
MANAGER
CHARLES W MOORE ASSOCIATES
ESSEX CT 06426

-PHONE-
203 7670101
SOLAR RESEARCH
HOMES

GUPTA DR MAKAM C
PROF MECH ENGR
INDIAN INSTITUTE OF TECHNOLOGY
MADRAS 600036 INDIA

-PHONE-
802742 X235
SOLAR RESEARCH
SOLAR SPACE HEATING & COOLING

GUTHRIE DAVID L
SISTEMAS DE ENERGIA SOLAR
AVENIDA MAGDALA 1151
CONDADO
SANTURCE PUERTO RICO
-PHONE-
809 7257825
SOLAR/ENERGY CONSERV INFO DISSEMINATION ORG
ALL SOLAR TECHNOLOGIES

HAACKE DR GOTTFRIED
AMERICAN CYANAMID CO
CHEM RESEARCH DIV
1937 WEST MAIN ST
STAMFORD CT 06904
-PHONE-
203 3487331 X329
SOLAR RESEARCH
INORGANIC SEMI-CONDUCTORS

HAGGARD KEN
CALIF POLYTECHNIC STATE UNIV
ARVH&ENVIRO DESGN&EN
SAN LUIS OBISPO CA 93401

-PHONE-
805 5462573
SOLAR RESEARCH
SOLAR SPACE HEATING & COOLING

HAGGART KEITH
PRESIDENT
NEW MEXICO SOLAR ENERGY ASSOC
602 1/2 CANYON RD
SANTA FE NM 87501
-PHONE-
505 9832861
SOLAR/ENERGY CONSERV INFO DISSEMINATION ORG
ALL SOLAR TECHNOLOGIES

INDEX OF INDIVIDUALS

HALL NORMAN W
ASHLAND OIL INC
RESEARCH & DEVELOP
P O BOX 391
ASHLAND KY 41101

-PHONE-
606 3293333 X8374
SOLAR RESEARCH
SOLAR SPACE HEATING & COOLING

HALLET JR RAYMON W
DIRECTOR
MCDONNELL DOUGLAS ASTRONAUTICS
ADV SOL&NVC ENGY SYS
5301 BOLSA AVE
HUNTINGTON BEACH CA 92646
-PHONE-
714 8963664
SOLAR RESEARCH
SOLAR ELECTRIC POWER GENERATION

HAMMOND JONATHAN
CALIF, UNIV OF-DAVIS
DEPT ENVIRO HORTCULT
DAVIS CA 95616

-PHONE-
916 7523697
ENERGY CONSERVATION RESEARCH
SOLAR SPACE HEATING & COOLING

HANSON GARRY D
MANAGER
LENNOX INDUSTRIES INC
HEATING EQUIP ENGR
PO BOX 250
MARSHALLTOWN IA 50158
-PHONE-
515 7544011
SOLAR INTEREST
SOLAR SPACE HEATING & COOLING

HARKNESS SARAH P
ARCHITECTS COLLABORATIVE INC
46 BRATTLE STREET
CAMBRIDGE MA 02138

-PHONE-
617 8684200 X333
SOLAR RESEARCH
ALL SOLAR TECHNOLOGIES

HARMAN PE GORDON E
ENVIROMT AFF OFFICER
WISCONSIN, STATE OF
FACLTY MGMT BUREAU
1 WEST WILSON ST
MADISON WI 53702
-PHONE-
608 2662885
SOLAR RESEARCH
SOLAR SPACE HEATING

HARRIS DWIGHT J
ROUTE 1, BOX 249A
CHESNEE SC 29323

-PHONE-
803 5780768

SOLAR RESEARCH
SOLAR SPACE HEATING & COOLING

HARRIS W B
TEXAS A&M UNIV
CHEMICAL ENGR DEPT
COLLEGE STATION TX 77843

-PHONE-
713 8453361

SOLAR RESEARCH
MEDIUM TEMPERATURE SOLAR COLLECTOR SYSTEMS

HARRISON KEITH
DIVISION MGR
KALWALL CORPORATION
P O BOX 237
MANCHESTER NH 03105
-PHONE-
603 6273861
603 6277887
SOLAR MANUFACTURERS
SOLAR SPACE HEATING

HARRISON RAYMOND
SANDIA LABS
ORGANIZATION 5717
ALBUQUERQUE NM 87110

-PHONE-
505 2648669
SOLAR RESEARCH
TOTAL SYSTEMS INTEGRATION

HAUER * CHARLES R
DIRECTOR
NATIONAL SCIENCE FOUNDATN
EXPLOR RESCH&PROB AS
WASHINGTON DC 20550
-PHONE-
202 6324175
SOLAR RESEARCH
SOLAR SPACE HEATING & COOLING

HAY HAROLD R
SKY THERM PROCESS& ENGINEERING
945 WILSHIRE BLVD
LOS ANGELES CA 90017

-PHONE-
213 6247261 X306
SOLAR POTENTIAL MFG,ESTABLISHED BUSINESS
SOLAR SPACE HEATING & COOLING

h

INDEX OF INDIVIDUALS

HEAD THE SECTION
ENERGY SECTION ESA
UNITED NATIONS
RESOURCES& TRANSPORT
NEW YORK NY 10017

-PHONE-
212 7541234 X4339
SOLAR RESEARCH
ALL SOLAR TECHNOLOGIES

HEIDRICH WILLIAM J
REVERE COPPER & BRASS INC
BLDG PROD DEPT
P O BOX 151
ROME NY 13440

-PHONE-
315 3382022

SOLAR MANUFACTURERS
SOLAR SPACE HEATING & COOLING

HENRY JR JOHN P
DIR ENERGY TECH DEPT
STANFORD RESEARCH INSTITUTE
ECONOMICS DIVISION
333 RAVENSWOOD AVE
MENLO PARK CA 94025
-PHONE-
415 3266200 X2358
415 3266200 X2704
SOLAR RESEARCH
PUBLISH ORIGINAL RESEARCHED INFORMATION

HERENDEEN ROBERT A
ILLINOIS, UNIV OF
ADVANCED COMP CENTER
URBANA IL 61801

SOLAR/ENERGY CONSERV INFO DISSEMINATION ORG
HAVE&MAINTAIN ENERGY DATA BANK

HEDGER JACK
PO BOX 357
LAKESIDE CA 92040

SOLAR POTENTIAL MFG,NEW BUSINESS
SOLAR POWERED TRANSPORTATION

HEMPHILL RICHARD A
ENGINEER
LOS ALAMOS SCIENTIFIC LAB
ENGINEERING
ENG-9 MS 636
LOS ALAMOS NM 87544
-PHONE-
505 6676261
505 6721965
SOLAR INTEREST
NEW CONSTRUCTION, MECHANICAL SYSTEMS

HENSLEY JOHN R
2238 MOFFETT DRIVE
FORT COLLINS CO 80521

-PHONE-
303 4931688

SOLAR POTENTIAL MFG,NEW BUSINESS
SOLAR WATER HEATING

HERMAN STEWART W
INFORM
25 BROAD ST
NEW YORK NY 10004

-PHONE-
212 4253550
SOLAR/ENERGY CONSERV INFO DISSEMINATION ORG
PUBLISH ORIGINAL RESEARCHED INFORMATION

HEDGER JOHN H
PRESIDENT
PHYSICAL INDUSTRIES CORP
SOLAR DIVISION
P O BOX 357
LAKESIDE CA 92040
-PHONE-
714 5611266
SOLAR MANUFACTURERS
SOLAR HEAT,COOL,&ELECTRIC POWER GENERATION

HENKE GREGORY P
2190 GARFIELD
EUGENE OR 97405

-PHONE-
503 3448065

SOLAR RESEARCH
WIND CONVERSION

HERBERT * JAMES S
DIR CUST TECH SERV
ASG INDUSTRIES, INC
BOX 929
KINGSPORT TN 37662

-PHONE-
615 2450211 X325

SOLAR COMPONENTS MANUFACTURER
SOLAR SPACE HEATING & COOLING

HERMSDORF RONALD K
KALWALL CORPORATION
KAL-LITE DIVISION
P O BOX 237
MANCHESTER NH 03105
-PHONE-
603 6273861
SOLAR COMPONENTS MANUFACTURER
MEDIUM TEMPERATURE SOLAR COLLECTOR SYSTEMS

INDEX OF INDIVIDUALS

HERONEMUS W E
MASSACHUSETTS, UNIV OF
CIVIL ENGINEERING
AMHERST MA 01002

SOLAR RESEARCH
OCEAN THERMAL GRADIENT SYSTEMS

HERWIG * LLOYD O
DIRECTOR
NATIONAL SCIENCE FOUNDATION
ADVNCED ENGY RES&TEC
WASHINGTON DC 20550
-PHONE-
202 6327230
SOLAR RESEARCH
ALL SOLAR TECHNOLOGIES

HESS * ROBERT
PRESIDENT
ECO-SOLUTIONS,INC.
P O BOX 4117
BOULDER CO 80302
-PHONE-
303 4436798
SOLAR POTENTIAL MFG,NEW BUSINESS
SOLAR SPACE HEATING

HILDEBRANDT ALVIN
PROFESSOR
HOUSTON, UNIV OF
PHYSICS DEPT
HOUSTON TX 77004

-PHONE-
713 7492834
SOLAR RESEARCH
SOLAR ELECTRIC POWER GENERATION

HILL JAMES E
THERMAL ENGR SYS SEC
NATIONAL BUREAU OF STANDARDS
CTR FOR BLDG TECH
BLDG 226 RM B104
WASHINGTON DC 20234
-PHONE-
301 9213503
SOLAR RESEARCH
SOLAR SPACE HEATING & COOLING

HILL SAM
MGR SALES&MARKETING
RUDKIN-WILEY CORPORATION
POLARCH
760 HONEYSPOT
STRATFORD CT 06497
-PHONE-
202 3755966
SOLAR INTEREST
SOLAR SPACE HEATING & COOLING

HIRAYAMA TAKASHI
TOKYO UNIVERSITY
FAC OF ENGINEERING
BUNKYO-KU
TOKYO JAPAN
-PHONE-
03 8122111
SOLAR RESEARCH
SOLAR SPACE HEATING

HIRSCHMANN R JULIO
UNIV TECNICA FED SANTA MARIA
LAB DE ENERGIA SOLAR
CASILLA 110 V
VALPARAISO CHILE
-PHONE-
61268
SOLAR RESEARCH
MEDIUM TEMPERATURE SOLAR COLLECTOR SYSTEMS

HOFFMAN ALLAN R
MASSACHUSETTS, UNIV OF
PHYSICS & ASTRONOMY
AMHERST MA 01002

-PHONE-
413 5450933
SOLAR RESEARCH
SOLAR SPACE HEATING

HOFFMANN ERICH W
HOFLAR INDUSTRIES
5511 128TH ST
SURREY B C CANADA

-PHONE-
604 5962665
SOLAR MANUFACTURERS
SOLAR SPACE HEATING

HOGG * F G
CHIEF
C I R S O
DIV OF MECH ENGR
P O BOX 26
HIGHETT, VICTORIA AUSTRALIA
-PHONE-
03 950333
SOLAR RESEARCH
MEDIUM TEMPERATURE SOLAR COLLECTOR SYSTEMS

HOLLOWAY DENNIS R
ASSISTANT PROFESSOR
MINNESOTA, UNIVERSITY OF
ARCH&LANDSCAPE ARCH
110 ARCHITECTURE BLG
MINNEAPOLIS MN 55455
-PHONE-
612 3732198
SOLAR RESEARCH
SOLAR SPACE HEATING

INDEX OF INDIVIDUALS

HOLT JR WILLIAM C
UNIVERSAL OIL PRODUCTS COMPANY
10 UOP PLAZA
DES PLAINES IL 60016

-PHONE-
312 3913330
SOLAR POTENTIAL MFG,ESTABLISHED BUSINESS
MEDIUM TEMPERATURE SOLAR COLLECTOR SYSTEMS

HOROWITZ HAROLD
PROG MGR HEAT&COOL B
NATIONAL SCIENCE FOUNDATION
ADVNCED ENGY RES&TEC
WASHINGTON DC 20550
-PHONE-
202 6327364
SOLAR RESEARCH
SOLAR SPACE HEATING & COOLING

HOSKINS EDWARD
COGSWELL/HAUSLER ASSOC
P O BOX 2214
CHAPEL HILL NC 27514

-PHONE-
919 9425197
SOLAR/ENERGY CONSERV INFO DISSEMINATION ORG
SOLAR SPACE HEATING & COOLING

HOWE EVERETT D
CALIF, UNIV OF-RICHMOND FLD ST
177 RICHMOND FLD STA
1301 SOUTH 46TH ST
RICHMOND CA 94704

-PHONE-
415 6426352
415 5254990 X237
SOLAR RESEARCH
SOLAR WATER DISTILLATION

HORIGOME TAKASHI
ENERGY DIVISION
MINISTRY OF INT'L TRADE
ENERGY TRNSPT SECT
ELECTROTECHNICAL LAB
TOKYO JAPAN
-PHONE-
042 6612141
SOLAR RESEARCH
SOLAR ELECTRIC POWER GENERATION

HORSFIELD BRIAN
PURDUE UNIV
DEPT OF AG ENGR
W LAFAYETTE IN 47907

-PHONE-
317 7492971
SOLAR RESEARCH
SOLAR AGRICULTURAL DRYING

HOUSE DAVID
EARTHMIND
JOSEL SAUGUS CA 91350

-PHONE-
805 2513053
SOLAR/ENERGY CONSERV INFO DISSEMINATION ORG
WIND CONVERSION

HSU YIH-YUN
MAIL STOP 301-1
NASA LEWIS RESEARCH CTR
PHYSICAL SCIENCE DIV
21000 BROOKPARK ROAD
CLEVELAND OH 44135
-PHONE-
216 4334000 X205

SOLAR RESEARCH
BIO-CONVERSION

HORNING DON O
CALIF, UNIV OF-RICHMOND FLD ST
RICHMOND FIELD STA
1301 SOUTH 46TH ST
RICHMOND CA 94704

-PHONE-
415 2356000 X211
SOLAR RESEARCH
HIGH TEMPERATURE SOLAR COLLECTOR SYSTEMS

HORSTER DR HORST
PHILIPS FORSCHUNGSLABORATORIUM
AACHEN GMBH
D-51 AACHEN
WEISSHAUSSTRASSE WEST GERMANY
-PHONE-
024 1 62071
SOLAR RESEARCH
MEDIUM TEMPERATURE SOLAR COLLECTOR SYSTEMS

HOWARD * DR MICHAEL
ASST PROF
CONNECTICUT, UNIVERSITY OF
CHEMICAL ENGR
STORRS CT 06268
-PHONE-
203 4864020
ENERGY CONSERVATION RESEARCH
RETROFITTING, TECHNOLOGIES

HUBBARD JAMES O
ARCHITECT
NEVADA,STATE OF
HIGHWAY DEPARTMENT
1263 SO STEWART ST
CARSON CITY NV 89701
-PHONE-
702 8827485

SOLAR INTEREST
SOLAR SPACE HEATING

HUGHES THOMAS
BOARD OF COUNTY COMMISSIONERS
BROWARD COUNTY
CO COURTHOUSE RM 270
FT LAUDERDALE FL 33301

SOLAR INTEREST
SOLAR WATER HEATING

HUGHES WILLIAM L
HEAD SCHOOL OF E E
OKLAHOMA STATE UNIVERSITY
ELECTRICAL ENGG
ENGINEERING SOUTH
STILLWATER OK 74074
-PHONE-
405 3726211 X7581
SOLAR RESEARCH
WIND CONVERSION

HUMPHREY CLIFFORD C
ECOLOGY ACTION EDUCATION INST
BOX 3895
MODESTO CA 95352

-PHONE-
209 5291964
SOLAR/ENERGY CONSERV INFO DISSEMINATION ORG
PUBLISH ORIGINAL RESEARCHED INFORMATION

HYSOM ERVIN E
CONSULTING MECH ENG
PROF ENG CONSULTANTS
MECHANICAL
1440 E ENGLISH
WICHITA KS 67211
-PHONE-
316 2622691

SOLAR INTEREST
SOLAR SPACE HEATING & COOLING

IGNAZITO MARTIN D
SANDHU AND ASSOCIATES
309 S RANDOLPH
CHAMPAIGN IL 61820

-PHONE-
217 3522900

SOLAR RESEARCH
SOLAR SPACE HEATING & COOLING

ILSE JOHN
RESEARCH ASSISTANT
MINNESOTA, UNV OF
MECHANICAL ENGR
MINNEAPOLIS MN 55455

-PHONE-
612 3764916
612 4233701
SOLAR RESEARCH
SOLAR SPACE HEATING

INDUSTRIES ASSOC SOLAR ENERGY
SOLAR ENERGY RESEARCH&INFO CTR
1001 CONN AVE NW
WASHINGTON DC 20036

SOLAR/ENERGY CONSERV INFO DISSEMINATION ORG
ALL SOLAR TECHNOLOGIES

INGRAHAM ELIZABETH W
DIRECTOR
WRIGHT-INGRAHAM INSTITUTE
1228 TERRACE ROAD
COLORADO SPRINGS CO 80904

-PHONE-
303 6337011

SOLAR RESEARCH
SOLAR HEAT,COOL,&ELECTRIC POWER GENERATION

INMAN ROBERT E
PLANT BIOLOGIST
STANFORD RESEARCH INSTITUTE
FOOD&PLANT SCIENCES
333 RAVENSWOOD AVE
MENLO PARK CA 94025
-PHONE-
415 3266200 X2155
408 2558853
SOLAR RESEARCH
BIO-CONVERSION

IONNIDES MARIUS
PROD MGR
METALCO
NICOSIA CYPRUS GREECE

SOLAR MANUFACTURERS
SOLAR WATER HEATING

IRVINE P E R GERALD
1 TURNER LANE
LOUDONVILLE NY 12211

-PHONE-
518 4631334
SOLAR RESEARCH
SOLAR SPACE HEATING & COOLING

IVOJCIECHOIVSKI PAUL
ROCHESTER INST OF TECH
50 WEST MAIN
ROCHESTER NY 14614

-PHONE-
716 4642411
SOLAR RESEARCH
SOLAR SPACE HEATING

JAMES GEORGE
INFORMATION PROG MGR
NATIONAL SCIENCE FOUNDATION
ADVNCED ENGY RESETEC
WASHINGTON DC 20550
-PHONE-
202 6327398

SOLAR RESEARCH
ALL SOLAR TECHNOLOGIES

JENKINS ALFRED
PRESIDENT
ENERGEX CORP
418 TROPICANA
LAS VEGAS NV 89109
-PHONE-
702 7362994
SOLAR MANUFACTURERS
ALL SOLAR TECHNOLOGIES

JEWETT DAVID N
MANAGER
TYCO LABORATORIES INC
SILICON DEPARTMENT
16 HICKORY DRIVE
WALTHAM MA 02154
-PHONE-
617 8902400

SOLAR RESEARCH
INORGANIC SEMI-CONDUCTORS

JOHNSON ROBERT A
RESEARCH ASSOCIATE
MASSACHUSETTS, UNIVERSITY OF
PHYSICS DEPT
HASBROUCK LABORATORY
AMHERST MA 01002
-PHONE-
413 5450346

SOLAR INTEREST
NEW CONSTRUCTION, RESEARCHING

JANNEY * RAYMOND
ASSIST TO CHAIRMAN
AVCO EVERETT RESEARCH LAB INC
2385 REVERE BEACH PK
EVERETT MA 02149
-PHONE-
617 3893000

SOLAR RESEARCH
WIND CONVERSION

JENNINGS DANA C
RT 3 BOX 177
MADISON SD 57042

-PHONE-
605 2563377
SOLAR/ENERGY CONSERV INFO DISSEMINATION ORG
PROVIDE INFORMATION SEARCHES TO CLIENTS

JOHNSON DENNIS O
BIG OUTDOORS PEOPLE
P O BOX 936
MINNEAPOLIS MN 55440

-PHONE-
612 3315344
612 3315430
SOLAR POTENTIAL MFG,ESTABLISHED BUSINESS
MEDIUM TEMPERATURE SOLAR COLLECTOR SYSTEMS

JOHNSON STAN
JOHNSON DIVERSIFIED INC
2340 QUEEN ANN ST
MERRITT ISLAND FL 32952

-PHONE-
305 4525545

SOLAR MANUFACTURERS
SOLAR WATER HEATING

JEFFORDS EDD
OZARK ACCESS CENTER INC
55 SPRING ST BOX 506
EUREKA SPRINGS AR 72632

-PHONE-
501 2539616
501 2539601
SOLAR/ENERGY CONSERV INFO DISSEMINATION ORG
PUBLISH LAY PERIODICALS

JEWELL WILLIAM H
R 4
WEST BRATTLEBORO VT 05301

-PHONE-
802 2542403
SOLAR RESEARCH
PROVIDE INFORMATION SEARCHES TO CLIENTS

JOHNSON RALPH J
STAFF VICE PRESIDENT
NAHB RESEARCH FOUNDATION INC
PO BOX 1627
ROCKVILLE MD 20850

-PHONE-
301 7624200

ENERGY CONSERVATION RESEARCH
PUBLISH TECHNICAL BOOKS

JOHNSON VICTOR J
1380 55TH STREET
BOULDER CO 80303

-PHONE-
303 4421581
303 4991000 X3523
SOLAR RESEARCH
SOLAR REFRIGERATION

INDEX OF INDIVIDUALS

JONAS ZWEIG
ASSOCIATE PROFESSOR
WM PATTERSON COLLEGE
SECONDARY EDUCATION
300 POMPTON RD
WAYNE NJ 07470
-PHONE-
201 8812119
ENERGY CONSERVATION RESEARCH
SOLAR HEAT,COOL,&ELECTRIC POWER GENERATION

JONES JOSEPH M
DIRECTOR
NASA G C MARSHALL SPACE FLT CT
PUBLIC AFFAIRS
MARSHAL SPACE F C AL 35812

SOLAR RESEARCH
MEDIUM TEMPERATURE SOLAR COLLECTOR SYSTEMS

JONES WILLIAM E
CAMPUS ARCHITECT
MARYLAND, UNIV OF
CAMPUS PLANNING DEPT
SERVICE PLDG 2ND FL
COLLEGE PARK MD 20740
-PHONE-
301 4545770
SOLAR RESEARCH
SOLAR SPACE HEATING

JORDAN * R C
MINN, UNIV OF
DEPT OF MECH ENGR
MINNEAPOLIS MN 55455

-PHONE-
612 3733302
SOLAR RESEARCH
SOLAR ELECTRIC POWER GENERATION

JORDI EDOUARD
ESB INCORPORATED
19 WEST COLLEGE AVE
PA 19067

-PHONE-
215 4933659
SOLAR RESEARCH
SOLAR WATER DISTILLATION

JOSLYN DANFORTH
OPTICAL COATING LABORATORY INC
2789 GIFFEN AVENUE
SANTA ROSA CA 95403

-PHONE-
707 5456440
SOLAR COMPONENTS MANUFACTURER
MEDIUM TEMPERATURE SOLAR COLLECTOR SYSTEMS

KANTROWITZ ARTHUR
CHAIRMAN
AVCO EVERETT RESEARCH LAB INC
2385 REVERE.BEACH PK
EVERETT MA 02149
-PHONE-
617 3893000
SOLAR RESEARCH
WIND CONVERSION

KAPLAN GEORGE M
PROG MGR SOLAR THERM
NATIONAL SCIENCE FOUNDATION
ADVNCED ENGY RES&TEC
WASHINGTON DC 20550
-PHONE-
202 6327364
SOLAR RESEARCH
MEDIUM TEMPERATURE SOLAR COLLECTOR SYSTEMS

KAPLAN GEORGE M
NASA LEWIS RESEARCH CTR
POWER SYSTEMS DIV
21000 BROOKPARK ROAD
CLEVELAND OH 44135
-PHONE-
216 4334000 X6845
SOLAR RESEARCH
SOLAR ELECTRIC POWER GENERATION

KAPUR J C
KAPUR SOLAR FARMS
BIJWASAN-NAJAFGARH R
P O KAPAS HERA
NEW DELHI 110037 INDIA

-PHONE-
391747
SOLAR RESEARCH
MEDIUM TEMPERATURE SOLAR COLLECTOR SYSTEMS

KARAKI SUSUMU
ASSOC DIRECTOR
COLORADO STATE UNIVERSITY
SOLAR ENGY APPL LAB
ENGR RESEARCH CENTER
FT COLLINS CO 80521
-PHONE-
303 4911101
SOLAR RESEARCH
SOLAR SPACE HEATING & COOLING

KASABOV GEORGE
23 STRATFORD VILLAS
LONDON N W 1 ENGLAND

SOLAR RESEARCH
MEDIUM TEMPERATURE SOLAR COLLECTOR SYSTEMS

k

INDEX OF INDIVIDUALS

KAUZLARICH DR JAMES J
VIRGINIA, UNIV OF
DEPT OF MECH ENGR
CHARLOTTESVILLE VA 22901

-PHONE-
804 9243661

SOLAR RESEARCH
SOLAR SPACE HEATING & COOLING

KEGEL PE ROBERT A
ILLINOIS, STATE OF
CAPITAL DEVELOP BORD
1030 S LAGRANGE RD
LAGRANGE IL 60525

-PHONE-
312 3528134
SOLAR/ENERGY CONSERV INFO DISSEMINATION ORG
NEW CONSTRUCTION, TECHNOLOGIES

KENDRICK J D
SR LECTURER BLDG SCI
ADELAIDE, UNIV OF
ARCH & TOWN PLANNING
ADELAIDE, S A AUSTRALIA
-PHONE-
223 4333 X2444
SOLAR RESEARCH
ALL SOLAR TECHNOLOGIES

KERNOZEK FRANK
NUCLEAR TECHNOLOGY
P O BOX 1
AMSTON CN 06231

-PHONE-
203 5372387
SOLAR RESEARCH
MEDIUM TEMPERATURE SOLAR COLLECTOR SYSTEMS

KAY THOMAS P
PRESIDENT
T-K SOLAR DISTRIBUTORS
5 TH FLOOR
9 EAST 16TH STREET
NEW YORK NY 10003
-PHONE-
212 9894422

SOLAR PRODUCTS DISTRIBUTORS
SOLAR WATER HEATING

KEISTER DR DONALD L
CHARLES F KETTERING RESRCH LAB
150 E SO COLLEGE ST
YELLOW SPRINGS OH 45387

-PHONE-
513 7677271
SOLAR RESEARCH
BIO-CONVERSION

KEOUGH MARTIN E
MAIL CODE EZ
NASA JOHNSON SPACE CRAFT CTR
URBAN SYS PROJ OFF
HOUSTON TX 77053
-PHONE-
713 4835557
SOLAR RESEARCH
SOLAR SPACE HEATING & COOLING

KEYES ROBERT W
I B M
T WATSON RESERCH CTR
PO BOX 218
YORKTOWN HEIGHTS NY 10598
-PHONE-
914 9453000
SOLAR RESEARCH
INORGANIC SEMI-CONDUCTORS

KAYE DOUGLAS R
PRODUCER
GLEN-KAYE FILMS
100 EAST 21ST STREET
BROOKLYN NY 11226

-PHONE-
212 2872929
212 6939618
SOLAR/ENERGY CONSERV INFO DISSEMINATION ORG
PUBLISH AND/OR DISTRIBUTE FILMS

KELLY JR EDWARD J
INSTRUCTOR ENV TECH
MICHIGAN, UNIV OF
DEPT OF ARCH&DESIGN
TAPPAN & MONROE STS
ANN ARBOR MI 48104
-PHONE-
313 7641340
SOLAR RESEARCH
SOLAR SPACE COOLING

KERN DAVID F
STUDENT
COLORADO COLLEGE
PHYSICS
COLORADO SPRINGS CO 80903
-PHONE-
303 6328168
SOLAR RESEARCH
SOLAR SPACE HEATING

KIDDER ERNEST H
MICHIGAN STATE UNIV
211 AG ENGINEERING
EAST LANSING MI 48824

-PHONE-
517 3554720
SOLAR RESEARCH
SOLAR RADIATION MEASUREMENTS

INDEX OF INDIVIDUALS

KIMBALL CHRISTOPHER
ENVIRONMENTAL INFORMATION CTR
124 EAST 39TH STREET
NEW YORK NY 10016

-PHONE-
212 6790810

SOLAR/ENERGY CONSERV INFO DISSEMINATION ORG
PUBLISH LAY + TECHNICAL BOOKS

KIMURA KEN-ICHI
WASEDA UNIVERSITY
DEPT OF ARCHITECTURE
4-170 NISHIOKUBO
SHINJUKU-KU TOKYO JAPAN

-PHONE-
03 2093211 X420

SOLAR RESEARCH
SOLAR SPACE HEATING & COOLING

KINGMAN ELIZABETH A
SOLAR EXHIBIT PROG
COLORADO, UNIV OF-DENVER

1100 14TH STREET
DENVER CO 80202
-PHONE-
SEE PAGE 4.009 + 4.022
303 8921117 X471
SOLAR/ENERGY CONSERV INFO DISSEMINATION ORG
PROVIDE AUDIO & VISUAL SERVICES

KINNEY J F
GEORGIA INST OF TECHNOLOGY
ENGINEERING EXP STAT
ATLANTA GA 30332

-PHONE-
404 8943661
SOLAR RESEARCH
MEDIUM TEMPERATURE SOLAR COLLECTOR SYSTEMS

KINNEY WILL C
PORTABLE ALUMINUM IRRIG CO INC
PO BOX 878
VISTA CA 92083

-PHONE-
714 7242163
SOLAR POTENTIAL MFG,ESTABLISHED BUSINESS
SOLAR WATER DISTILLATION

KIRK GEORGE L
PRESIDENT
EPPLEY LABORATORY INC
12 SHEFFIELD AVE
NEWPORT RI 02840
-PHONE-
4 0184710 X20
SOLAR MANUFACTURERS
SOLAR RADIATION MEASUREMENTS

KIRKER N
TECH OPERATIONS DIR
UK ATOMIC ENERGY AUTHORITY
REACTOR GROUP
RISLEY
WARRINGTON LANCASHIR ENGLAND
-PHONE-
WAR 31244 X2290
SOLAR RESEARCH
INORGANIC SEMI-CONDUCTORS

KIRSCHKE CAPT ERNERST J
U S NAVAL AMMUNITION DEPOT
HAWTHORNE NV 89415

-PHONE-
702 9452451 X590
SOLAR RESEARCH
SOLAR SPACE HEATING

KITHIL RICHARD
VEXOCAVE COMPANY
STAR ROUTE BOX 400
N TURKEY CREEK
MORRISON CO 80465

-PHONE-
303 6978246
SOLAR RESEARCH
SOLAR SPACE HEATING

KLAUSNER MADELEINE S
DIR OF ADMIN
CTR FOR RESEARCH-ACTS OF MAN
4025 CHESTNUT STREET
PHILADELPHIA PA 19104

-PHONE-
215 5946241
SOLAR SOCIAL CONSIDERATIONS
ALL SOLAR TECHNOLOGIES

KLEIN DR WILLIAM H
SECRETARY/TREASURER
INTL SOLAR ENGY SOC/US SECT
C/O SMITHSONIAN RBL
12441 PARKLAWN DR
ROCKVILLE MD 20852

SOLAR/ENERGY CONSERV INFO DISSEMINATION ORG
ALL SOLAR TECHNOLOGIES

KLOPFENSTEIN KEITH
SYSTEMS RESEARCH LABS INC
2800 INDIAN RIPPLE
DAYTON OH 45440

-PHONE-
513 2528758
SOLAR RESEARCH
SOLAR LASERS

k

INDEX OF INDIVIDUALS

KOHIN ROGER P
CLARK UNIVERSITY
PHYSICS DEPT
WORCESTER MA 01610

-PHONE-
617 7937711
SOLAR RESEARCH
SOLAR SPACE HEATING

KRAUS MARJORIE P
DELAWARE, UNIV OF
CHEMISTRY DEPT
NEWARK DE 19711

-PHONE-
302 7382458
215 2748114
SOLAR RESEARCH
BIO-CONVERSION

KRONER WALTER
RENSSELAER POLYTECHNIC INST
ARCHITECT RESERCH CT
TROY NY 12181

-PHONE-
518 2706461

SOLAR RESEARCH
ALL SOLAR TECHNOLOGIES

KUO DR SIMION C
PRINCIPAL SYS ENGR
UNITED AIRCRAFT RESEARCH LABS
ENERGY RESEARCH DEPT
SILVER LANE
EAST HARTFORD CT 06108
-PHONE-
203 5658758

SOLAR RESEARCH
SOLAR SPACE HEATING & COOLING

KONHEIM CAROLYN S
EXECUTIVE DIR
N Y SCI COMM FOR PUBLIC INFO
30 EAST 68TH ST
NEW YORK NY 10021
-PHONE-
212 7377302
SOLAR/ENERGY CONSERV INFO DISSEMINATION ORG
ALL SOLAR TECHNOLOGIES

KREIDER DR JAN F
CHIEF ENGR
ENVIRONMENTAL CONSULT SERV INC
1485 SIERRA DRIVE
BOULDER CO 80302
-PHONE-
303 4472218
303 4431235
SOLAR RESEARCH
SOLAR HEAT,COOL,&ELECTRIC POWER GENERATION

KUEHNER RICHARD L
STAFF SCIENTIST
BORG WARNER CORP.
RESEARCH CENTER
ALGONQUIN&WOLF RDS
DES PLAINES IL 60018
-PHONE-
312 8273131

SOLAR RESEARCH
SOLAR SPACE HEATING & COOLING

KUSUDA T
NATIONAL BUREAU OF STNDRDS
CTR FOR BLDG TECH
ROOM 1008 BLDG 226
WASHINGTON DC 20234

-PHONE-
301 9213522
301 6566556
SOLAR RESEARCH
SOLAR SPACE HEATING & COOLING

KONOPKA EDWARD J
PRESIDENT
FUN & FROLIC INC
P O BOX 277
MADISON HEIGHTS MI 48071

SOLAR MANUFACTURERS
SOLAR WATER HEATING

KREMSER DR THURMAN R
ALBRIGHT COLLEGE
DEPT OF PHYSICS
READING PA 19604

-PHONE-
215 9212381 X266

SOLAR INTEREST
SOLAR SPACE HEATING

KUHARICH RODNEY F
327 E. PLATTE
COLORADO SPRINGS CO 80902

-PHONE-
303 4711111
303 6337011
SOLAR RESEARCH
TOTAL SYSTEMS INTEGRATION

KYRYLUK WILLIAM F
1578 WEST 71ST AVE
VANCOUVER 14 B C CANADA

-PHONE-
2638315

SOLAR RESEARCH
MEDIUM TEMPERATURE SOLAR COLLECTOR SYSTEMS

INDEX OF INDIVIDUALS

LA ROSA RICHARD
317 OAK STREET
SOUTH HEMPSTEAD NY 11550

-PHONE-
516 2617000 X321

SOLAR RESEARCH
SOLAR SPACE HEATING & COOLING

LANDERS TED
FARMER
NEW LIFE FARM
DRURY MO 65638

-PHONE-
417 2612553
SOLAR RESEARCH
TOTAL SYSTEMS INTEGRATION

LANSFORD * HENRY H
INFORMATION OFFICER
NATL CTR FOR ATMOS RESEARCH
BUDGET PLAN & INFO
BOX 3000
BOULDER CO 80303
-PHONE-
303 4945151
SOLAR RESEARCH
SOLAR SPACE HEATING & COOLING

LAWRENCE ROBERT S
PHYSICIST
SALINA STAR ROUTE
BOULDER CO 80302

-PHONE-
303 4991000 X6353
303 4427534
SOLAR RESEARCH
SOLAR WATER HEATING

LAIRD CAMERON B
R R 3
RANSSELAER IN 47906

-PHONE-
317 4932456
219 8663628
SOLAR RESEARCH
TOTAL SYSTEMS INTEGRATION

LANDIS * LARRY
PERSONNEL&PUBLIC REL
BABCOCK & WILCOX CO
RESEARCH CENTER
1562 BEESON
ALLIANCE OH 44601
-PHONE-
216 8219110 X387
SOLAR RESEARCH
SOLAR ELECTRIC POWER GENERATION

LARSON RONAL W
OFFICE OF TECH ASSESSMENT
119 'D' STREET NE
WASHINGTON DC 20510

-PHONE-
202 2258996
SOLAR LEGAL CONSIDERATIONS
LEGISLATIVE, FEDERAL

LEE BRYAN
840 37TH
BOULDER CO 80303

-PHONE-
303 4496278

SOLAR RESEARCH
SOLAR SPACE HEATING

LAMBERT LLOYD M
VERMONT, UNIV OF
ELECT ENGR DEPT
BURLINGTON VT 05401

-PHONE-
802 6563332

SOLAR RESEARCH
INORGANIC SEMI-CONDUCTORS

LANGDON CARL M
EDITOR-PUBLISHER
ADV SOLAR ENGY TECH NEWSLETTER
1609 WEST WINDROSE
PHOENIX AZ 85029

-PHONE-
602 9423796
SOLAR/ENERGY CONSERV INFO DISSEMINATION ORG
PUBLISH LAY + TECHNICAL PERIODICALS

LAWAND * T A
DIR FIELD OPERATIONS
MCGILL UNIV MACDONALD COLLEGE
BRACE RESEARCH INST
ST ANN DE BELEVUE800
QUEBEC H9X 3MI CANADA
-PHONE-
514 4576580 X341
SOLAR RESEARCH
ALL SOLAR TECHNOLOGIES

LEE ORVILLE
SOLAR ENGY PROG MGR
DEPT OF HOUSING & URBAN DEVEL
BUILDING TECH DIV
451 7TH STREET S W
WASHINGTON DC 20410
-PHONE-
202 7575356

SOLAR RESEARCH
SOLAR SPACE HEATING & COOLING

INDEX OF INDIVIDUALS

LEE ROBERT A
STEAM GEN TECH SECT
BABCOCK & WILCOX CO
RESEARCH CENTER
1562 BEESON
ALLIANCE OH 44601
-PHONE-
2.5 8219110 X376
SOLAR RESEARCH
SOLAR ELECTRIC POWER GENERATION

LEFEVER H R
SONNEWALD SERVICE
RD 1 BOX 457
SPRING GROVE PA 17362

-PHONE-
717 2253456
SOLAR RESEARCH
SOLAR SPACE HEATING

LEHR VALENTINE A
LEHR ASSOC,CONSULTING ENGRS
ONE PENN PLAZA
NEW YORK NY 10001

-PHONE-
212 9478050
SOLAR/ENERGY CONSERV INFO DISSEMINATION ORG
SOLAR WATER HEATING

LENTZ DANIEL M
AMERICAN INST AERONAUT&ASTRONT
MILWAUKEE STUDENT BR
3228 N 26TH STREET
MILWAUKEE WI 53206

-PHONE-
414 8730133
414 9635168
SOLAR/ENERGY CONSERV INFO DISSEMINATION ORG
INORGANIC SEMI-CONDUCTORS

LESK DR ARNOLD
HEAD
MOTOROLA INC
SOLAR ENERGY GROUP
8201 E MCDOWELL ROAD
SCOTSDALE AZ 85252
-PHONE-
602 9493355
SOLAR RESEARCH
SOLAR ELECTRIC POWER GENERATION

LESSMANN DR RICHARD C
RHODE ISLAND, UNIV OF
MECHANICAL ENGR DEPT
KINGSTON RI 02881

SOLAR RESEARCH
TOTAL SYSTEMS INTEGRATION

LESTER NANCY C
GENERAL RESEARCH CORPORATION
SYSTEMS RESEARCH DIV
1501 WILSON BLVD
ARLINGTON VA 22209

-PHONE-
703 5347206
SOLAR/ENERGY CONSERV INFO DISSEMINATION ORG
PUBLISH ORIGINAL RESEARCHED INFORMATION

LEVY STEVE
MANAGER
MOTOROLA INC
NEW VENTURES LABS
8201 E MCDOWELL ROAD
SCOTSDALE AZ 85252
-PHONE-
602 9493355
SOLAR RESEARCH
SOLAR ELECTRIC POWER GENERATION

LEWCHUK R R
PROJECT MGR
PPG INDUSTRIES INC
NEW PROD DEVELOPMENT
ONE GATEWAY CENTER
PITTSBURGH PA 15222
-PHONE-
412 4342645
SOLAR MANUFACTURERS
MEDIUM TEMPERATURE SOLAR COLLECTOR SYSTEMS

LEWIS WILLIAM S
MARKET DEV MANAGER
ALCOA ALUMINUM
1501 ALCOA BLDG
PITTSBURGH PA 15219

-PHONE-
412 5532748
SOLAR COMPONENTS MANUFACTURER
MEDIUM TEMPERATURE SOLAR COLLECTOR SYSTEMS

LICHTIN N N
CHAIRMAN
BOSTON UNIVERSITY
CHEMISTRY DEPT
685 COMMONWEALTH AVE
BOSTON MA 02215
-PHONE-
617 3532493
SOLAR RESEARCH
INORGANIC SEMI-CONDUCTORS

LICHTY LYALL J
TUBAC AZ 85640

-PHONE-
602 3982439
ENERGY CONSERVATION RESEARCH
ALL SOLAR TECHNOLOGIES

INDEX OF INDIVIDUALS

LIECHTY MORRIS
R R 1
GRABILL IN 46741

-PHONE-
219 6272780
219 4858095
SOLAR RESEARCH
SOLAR HEAT,COOL,&ELECTRIC POWER GENERATION

LINDSAY HARRY
ZURN INDUSTRIES INC
5533 PERRY HWY
ERIE PA 16509

-PHONE-
814 8644041
SOLAR PRODUCTS DISTRIBUTORS
INORGANIC SEMI-CONDUCTORS

LISTON THOMAS L
LISTON & ASSOC,MECH ENGRS
890 SARATOGA AVE
SAN JOSE CA 95129

-PHONE-
408 2466281
SOLAR RESEARCH
SOLAR SPACE HEATING

LOF DR GEORGE
DIRECTOR
COLORADO STATE UNIVERSITY
SOLAR ENGY APPL LAB
ENGR RESEARCH CENTER
FORT COLLINS CO 80521
-PHONE-
303 4918632
303 4918325
SOLAR RESEARCH
SOLAR SPACE HEATING & COOLING

LILLYWHITE MALCOLM A
COORDINATOR
MTN AREA DOMESTIC SOLAR ENGY
TECH CONSORTIUM
RR 2 BOX 274
EVERGREEN CO 80439
-PHONE-
303 6746633

SOLAR/ENERGY CONSERV INFO DISSEMINATION ORG
ALL SOLAR TECHNOLOGIES

LIOR NOAM
ASSISTANT PROFESSOR
PENN, UNIV OF
N CTR ENGY MGT&POWER
260 TOWNE BLDG D3
PHILADELPHIA PA 19174
-PHONE-
215 5944803
SOLAR RESEARCH
SOLAR SPACE HEATING & COOLING

LODWIG ERWIN
NIAGARA BLOWER CO
405 LEXINGTON AVE
NEW YORK NY 10017

-PHONE-
212 6976151
SOLAR COMPONENTS MANUFACTURER
SOLAR SPACE COOLING

LOF GEORGE O
CONSULTING ENGINEER
GEORGE O G LOF CONS ENGR
158 FILLMORE ST R204
DENVER CO 80206

-PHONE-
303 3220446
303 4918632
SOLAR RESEARCH
SOLAR SPACE HEATING & COOLING

LILLYWHITE MALCOM
V P RESEARCH&DEVELOP
SOLAR POWER SUPPLY
RT 3 BOX A10
EVERGREEN CO 80439

-PHONE-
303 6744734

SOLAR POTENTIAL MFG,NEW BUSINESS
SOLAR HEAT,COOL,&ELECTRIC POWER GENERATION

LISHNESS ALLEN
ZEPHER WIND DYNAMO
P O BOX 241
BRUNSWICK MA 04011

-PHONE-
207 7256534
SOLAR POTENTIAL MFG,NEW BUSINESS
WIND CONVERSION

LOEB MILTON
COLORADO, UNIV OF-BOULDER
ATIP
RM 331A UMC
BOULDER CO 80302
-PHONE-
303 4926659
SOLAR/ENERGY CONSERV INFO DISSEMINATION ORG
HAVE&MAINTAIN ENERGY DATA BANK

LOF L
LOF BROS SOLAR APPLIANCES
P O BOX 10594
DENVER CO 80210

-PHONE-
303 7446852
303 7582585
SOLAR MANUFACTURERS
SOLAR WATER HEATING

LOFERSKI J J
BROWN UNIVERSITY
ENGINEERING DEP
UNIV HALL BOX 1867
PROVIDENCE RI 02912

SOLAR RESEARCH
INORGANIC SEMI-CONDUCTORS

LOGAN J C
GRANT RD
NEWMARKET NH 03857

-PHONE-
603 6595774
SOLAR POTENTIAL MFG,NEW BUSINESS
SOLAR WATER HEATING

LOKMANHEKIM DR METIN
9190 RED BRANCH RD
COLUMBIA MD 21045

-PHONE-
301 7307800
ENERGY CONSERVATION RESEARCH
RETROFITTING, MECHANICAL SYSTEMS

LOOMIS MILDRED J
SCHOOL OF LIVING
RT 1 BOX 129
FREELAND MD 21053

-PHONE-
301 3575619
301 3296549
SOLAR/ENERGY CONSERV INFO DISSEMINATION ORG
PUBLISH LAY PERIODICALS

LOONEY TEMPLE
MCFALL AND KONKEL CONSULT ENGR
2160 S CLERMONT ST
DENVER CO 80222

-PHONE-
303 7574976
SOLAR RESEARCH
SOLAR SPACE HEATING & COOLING

LORD DAVID
ASST PROF
HARVARD UNIV
GRADUATE SCH DESIGN
DEPT OF ARCHITECTURE
CAMBRIDGE MA 02138
-PHONE-
617 4952570

SOLAR RESEARCH
ALL SOLAR TECHNOLOGIES

LORSCH HAROLD G
PENN, UNIV OF
N CTR ENGY MGT&POWER
260 TOWNE BLDG D3
PHILADELPHIA PA 19174
-PHONE-
215 5947181
SOLAR RESEARCH
SOLAR SPACE HEATING & COOLING

LUMSDAINE DR EDWARD
ASSOCIATE PROFESSOR
TENNESSEE, UNIV OF
MECH&AERO ENGR DEPT
KNOXVILLE TN 37916
-PHONE-
615 9743035
SOLAR RESEARCH
SOLAR SPACE HEATING & COOLING

MAC DOUGALL R TODD
ENGINEER
STONEWALLEN ENGINEERING
RT 2 BOX 381A
DELTA CO 81416

SOLAR RESEARCH
SOLAR ENERGY STORAGE

MACDONALD BRUCE W
SYSTEM PLANNING CORP
1500 WILSON BLVD
ARLINGTON VA 22209

- NE-
7 5254904
SL AR RESEARCH
PROVIDE ECONOMIC ANALYSIS OF SOLAR TECH

MACKILLOP ANDREW
DIRECTOR
LOW IMPACT TECHNOLOGY LTD
73 MOLESWORTH ST
CORNWALL ENGLAND

SOLAR COMPONENTS MANUFACTURER
SOLAR SPACE HEATING & COOLING

MACRON NICHOLAS S
SOLAR ENERGY DEVELOPMENT INC
1457 ALAMEDA AVE
LAKEWOOD OH 44107

-PHONE-
216 2213500
SOLAR RESEARCH
MEDIUM TEMPERATURE SOLAR COLLECTOR SYSTEMS

MADLEY DR ROBERT
MGR RESEARCH&DEVEL
GRUMMAN AEROSPACE CORPORATION
ENERGY SYSTEMS DEPT
PLANT 30
BETHPAGE NY 11714
-PHONE-
516 5751573
SOLAR POTENTIAL MFG,ESTABLISHED BUSINESS
MEDIUM TEMPERATURE SOLAR COLLECTOR SYSTEMS

MAHONE * DOUGLAS E
TOTAL ENVIRONMENTAL ACTION
DESIGN
BOX 47
HARRISVILLE NH 03450
-PHONE-
603 8273087
SOLAR/ENERGY CONSERV INFO DISSEMINATION ORG
ALL SOLAR TECHNOLOGIES

MANDT LOLLAR
IBM
ELECTRONICS SYSTEMS
HUNTSVILLE AL 35804
SOLAR MANUFACTURERS
SOLAR RADIATION MEASUREMENTS

MANFREDI FRED
PRODUCER
KNME CHANNEL 5
1130 UNIV BLVD NE
ALBUQUERQUE NM
-PHONE-
505 2772121
SOLAR/ENERGY CONSERV INFO DISSEMINATION ORG
PUBLISH AND/OR DISTRIBUTE FILMS

MAPHAM NEVILLE
GENERAL MANAGER
ENERGY DEVELOPMENT ASSOC
1100 W WHITCOMB AVE
MADISON HTS MI 48071
-PHONE-
313 5664000
313 5839434
SOLAR COMPONENTS MANUFACTURER
SOLAR ENERGY STORAGE

MARINER DONALD
ALTERNATIVE SOURCES OF ENERGY
RT 2, BOX 90A
MILACA MN 56353
-PHONE-
612 9836892
SOLAR/ENERGY CONSERV INFO DISSEMINATION ORG
PUBLISH LAY PERIODICALS

MARMOR ALFRED C
DIRECTOR
U S PATENT OFFICE
TECH ASSESMNT&FORCST
WASHINGTON DC 20231
-PHONE-
703 5573051
SOLAR/ENERGY CONSERV INFO DISSEMINATION ORG
PROVIDE INFORMATION SEARCHES TO CLIENTS

MAROVSH DAN
CONTOUR CO INC
9300 WHITMORE ST
EL MONTE CA 91731
-PHONE-
212 2834101
SOLAR POTENTIAL MFG,ESTABLISHED BUSINESS
SOLAR SPACE HEATING & COOLING

MARTIN ERNESTO Q
DIRECTOR
PUERTO RICO,UNIV OF
CENTRO DE INFO TECNI
RECINTO UNIV MAYAGUZ
MAYAGUEZ PUERTO RICO
SOLAR/ENERGY CONSERV INFO DISSEMINATION ORG
PROVIDE INFORMATION SEARCHES TO CLIENTS

MARTIN J H
TEXAS A&M UNIV
CHEMICAL ENGRG DEPT
COLLEGE STATION TX 77843
-PHONE-
713 8453361
SOLAR ECONOMIC CONSIDERATIONS
SOLAR SPACE HEATING & COOLING

MARTIN JOHN H
SENIOR PHYSICIST
ARGONNE NATIONAL LABORATORY
ACCELERATOR FACILITY
9700 SOUTH CASS AVE
ARGONNE IL 60439
-PHONE-
312 7397711 X4881
312 8529714
SOLAR/ENERGY CONSERV INFO DISSEMINATION ORG
SOLAR HEAT,COOL,&ELECTRIC POWER GENERATION

MASSICOT DR PAUL
MARYLAND DEPT OF NAT RESOURCES
PWR PLT SITING PROG
TAWES ST OFFICE BLDG
ANNAPOLIS MD 21401
-PHONE-
301 2675384
SOLAR RESEARCH
SOLAR ELECTRIC POWER GENERATION

MATHER DR KEITH
DIRECTOR
ALASKA,UNIV OF
GEOPHYSICAL INSTITUT
FAIRBANKS AK 99701
-PHONE-
907 4797607
907 4797558
SOLAR RESEARCH
WIND CONVERSION

MAUNG U MAUNG
CENTRAL RESEARCH ORGANIZATION
PHYSICS & ENGR RESCH
KANBE
YANKIN P O RANGOON BURMA
-PHONE-
50544

SOLAR RESEARCH
ALL SOLAR TECHNOLOGIES

MCAULIFFE D P
PHYSICS & ENGR LAB D S I R
PRIVATE BAG
LOWER HUTT NEW ZEALAND

-PHONE-
WEL 699199 X818

SOLAR RESEARCH
SOLAR WATER HEATING

MCCAULEY * ELLEN
ARTHUR D LITTLE,INC
PUBLIC REALTIONS
20 ACORN PARK
CAMBRIDGE MA 01240
-PHONE-
617 8645770
SOLAR RESEARCH
SOLAR SPACE HEATING & COOLING

MCCRACKEN HORACE
SUNWATER CO
1112 PIONEER WAY
EL CAJON CA 92020

-PHONE-
714 2830454
SOLAR MANUFACTURERS
SOLAR WATER HEATING

MCMAHON THOMAS J
EDUCATIONAL MATERIALS&EQUIP CO
BOX 17
PELHAM NY 10803

-PHONE-
914 5761121
SOLAR MANUF, EDUCATIONAL EQUIPMENT
ALL SOLAR TECHNOLOGIES

MCQUILKIN GEORGE H
PRODUCER-DIRECTOR
CHURCHILL FILMS
662 NO ROBERTSON
LOS ANGELES CA 90069

-PHONE-
213 6575110

SOLAR/ENERGY CONSERV INFO DISSEMINATION ORG
PUBLISH AND/OR DISTRIBUTE FILMS

MEADOWS BETTY JANE
ENV EDUC SPECIALIST
JEFFERSON CO PUBLIC SCHOOLS
ENVIRONMENTAL EDUC
2192 COORS DRIVE
GOLDEN CO 80401
-PHONE-
303 2793346
303 2793347
SOLAR ENVIRONMENTAL CONSIDERATIONS
INCENTIVES FOR ENERGY CONSERVATION

MEAGHER ANTHONY
SUNDU COMPANY
3319 KEYES LANE
ANAHEIM CA 92804

-PHONE-
714 8282873

SOLAR MANUFACTURERS
SOLAR WATER HEATING

MEEK IVAN R
PRES
MEEK'S
6520 E 25 PLACE
TULSA OK 74129
-PHONE-
918 8356081
918 8356638
SOLAR POTENTIAL MFG,NEW BUSINESS
SOLAR SPACE HEATING & COOLING

MEHROTRA P K
ROORKEE, UNIV OF
DEPT OF MECH ENGR
ROORKEE 247667 INDIA

SOLAR RESEARCH
SOLAR SPACE HEATING & COOLING

MEINEL ADEN B
ARIZONA, UNIV OF
OPTICAL SCIENCES CTR
TUCSON AZ 85721

-PHONE-
602 7493322
602 8843138
SOLAR RESEARCH
HIGH TEMPERATURE SOLAR COLLECTOR SYSTEMS

MEINEL MARJORIE P
ARIZONA, UNIV OF
OPTICAL SCIENCES CTR
TUCSON AZ 85721

-PHONE-
602 7493322
602 8843138
SOLAR RESEARCH
HIGH TEMPERATURE SOLAR COLLECTOR SYSTEMS

MERRIAM MARSHAL F
CALIF,UNIV OF--BERKELEY
MATERIALS SCI DEPT
HEARST MINING BLDG
BERKELEY CA 94720
-PHONE-
415 6423815

SOLAR RESEARCH
MEDIUM TEMPERATURE SOLAR COLLECTOR SYSTEMS

MERRILL RICHARD
DIRECTOR
NEW ALCHEMY INSTITUTE WEST
BOX 376
PESCADERO CA 94040
-PHONE-
805 9690101

SOLAR RESEARCH
MEDIUM TEMPERATURE SOLAR COLLECTOR SYSTEMS

MERRILL YEDIDA
NEW ALCHEMY INSTITUTE WEST
BOX 376
PESCADERO CA 94040

-PHONE-
805 9690101
SOLAR/ENERGY CONSERV INFO DISSEMINATION ORG
PUBLISH LAY PERIODICALS

MEYER HANS
WINDWORKS
BOX 329 RTE 3
MUKWONAGO WI 53149

SOLAR COMPONENTS MANUFACTURER
WIND CONVERSION

MEYER JAMES H
SR PROJ ENGR
SUNDSTRAND HYDRAULICS
2210 HARRISON AVE
ROCKFORD IL 61101
-PHONE-
815 2267246
SOLAR POTENTIAL MFG,ESTABLISHED BUSINESS
SOLAR SPACE HEATING & COOLING

MICHEL FREDERICK J
MGR INT'L SERVICES
WESTINGHOUSE ELECTRIC CORP
MFG DEVEL. LAB
R & D CENTER
PITTSBURGH PA 15235
-PHONE-
412 2563548
412 2564724
SOLAR POTENTIAL MFG,ESTABLISHED BUSINESS
SOLAR HEAT,COOL,&ELECTRIC POWER GENERATION

MIDDLETON DIRECTOR ARTHUR E
ELECT MAT&DEVICE LAB
OHIO STATE UNIV
ELECT ENGR LAB
2015 NEIL RM300 CL
COLUMBUS OH 43210
-PHONE-
614 4226210
614 4869561
ENERGY CONSERVATION RESEARCH
INORGANIC SEMI-CONDUCTORS

MIHALOEW JAMES R
13463 ATLANTIC RD
STRONGSVILLE OH 44136

-PHONE-
216 2384184
216 4334000 X6819
SOLAR RESEARCH
SOLAR HEAT,COOL,&ELECTRIC POWER GENERATION

MILLER ASHRAE ASME RICHARD D
CHIEF ENGINEER
FEUER CORPORATION
ENGINEERING
5401 MCCONNELL AVE
LOS ANGELES CA 90066
-PHONE-
213 3904046
213 8217209
SOLAR INTEREST
SOLAR SPACE HEATING & COOLING

MILNES PROF A G
CARNEGIE MELLON UNIVERSITY
ELECTRICAL ENGR DEPT
PITTSBURG PA 15213

-PHONE-
412 6212600

SOLAR RESEARCH
INORGANIC SEMI-CONDUCTORS

MINARDI JOHN E
DAYTON UNIV OF
DEPT OF MECH ENGR
COLLEGE PK AV
DAYTON OH 45469

-PHONE-
513 2290123

SOLAR RESEARCH
MEDIUM TEMPERATURE SOLAR COLLECTOR SYSTEMS

MINGENBACH WILLIAM A
PRESIDENT
GEOCENTRIC SYSTEMS INC
ARCHITECTURAL
PO BOX 49
TAOS NM 87571
-PHONE-
505 7584111
505 7582864
SOLAR RESEARCH
NEW CONSTRUCTION, PRACTICING

MINNEMAN JACK
V P
UNITED BANK OF BOULDER
1300 WALNUT ST
BOULDER CO 80302
-PHONE-
303 4423734
SOLAR FINANCIAL CONSIDERATIONS
BANKERS

MISSAR CHARLES E
EDITOR
PORTOLA INSTITUTE
ENERGY PRIMER
558 SANTA CRUZ AVE
MENLO PARK CA 94025
-PHONE-
415 3235535
415 3230313
SOLAR/ENERGY CONSERV INFO DISSEMINATION ORG
PUBLISH LAY + TECHNICAL BOOKS

MIZOJIRI HIROSHI
PRESIDENT
TOKIWA OPTIC&ELECT MANUFACTORY
2-9 1CHOME KOYAMADAI
SHINAGAWA-KU TOKYO JAPAN
-PHONE-
03 7126212
03 7149947
SOLAR MANUFACTURERS
HIGH TEMPERATURE SOLAR COLLECTOR SYSTEMS

MLAVSKY A I
TYCO LABORATORIES INC
16 HICKORY DRIVE
WALTHAM MA 02154
-PHONE-
617 8902400
SOLAR RESEARCH
INORGANIC SEMI-CONDUCTORS

MOCKOVCIAK, JR * JOHN
MGR ENERGY SYS DEPT
GRUMMAN AEROSPACE CORPORATION
PLANT 30
BETHPAGE NY 11714
-PHONE-
516 5753785
SOLAR POTENTIAL MFG,ESTABLISHED BUSINESS
MEDIUM TEMPERATURE SOLAR COLLECTOR SYSTEMS

MOKADAM R G
SUNSTRAND AVIATION
RESEARCH DEPT
4747 HARRISON AVE
ROCKFORD IL 61101
-PHONE-
815 2266780
SOLAR POTENTIAL MFG,ESTABLISHED BUSINESS
SOLAR HEAT,COOL,&ELECTRIC POWER GENERATION

MONROE JAMES I
SCIENTIFIC ADVISOR
LEGISLATIVE ENERGY COMMISSION
NEW YORK STATE
ROOM 718 L.O.B.
ALBANY NY 12207
-PHONE-
518 4722700
SOLAR RESEARCH
LEGISLATIVE, STATE

MOORE JAMES G
LIBRARY OF CONGRESS
SCIENCE POLICY DIV
WASHINGTON DC 20540
-PHONE-
202 4266040
SOLAR/ENERGY CONSERV INFO DISSEMINATION ORG
PROVIDE INFORMATION SEARCHES TO CLIENTS

MOORE JOHN P
NEWCASTLE UNIV OF
DEPT OF ELECT ENGR
NEW SOUTH WALES AUSTRALIA
SOLAR RESEARCH
SOLAR SPACE HEATING & COOLING

MOORE L PAUL
PLANT ENGINEER
NATL CNTR FOR ATMOS RESEARCH
PLANT FACILITIES
BOX 3000
BOULDER CO 80303
-PHONE-
303 4945151 X547
SOLAR RESEARCH
SOLAR SPACE HEATING & COOLING

MORITZ * KEN
TRW SYSTEMS
ONE SPACE PARK
REDONDO BEACH CA 90278
-PHONE-
213 5362253
SOLAR POTENTIAL MFG,ESTABLISHED BUSINESS
SOLAR SPACE HEATING & COOLING

MORRIS D N
RAND COPP
1700 MAIN ST
SANTA MONICA CA 90406

-PHONE-
213 3930411
ENERGY CONSERVATION RESEARCH
INCENTIVES FOR ENERGY CONSERVATION

MORRISS JOHN
BIOMASS ENERGY INST INC
PO BOX 129 P STA C
WINNIPEG MANITOBA CANADA

-PHONE-
204 2840472
SOLAR RESEARCH
BIO-CONVERSION

MUELLER JOHN J
MARKETING COORD
3 M CO
MARKETING DEPT
3 M CENTER
ST PAUL MN 55101
-PHONE-
612 7337201
SOLAR COMPONENTS MANUFACTURER
MEDIUM TEMPERATURE SOLAR COLLECTOR SYSTEMS

MUELLER OTWARD M
ELECTRONICS ENGINEER
GENERAL ELECTRIC COMPANY
TPD
MOUNTAIN VIEW ROAD
LYNCHBURG VA 24502
-PHONE-
804 8467311 X2765
804 2371161
SOLAR INTEREST
SOLAR HEAT,COOL,&ELECTRIC POWER GENERATION

MUKHERJEE MANISH KR
PROFESSOR
JADAVPUR UNIVERSITY
ELEC & TELECOMM ENGR
CALCUTTA 700032 INDIA

SOLAR RESEARCH
INORGANIC SEMI-CONDUCTORS

NALLE JR GEORGE S
NALLE PLASTICS INC
203 COLORADO STREET
AUSTIN TX 78701

-PHONE-
512 4776168

SOLAR COMPONENTS MANUFACTURER
SOLAR WATER DISTILLATION

NASH ROBERT
ASSOC PROF
VANDERBILT UNIV
ELECT ENGR
21ST AV S
NASHVILLE TN 37203
-PHONE-
615 3227311
SOLAR RESEARCH
SOLAR SPACE HEATING

NATHAN RICHARD A
PRINCIPAL CHEMIST
BATTELLE COLUMBUS LABS
505 KING AVENUE
COLUMBUS OH 43201

-PHONE-
614 2993151 X2049
SOLAR RESEARCH
SOLAR ENERGY STORAGE

NEBBIA GIORGIO
BARI, UNIV OF
ENERGY SOURCES LAB
LARGO FRACCACRETA 1
1 70122 BARY ITALY

-PHONE-
080 214495
SOLAR RESEARCH
SOLAR WATER DISTILLATION

NEILANDS J B
CALIF, UNIV OF-BERKELEY
BIOCHEM DEPT
BERKELEY CA 94720

-PHONE-
415 6427460
SOLAR RESEARCH
BIO-CONVERSION

NELSON DR VAUGHN
WEST TEXAS STATE UNIV
BOX 248
CANYON TX 79016

-PHONE-
806 6563904
SOLAR RESEARCH
WIND CONVERSION

NELSON GARY
ITEK CORPORATION
OPTICAL SYSTEMS DIV
10 MAGUIRE ROAD
LEXINGTON MA 02173
-PHONE-
617 2763411
SOLAR POTENTIAL MFG,ESTABLISHED BUSINESS
SOLAR SPACE HEATING & COOLING

NELSON * JAMES F
PRODUCT MANAGER
SUNDSTRAND HYDRAULICS
2210 HARRISON AVE
ROCKFORD IL 61101
-PHONE-
815 2267212

SOLAR POTENTIAL MFG,ESTABLISHED BUSINESS
SOLAR SPACE HEATING & COOLING

NIEMANN RALPH C
MECH ENGR
ARGONNE NATIONAL LAB
ENERGY & ENVIRONMENT
9500 SOUTH CASS AVE
ARGONNE IL 60439
-PHONE-
312 7397711 X5007
SOLAR RESEARCH
SOLAR HEAT,COOL,&ELECTRIC POWER GENERATION

NORDONE JOSEPH A
70 N W 94TH ST
MIAMI SHORES FL 33150

-PHONE-
305 7572194
SOLAR POTENTIAL MFG,NEW BUSINESS
SOLAR WATER HEATING

NUNEZ JR ALBERT
SOLAR PROJECT COORD

ENVIRONMTL ACTION-CO

DENVER CO 80206
-PHONE-
303 5341602
SEE PAGE 4.009.
SOLAR/ENERGY CONSERV INFO DISSEMINATION ORG
PROVIDE INFORMATION SEARCHES TO CLIENTS

NEUBAUER DR L W
CALIF, UNIV OF-DAVIS
AGR ENGR DEPT
DAVIS CA 95616

-PHONE-
916 7520102

ENERGY CONSERVATION RESEARCH
SOLAR SPACE HEATING & COOLING

NOGUCHI TETSUO
GOVT INDUST RESCH INST NAGOYA
SOLAR RESEARCH LAB
1 HIRATE-MACHI
KITA-KU NAGOYA JAPAN

-PHONE-
052 9112111
SOLAR RESEARCH
SOLAR FURNACE

NORTHRUP LYNN L
NORTHRUP INC
2208 CANTON ST
DALLAS TX 75201

-PHONE-
214 7415631
SOLAR MANUFACTURERS
MEDIUM TEMPERATURE SOLAR COLLECTOR SYSTEMS

OLSEN JOHN M
BIOPHYSICIST
BROOKHAVEN NATIONAL LABORATORY
BIOLOGY DEPT
50 BELL AVENUE
UPTON NY 11973
-PHONE-
516 3453382
SOLAR RESEARCH
WIND CONVERSION

NEWTON ALWIN B
CONSULTANT
136 SHELBOURNE DR
YORK PA 17403

-PHONE-
717 8435405
717 8430731 X223
SOLAR RESEARCH
SOLAR SPACE HEATING & COOLING

NOLAND MICHAEL C
MIDWEST RESEARCH INST
ENGR SCIENCES DIV
425 VOULKER BLVD
KANSAS CITY MO 64110

-PHONE-
816 5610202
SOLAR RESEARCH
ALL SOLAR TECHNOLOGIES

NOZIK A J
ALLIED CHEMICAL CORP
MATERIAL RESERCH CTR
P O BOX 1021R
MORRISTOWN NJ 07960

SOLAR RESEARCH
SOLAR ENERGY STORAGE

OLSON CURTIS L
DIR OF MARKETING
CARSON ASTRONOMICAL INSTRUMENT
P O BOX 5566
VALENCIA CA 91355

-PHONE-
805 2551234

SOLAR POTENTIAL MFG,ESTABLISHED BUSINESS
HIGH TEMPERATURE SOLAR COLLECTOR SYSTEMS

ORONA FIDEL
3417 MEMPHIS DR
EL PASO TX 79902
-PHONE-
915 7793877
SOLAR RESEARCH
INORGANIC SEMI-CONDUCTORS

ORTAEASI UGUR
SENIOR PHYSIST
CORNING GLASS WORKS
PHYSICAL RESEARCH
CORNING NY 14830
-PHONE-
607 9743370
SOLAR RESEARCH
MEDIUM TEMPERATURE SOLAR COLLECTOR SYSTEMS

OSHIDA ISAO
FACULTY OF SCI & TEC
SOPHIA UNIVERSITY
DEPT OF PHYSICS
7 KIOI-CHO CHIYODA-K
TOKYO JAPAN
-PHONE-
03 2659211 X567
SOLAR RESEARCH
HIGH TEMPERATURE SOLAR COLLECTOR SYSTEMS

OSWALD W J
CALIF, UNIV OF-BERKELEY
SCH OF PUBLIC HEALTH
BERKELEY CA 94720
SOLAR RESEARCH
BIO-CONVERSION

OTTMER JAMES H
MECHANICAL ENGINEER
U S GOVT GSA
CONSTRUCTION MGMT
BLDG#41 DFC
DENVER CO 80225
-PHONE-
303 2342641
SOLAR INTEREST
PLANNING OFFICES, FEDERAL

OTTUM AL
UNITSPAN ARCH SYS INC
9419 MASON AVE
CHATSWORTH CA 91311
-PHONE-
213 9981131
SOLAR MANUFACTURERS
MEDIUM TEMPFRATURE SOLAR COLLECTOR SYSTEMS

OUTTERSON G GILBERT
DIRECTOR
ENVIRONMENTAL ENERGY FOUNDATN
ROOM 1001
12000 EDGEWATER DR
LAKEWOOD OH 44107
-PHONE-
216 2218130
SOLAR RESEARCH
SOLAR HEAT,COOL,&ELECTRIC POWER GENERATION

OVSHINSKY STANFORD R
ENERGY CONVERSION DEVICES INC
TROY MI 48084
-PHONE-
313 5497300
SOLAR MANUFACTURERS
INORGANIC SEMI-CONDUCTORS

OWEN,PE BRAD
COBALT ENGINEERING CO
164 PENINGTN-HBOURTN
PENNINGTON NJ 08534
-PHONE-
609 7371376
609 8821414 X360
SOLAR POTENTIAL MFG,ESTABLISHED BUSINESS
SOLAR SPACE HEATING

PADOVANI * ANN
CURRENT PRODUCTS
JOHNS-MANVILLE CORP
PRODUCT INFO CENTER
GREENWOOD PLAZA
DENVER CO 80217
-PHONE-
303 7701000
ENERGY CONSERVATION RESEARCH
NEW CONSTRUCTION, PRACTICING

PAGE J K
CHAIRMAN
INTL SOLAR ENGY SOC/UK SECT
21 ALBEMARLE ST
LONDON ENGLAND
SOLAR/ENERGY CONSERV INFO DISSEMINATION ORG
ALL SOLAR TECHNOLOGIES

PALMER HOWARD B
PROF OF FUEL SCIENCE
PENNSYLVANIA STATE UNIV
MATERIAL SCIENCE DPT
320 M I BLDG
UNIVERSITY PARK PA 16802
-PHONE-
814 8656512
SOLAR RESEARCH
HIGH TEMPERATURE SOLAR COLLECTOR SYSTEMS

PALZ WOLFGANG
C N E S
91 BRETIGNY BOX 04
FRANCE

-PHONE-
4909220
SOLAR RESEARCH
INORGANIC SEMI-CONDUCTORS

PANAGOS PAUL T
PUBLIC HEALTH ENGR
MINNESOTA STATE HEALTH DEPT
ENVIRO HEALTH DIV
5900 PENN AVE SOUTH
MINNEAPOLIS MN 55419
-PHONE-
614 9229225
SOLAR RESEARCH
MEDIUM TEMPERATURE SOLAR COLLECTOR SYSTEMS

PAPINEAU ROGER
PRESIDENT
SOLAR ENERGY CO
DEERWOOD DR
MERRIMACK NH 03054

-PHONE-
603 4245168
SOLAR MANUFACTURERS
SOLAR WATER HEATING

PARKER ANDREW J
MANAGER
RICHARD P MUELLER & ASSOC INC
ENVTL ENGRG DEPT
1900 SULPHUR SPRG RD
BALTIMORE MD 21227
-PHONE-
301 2745666
202 7839159
SOLAR RESEARCH
SOLAR SPACE HEATING & COOLING

PAUL SAMUEL
SAMUEL PAUL ARCHITECT, OFF OF
107-40 QUEENS BLVD
FOREST HILLS NY 11375

-PHONE-
212 2615100

SOLAR INTEREST
SOLAR SPACE HEATING

PAYNTON WILLIAM C
MANAGER
TEXAS INSTRUMENTS INC
GOVT PROD MS10-10
ATTLEBORO MA 02703

-PHONE-
617 2222800 X6654
617 2222800 X484
SOLAR POTENTIAL MFG,ESTABLISHED BUSINESS
INORGANIC SEMI-CONDUCTORS

PEACORE GEORGE R
PRESIDENT
WEATHER ENGINEERING & MFG INC
4703 N PARK DR
COLORADO SPRINGS CO 80907
-PHONE-
303 5988228
SOLAR PRODUCTS DISTRIBUTORS
SOLAR HEAT,COOL,&ELECTRIC POWER GENERATION

PECK DR JOHN F
ARIZONA, UNIVERSITY OF
ENVIRONMENT RES LAB
TUCSON INT'L AIRPORT
TUCSON AZ 85706
-PHONE-
602 8842931
SOLAR RESEARCH
HIGH TEMPERATURE SOLAR COLLECTOR SYSTEMS

PEI DR Y K
LIBBEY OWENS FORD CO
811 MADISON AVE
TOLEDO OH 43695

-PHONE-
419 2473731
SOLAR POTENTIAL MFG,ESTABLISHED BUSINESS
MEDIUM TEMPERATURE SOLAR COLLECTOR SYSTEMS

PELLETT MICHAEL P
PRES
PELCASP SOLAR DESIGNERS
240 LANGDON ST
MADISON WI 53703
-PHONE-
608-2519967
SOLAR INTEREST
SOLAR SPACE HEATING

PENCE RICHARD
VANTAGE HEATING&AIR COND INC
5611 KENDALL CT
ARVADA CO 80002

-PHONE-
303 4204410
SOLAR PRODUCTS DISTRIBUTORS
SOLAR SPACE HEATING & COOLING

PENNOYER RAY
V P
INNOTECH CORP
181 MAIN ST
NORWALK CN 06851
-PHONE-
203 8462041
SOLAR RESEARCH
INORGANIC SEMI-CONDUCTORS

PERI GEORGES
FACULTY OF SCIENCE
ST JEROME UNIVERSITY
DEPT D'HELIOPHYSIQUE
13013 MARSEILLE FRANCE

-PHONE-
93 98 0901
SOLAR RESEARCH
SOLAR RADIATION MEASUREMENTS

PERKINS PHILIP J
ENVIRONMENTAL ED
UNITED AUTO WORKERS
CONSERVATION DEPT
UAW FAMILY ED CENTER
ONAWAY MI 49765
-PHONE-
517 7338521
SOLAR/ENERGY CONSERV INFO DISSEMINATION ORG
PROVIDE AUDIO & VISUAL SERVICES

PERTHERBRIDGE P
BUILDING RESCH STATION
GARSTON WATFORD
WD2 7JR UNITED KINGDOM

-PHONE-
092 7 3 740 X40
ENERGY CONSERVATION RESEARCH
SOLAR RADIATION MEASUREMENTS

PESKO CAROLYN M
SOLAR DIRECT PROJ DR
COLORADO, UNIV OF-DENVER

1100 14TH STREET
DENVER CO 80202
-PHONE-
SEE PAGE 4.009
303 8921117 X471
SOLAR/ENERGY CONSERV INFO DISSEMINATION ORG
HAVE&MAINTAIN ENERGY DATA BANK

PETERSON CRAIG H
ASST PROF ECONOMICS
UTAH STATE UNIVERSITY
DEPT OF ECONOMICS
LOGAN UT 84322

-PHONE-
801 7524100 X7307
SOLAR ECONOMIC CONSIDERATIONS
PROVIDE ECONOMIC ANALYSIS OF SOLAR TECH

PETERSON MARK
DENVER MUSEUM OF NATURAL HIST
GATES PLANETARIUM
CITY PARK
DENVER CO 80205

-PHONE-
303 3884201

SOLAR/ENERGY CONSERV INFO DISSEMINATION ORG
PROVIDE AUDIO & VISUAL SERVICES

PETROCELLI J V
FORD MOTOR COMPANY
SCIENTIFIC RES STAFF
ADVANCED TECH APP/RA
DEARBORN MI 48121

SOLAR RESEARCH
SOLAR ENERGY STORAGE

PFEFFER J T
ILLINOIS, UNIV OF
SCHOOL OF ENGINEERNG
URBANA IL 61801

SOLAR RESEARCH
BIO-CONVERSION

PIERRE YVON F
8709 SOUTHERN PINES
VIENNA VA 22180

-PHONE-
703 9388206
SOLAR RESEARCH
INORGANIC SEMI-CONDUCTORS

PINNOW LANE A
MGR ENGR & SALES
WEATHER ENGINEERING & MFG INC
4703 N PARK DR
COLORADO SPRINGS CO 80907
-PHONE-
303 5988228
SOLAR PRODUCTS DISTRIBUTORS
SOLAR HEAT,COOL,&ELECTRIC POWER GENERATION

POLLARD HOWARD E
SR ENG SPECIALIST
PHILCO-FORD CORP
3939 FABIAN WAY
PALO ALTO CA 94303
-PHONE-
415 3264350 X6329
SOLAR RESEARCH
INORGANIC SEMI-CONDUCTORS

POLLUCK ROBERT
METROPOLITAN ECOLOGY WORKSHOP
74 JOY STREET
BOSTON MA 02114

-PHONE-
617 7234699
SOLAR/ENERGY CONSERV INFO DISSEMINATION ORG
ALL SOLAR TECHNOLOGIES

POPE JOSEPH H
7328 MT SHERMAN RD
LONGMONT CO 80501

-PHONE-
303 4428452
SOLAR RESEARCH
SOLAR WATER HEATING

PORTER KENNETH A
FEDERAL ENGY OFFICE REGION II
26 FEDERAL PLAZA
NEW YORK NY 10017

-PHONE-
212 2641041
SOLAR RESEARCH
ALL SOLAR TECHNOLOGIES

POTTER JAMES A
12 GREEN HOUSE BLVD
WEST HARTFORD CT 06110

-PHONE-
203 5611637
SOLAR POTENTIAL MFG,NEW BUSINESS
SOLAR HEAT,COOL,&ELECTRIC POWER GENERATION

POWE R E
MONTANA STATE UNIVESITY
SCHOOL OF ENGINEERG
BOZEMAN MT 59715

SOLAR RESEARCH
WIND CONVERSION

PRAKASH RAJENDRA
ROORKEE, UNIV OF
DEPT OF MECH ENGR
ROORKEE 247667 INDIA

-PHONE-
289
SOLAR RESEARCH
MEDIUM TEMPERATURE SOLAR COLLECTOR SYSTEMS

PULFORD A
HEAD NEW VENTURES
PILKINGTON BROTHERS LTD
R&D LABS
LATHOM ORMSHIRE
LANCASHIRE ENGLAND

SOLAR RESEARCH
MEDIUM TEMPERATURE SOLAR COLLECTOR SYSTEMS

PULLER ROBLEY S
ARIES CONSULTING ENGRS INC
SUITE D
1810 N W 6TH STREET
GAINESVILLE FL 32601
-PHONE-
904 3726687

SOLAR RESEARCH
SOLAR SPACE HEATING & COOLING

PULVARI CHARLES F
CATHOLIC UNIVERSITY OF AMERICA
DEPT OF ELECT ENGR
4 ST ON MICH AVE N E
WASHINGTON DC 20017
-PHONE-
202 6355193
202 5262400
SOLAR RESEARCH
INORGANIC SEMI-CONDUCTORS

PUTHOFF RICHARD L
NASA LEWIS RESEARCH CTR
PLUM BROOK STATION
21000 BROOKPARK ROAD
CLEVELAND OH 44135
-PHONE-
216 4334000

SOLAR RESEARCH
WIND CONVERSION

RAGSDALE ROBERT
ASST TO CHIEF
NASA LEWIS RESCH CTR
POWER APPLICATIONS
MAIL DROP 500 201
CLEVELAND OH 44142
-PHONE-
216 4334000 X6943
SOLAR RESEARCH
MEDIUM TEMPERATURE SOLAR COLLECTOR SYSTEMS

RAHL ARI
OHIO ST UNIV
DEPT OF PHYSICS
COLUMBUS OH 43210

-PHONE-
614 4226446
SOLAR RESEARCH
SOLAR PONDS

RAIN LLOYD
MANAGER-PROD DEVEL
IRVIN INDUSTRIES INC
STRUCTURES GROUP
555 S BROADWAY
LEXINGTON KY 40508
-PHONE-
606 2546406
SOLAR POTENTIAL MFG,ESTABLISHED BUSINESS
SOLAR SPACE HEATING

RAMAKUMAR RAMACHANDRA C
ASSOC PROF ELEC ENGG
OKLAHOMA STATE UNIVERSITY
ELECTRICAL ENGG
ENGINEERING SOUTH
STILLWATER OK 74074
-PHONE-
405 3726211 X6476
SOLAR RESEARCH
WIND CONVERSION

RAMSTETTER J K
PRODUCT DESIGNER
1520 W ALASKA PLACE
DENVER CO 80223

SOLAR/ENERGY CONSERV INFO DISSEMINATION ORG
SOLAR SPACE HEATING

RANKEN WILLIAM A
GROUP LEADER Q-25
CALIF, UNIV OF--LOS ALAMOS
LOS ALAMOS SCI LAB
P O BOX 1663
LOS ALAMOS NM 87544
-PHONE-
505 6676578
SOLAR RESEARCH
SOLAR SPACE HEATING & COOLING

RAPPAPORT PAUL
DIRECTOR M&P LAB
RCA CORPORATION
D SARNOFF RESCH CTR
PRINCETON NJ 08540
-PHONE-
609 4522700
SOLAR RESEARCH
INORGANIC SEMI-CONDUCTORS

RASEMAN CHAD J
SOLAR SUNSTILL INC
SETAUKET NY 11733

-PHONE-
516 9414078
SOLAR COMPONENTS MANUFACTURER
GREENHOUSES

RASMUSSEN WILLIAM
BOX D
SUGAR HILL NH 03583

-PHONE-
603 8237713
SOLAR RESEARCH
SOLAR SPACE HEATING

RATTIN EGAN J
SOLERG ASSOCIATES
PO BOX 90691
LOS ANGELES CA 90009

-PHONE-
213 8364200
SOLAR MANUFACTURERS
MEDIUM TEMPERATURE SOLAR COLLECTOR SYSTEMS

RAVIN * JERRY W
MARKETING MGR
SPECTROLAB DIV OF TEXTRON
SOLAR POWER SYSTEMS
12484 GLADSTONE AVE
SYLMAR CA 91342
-PHONE-
213 3654611
SOLAR MANUFACTURERS
INORGANIC SEMI-CONDUCTORS

REDMOND THOMAS B
MGR RESEARCH & DEVEL
NATL CONCRETE MASONRY ASSOC
PO BOX 9185 ROSLYN S
ARLINGTON VA 22209

-PHONE-
703 5240815
SOLAR RESEARCH
SOLAR SPACE HEATING & COOLING

REED JOHN C
PROF EMERITUS
FLORIDA, UNIV OF
126 MECH ENGR BLDG
GAINESVILLE FL 32611
-PHONE-
904 3920805

SOLAR RESEARCH
ALL SOLAR TECHNOLOGIES

REIFSNYDER WILLIAM E
YALE UNIV
FORESTRY&ENVIRO STDY
360 PROSPECT ST
NEW HAVEN CT 06511
-PHONE-
203 4360020

SOLAR RESEARCH
SOLAR RADIATION MEASUREMENTS

REYNOLDS JOHN K
CONSULTANT
ARTHUR D LITTLE, INC.
3601 S WADSWORTH
DENVER CO 80235
-PHONE-
303 9854505
303 9854741
SOLAR RESEARCH
ALL SOLAR TECHNOLOGIES

r

INDEX OF INDIVIDUALS

REYNOLDS JR   CHARLES F
VICE PRES
PRUDENTIAL INS CO OF AMERICA
REAL ESTATE INVESTNT
1180 RAYMOND BLVD
NEWARK   NJ   07102
-PHONE-
201 3364578

ENERGY CONSERVAT'N,FINANCIAL CONSIDERATIONS
NEW CONSTRUCTION, RESOURCES

RHODES   LINDA
JEFFERS POND
HOP BOTTOM   PA   18824

-PHONE-
717 2894743

SOLAR RESEARCH
SOLAR SPACE HEATING

RICE * CONE
V P SALES
SOLARSYSTEMS INC
1515 WSW LOOP 323
TYLER   TX   75701

-PHONE-
214 5920945
214 5975534
SOLAR MANUFACTURERS
SOLAR SPACE HEATING & COOLING

RICE   FRED
PRESIDENT
FRED RICE PRODUCTIONS INC
6313 PEACH AVE
VAN NUYS   CA   91401
-PHONE-
213 7863860
SOLAR MANUFACTURERS
SOLAR WATER HEATING

RIEL * R K
WESTINGHOUSE RESEARCH LABS
BEULAH RD
PITTSBURG   PA   15235

-PHONE-
412 2563614
SOLAR RESEARCH
INORGANIC SEMI-CONDUCTORS

RILEY   THOMAS J
NASA LEWIS RESEARCH CTR
POWER ELECTRONICS BR
21000 BROOKPARK RD
CLEVELAND   OH   44135
-PHONE-
216 4334000 X288
SOLAR RESEARCH
INORGANIC SEMI-CONDUCTORS

RITTNER   DR EDMUND S
DIRECTOR
COMSTAT LABORATORIES
APPLIED SCI DIV
P O BOX 115
CLARKSBURG   MD   20734
-PHONE-
301 4284541

SOLAR RESEARCH
INORGANIC SEMI-CONDUCTORS

ROBINSON   DAVID A
DESIGN ENGINEER
DOME EAST CORPORATION
325 DUFFY AVENUE
HICKSVILLE   NY   11801

-PHONE-
516 9380545

SOLAR MANUFACTURERS
MEDIUM TEMPERATURE SOLAR COLLECTOR SYSTEMS

RODRIGUEZ PERAZZA   DR MANUEL
PUERTO RICO, UNIV OF
GEN ENGR DEPT
MAYAGUEZ   PUERTO RICO

-PHONE-
809 8324040 X209
809 8324040 X455
SOLAR RESEARCH
MEDIUM TEMPERATURE SOLAR COLLECTOR SYSTEMS

ROGERS   BENJAMIN T
CONSULTING ENGINEER
PO BOX 2
EMBUDO   NM   87531

-PHONE-
505 5794355
505 6675010
SOLAR RESEARCH
SOLAR SPACE HEATING

ROGERS   CHARLES F
PARTNER
PERRY-DEAN-STEWART ARCHITECTS
955 PARK SQ BLDG
BOSTON   MA   02116
-PHONE-
617 4829160

SOLAR INTEREST
SOLAR HEAT,COOL,&ELECTRIC POWER GENERATION

ROMER   ROBERT H
AMHERST COLLEGE
PHYSICS DEPT
AMHERST   MA   01002

-PHONE-
413 5422258
413 5422251
SOLAR RESEARCH
SOLAR SPACE HEATING

ROOS WILLIAM C
318 RENA DRIVE
LAFAYETTE LA 70501

-PHONE-
318 9843430

SOLAR RESEARCH
SOLAR ENGINES

RUSH WILLIAM F
INSTITUTE OF GAS TECHNOLOGY
3424 S STATE
CHICAGO IL 60616

-PHONE-
312 2259600 X685
SOLAR RESEARCH
SOLAR SPACE HEATING & COOLING

SAFDARI Y B
BRADLEY UNIV
MECH ENGR DEPT
PEORIA IL 61606

-PHONE-
309 6767611
SOLAR RESEARCH
SOLAR SPACE HEATING & COOLING

SALES DIVISION SOLAR EQUIP
SOLAR ENERGY DIGEST
P O BOX 17776
SAN DIEGO CA 92117

-PHONE-
714 2772980
SOLAR PRODUCTS DISTRIBUTORS
SOLAR WATER HEATING

ROST DR D F
DIR SOLAR ENGY TSK G
YOUNGSTOWN STATE UNIV
ELECT ENGR DEPT
YOUNGSTOWN OH 44503
-PHONE-
216 7461851 X425
216 5336128
SOLAR RESEARCH
TOTAL SYSTEMS INTEGRATION

RUSSELL DR JOHN
HEAD,SOLAR RESEARCH
GENERAL ATOMIC COMPANY
P O BOX 81608
SAN DIEGO CA 92138
-PHONE-
714 4531000
SOLAR RESEARCH
SOLAR ELECTRIC POWER GENERATION

SAHA HIRANMAY
KALYANI, UNIV OF
DEPT OF PHYSICS
KALYANI NADIA
WEST BENGAL INDIA

SOLAR RESEARCH
INORGANIC SEMI-CONDUCTORS

SAN MARTIN DR ROBERT L
BOX 3450
NEW MEXICO STATE UNIV
MECHANICAL ENGR DEPT
LAS CRUCES NM 88003
-PHONE-
505 6463501
SOLAR RESEARCH
SOLAR SPACE HEATING & COOLING

RUCH JOHN W
447 E DRIVE
OAK RIDGE TN 37830

-PHONE-
615 4836436

SOLAR RESEARCH
SOLAR SPACE HEATING & COOLING

SADLER GERALD W
ALBERTA, UNIV OF
DEPT OF MECH ENGR
EDMONTON 7 ALBERTA CANADA

-PHONE-
403 4323450
SOLAR RESEARCH
SOLAR RADIATION MEASUREMENTS

SAHLY SAM
DO LITTLE INDUSTRIES
23 TERRACE PLACE
DANBURY CT 06810

SOLAR RESEARCH
HOMES

SANTALA TUEVO
TEXAS INSTRUMENTS
34 FOREST ST
ATTLEBORO MA 02703

-PHONE-
217 2222800 X6150
SOLAR RESEARCH
MEDIUM TEMPERATURE SOLAR COLLECTOR SYSTEMS

SARKAR * PROF D C
HEAD
KALYANI, UNIV OF
DEPT OF PHYSICS
KALYANI NADIA
WEST BENGAL INDIA

SOLAR RESEARCH
INORGANIC SEMI-CONDUCTORS

SATCHWELL DAVID
PRESIDENT
SOLAR ENERGY SOC OF AMERICA
2780 SEPULA BLVD
TORRANCE CA 90510

-PHONE-
213 3263283
SOLAR/ENERGY CONSERV INFO DISSEMINATION ORG
PUBLISH LAY + TECHNICAL PERIODICALS

SATER BERNARD L
NASA LEWIS RESEARCH CTR
M S 54-4
21000 BROOKPARK ROAD
CLEVELAND OH 44135

-PHONE-
216 4334000 X220
SOLAR RESEARCH
INORGANIC SEMI-CONDUCTORS

SAVOLAINEN UNTO
TEXAS INSTRUMENTS INC
MS10-16
ATTLEBORO MA 02703

-PHONE-
617 2222800 X272
SOLAR RESEARCH
MEDIUM TEMPERATURE SOLAR COLLECTOR SYSTEMS

SAYLER J P
AMERICAN ENERGY ALTERNATVS INC
P O BOX 905
BOULDER CO 80302

-PHONE-
303 4471865
SOLAR PRODUCTS DISTRIBUTORS
WIND CONVERSION

SCHAEPER DR HERBERT
FLORIDA, UNIV OF
DEPT MECH ENGR
GAINESVILLE FL 32601

-PHONE-
904 3920809
SOLAR RESEARCH
WIND CONVERSION

SCHARFF JACK
DIR MKT & ADV
EDMUND SCIENTIFIC CO
ADV & MKT
101 E GLOUCESTER PK
BARRINGTON NJ 08007
-PHONE-
609 7678483
SOLAR MANUF, EDUCATIONAL EQUIPMENT
SOLAR SPACE HEATING & COOLING

SCHECHTER J
DIRECTOR
NEGEV, UNIV OF THE
R & D AUTHORITY
P O B 1025
BEER SHEVA ISRAEL
-PHONE-
0573382
SOLAR RESEARCH
SOLAR PONDS

SCHENEEBERG FLOYD H
VICE PRES
LEAR SIEGLER INC
MAMMOTH DIVISION
13120 B COUNTY RD 6
MINNEAPOLIS MN 55441
-PHONE-
612 5442711
SOLAR RESEARCH
SOLAR SPACE HEATING

SCHIEFER D
ISRAEL INSTITUTE OF TECHNOLOGY
TECHNION
HAIFA ISRAEL

-PHONE-
04 235151
SOLAR RESEARCH
INORGANIC SEMI-CONDUCTORS

SCHLESINGER P E ROBERT J
RHO SIGMA UNLIMITED
5108 MELVIN AVE
TARZANA CA 91356

-PHONE-
213 3424376
SOLAR COMPONENTS MANUFACTURER
SOLAR SPACE HEATING

SCHMIDT ROGER N
HONEYWELL INC
SYSTEMS&RESEARCH CTR
2600 RIDGWAY PARKWAY
MINNEAPOLIS MN 55413
-PHONE-
612 3314141 X4078
SOLAR RESEARCH
SOLAR ELECTRIC POWER GENERATION

SCHNEIDER RALPH
MARKETING MGR
FAFCO INC
2860 SPRING ST
REDWOOD CITY CA 94063
-PHONE-
415 3646772
SOLAR MANUFACTURERS
SOLAR WATER HEATING

SCHOENEALL WALTER
45 CH M DUBOULE
CH-1211
GENEVA 19 SWITZERLAND

SOLAR MANUFACTURERS
WIND CONVERSION

SCHURR SAM H
RESOURCES FOR THE FUTURE INC
ENERGY RESEARCH&TECH
1755 MASS AVE NW
WASHINGTON DC 20036

ENERGY CONSERVATION RESEARCH
FOUNDATIONS PROVIDING GRANTS/FUNDS

SCOTT * JOHN L
DESERT SUNSHINE EXPOSURE TESTS
CLIENT SERVICES
BOX185 BLACKCAN STGE
PHOENIX AZ 85020
-PHONE-
602 4657525
602 4657521
SOLAR ENVIRONMENTAL CONSIDERATIONS
SOLAR,MATERIAL WEATHERABILITY TESTS

SCHNEPP BARBARA J
N Y STATE COUNCIL ON ARCHITECT
810 SEVENTH AVE
NEW YORK NY 10019

-PHONE-
212 7657630
SOLAR/ENERGY CONSERV INFO DISSEMINATION ORG
NEW CONSTRUCTION, TECHNOLOGIES

SCHRENK GEORGE L
PENN, UNIVERSITY OF
N CTR ENGY MGT&POWER
TOWNE BUILDING
PHILADELPHIA PA 19104

-PHONE-
215 5945121
SOLAR RESEARCH
HIGH TEMPERATURE SOLAR COLLECTOR SYSTEMS

SCHWARTZ CHAS
DIRECTOR
TR PRODUCTIONS
SOLAR ENERGY PROJECT
1031 COMMONWEALTH AV
BOSTON MA 02215
-PHONE-
617 7830200

SOLAR/ENERGY CONSERV INFO DISSEMINATION ORG
PUBLISH AND/OR DISTRIBUTE FILMS

SCOTT PAUL B
PRINCIPAL SCIENTIST
XONICS INC
6837 HAYVENHURST AVE
VAN NUYS CA 91406
-PHONE-
213 7877380

SOLAR RESEARCH
SOLAR SPACE HEATING

SCHOEN DAVID A
EXECUTIVE SECRETARY
ECO-SOLUTIONS,INC.
PO BOX 4117
BOULDER CO 80302
-PHONE-
303 4436798
SOLAR POTENTIAL MFG,NEW BUSINESS
SOLAR SPACE HEATING

SCHULZE KARL H
RESEARCH PHYSICIST
BABCOCK & WILCOX CO
RESEARCH CENTER
1562 BEESON STREET
ALLIANCE OH 44601
-PHONE-
216 8219110 X227
SOLAR RESEARCH
SOLAR ELECTRIC POWER GENERATION

SCHWERZEL DR ROBERT E
RESEARCH CHEMIST
BATTELLE COLUMBUS LABS
ORGANIC CHEMISTRY
505 KING AVENUE
COLUMBUS OH 43201
-PHONE-
614 2993151 X3457
614 2993151 X3154
SOLAR RESEARCH
ALL SOLAR TECHNOLOGIES

SEALANDER DAVID
ROUTE 4 BOX 371
IDAHO FALLS ID 83401

SOLAR RESEARCH
MEDIUM TEMPERATURE SOLAR COLLECTOR SYSTEMS

SECRETARY THE
INTL SOLAR ENERGY SOCIETY
PO BOX 52
PARKVILLE VIC AUSTRALIA

SOLAR/ENERGY CONSERV INFO DISSEMINATION ORG
ALL SOLAR TECHNOLOGIES

SEICLEL PETER
ECOTEC FOUNDATION INC
2923 WOR
LD AVENUE
CINCINNATI OH 45206
-PHONE-
513 8618324
SOLAR RESEARCH
HOMES

SELCH GLENN
B F SALES ENGR CO
4770 FOX ST NO 9
DENVER CO 80216

SOLAR PRODUCTS DISTRIBUTORS
MEDIUM TEMPERATURE SOLAR COLLECTOR SYSTEMS

SELCUK DR M K
HONEYWELL INC
SYSTEMS&RESEARCH CTR
2600 RIDGWAY PARKWAY
MINNEAPOLIS MN 55413

-PHONE-
612 3314141
SOLAR RESEARCH
SOLAR SPACE HEATING & COOLING

SELKOWITZ STEPHEN
INSTRUCTOR
CALIF INST OF THE ARTS
SCHOOL OF DESIGN
24700 MCBEAN PKWY
VALENCIA CA 91355
-PHONE-
805 2551050 X328
SOLAR RESEARCH
SOLAR WATER HEATING

SEPSY CHARLES F
OHIO STATE UNIVERSITY
NORTH HIGH ST
COLUMBUS OH 43210

-PHONE-
614 4226898
SOLAR RESEARCH
SOLAR SPACE HEATING & COOLING

SERAPHIN BERNHARD O
ARIZONA, UNIV OF
OPTICAL SCIENCE CTR
TUCSON AZ 85721

-PHONE-
602 8842263
SOLAR RESEARCH
INORGANIC SEMI-CONDUCTORS

SEVCIK VACLAV J
MECH ENGR
ARGONNE NATIONAL LABORATORY
ARF
9700 SOUTH CASS AVE
ARGONNE IL 60439
-PHONE-
312 7397711 X4182
SOLAR RESEARCH
SOLAR SPACE HEATING & COOLING

SHANNON * WAYNE
LOCKHEED PALO ALTO RESRCH LAB
DEPT 52-01 BLDG 201
3251 HANOVER ST
PALO ALTO CA 94304

-PHONE-
415 4934411
SOLAR RESEARCH
SOLAR SPACE HEATING & COOLING

SHELTON JAY
WILLIAMS COLLEGE
PHYSICS DEPT
WILLIAMSTOWN MA 01267

-PHONE-
413 5972123
SOLAR RESEARCH
MEDIUM TEMPERATURE SOLAR COLLECTOR SYSTEMS

SHER NEIL C
HONEYWELL INCORP
URBAN & ENVIRO SYS
2700 RIDGWAY PARKWAY
MINNEAPOLIS MN 55413
-PHONE-
612 3314141 X4033
SOLAR RESEARCH
SOLAR HEAT,COOL,&ELECTRIC POWER GENERATION

SHERIDAN NORMAN R
QUEENSLAND, UNIV OF
DEPT OF MECH ENGR
ST LUCIA, QUEENSLAND AUSTRALIA

-PHONE-
072 700111
SOLAR RESEARCH
MEDIUM TEMPERATURE SOLAR COLLECTOR SYSTEMS

SHERMAN JOSEPH
DIR BLDG TECH&SAFETY
U S DEPT HOUSING & URBAN DEV
OFF POLICY DEV&RESCH
451 7TH ST S W
WASHINGTON DC 20410
-PHONE-
202 7555364
SOLAR ECONOMIC CONSIDERATIONS
GOVERNMENT AGENCIES PROVIDING GRANTS

SHIRLAND F A
WESTINGHOUSE RESEARCH LABS
SOLID STATE RES
BEULAH RD
PITTSBURGH PA 15235
-PHONE-
412 2563613

SOLAR RESEARCH
INORGANIC SEMI-CONDUCTORS

SHROPSHIRE JR WALTER
SMITHSONIAN RADIATION BIO LAB
12441 PARKLAWN DR
ROCKVILLE MD 20852

-PHONE-
301 4432333

SOLAR RESEARCH
BIO-CONVERSION

SILVER JOHN
JOHN SIMKINS SCHOOL
MAIN ST
SOUTH YARMOUTH MA 02664

-PHONE-
617 3982412
SOLAR RESEARCH
MEDIUM TEMPERATURE SOLAR COLLECTOR SYSTEMS

SHICK WAYNE L
ILLINOIS, UNIVERSITY OF
SMALL HOMES COUNCIL
ONE EAST ST MARYS RD
CHAMPAIGN IL 61820

-PHONE-
217 3331801 X457
SOLAR/ENERGY CONSERV INFO DISSEMINATION ORG
PROVIDE INFORMATION SEARCHES TO CLIENTS

SHITZER AVRAHAM
TECHNION-ISRAEL INST OF TECH
DEPT MECH ENGR
TECHNION CITY
HAIFA ISRAEL
-PHONE-
04 235105

SOLAR RESEARCH
SOLAR SPACE HEATING & COOLING

SHURCLIFF WILLIAM A
19 APPLETON ST
CAMBRIDGE MA 02138

-PHONE-
617 4952816
617 8760764
SOLAR RESEARCH
SOLAR SPACE HEATING

SIMON FREDERICK
M S 500-201
NASA LEWIS RESEARCH CTR
SOLAR SYS SECTION
21000 BROOKPARK ROAD
CLEVELAND OH 44135
-PHONE-
216 4334000 X726
SOLAR RESEARCH
SOLAR SPACE HEATING & COOLING

SHIPMAN ROBERT H
KANSAS STATE UNIV
CHEM ENGR DEPT
MANHATTAN KS 66506

-PHONE-
913 5325584
SOLAR RESEARCH
BIO-CONVERSION

SHORT W LEIGH
PROF
MASSACHUSETTS, UNIV OF
DEPT CHEM ENGR
AMHERST MA 01002
-PHONE-
413 5451588
413 5452507
ENERGY CONSERVATION RESEARCH.
BIO-CONVERSION

SILBERT MD JEROME A
444 EAST 84TH ST
NEW YORK NY 10028

-PHONE-
212 2496487

SOLAR/ENERGY CONSERV INFO DISSEMINATION ORG
OCEAN THERMAL GRADIENT SYSTEMS

SINGH RAM BUX
INCHARGE
GOBAR GAS RESEARCH STATION
AJITMAL ETAWAH U P INDIA

SOLAR/ENERGY CONSERV INFO DISSEMINATION ORG
BIO-CONVERSION.

S

INDEX OF INDIVIDUALS

SITTENFELD ITAMAR
VICE PRESIDENT
SOL-THERM CORP
7 WEST 14TH STREET
NEW YORK NY 10011

-PHONE-
212 6914632

SOLAR MANUFACTURERS
SOLAR WATER HEATING

SKINROOD ALAN C
SUPERVISOR
SANDIA LABORATORIES
DIVISION 8184
LIVERMORE CA 94550

-PHONE-
415 4552501
SOLAR RESEARCH
TOTAL SYSTEMS INTEGRATION

SMITH THOMAS W
RESEARCH SPECIALIST
WISCONSIN, UNIV OF
MARINE STUDIES CENTE
1225 WEST DAYTON ST
MADISON WI 53706
-PHONE-
608 2634578
ENERGY CONSERVATION RESEARCH
NEW CONSTRUCTION, RESEARCHING

SOLERI PAOLO
PRESIDENT
COSANTI FOUNDATION
6433 DOUBLETREE ROAD
SCOTTSDALE AZ 85253
-PHONE-
602 9486145
SOLAR RESEARCH
ALL SOLAR TECHNOLOGIES

SKELLY CHARLES
BOX 971
USASAFSB APO NY 09742

SOLAR RESEARCH
SOLAR SPACE HEATING & COOLING

SMITH DEAN L
ASSISTANT PROFESSOR
LOUISIANA TECH UNIVERSITY
ELECTRICAL ENGINEER
PO BOX 4967,TECH STA
RUSTON LA 71270
-PHONE-
318 2572262
SOLAR RESEARCH
INORGANIC SEMI-CONDUCTORS

SMITH WILLIAM
P O BOX 281
JAMESTOWN RI 02835

SOLAR RESEARCH
WIND CONVERSION

SOLOMON WYNN
NORTH MOUNTAIN COMMUNITY
RR 2 BOX 207
LEXINGTON VA 24450

-PHONE-
703 4637095
SOLAR RESEARCH
TOTAL SYSTEMS INTEGRATION

SKINNER * WILLIAM
MARKETING MGR
SOUTHERN CALIF, UNIV OF
WESRAC
809 WEST 34TH ST
LOS ANGELES CA 90007
-PHONE-
213 7466132
213 7463675
SOLAR/ENERGY CONSERV INFO DISSEMINATION ORG
PROVIDE INFORMATION SEARCHES TO CLIENTS

SMITH RICHARD R
5325 LUDLOW DRIVE
CAMP SPRINGS MD 20031

-PHONE-
301 4235324
SOLAR POTENTIAL MFG,NEW BUSINESS
INORGANIC SEMI-CONDUCTORS

SOLAN ALEXANDER
ASSOCIATE PROFESSOR
TECHNION-ISRAEL INST OF TECH
MECHANICAL ENGINEERG
TECHNION CITY
HAIFA ISRAEL

ENERGY CONSERVATION RESEARCH
MEDIUM TEMPERATURE SOLAR COLLECTOR SYSTEMS

SORKORAM PAUL
RAM PRODUCTS INC
BOX 340
STURGIS MI 49091

-PHONE-
616 6519351
SOLAR POTENTIAL MFG,ESTABLISHED BUSINESS
HIGH TEMPERATURE SOLAR COLLECTOR SYSTEMS

SOUTH PETER
ASSOC RESCH OFFICER
NATIONAL RESEARCH COUNCIL
OTTAWA CANADA

SOLAR RESEARCH
WIND CONVERSION

SPARROW E M
MINN, UNIV OF
DEPT OF MECH ENGR
MINNEAPOLIS MN 55455

-PHONE-
612 3733034
SOLAR RESEARCH
SOLAR ELECTRIC POWER GENERATION

SPEED RICHARD S
PRESIDENT
SOLTEC
PO BOX 6844
DENVER CO 80206
-PHONE-
303 3338869
SOLAR POTENTIAL MFG,NEW BUSINESS
GREENHOUSES

SPETZER HARTMUT
COLORADO UNIV OF-BOULDER
CIRES
BOULDER CO 80302

-PHONE-
303 4928028

SOLAR RESEARCH
HIGH TEMPERATURE SOLAR COLLECTOR SYSTEMS

STANFIELD LYNN M
PRESIDENT
STANFIELD AIR SYSTEMS
110 TARA WAY
ATHENS GA 30601
-PHONE-
404 5494767
404 5434202
SOLAR POTENTIAL MFG,ESTABLISHED BUSINESS
RETROFITTING, INTEREST

STANFORD GEOFFREY
AGRO-CITY PLANNING CONSULTANTS
3410 GARROTT RD
HOUSTON TX 77006

-PHONE-
713 5297266
SOLAR RESEARCH
BIO-CONVERSION

STEIGLEMANN MANAGER W H
ENERGY SYS LABS
FRANKLIN INST RESEARCH LABS
MECH&NUCLEAR ENGR DP
20TH & RACE STREETS
PHILADELPHIA PA 19103
-PHONE-
215 4481138
SOLAR RESEARCH
SOLAR SPACE HEATING & COOLING

STERN CARLOS D
ASSISTANT PROFESSOR
CONNECTICUT, UNIV OF
ENVIRONMENTAL ECON.
BOX U-21 UNIV. OF CT
STORRS CT 06268
-PHONE-
203 4862740
ENERGY CONSERVATION RESEARCH
RETROFITTING, TECHNOLOGIES

STERN MARTIN O
INT'L RESEARCH&TECHNOLOGY CORP
1225 CONN AVENUE NW
WASHINGTON DC 20036

SOLAR RESEARCH
SOLAR ENGINES

STERNBERG * JUDY
DIRECTOR INFO EXCHNG
SOLAR ENERGY SOC OF AMERICA
2780 SEPULA BLVD
TORRANCE CA 90510
-PHONE-
213 3263283
SOLAR/ENERGY CONSERV INFO DISSEMINATION ORG
PUBLISH LAY + TECHNICAL PERIODICALS

STEVENSON ROBERT H
MEGATHERM CORP
TAUNTON AVE
EAST PROVIDENCE RI 02914

-PHONE-
401 4383800
SOLAR COMPONENTS MANUFACTURER
SOLAR SPACE HEATING

STICHER JAKE
D&J SHEET METAL CO
10555 NW 7TH AVE
MIAMI FL 33150

-PHONE-
305 7577033
SOLAR MANUFACTURERS
SOLAR WATER HEATING

STREB J A
V.P.MARKETING
DYNATHERM CORP
MARBLE CT OFF IND LN
COCKEYSVILLE MD 21030
-PHONE-
301 6669151
SOLAR POTENTIAL MFG,ESTABLISHED BUSINESS
SOLAR SPACE HEATING & COOLING

SUHAY FRANK L
SUHAY ENTERPRISES
1505 E WINDSOR RD
GLENDALE CA 91204

-PHONE-
213 2469352

SOLAR MANUFACTURERS
SOLAR WATER HEATING

SUSEY PETER E
MANAGER
AMERICAN GAS ASSOCIATION
ENGY&ENVIROMENT SYS
1515 WILSON BLVD
ARLINGTON VA 22209
-PHONE-
703 5242000
SOLAR RESEARCH
SOLAR SPACE HEATING & COOLING

STONIER TOM T
PROF OF PEACE STUD
BRADFORD,UNIV OF
BRADFORD YORKSHIRE ENGLAND

-PHONE-
33466
SOLAR POLITICAL CONSIDERATIONS
LOBBY TO PROMOTE SOLAR LEGISLATION

STREED ELMER R
LOCKHEED PALO ALTO RESRCH LAB
DEPT 52/21 BLDG 205
3251 HANOVER ST
PALO ALTO CA 94304
-PHONE-
415 4934411
SOLAR RESEARCH
SOLAR SPACE HEATING & COOLING

SUMMERS GEORGE D
DIRECTOR
ATLANTIC RESEARCH CORPORATION
5390 CHEROKEE AVENUE
ALEXANDRIA VA 22314
-PHONE-
703 3543400 X315
703 3543400 X312
SOLAR POTENTIAL MFG,ESTABLISHED BUSINESS
ALL SOLAR TECHNOLOGIES

SWANSEN PE T L
RT 2 BOX 345
EAST TROY WI 53120

-PHONE-
414 6427487
SOLAR RESEARCH
SOLAR ELECTRIC POWER GENERATION

STRADER NANCY
NATIONAL CAPITAL PARKS
KLINGLE URBAN ENV CT
1100 OHIO DRIVE S W
WASHINGTON DC 20242
-PHONE-
212 2827020
SOLAR/ENERGY CONSERV INFO DISSEMINATION ORG
PROVIDE AUDIO & VISUAL SERVICES

STROMBERG R P
SANDIA LABS
DIV 4736 BOX 5800
ALBUQUERQUE NM 87115

-PHONE-
505 2648170
SOLAR RESEARCH
TOTAL SYSTEMS INTEGRATION

SUN THE
SOLAR ENERGY SERIES
SUN PUBLISHING CO
P O BOX 4383
ALBUQUERQUE NM 87106
-PHONE-
505 2556550

SOLAR/ENERGY CONSERV INFO DISSEMINATION ORG
PUBLISH LAY + TECHNICAL PERIODICALS

SWANSON EDWIN K
5615 S JOLLY ROGER
TEMPE AZ 85283

-PHONE-
602 8383727
SOLAR POTENTIAL MFG,NEW BUSINESS
SOLAR SPACE HEATING

SWARTMAN ROBERT K
WESTERN ONTARIO, UNIV OF
FAC OF ENGINEERING
LONDON N6A3K7 CANADA

-PHONE-
519 6793332
SOLAR RESEARCH
SOLAR REFRIGERATION

SWENSON * CARL
CONST ENGR SUPERVISR
U S GOVT GSA
CONSTRUCTION MGMT
BLDG#41 DFC
DENVER CO 80225
-PHONE-
303 2342641
SOLAR INTEREST
PLANNING OFFICES, FEDERAL

SWET CHARLES J
JOHNS HOPKINS UNIV
APPLIED PHYSICS LAB
8621 GEORGIA AVE
SILVER SPRING MD 20910

-PHONE-
301 9537100
SOLAR RESEARCH
ALL SOLAR TECHNOLOGIES

SWITZEN ROSALYN
SOLR ENGY OMBUDSMAN
BOX 1162
SAN FERNANDO CA 91341

SOLAR/ENERGY CONSERV INFO DISSEMINATION ORG
LOBBY TO PROMOTE SOLAR LEGISLATION

SYNDER RACHEL
WRITER
SOLAR ENERGY RESEARCH&INFO CTR
1001 CONN AVE NW
WASHINGTON DC 20036
-PHONE-
202 2931000

SOLAR/ENERGY CONSERV INFO DISSEMINATION ORG
PUBLISH LAY PERIODICALS

SZEGO GEORGE C
PRESIDENT
INTERTECHNOLOGY CORPORATION
PO BOX 340
WARRENTON VA 22186
-PHONE-
703 3477900
703 3471113 X4121
SOLAR RESEARCH
SOLAR SPACE HEATING & COOLING

SZYEALSKI S
ENERGY CONSV SECT
CALIF, UNIV OF-LIVERMORE
LAWRENCE LIVER LAB
P O BOX 808
LIVERMORE CA 94550
-PHONE-
415 4471100 X7721
SOLAR RESEARCH
SOLAR HEAT,COOL,&ELECTRIC POWER GENERATION

TABOR DR H
SCIENTIFIC RESEARCH FOUNDATION
P O B 3745
JERUSALEM ISRAEL

-PHONE-
02 534515
SOLAR RESEARCH
MEDIUM TEMPERATURE SOLAR COLLECTOR SYSTEMS

TABOR DR HARRY
NATIONAL PHYSICAL LAB ISRAEL
HEBREW UNIV CAMPUS
DANZIGER BLDG A
JERUSALEM ISRAEL

-PHONE-
30211 X475
SOLAR RESEARCH
MEDIUM TEMPERATURE SOLAR COLLECTOR SYSTEMS

TAFF DR DOUGLAS C
GARDEN WAY LABORATORIES
P O BOX 66
CHARLOTTE VT 05445

-PHONE-
802 8626501
SOLAR MANUFACTURERS
WIND CONVERSION

TALBERT SHERWOOD G
SR MECH ENGR
BATTELLE COLUMBUS LABS
THERM&MECH ENGY SYS
505 KING AVENUE
COLUMBUS OH 43201
-PHONE-
614 2993151
SOLAR RESEARCH
MEDIUM TEMPERATURE SOLAR COLLECTOR SYSTEMS

TANI TATSU
ELECTROTECHNICAL LAB
ENGY SYSTEM SECTION
TANASHI
TOYKO JAPAN

SOLAR RESEARCH
HIGH TEMPERATURE SOLAR COLLECTOR SYSTEMS

t

INDEX OF INDIVIDUALS

TAPPERT GEORGE
DEPT OF HOUSING & URBAN DEVEL
ENERGY AFF RM 10230
451 7TH STREET S W
WASHINGTON DC 20410
-PHONE-
202 7556480
SOLAR FINANCIAL CONSIDERATIONS
ALL SOLAR TECHNOLOGIES

TEAGUE W D
PRES
DARWIN TEAGUE INC
375 SYLVAN AVE
ENGLEWOOD CLIFFS NJ 07632
-PHONE-
201 5685124
SOLAR RESEARCH
SOLAR SPACE HEATING & COOLING

THEKAEKARA DR MATTHEW P
NASA GODDARD SPACE FLIGHT CTR
CODE 912
GREENBELT MD 20770

-PHONE-
301 9825034
SOLAR RESEARCH
SOLAR RADIATION MEASUREMENTS

THOMASON DR HARRY E
THOMASON SOLAR HOMES INC
6802 WALKER MILL SE
WASHINGTON DC 20027

-PHONE-
301 3364042
301 3365329
SOLAR MANUFACTURERS
SOLAR SPACE HEATING & COOLING

TAUSSIG PROF ROBERT
COLUMBIA UNIVERSITY
MECHANICAL ENGR
214 MUDD
NEW YORK NY 10027
-PHONE-
212 2802961
SOLAR RESEARCH
MEDIUM TEMPERATURE SOLAR COLLECTOR SYSTEMS

TELKES MARIA
ADJUNCT PROFESSOR
DELAWARE, UNIV OF
ENERGY CONVERSION
NEWARK DE 19711
-PHONE-
302 7388481
SOLAR RESEARCH
SOLAR ENERGY STORAGE

THIEN * CARL R
INFORMATION OFFICER
BROOKHAVEN NATIONAL LAB
PUBLIC RELATIONS
40 BROOKHAVEN AVE
UPTON NY 11973
-PHONE-
516 3452345
SOLAR RESEARCH
BIO-CONVERSION

THORNE BRINKLEY
140 MAIN STREET
NORTHAMPTON MA 01060

-PHONE-
413 5863212

SOLAR RESEARCH
SOLAR SPACE HEATING

TAYLOR THEODORE B
INT'L RESEARCH&TECH CORP
1501 WILSON BLVD
ARLINGTON VA 22209

-PHONE-
703 5245834
SOLAR RESEARCH
HIGH TEMPERATURE SOLAR COLLECTOR SYSTEMS

TEMPLE TRUMAN D
PRES
HELIO-DYNAMICS INC
518 SOUTH VAN NESS A
LOS ANGELES CA 90020
-PHONE-
213 3849853
SOLAR MANUFACTURERS
MEDIUM TEMPERATURE SOLAR COLLECTOR SYSTEMS

THOMAS RONALD
500-201
NASA LEWIS RESEARCH CTR
SOLAR SYS SECTION
2100 BROOKPARK RD
CLEVELAND OH 44135
-PHONE-
216 4334000
SOLAR RESEARCH
SOLAR SPACE HEATING & COOLING

THORNE JOHN K
MGR PRODUCT DEVELOP
UNIVERSAL OIL PRODUCTS CO
WOLVERINE TUBE DIV
P O BOX 2202
DECATUR AL 35601
-PHONE-
205 3531310 X290

SOLAR POTENTIAL MFG,ESTABLISHED BUSINESS
MEDIUM TEMPERATURE SOLAR COLLECTOR SYSTEMS

TILMANN RICHARD
NEW LIFE ENVIRO DESIGNS INST
P O BOX 648
KALAMAZOO MI 49005

-PHONE-
616 3452948
SOLAR RESEARCH
TOTAL SYSTEMS INTEGRATION

TRATNER ALLAN
ENVIRONMENTAL EDUCATION GROUP
67 LURLINE AVENUE
CANOGO CA 91306

SOLAR/ENERGY CONSERV INFO DISSEMINATION ORG
PUBLISH ORIGINAL RESEARCH + REPRINTS

TROYER LEROY S
LEROY S TROYER & ASSOC
112 1/2 LINCOLNWAY E
MISHAWAKA IN 46544

-PHONE-
219 2599976
SOLAR INTEREST
ALL SOLAR TECHNOLOGIES

ULVILD, FAIC GEORGE B
OWNER
GROUP SEVEN CHEMICALS CO
1210 DONA DRIVE
CORPUS CHRISTI TX 78407
-PHONE-
512 8886191
512 8835809
SOLAR RESEARCH
HIGH TEMPERATURE SOLAR COLLECTOR SYSTEMS

TLEIMAT BADAWI W
CALIF, UNIV OF-SEA H2O CON LAB
SEA WATER CONV LAB
1301 SOUTH 46TH ST
RICHMOND CA 94804
-PHONE-
415 2356000 X231
SOLAR RESEARCH
SOLAR SPACE HEATING & COOLING

TRAUTMAN PE FRED L
OWNER
MECHANICAL ENGINEER
3881 S GRAPE ST
DENVER CO 80237
-PHONE-
303 7570416
SOLAR RESEARCH
SOLAR SPACE HEATING & COOLING

UBEROI MAHINDER
COLORADO UNIV OF-BOULDER
AEROSPACE ENG SCI
BOULDER CO 80302

-PHONE-
303 4926612
SOLAR RESEARCH
HIGH TEMPERATURE SOLAR COLLECTOR SYSTEMS

URQUHART * DONALD A
MANAGER SPECIAL PROJ
CORNING GLASS WORKS
LIGHTING PROD DIV
CORNING NY 14830
-PHONE-
607 9747306

SOLAR POTENTIAL MFG,ESTABLISHED BUSINESS
MEDIUM TEMPERATURE SOLAR COLLECTOR SYSTEMS

TODD JOHN
PRES
NEW ALCHEMY INSTITUTE EAST
P O BOX 432
WOODS HOLE MA 02543
-PHONE-
617 5632655
SOLAR RESEARCH
TOTAL SYSTEMS INTEGRATION

TRENT JACK
OWNER
NATIONAL SUN CONTROL
PO BOX 330205
MIAMI FL 33133
-PHONE-
305 4447494
SOLAR COMPONENTS MANUFACTURER
SOLAR SPACE COOLING

ULRICH RICHARD D
CODE 45701
US NAVAL WEAPONS CENTER
CHINA LAKE CA 93555

-PHONE-
714 9397384
SOLAR RESEARCH
TOTAL SYSTEMS INTEGRATION

VAN BUREN TEAGUE
SOLAR HYDRONICS
P O BOX 4484
BOULDER CO 80302

-PHONE-
303 4479828

SOLAR RESEARCH
SOLAR SPACE HEATING

V

INDEX OF INDIVIDUALS

VAN DRESSER PETER
EL RITO NM 87530

SOLAR RESEARCH
HOMES

VARANI FREDERICK T
ADRIAN MANUF & RESEARCH
619 19TH STREET
GOLDEN CO 80401

-PHONE-
303 2794536
SOLAR RESEARCH
SOLAR SPACE HEATING

VERNON RICHARD W
NASA LEWIS RESEARCH CTR
SOLAR SYS SECTION
2100 BROOKPARK ROAD
CLEVELAND OH 44135

-PHONE-
216 4334000 X6857

SOLAR RESEARCH
SOLAR SPACE HEATING & COOLING

VITOUSEK MARTIN
HAWAII, UNIV OF
HI INST OF GEOPHYSIC
2525 CORREA ROAD
HONOLULU HI 96822
-PHONE-
808 9487668
SOLAR RESEARCH
OCEAN THERMAL GRADIENT SYSTEMS

VANT-HULL DR LORIN L
HOUSTON, UNIV OF
PHYSICS DEPT
HOUSTON TX 77004

-PHONE-
713 7493809
SOLAR RESEARCH
SOLAR ELECTRIC POWER GENERATION

VENDITTI FRED P
SP RES ENGR
DENVER, UNIV OF
DENVER RESEARCH INST
UNIVERSITY OF DENVER
DENVER CO 80210
-PHONE-
303 7532241
SOLAR RESEARCH
SOLAR ELECTRIC POWER GENERATION

VERSCHOOR JACK D
MANAGER-PLANNING
JOHNS-MANVILLE CORP
GROWTH PLAN&DEVELOP
GREENWOOD PLAZA
DENVER CO 80217
-PHONE-
303 7701000 X2879
303 9865757
ENERGY CONSERVATION RESEARCH
NEW CONSTRUCTION, MECHANICAL SYSTEMS

VOGENTHALER THOMAS
DIRECTOR
COLORADO ENERGY RESEARCH INST
COLO SCHOOL OF MINES
GOLDEN CO 80401
-PHONE-
303 2790300
SOLAR RESEARCH
GOVERNMENT AGENCIES PROVIDING GRANTS

VARADI PETER F
EXEC V P
SOLAREX CORP
1335 PICCARD DR
ROCKVILLE MD 20850
-PHONE-
301 9480202
SOLAR MANUFACTURERS
INORGANIC SEMI-CONDUCTORS

VERMEERS HELEN K
358 SW 27TH AVE #12
MIAMI FL 33135

-PHONE-
305 6423494
SOLAR SOCIAL CONSIDERATIONS
ALL SOLAR TECHNOLOGIES

VILLECCO MARGOT
SENIOR EDITOR
ARCHITECTURE PLUS
1345 SIXTH AVE
NEW YORK NY 10019

-PHONE-
212 4898697

SOLAR/ENERGY CONSERV INFO DISSEMINATION ORG
PUBLISH LAY + TECHNICAL PERIODICALS

VON BREIEN JR HANS
COORDINATOR
LAMA FOUNDATION
PO BOX 444
SAN CRISTOBAL NM 87564

SOLAR RESEARCH
MEDIUM TEMPERATURE SOLAR COLLECTOR SYSTEMS

WAHIG MICHAEL
ENERGY&ENVIRO PROG
CALIF, UNIV OF-BERKELEY
LAWRENCE BERKELEY LB
BERKELEY CA 94720
-PHONE-
415 8432740 X5787
415 8432740 X5001
SOLAR RESEARCH
INORGANIC SEMI-CONDUCTORS

WALCK * DAVE
GENERAL ATOMIC COMPANY
P O BOX 81608
SAN DIEGO CA 92138

-PHONE-
714 4531000

SOLAR RESEARCH
SOLAR ELECTRIC POWER GENERATION

WALKER * WAYLAND
CITY PLANNER
DENVER PLANNING OFFICE
1445 CLEVELAND PLACE
DENVER CO 80202
-PHONE-
303 2972736

SOLAR INTEREST
PLANNING OFFICES, LOCAL

WALTON JR JESSE D
GEORGIA INST OF TECHNOLOGY
ENGINEERING EXP STAT
ATLANTA GA 30332

-PHONE-
404 8943661
SOLAR RESEARCH
SOLAR ELECTRIC POWER GENERATION

WARBURTON RALPH
ASSOCIATE DEAN
MIAMI, UNIV OF
ARCHITECT& PLANNING
CORAL GABLES FL 33124
-PHONE-
305 2843438
SOLAR RESEARCH
SOLAR SPACE HEATING & COOLING

WARD DAN S
COLORADO STATE UNIV
SOLAR ENGY APPL LAB
FORT COLLINS CO 80521

-PHONE-
303 4918211
SOLAR RESEARCH
SOLAR SPACE HEATING & COOLING

WARD-SWANN WILLIAM
HELTER-SKELTER ENGINEERING
BOX 479
FOREST KNOLLS CA 94933

-PHONE-
415 4572986

SOLAR MANUFACTURERS
SOLAR SPACE HEATING

WARING JOHN A
RES WRITER&CONSULTNT
8502 FLOWER AVENUE
TAKOMA PARK MD 20012

-PHONE-
202 6938675
301 5887738
SOLAR RESEARCH
SOLAR ELECTRIC POWER GENERATION

WARREN ROGER W
MANAGER
WESTINGHOUSE RESEARCH LAB
CHURCHILL BORO
PITTSBURGH PA 15235
-PHONE-
412 2565187

SOLAR RESEARCH
HIGH TEMPERATURE SOLAR COLLECTOR SYSTEMS

WASON ROBERT G
CONSULTING ENGR
ENVIRONMENTAL SYS FOR PEOPLE
213 SUN 2030 E SPDWY
TUCSON AZ 85719
-PHONE-
602 7955437
SOLAR INTEREST
SOLAR SPACE HEATING & COOLING

WATSON JOHN
1502 1/2 WEST 9TH
AUSTIN TX 78703

-PHONE-
512 4772042
SOLAR RESEARCH
SOLAR SPACE HEATING & COOLING

WATSON * SKIP
EDITOR
SUN PUBLISHING CO
P O BOX 4383
ALBUQUERQUE NM 87106
-PHONE-
505 2556550
SOLAR/ENERGY CONSERV INFO DISSEMINATION ORG
PUBLISH LAY + TECHNICAL PERIODICALS

W

INDEX OF INDIVIDUALS

WATSON THOMAS
1144 CANYON RD
SANTA FE NM 87501

-PHONE-
505 9881800
SOLAR RESEARCH
TOTAL SYSTEMS INTEGRATION

WEDGE A C
EXECUTIVE VP
SENSENICH CORPORATION
PO BOX 1168
LANCASTER PA 17604

SOLAR COMPONENTS MANUFACTURER
WIND CONVERSION

WEINSTEIN ALBERT
MANAGER
WESTINGHOUSE ELECTRIC CORP.
SPECIAL ENERGY SYST
PO BOX 1693 MS-999
BALTIMORE MD 21203
-PHONE-
301 7653454
SOLAR RESEARCH
SOLAR SPACE HEATING & COOLING

WELCH BRUCE L
FRIENDS RESEARCH CENTER
ENVIRONMENTAL STUDYS
308 HILTON AVE
CATONSVILLE MD 21228
-PHONE-
301 7472773
SOLAR POLITICAL CONSIDERATIONS
ALL SOLAR TECHNOLOGIES

WATT ARTHUR D
WATT ENGR LTD
8395 BASELINE RD
BOULDER CO 80303

-PHONE-
303 4946734
SOLAR/ENERGY CONSERV INFO DISSEMINATION ORG
SOLAR SPACE HEATING & COOLING

WEHNER G K
MINN, UNIV OF
ELECT ENGINEERING
MINNEAPOLIS MN 55455

-PHONE-
612 3737831
SOLAR RESEARCH
SOLAR ELECTRIC POWER GENERATION

WEINSTEIN FRANK
FRANKLIN INST RESEARCH LABS
SOLAR APPLC INFO CTR
20TH & RACE STREETS
PHILADELPHIA PA 19103

-PHONE-
215 4481500
SOLAR/ENERGY CONSERV INFO DISSEMINATION ORG
HAVE&MAINTAIN ENERGY DATA BANK

WERDEN ROBERT G
ROBERT G WERDEN&ASSOCIATES INC
PO BOX 414
JENKINTOWN PA 19046

-PHONE-
215 8852500
SOLAR RESEARCH
SOLAR SPACE HEATING & COOLING

WEBER MARTIN
67-29-218 STREET
BAYSIDE NY 11364

-PHONE-
212 2249352
SOLAR RESEARCH
ALL SOLAR TECHNOLOGIES

WEINGART JEROME
SOLAR CONSULTANT
AYRES&HAYAKAWA ENERGY MGT
CONSULTING ENGINEERS
1190 SOUTH BEVERLY
LOS ANGELES CA 90035
-PHONE-
213 8794477
SOLAR RESEARCH
SOLAR SPACE HEATING & COOLING

WEITZEL DAN
CONSULTANT
3040 ASH AVE
BOULDER CO 80303

SOLAR RESEARCH
WIND CONVERSION

WERLEIN PHILIP
7927 ST CHARLES AVE
NEW ORLEANS LA 70118

-PHONE-
504 8610544
SOLAR RESEARCH
SOLAR COOKING

WHETSTONE GEORGE A
TEXAS TECH UNIVERSITY
DEPT OF CIVIL ENGRG
PO BOX 4089
LUBBOCK TX 79409
-PHONE-
806 7421234
SOLAR RESEARCH
SOLAR ELECTRIC POWER GENERATION

WHILLIER AUSTIN
CHAMBER OF MINES OF S AFRICA
RESEARCH LABS
PO BOX 61809
JOHANNESBURG SOUTH AFRICA
-PHONE-
JOH 31161
SOLAR RESEARCH
MEDIUM TEMPERATURE SOLAR COLLECTOR SYSTEMS

WHITE CHARLES R
AUBURN UNIVERSITY
DEPT OF INDUST ENGR
AUBURN AL 36830
-PHONE-
205 8264340
SOLAR RESEARCH
MEDIUM TEMPERATURE SOLAR COLLECTOR SYSTEMS

WHITE * ROGER
ASST TO PRESIDENT
COLORADO NATIONAL BANK
P O BOX 5168TA
DENVER CO 80217
-PHONE-
303 8931862
SOLAR FINANCIAL CONSIDERATIONS
BANKERS

WHITLOW DR EUGENE P
WHIRLPOOL CORP
R&E
MONTE ROAD
BENTON HARBOR MI 49022
-PHONE-
616 9265341
SOLAR RESEARCH
MEDIUM TEMPERATURE SOLAR COLLECTOR SYSTEMS

WHITMER AIA ROGER G
ASSOC PROF ARCH DEPT
ILLINOIS, UNIV OF-CHICAGO
ARCHITECTURE&ART CLG
BOX 4348
CHICAGO IL 60680
-PHONE-
312 9963335
SOLAR RESEARCH
NEW CONSTRUCTION, TECHNOLOGIES

WHITNEY EDWARD N
PE
1978 E 2ND AVE
DURANGO CO 81301
-PHONE-
303 2473584
SOLAR RESEARCH
SOLAR HEAT,COOL,&ELECTRIC POWER GENERATION

WIEGAND JAMES B
PRESIDENT
SOLAR ENERGY RESEARCH CORP
RT 4 BOX 26B
LONGMONT CO 80501
-PHONE-
303 7724522
SOLAR MANUFACTURERS
SOLAR SPACE HEATING & COOLING

WILCOX HOWARD
NAVAL UNDERSEAS CENTER
SAN DIEGO CA 92132
-PHONE-
714 2257957
SOLAR RESEARCH
BIO-CONVERSION

WILDERMAN RICHARD F
SEA PINES COMPANY
ENVIRONMENTAL SERV
PO BOX 5608
HILTON HEAD ISLAND SC 29928
-PHONE-
803 7853333 X3144
SOLAR RESEARCH
SOLAR SPACE HEATING & COOLING

WILEY * HARRY T
ASHLAND OIL INC
P O 391
ASHLAND KY 41101
-PHONE-
606 3293333
SOLAR RESEARCH
SOLAR SPACE HEATING & COOLING

WILLIS ROBERT W
VICE PRESIDENT
SOLAR POWER CORP
186 FORBES ROAD
BRAINTREE MA 02184
-PHONE-
617 8486877
SOLAR MANUFACTURERS
INORGANIC SEMI-CONDUCTORS

W,X,Y

INDEX OF INDIVIDUALS

WINSLOW FRANK
MONSANTO RESEARCH CORP
1515 NICHOLAS ROAD
DAYTON OH 45407

-PHONE-
513 2683411
SOLAR POTENTIAL MFG,ESTABLISHED BUSINESS
INORGANIC SEMI-CONDUCTORS

WOLF RICHARD W
SERVICE UNLIMITED INC
SECOND & WALNUT STS
WILMINGTON DE 19801

-PHONE-
302 6551568

SOLAR POTENTIAL MFG,ESTABLISHED BUSINESS
SOLAR SPACE HEATING

WORTES L L
550 E 12TH AVE #1801
DENVER CO 80203

-PHONE-
303 2556062
SOLAR RESEARCH
SOLAR SPACE HEATING & COOLING

YAROSH MARVIN
FLORIDA ENERGY COMMISSION
SENATE OFFICE BLDG
TALLAHASSEE FL 32304

-PHONE
904 4881078
904 4881167
SOLAR POLITICAL CONSIDERATIONS
ALL SOLAR TECHNOLOGIES

WINSTON PROF ROLAND
ARGONNE NATIONAL LABORATORY
9700 SOUTH CASS AVE
ARGONNE IL 60439

SOLAR RESEARCH
HIGH TEMPERATURE SOLAR COLLECTOR SYSTEMS

WOLFSON MOREY
DIRECTOR

ENVIRONMTL ACTION-CO

DENVER CO 80206
-PHONE-
303 5341602
SEE PAGE 4.009 + 4.021
SOLAR/ENERGY CONSERV INFO DISSEMINATION ORG
ALL SOLAR TECHNOLOGIES

WRIGHT * JOHN E
SUPERVISOR PUB REL
ALCOA ALUMINUM
1501 ALCOA BLDG
PITTSBURGH PA 15219
-PHONE-
412 5534751
SOLAR COMPONENTS MANUFACTURER
MEDIUM TEMPERATURE SOLAR COLLECTOR SYSTEMS

YATABE YOSHIO
YUP COMPANY LTD
1-10-3 JUJONAKAHARA
KITA-KU TOKYO JAPAN

-PHONE-
03 9008987

SOLAR POTENTIAL MFG,ESTABLISHED BUSINESS
HIGH TEMPERATURE SOLAR COLLECTOR SYSTEMS

WINTER FRANCIS DE
2940 THORNDIKE ROAD
PASADENA CA 91107

-PHONE-
213 7966414
SOLAR RESEARCH
SOLAR WATER HEATING

WORMSER ERIC M
PRESIDENT
WORMSER SCIENTIFIC CORP
88 FOXWOOD RD
STAMFORD CT 06903

-PHONE-
203 3221981
203 3226200
SOLAR/ENERGY CONSERV INFO DISSEMINATION ORG
SOLAR SPACE HEATING & COOLING

YANNONI NICHOLAS F
US AFCRL-PHF
HANSCOM FIELD
BEDFORD MA 01730

-PHONE-
617 8612265
SOLAR RESEARCH
INORGANIC SEMI-CONDUCTORS

YELLOTT JOHN I
ARIZONA STATE UNIVERSITY
COLL OF ARCHITECTURE
TEMPE AZ 85281

-PHONE-
602 9655562
602 9653216
SOLAR RESEARCH
SOLAR SPACE HEATING & COOLING

YERS JACK M
TEXAS, UNIV OF
ZOOLOGY DEPT
AUSTIN TX 78712

-PHONE-
512 4711686
SOLAR RESEARCH
BIO-CONVERSION

ZAFIRIOU OLIVER C
WOODS HOLE OCEANOGRAPHIC INST
CHEMISTRY DEPT
WOODS HOLE MA 02543

-PHONE-
617 5481400 X302

SOLAR RESEARCH
BIO-CONVERSION

ZOSCHAK R J
MANAGER
FOSTER WHEELER CORP
APP THERMODYNAMICS
110 S ORANGE AVE
LIVINGSTON NJ 07039
-PHONE-
201 5331100
SOLAR RESEARCH
BIO-CONVERSION

YOKOMIZO CLIFFORD T
TECHNICAL STAFF
SANDIA LABORATORIES
P O BOX 969
LIVERMORE CA 94550
-PHONE-
415 4557011 X2668
SOLAR RESEARCH
HIGH TEMPERATURE SOLAR COLLECTOR SYSTEMS

ZENER CLARENCE
CARNEGIE MELLON UNIV
6123 SCIENCE HALL
PITTSBURG PA 15213

-PHONE-
412 6212600 X229

SOLAR RESEARCH
OCEAN THERMAL GRADIENT SYSTEMS

ZUCKERMAN RONALD A
EDITOR
ENERGY RESEARCH CORP
ENERGY REVIEW
6 EAST VALERIO
SANTA BARBARA CA 93101
-PHONE-
805 9631388
SOLAR RESEARCH
OCEAN THERMAL GRADIENT SYSTEMS

YOUNG CHESTER W
WALTER V STERLING INC
TECH&MANAGMT CONSULT
16262 E WHITTIER B
WHITTIER CA 90603
-PHONE-
213 9438703
SOLAR RESEARCH
SOLAR ELECTRIC POWER GENERATION

ZERLAUT GENE A
PRESIDENT
DESERT SUNSHINE EXPOSURE TESTS
BOX185 BLACKCAN STGE
PHOENIX AZ 85020
-PHONE-
602 4657525
602 4657521
SOLAR RESEARCH
SOLAR,MATERIAL WEATHERABILITY TESTS

yancy

. . . . some day soon